CANCER AND INFLAMMATION MECHANISMS

CANCER AND INFLAMMATION MECHANISMS
Chemical, Biological, and Clinical Aspects

Edited by

YUSUKE HIRAKU

Department of Environmental and Molecular Medicine
Mie University Graduate School of Medicine, Tsu, Mie, Japan

SHOSUKE KAWANISHI

Faculty of Pharmaceutical Sciences
Suzuka University of Medical Science, Suzuka, Mie, Japan

HIROSHI OHSHIMA

Department of Nutritional and Environmental Sciences
Graduate School of Integrated Pharmaceutical and Nutritional Sciences
University of Shizuoka, Shizuoka, Japan

Published by John Wiley & Sons, Inc., Hoboken, New Jersey.
Published simultaneously in Canada.

For general information on our other products and services or for technical support, please contact our
Customer Care Department within the United States at (800) 762-2974, outside the United States at
(317) 572-3993 or fax (317) 572-4002.

Wiley also publishes its books in a variety of electronic formats. Some content that appears in print may
not be available in electronic formats. For more information about Wiley products, visit our web site at
www.wiley.com.

Library of Congress Cataloging-in-Publication Data:

Cancer and inflammation mechanisms : chemical, biological, and clinical aspects /
edited by Dr. Yusuke Hiraku, Dr. Shosuke Kawanishi, Dr. Hiroshi Ohshima.
    1 online resource.
  Includes index.
    ISBN 978-1-118-82655-3 (ePub) – ISBN 978-1-118-82667-6 (Adobe PDF)
– ISBN 978-1-118-16030-5 (cloth) 1. Carcinogenesis. 2. Inflammation–Mediators.
3. Inflammation–Immunological aspects. 4. Cancer–Immunological aspects.
I. Hiraku, Yusuke, editor of compilation. II. Kawanishi, Shosuke, 1942– editor of compilation.
III. Oshima, Hiroshi, 1949– editor of compilation.
  RC268.5
  616.99′401–dc23

                                        2013039801

Printed in the United States of America

10 9 8 7 6 5 4 3 2 1

CONTENTS

CONTRIBUTORS

Helmut Bartsch Division of Toxicology and Cancer Risk Factors, German Cancer Research Center, Heidelberg, Germany

Michele Carbone University of Hawai'i Cancer Center, University of Hawai'i, Honolulu, HI, United States

Department of Pathology, John. A. Burns School of Medicine, University of Hawai'i, Honolulu, HI, United States

Fu Chen Department of Radiation Oncology, Eye Ear Nose & Throat Hospital of Fudan University, Shanghai, China

Kaiwen W. Chen Institute for Molecular Bioscience, The University of Queensland, St Lucia, Australia

Tsutomu Chiba Department of Gastroenterology and Hepatology, Graduate School of Medicine, Kyoto University, Kyoto, Japan

Pei-Hsin Chou Department of Environmental Engineering, National Cheng Kung University, Tainan, Taiwan

Kyung-Soo Chun College of Pharmacy, Keimyung University, Daegu, South Korea

Jonas Fuxe Department of Medical Biochemistry and Biophysics, Division of Vascular Biology, Karolinska Institute, Stockholm, Sweden

Giovanni Gaudino University of Hawai'i Cancer Center, University of Hawai'i, Honolulu, HI, United States

Alexandros G. Georgakilas Physics Department, School of Applied Mathematical and Physical Sciences, National Technical University of Athens, Zografou Campus, Athens, Greece

Zhaojian Gong Hunan Key Laboratory of Nonresolving Inflammation and Cancer, Disease Genome Research Center, The Third Xiangya Hospital, Central South University, Changsha, China

Cancer Research Institute, Key Laboratory of Carcinogenesis of Ministry of Health, Key Laboratory of Carcinogenesis and Cancer Invasion of Ministry of Education, Central South University, Changsha, China

Department of Stomatology, The Second Xiangya Hospital, Central South University, Changsha, China

Yusuke Hiraku Department of Environmental and Molecular Medicine, Mie University Graduate School of Medicine, Tsu, Mie, Japan

Lorne J. Hofseth Department of Pharmaceutical and Biomedical Sciences, South Carolina College of Pharmacy, University of South Carolina, Columbia, SC, United States

Tingting Huang Department of Radiation Oncology, First Affiliated Hospital of Guangxi Medical University, Nanning, China

Sandro Jube University of Hawai'i Cancer Center, University of Hawai'i, Honolulu, HI, United States

Mikael C. I. Karlsson Department of Medicine, Translational Immunology Unit, Karolinska Institutet, Karolinska Hospital, Stockholm, Sweden

Shosuke Kawanishi Faculty of Pharmaceutical Sciences, Suzuka University of Medical Science, Suzuka, Mie, Japan

Ching-Lung Lai Department of Medicine, the University of Hong Kong, Queen Mary Hospital, Hong Kong

State Key Laboratory for Liver Research, the University of Hong Kong, Queen Mary Hospital, Hong Kong

Danae A. Laskaratou Physics Department, School of Applied Mathematical and Physical Sciences, National Technical University of Athens, Zografou Campus, Athens, Greece

Guiyuan Li Hunan Key Laboratory of Nonresolving Inflammation and Cancer, Disease Genome Research Center, the Third Xiangya Hospital, Central South University, Changsha, China

Cancer Research Institute, Key Laboratory of Carcinogenesis of Ministry of Health, Key Laboratory of Carcinogenesis and Cancer Invasion of Ministry of Education, Central South University, Changsha, China

Xiaoling Li Hunan Key Laboratory of Nonresolving Inflammation and Cancer, Disease Genome Research Center, The Third Xiangya Hospital, Central South University, Changsha, China

Cancer Research Institute, Key Laboratory of Carcinogenesis of Ministry of Health, Key Laboratory of Carcinogenesis and Cancer Invasion of Ministry of Education, Central South University, Changsha, China

Xiayu Li Hunan Key Laboratory of Nonresolving Inflammation and Cancer, Disease Genome Research Center, the Third Xiangya Hospital, Central South University, Changsha, China

Cancer Research Institute, Key Laboratory of Carcinogenesis of Ministry of Health, Key Laboratory of Carcinogenesis and Cancer Invasion of Ministry of Education, Central South University, Changsha, China

Yong Li Department of Biochemistry and Molecular Biology, Center for Genetics and Molecular Medicine, School of Medicine, University of Louisville, Louisville, KY, United States

Jian Ma Hunan Key Laboratory of Nonresolving Inflammation and Cancer, Disease Genome Research Center, The Third Xiangya Hospital, Central South University, Changsha, China

Cancer Research Institute, Key Laboratory of Carcinogenesis of Ministry of Health, Key Laboratory of Carcinogenesis and Cancer Invasion of Ministry of Education, Central South University, Changsha, China

Hiroyuki Marusawa Department of Gastroenterology and Hepatology, Graduate School of Medicine, Kyoto University, Kyoto, Japan

Tomonari Matsuda Research Center for Environmental Quality Management, Kyoto University, Otsu, Shiga, Japan

William Matsui Department of Oncology, Sidney Kimmel Comprehensive Cancer Center, Johns Hopkins University School of Medicine, Baltimore, MD, United States

Ifigeneia V. Mavragani Physics Department, School of Applied Mathematical and Physical Sciences, National Technical University of Athens, Zografou Campus, Athens, Greece

Pranab Behari Mazumder Department of Biochemistry and Molecular Biology, Center for Genetics and Molecular Medicine, School of Medicine, University of Louisville, Louisville, KY, United States

Noriyuki Miyoshi Department of Nutritional and Environmental Sciences, Graduate School of Integrated Pharmaceutical and Nutritional Sciences, University of Shizuoka, Shizuoka, Japan

Yiqun Mo Department of Environmental and Occupational Health Sciences, School of Public Health and Information Sciences, University of Louisville, Louisville, KY, United States

Akira Murakami Division of Food Science and Biotechnology, Graduate School of Agriculture, Kyoto University, Kyoto, Japan

Michihiro Mutoh Division of Cancer Prevention Research, National Cancer Center Research Institute, Tokyo, Japan

Urmila Jagadeesan Nair Division of Toxicology and Cancer Risk Factors, German Cancer Research Center, Heidelberg, Germany

Unit of Cancer Prevention and WHO Collaborating Center, German Cancer Research Center, Heidelberg, Germany

Andrea Napolitano University of Hawai'i Cancer Center, University of Hawai'i, Honolulu, HI, United States

Department of Molecular Biosciences and Bioengineering, University of Hawai'i, Honolulu, HI, United States

Chikako Nishigori Division of Dermatology, Department of Internal Related, Kobe University Graduate School of Medicine, Kobe, Japan

Hiroshi Ohshima Department of Nutritional and Environmental Sciences, Graduate School of Integrated Pharmaceutical and Nutritional Sciences, University of Shizuoka, Shizuoka, Japan

Futoshi Okada Division of Pathological Biochemistry, Tottori University Faculty of Medicine, Yonago, Tottori, Japan

Chromosome Engineering Research Center, Tottori University, Yonago, Tottori, Japan

Harvey I. Pass Department of Cardiothoracic Surgery, NYU Langone Medical Center, New York, NY, United States

Angela Poehlmann Department of Pathology, Otto-von-Guericke University Magdeburg, Magdeburg, Germany

Deepak Poudyal Department of Pharmaceutical and Biomedical Sciences, South Carolina College of Pharmacy, University of South Carolina, Columbia, SC, United States

Ayanthi A. Richards Institute for Molecular Bioscience, The University of Queensland, St Lucia, Australia

Albert Roessner Department of Pathology, Otto-von-Guericke University Magdeburg, Magdeburg, Germany

Kurt J. Sales Institute of Infectious Disease and Molecular Medicine and Division of Medical Biochemistry, University of Cape Town Faculty of Health Sciences, Observatory, Cape Town, South Africa

Kate Schroder Institute for Molecular Bioscience, The University of Queensland, St Lucia, Australia

Australian Infectious Disease Research Centre, The University of Queensland, St Lucia, Australia

Wai-Kay Seto Department of Medicine, the University of Hong Kong, Queen Mary Hospital, Hong Kong

Haruhiko Sugimura Department of Pathology, Hamamatsu University School of Medicine, Hamamatsu, Shizuoka, Japan

Young-Joon Surh Tumor Microenvironment Global Core Research Center, College of Pharmacy, Seoul National University, Seoul, South Korea

Mami Takahashi Central Animal Division, National Cancer Center Research Institute, Tokyo, Japan

Osamu Takeuchi Laboratory of Infection and Prevention, Institute for Virus Research Centre, Kyoto University, Kyoto, Japan

Toshihiko Tanno Department of Oncology, Sidney Kimmel Comprehensive Cancer Center, Johns Hopkins University School of Medicine, Baltimore, MD, United States

Sarang Tartey Laboratory of Infection and Prevention, Institute for Virus Research Centre, Kyoto University, Kyoto, Japan

Department of Host Defense, WPI Immunology Frontier Research Centre, Osaka University, Suita, Osaka, Japan

David J. Tollerud Department of Environmental and Occupational Health Sciences, School of Public Health and Information Sciences, University of Louisville, Louisville, KY, United States

Susumu Tomono Department of Nutritional and Environmental Sciences, Graduate School of Integrated Pharmaceutical and Nutritional Sciences, University of Shizuoka, Shizuoka, Japan

Keiji Wakabayashi Division of Nutritional and Environmental Sciences, University of Shizuoka, Shizuoka, Japan

Rong Wan Department of Environmental and Occupational Health Sciences, School of Public Health and Information Sciences, University of Louisville, Louisville, KY, United States

Xue Xiao Department of Otolaryngology-Head & Neck Surgery, First Affiliated Hospital of Guangxi Medical University, Nanning, China

Wei Xiong Hunan Key Laboratory of Nonresolving Inflammation and Cancer, Disease Genome Research Center, The Third Xiangya Hospital, Central South University, Changsha, China

Cancer Research Institute, Key Laboratory of Carcinogenesis of Ministry of Health, Key Laboratory of Carcinogenesis and Cancer Invasion of Ministry of Education, Central South University, Changsha, China

Haining Yang University of Hawai'i Cancer Center, University of Hawai'i, Honolulu, HI, United States

Department of Pathology, John. A. Burns School of Medicine, University of Hawai'i, Honolulu, HI, United States

Man-Fung Yuen Department of Medicine, the University of Hong Kong, Queen Mary Hospital, Hong Kong

State Key Laboratory for Liver Research, the University of Hong Kong, Queen Mary Hospital, Hong Kong

Alina Zamoshnikova Institute for Molecular Bioscience, The University of Queensland, St Lucia, Australia

Zhaoyang Zeng Hunan Key Laboratory of Nonresolving Inflammation and Cancer, Disease Genome Research Center, The Third Xiangya Hospital, Central South University, Changsha, China

Cancer Research Institute, Key Laboratory of Carcinogenesis of Ministry of Health, Key Laboratory of Carcinogenesis and Cancer Invasion of Ministry of Education, Central South University, Changsha, China

Qunwei Zhang Department of Environmental and Occupational Health Sciences, School of Public Health and Information Sciences, University of Louisville, Louisville, KY, United States

Zhe Zhang Department of Otolaryngology-Head & Neck Surgery, First Affiliated Hospital of Guangxi Medical University, Nanning, China

Ming Zhou Hunan Key Laboratory of Nonresolving Inflammation and Cancer, Disease Genome Research Center, The Third Xiangya Hospital, Central South University, Changsha, China

Cancer Research Institute, Key Laboratory of Carcinogenesis of Ministry of Health, Key Laboratory of Carcinogenesis and Cancer Invasion of Ministry of Education, Central South University, Changsha, China

Xiaoying Zhou Department of Otolaryngology-Head & Neck Surgery, First Affiliated Hospital of Guangxi Medical University, Nanning, China

PREFACE

Cancer is a life-threatening human disease, and it is estimated that approximately 25% of cancer cases worldwide are attributed to chronic inflammation. In the nineteenth century, Rudolf Virchow noted leucocytes in neoplastic tissues and suggested that the "lymphoreticular infiltrate" reflected the origin of cancer at sites of chronic inflammation. Since then, numerous epidemiological and experimental studies have provided evidence of linkage between inflammation and cancer. Chronic inflammation can be induced by a wide variety of environmental factors, such as infectious agents, inflammatory diseases, and physicochemical factors.

The aim of this book is to review current knowledge on the linkage between chronic inflammation and cancer, and to discuss comprehensively the mechanisms of carcinogenesis in terms of its chemical, biological, and clinical aspects. Future perspectives of chemoprevention of inflammation-related cancer are included. The book consists of the following sections:

Section I: General overview of inflammation-related cancer. The book begins with a general overview of the mechanisms of carcinogenesis mediated by chronic infection and inflammation. This is followed by topics in emerging fields of cancer biology, such as stem cell theory and epithelial–mesenchymal transition, and their roles in inflammation-related carcinogenesis are discussed.

Section II: Biochemistry in inflammation-related cancer. During chronic inflammation, reactive oxygen and nitrogen species are generated in biological systems and attack various biomolecules, including DNA. In this section the role of DNA damage mediated by reactive species in inflammation-related carcinogenesis and a comprehensive analytical method for DNA adducts (the "adductome" approach) are discussed.

Section III: Molecular biology in inflammation-related cancer. The main topics in this section are current knowledge on inflammation-related molecules, such as Toll-like receptors, inflammasome, activation-induced cytidine deaminase, and microRNAs, and their roles in carcinogenesis. An experimental animal model to investigate the role of inflammation in tumor progression is also reviewed.

Section IV: Inflammation-related cancer induced by specific causes. This section covers the mechanism of carcinogenesis mediated by oncogenic viruses such as human papillomavirus, hepatitis viruses, and Epstein–Barr virus, and inflammatory diseases such as Barrett's esophagus. Also covered are the mechanisms of carcinogenesis induced by physicochemical factors such as asbestos, nanomaterials, ultraviolet light, and ionizing radiation.

Section V: Prevention of inflammation-related carcinogenesis. While the first four sections deal primarily with the mechanisms of inflammation-related carcinogenesis, this section includes a perspective and strategy of cancer chemo-prevention using anti-inflammatory agents and natural components.

The target audience for this book includes researchers in the fields of medical, biological, and pharmacological sciences and clinical medicine. We also expect undergraduate and postgraduate students of these fields to be interested in the book. We are extremely grateful to all authors who contributed to the book. Finally, we thank Jonathan Rose and Amanda Amanullah for providing us with the opportunity to edit this book and for helpful advise throughout the editorial process.

YUSUKE HIRAKU
SHOSUKE KAWANISHI
HIROSHI OHSHIMA

CHAPTER 1

INFECTION, INFLAMMATION, AND CANCER: OVERVIEW

Hiroshi Ohshima, Noriyuki Miyoshi, and Susumu Tomono

It has been estimated that about 2 million (16.1%) of the total 12.7 million new cancer cases in 2008 were attributable to infections (1). This percentage was higher in less-developed (22.9%) than in more-developed (7.4%) countries, and varied 10-fold by region from 3.3% in Australia and New Zealand to 32.7% in sub-Saharan Africa. Four major infections with *Helicobacter pylori*, hepatitis B and C viruses, and human papillomavirus are estimated to be responsible for 1.9 million cases of gastric, liver, and cervical cancer. Cervical cancer accounts for about half of the infection-related burden of cancer in women, and in men liver and gastric cancers account for over 80%. In addition, as shown in Table 1.1, chronic infection by a variety of viruses, bacteria, or parasites and tissue inflammation such as gastritis and hepatitis, which are often caused by chronic infection, are recognized risk factors for human cancers at various sites. Furthermore, the chronic inflammation induced by chemical and physical agents such as tobacco smoke and asbestos is also associated with an increased risk of cancer. Thus, chronic bronchitis and emphysema lead to increased risks of lung cancer. Inhalation of asbestos causes chronic lung and pleural inflammation and increases the risk of mesothelioma. Gastroesophageal reflux disease and Barrett's esophagus, which are caused by abdominal obesity, gastroesophageal reflux, and cigarette smoking, induce chronic inflammation and increase the risk of esophageal adenocarcinoma. Autoimmune and inflammatory diseases of uncertain etiology are also associated with an increased risk of cancer. For example, inflammatory bowel diseases such as Crohn's disease and ulcerative colitis are associated with an increased risk of colon cancer. There is an increased risk of pancreatic cancer in chronic pancreatitis. Thus, a significant fraction of the global cancer burden is attributable to chronic infection and inflammation. It is estimated that there would be about 21% fewer cases of cancer in developing countries and 9% fewer cases in developed countries if these known infectious diseases were prevented (2).

Cancer and Inflammation Mechanisms: Chemical, Biological, and Clinical Aspects, First Edition.
Edited by Yusuke Hiraku, Shosuke Kawanishi, and Hiroshi Ohshima.
© 2014 John Wiley & Sons, Inc. Published 2014 by John Wiley & Sons, Inc.

TABLE 1.1   Infection and Inflammatory Conditions as Risk Factors for Human Cancers

Cancer site	Infection/inflammation
Breast	Inflammatory breast cancer
Cervix	Human papillomaviruses, herpes simplex virus
Esophagus	Barrett's esophagitis, gastroesophageal reflux
Gallbladder and extrahepatic biliary ducts	Stone/cholecystitis, *Salmonera typhimurium*
Kaposi's sarcoma	Human immunodeficiency viruses
Large intestine (colon/rectum)	Inflammatory bowel diseases, *Schistosomiasis japonicum*
Leukemia/lymphoma	Human T-cell leukemia virus, Epstein–Barr virus, malaria
Liver /intrahepatic biliary ducts	Hepatitis viruses B and C, cirrhosis, *Opistorchis viverrini, Clonorchis sinensis, Schistosomiasis japonicum*
Lung	Cigarette smoke, particles (asbestos, silica dust, nanomaterials, etc.)
Nasopharynx	Epstein–Barr virus
Oral cavity	Leukoplakia
Pancreas	Pancreatitis
Pleura (mesothelioma)	Asbestos
Prostate	Proliferative inflammatory atrophy
Skin	Ultraviolet radiation, sunburn, human papillomaviruses
Stomach	*Helicobacter pylori*, chronic atrophic gastritis, Epstein–Barr virus
Thyroid	Thyroiditis
Urinary bladder	Stones, bacterial infections, *Schistosomiasis haematobium*

INFECTION, INFLAMMATION, AND CANCER: POSSIBLE MECHANISMS

Although various mechanisms have been proposed for infection- and inflammation-associated carcinogenesis, at many sites carcinogenic mechanisms associated with infection and inflammation have not been fully elucidated. Both direct and indirect mechanisms may be involved in carcinogenesis associated with infection. Direct mechanisms include integration of viral DNA into the human genome, which often results in alterations of host DNA (insertion, deletion, translocation, and amplification). Products of integrated viral DNA (e.g., the X protein of hepatitis B virus and the E6 and E7 proteins of human papillomavirus) interact with tumor suppressor gene products such as pRB, p53, and Bax, inactivating these proteins in host cells (see Chapters 12 and 13). Viral products such as the E6 and E7 proteins of human papillomavirus may also immortalize infected cells (e.g., human genital keratinocytes) and interact with transcription factors of host genes (e.g., activation of c-myc by the X protein of hepatitis B virus), deregulating the cell cycle, or cell growth and death. In contrast, indirect mechanisms include inflammation-related cellular and genetic alterations and viral-infection-induced immunosuppression (e.g., human immunodeficiency virus), which can increase the risks of some types of malignancy (e.g., Kaposi's sarcoma). It is likely that both direct (integration of viral DNA into host genome) and indirect (immunosuppression and inflammatory responses) mechanisms cooperate in various cases. This is evident because (1) many infectious agents associated with human cancer are ubiquitous and widely distributed, but only a small fraction of infected subjects develop cancer; (2) there is a long latency period between initial infection and cancer appearance; and (3) other lifestyle factors, such as smoking and dietary habits, are known to modify cancer risks associated with infection and inflammation.

PRODUCTION OF INFLAMMATORY MEDIATORS AND REACTIVE OXIDANTS

Inflammation is the normal physiological response to tissue injury. The cellular and tissue responses to injury can increase the blood supply and enhance vascular permeability and migration of white blood cells to damaged sites. Granulocytes, monocytes, and lymphocytes are recruited to the injured area and concomitantly produce soluble mediators such as acute-phase proteins, eicosanoids, interleukins, and cytokines. Cytokines can be divided into pro-inflammatory [interleukin (IL)-1, IL-6, IL-15, IL-17, and IL-23 and tumor necrosis factor (TNF)-α] and anti-inflammatory [IL-4, IL-10, IL-13, transforming growth factor (TGF)-β, and interferon (IFN)-α]. Pro-inflammatory cytokines such as IL-1β, TNF-α, and IL-6 play an important role in inflammation and cancer development. These cytokines activate various transcription factors, such as nuclear factor (NF)-κB and signal transducer and activator of transcription (STAT) 3, which promote cell growth, suppress apoptotic cell death, and stimulate production of growth factors, cytokines, and a variety of oxidant-generating enzymes. Furthermore, these pro-inflammatory cytokines also activate multiple oncogenic pathways, such as the mitogen-activated protein kinase (MAPK) cascade, an important signaling pathway involved in various processes of carcinogenesis, such as cell proliferation and migration, and angiogenesis. Similarly, infection and inflammation activate multiple oncogenic pathways, such as the PI3K/AKT/GSK3β/STAT3 and β-catenin pathways (3) and induce aberrant expression of activation-induced cytidine deaminase (AID) (see Chapter 9)—all of which are important in promoting carcinogenesis. Upon activation, AID can deaminate cytosine to produce uracil in DNA and thus can induce C:G-to-T:A transition and other types of mutations. The enzyme cyclooxygenase-2, which plays pivotal roles in the progression of a variety of cancers through prostaglandin synthesis, can be induced by NF-κB and can be activated by excess nitric oxide (NO), produced by inducible NO synthase (iNOS) in inflamed tissues.

NF-κB activation by pro-inflammatory cytokines stimulates various inflammatory cells (e.g., macrophages, neutrophils, basophils, eosinophils) to produce potent reactive oxygen species (ROSs) and reactive nitrogen species (RNSs), primarily to attack and destroy invading microorganisms and foreign bodies. However, if foreign agents are not eliminated rapidly, inflammation becomes chronic, which often causes extensive tissue damage, due to continuous production of excessive active proteolytic enzymes, inflammatory mediators, and potent oxidants.

During infection or inflammation, various oxidant-generating enzymes are activated, including NADPH oxidase and xanthine oxidase, which produce a superoxide anion, iNOS, which produces excess NO, and peroxidases such as myeloperoxidase and eosinophil peroxidase, which generate hypochlorous acid (HOCl) and hypobromous acid (HOBr), respectively. Furthermore, these oxidants can react with lipids to generate lipid peroxidation products, which may react further with nucleobases in DNA and RNA and amino acid residues in proteins to form adducts (see Chapters 5 and 6).

DNA AND PROTEIN DAMAGE BY ROSs AND RNSs

The oxidants can cause oxidative damage to nucleobases and sugar moieties of DNA and RNA. Many different products resulting from oxidative DNA damage have been identified. The best studied nucleobase modification includes 8-oxo-7,8-dihydro-2′-deoxyguanosine (8-oxodG), thymidine glycol, and 5-hydroxymethyl-2′-deoxyuridine. Recent advances in mass spectrometric analysis of DNA damage present in inflamed tissues have allowed us to detect numerous DNA adducts in human tissues (see Chapter 6). Especially significant increases in nucleobase modifications induced by inflammatory oxidants have been observed, including 5-chlorocytosine (halogenation damage by HOCl) in both DNA and RNA, and increased deamination product hypoxanthine in DNA in the colon of *Helicobacter hepaticus*–infected mice, an animal model for inflammatory bowel diseases and colon carcinogenesis (4). RNSs such as dinitrogen trioxide (N_2O_3), formed by oxidation of NO, can deaminate nucleobases (e.g., adenine to hypoxanthine, guanine to xanthine and cytosine to uracil, 5-methylcytosine to thymine). NO also reacts with superoxide anion to form a highly reactive nitrating and oxidizing species, peroxynitrite anion ($ONOO^-$), which can damage DNA and RNA to induce DNA strand breaks and form oxidation and nitration products of nucleobases such as 8-oxodG and 8-nitroguanine (see Chapter 4). Thus, oxidative and nitrative damage can lead to point and frameshift mutations, to single- and double-stranded breaks, and to chromosome abnormalities.

ROSs and RNSs react with proteins to modify amino acid residues by oxidation, nitrosation, nitration, and halogenation. During aging, oxidative stress, and some pathological conditions, modified forms of proteins accumulate, resulting in alterations of protein structure and function. Tyrosine residues in proteins are modified by various RNSs to form 3-nitrotyrosine and react with ROSs such as HOCl and HOBr to form 3-chloro- or 3-bromotyrosines. Thiols, metals, and radical residues in protein are also modified by ROSs and RNSs. For example, the p53 tumor suppressor protein may be inactivated with excess RNSs through formation of disulfide bonds via S-nitrosation (5) and/or nitration of tyrosine residues (6). Similarly, various DNA repair enzymes such as 8-oxoguanine DNA glycosylase 1 (7) and O^6-methylguanine transferase (8) are inactivated by ROSs and RNSs, which may increase further mutations induced by ROSs and RNSs. Caspases and other pro-apoptotic enzymes have been reported to be inhibited by ROSs and RNSs, resulting in prevention of apoptotic cell death (9).

In contrast, NO and/or other reactive species are capable of activating the proto-oncogene *c-Ha-ras* product p21 protein via S-nitrosation (10). The activated p21 protein leads to an escape of transformed cells from cell-cycle control, rendering them independent to stimulation by growth factors, giving them almost unlimited proliferative capacity. ROSs and RNSs can also activate other enzymes, such as telomerase, by whose action cells acquire replicative potential (11), and metalloproteases, which facilitate invasion by cancer cells surrounding tissues (12).

INFLAMMATION, EPIGENETIC MODIFICATION, AND microRNA

Epigenetic modifications are DNA-associated modifications that are inherited upon somatic cell replication, which include DNA methylation and histone modifications (13,14). Cytosine methylation in promoter CpG islands is associated with gene silencing. In cancer cells, the presence of regional hypermethylation (aberrant DNA methylation) and global hypomethylation has been reported. In particular, many tumor suppressor genes that have promoter CpG islands [e.g., *CDKN2A*, mutL homolog *(MLH1)*, and cadherin-1] are hypermethylated and thus silenced permanently. It has been hypothesized that inflammatory signals mainly from macrophages, such as IL-1β and IL-6 and oxidative stress, possibly produced by iNOS, probably recruit a complex with DNA methyltrasferase (DNMT) 1 and histone methyltransferase (EZH2) to promoter CpG islands, which aberrantly methylate DNA at scattered CpG sites within a CpG island (14). If a promoter CpG island of a tumor suppressor gene is hypermethylated and silenced, such methylation promotes carcinogenesis (13).

Similarly, microRNAs (miRNAs) play an important role in inflammation and cancer (see Chapter 10). miRNAs are small, noncoding RNAs that regulate the translation of specific genes by base pairing with target RNAs. Inflammation signals lead to altered miRNAs expression, which may contribute to inflammation and cancer. During inflammation, miRNA expression in epithelial cells can be altered through various mechanisms, such as activations of NF-κB and/or activator protein-1 or stimulation with cytokines. For example, increased levels of *miR-21*, which targets a number of tumor suppressor genes, such as programmed cell death 4, are found in several chronic inflammatory diseases (e.g., ulcerative colitis), and various cancers and inflammatory stimuli can increase the expression of *miR-21* (15). In contrast, the expression of *miR-7*, which targets epidermal growth factor receptor *(Egfr)* and other signaling pathways and thus acts as a tumor suppressor, is inhibited by activated macrophages in *Helicobacter*-infected gastritis (16). Upon activation with a variety of inflammatory stimuli, the aberrant expression of many different miRNAs (e.g., *let-7*, *miR-9*, *miR-98*, *miR-214*) has been shown to occur (15).

CONCLUDING REMARKS

Inflammation facilitates the initiation of normal cells and their growth and progression to malignancy. Possible mechanisms include production of pro-inflammatory cytokines and oxidants, such as ROSs and RNSs, activation of signaling and oncogenic pathways associated with inflammation and carcinogenesis, and inactivation of

TABLE 1.2   Roles of Infection and Inflammation in Various Stages of Carcinogenesis

DNA damage or mutation by ROSs and RNSs	
Inhibition of antioxidant enzymes	
Inhibition of DNA repair enzymes	
Inhibition of apoptosis	(Evading apoptosis)[a]
Inactivation of tumor-suppressor functions	(Insensitivity to antigrowth signals)
Activation of oncogenic pathways	(Self-sufficiency in growth signals)
Production of growth factors	(Self-sufficiency in growth signals)
Telomerase activation	(Limitless replicative potential)
Increased vascular permeability	
Angiogenesis induction	(Sustained angiogenesis)
Metalloproteinase activation	(Tissue invasion)
Subversion of host immune system	

[a]Six hallmark capabilities necessary for carcinogenesis proposed by Hanahan and Weinberg (17,18) are shown in parentheses.

tumor-suppressor genes and their products, through genetic alterations and epigenetic mechanisms such as aberrant DNA methylation, aberrant expression of miRNAs, and post-translational modifications of gene products. Hanahan and Weinberg (17,18) recently proposed six major characteristics (self-sufficiency in growth signals, insensitivity to antigrowth signals, evading apoptosis, limitless replicative potential, sustained angiogenesis, and tissue invasion and metastasis) that are required for a normal cell to become a tumor cell. These pathways could be disrupted by genetic alterations in genes involved in carcinogenesis (oncogenes and tumor suppressor genes) or by epigenetic processes (e.g., gene methylation; miRNAs; post-translational modifications of proteins, including histones; DNA repair enzymes), and modification of the gene expression pattern. Diverse pro-inflammatory cytokines, ROSs, and RNSs are generated in inflamed tissues and can cause genetic and epigenetic changes, affecting the six major characteristics noted above (Table 1.2).

Better understanding of the molecular mechanisms by which chronic infection and inflammation increases cancer risks will lead to the development of new strategies for cancer prevention at various sites (19). In the following chapters, cancers associated with infection and inflammation are reviewed comprehensively, and possibilities for cancer prevention by modulating inflammatory processes are discussed.

REFERENCES

1. de Martel C, Ferlay J, Franceschi S, Vignat J, Bray F, Forman D, et al. Global burden of cancers attributable to infections in 2008: a review and synthetic analysis. Lancet Oncol 2012;13:607–615.
2. Pisani P, Parkin DM, Muñoz N, Ferlay J. Cancer and infection: estimates of the attributable fraction in 1990. Cancer Epidemiol Biomark Prev 1997;6:387–400.
3. Ding SZ, Goldberg JB, Hatakeyama M. *Helicobacter pylori* infection, oncogenic pathways and epigenetic mechanisms in gastric carcinogenesis. Future Oncol. 2010;6:851–862.
4. Mangerich A, Knutson CG, Parry NM, Muthupalani S, Ye W, Prestwich E, et al. Infection-induced colitis in mice causes dynamic and tissue-specific changes in stress response and DNA damage leading to colon cancer. Proc Natl Acad Sci USA 2012;109:E1820–E1829.

5. Calmels S, Hainaut P, Ohshima H. Nitric oxide induces conformational and functional modifications of wild-type p53 tumor suppressor protein. Cancer Res 1997;57:3365–3369.

6. Chazotte-Aubert L, Hainaut P, Ohshima H. Nitric oxide nitrates tyrosine residues of tumor-suppressor p53 protein in MCF-7 cells. Biochem Biophys Res Commun 2000;267:609–613.

7. Jaiswal M, LaRusso NF, Nishioka N, Nakabeppu Y, Gores GJ. Human Ogg1, a protein involved in the repair of 8-oxoguanine, is inhibited by nitric oxide. Cancer Res 2001;61:6388–6393.

8. Liu L, Xu-Welliver M, Kanugula S, Pegg AE. Inactivation and degradation of O(6)-alkylguanine-DNA alkyltransferase after reaction with nitric oxide. Cancer Res 2002;62:3037–3043.

9. Cauwels A, Brouckaert P. Survival of TNF toxicity: dependence on caspases and NO. Arch Biochem Biophys 2007;462:132–139.

10. Lander HM, Hajjar DP, Hempstead BL, Mirza UA, Chait BT, Campbell S, et al. A molecular redox switch on p21(ras): structural basis for the nitric oxide–p21(ras) interaction. J Biol Chem 1997;272:4323–4326.

11. Vasa M, Breitschopf K, Zeiher AM, Dimmeler S. Nitric oxide activates telomerase and delays endothelial cell senescence. Circ Res 2000;87:540–542.

12. Kar S, Subbaram S, Carrico PM, Melendez JA. Redox-control of matrix metalloproteinase-1: a critical link between free radicals, matrix remodeling and degenerative disease. Respir Physiol Neurobiol 2010;174:299–306.

13. Jones PA, Baylin SB. The epigenomics of cancer. Cell 2007;128:683–692.

14. Chiba T, Marusawa H, Ushijima T. Inflammation-associated cancer development in digestive organs: mechanisms and roles for genetic and epigenetic modulation. Gastroenterology 2012;143:550–563.

15. Schetter AJ, Heegaard NH, Harris CC. Inflammation and cancer: interweaving microRNA, free radical, cytokine and p53 pathways. Carcinogenesis 2010;31:37–49.

16. Kong D, Piao YS, Yamashita S, Oshima H, Oguma K, Fushida S et al. Inflammation-induced repression of tumor suppressor miR-7 in gastric tumor cells. Oncogene 2012;31:3949–3960.

17. Hanahan D, Weinberg RA. The hallmarks of cancer. Cell 2000;100:57–70.

18. Hanahan D, Weinberg RA. Hallmarks of cancer: the next generation. Cell 2011;144:646–674.

19. Coussens LM, Zitvogel L, Palucka AK. Neutralizing tumor-promoting chronic inflammation: A magic bullet? Science 2013;339:286–291.

CHAPTER 2

STEM CELL THEORY AND INFLAMMATION-RELATED CANCER

Toshihiko Tanno and William Matsui

Several chronic inflammatory conditions are associated with increased cancer risk. These inflammatory conditions involve a myriad of factors that have been implicated in the initiation, maintenance, and progression of cancer. Some of these factors, such as the production reactive oxygen species (ROSs) leading to DNA damage and inflammatory cytokines supporting tumor cell proliferation and survival, act in an intrinsic manner to promote tumorigenesis. Furthermore, changes in the extracellular environment, including relative tissue hypoxia, may also contribute to tumor formation.

Emerging data have suggested that the ability to propagate cancer growth is not a property shared universally by all cells within an individual tumor but, rather, is restricted to specialized and distinct subpopulations commonly referred to as *cancer stem cells* (1). Cancer stem cells have been defined primarily by two functional attributes: the ability to give rise to differentiated progeny, which phenotypically recapitulate the original tumor, and self-renewal, which maintains tumor growth over time (2). The factors responsible for the development of cancer stem cells and their subsequent regulation are unclear, but many of the processes induced by chronic inflammatory states are important regulators of normal stem cells. Thus, inflammation and the cellular responses required to adapt to the inflammatory state may lead to the generation of cancer stem cells and/or regulate their functional properties. In this chapter we discuss how these processes may become subverted to enhance the development and function of cancer stem cells.

CHRONIC INFLAMMATION AND CANCER

Chronic inflammatory states have long been associated with cancer. A major cause of chronic inflammation is persistent infection by bacteria, viruses, and parasites, leading to increased cancer risk in a wide range of tissues and organs. In some cases,

Cancer and Inflammation Mechanisms: Chemical, Biological, and Clinical Aspects, First Edition.
Edited by Yusuke Hiraku, Shosuke Kawanishi, and Hiroshi Ohshima.
© 2014 John Wiley & Sons, Inc. Published 2014 by John Wiley & Sons, Inc.

the infectious agent itself may lead directly to cellular transformation. For example, cervical epithelial carcinoma is virtually always associated with human papillomavirus (HPV) infection, and viral gene products can act as oncogenes to transform or immortalize epithelial cells. Similarly, a high proportion of head and neck, vaginal, and anal squamous cell carcinomas have been found to be associated with HPV infections (3). Human Epstein–Barr virus (EBV) represents another oncogenic virus, and infections may lead to a wide variety of cancers, including Hodgkin's and Burkitt's lymphomas and nasopharyngeal carcinoma (4). Several other viruses have also been linked causally to cancer, including human T-lymphotropic virus (HTLV), Kaposi's sarcoma–associated herpesvirus/human herpesvirus-8 (HKSV/HHV-8), and Merkel cell polyoma virus (5). In each of these cases, the risk of cancer development is relatively low compared to the overall infection rate. Therefore, additional host factors are required for cancer formation.

In addition to viral carcinogenesis, both bacterial and parasitic inflections have also been associated with increased cancer risk. These include *Helicobacter pylori* and gastric cancer, *Schistosoma haematobium* and bladder carcinoma, and liver flukes and cholangiocarcinoma (6). In these cases, as well as hepatitis viral infections that increase the risk of hepatocellular cancer, persistent infections probably induce chronic immune and inflammatory responses, leading to environmental changes at the tissue and cellular levels that are ultimately responsible for carcinogenesis. The association of other chronic inflammatory states occurring in the absence of infections supports this concept. Inflammatory bowel disease, primarily ulcerative colitis, has been associated with increased rates of colorectal adenocarcinoma (7). Similarly, chronic inflammation resulting from persistent gastroesophageal reflux disease may produce Barrett's esophagus and epithelial metaplasia, which carries an increased risk of esophageal carcinoma (8). These sustained inflammatory states, arising from chronic infections or from autoimmune or mechanical dysfunction, clearly increase cancer risk, but how the vast array of factors mediating both tissue injury and repair influence the development of cancer are not precisely understood.

CANCER STEM CELLS

Sustained and increased cellular proliferation is required to generate a macroscopic tumor; thus, it is possible that chronic inflammatory states leading to cancer involve the generation of cancer stem cells. Cancer stem cells represent an emerging concept in cancer biology, suggesting that not all malignant cells are equally capable of tumor growth but are restricted to specific cell subpopulations within an individual tumor (9). This hypothesis has arisen from two basic observations regarding normal and malignant tissues. The first is that tumors are typically composed of phenotypically heterogeneous cells that display variable levels of maturation. For example, in most solid malignancies, individual tumors are not composed of monotonous and identical tumor cells but, instead, recapitulate many of the histological features of the tissue of origin, such as gland formation in adenocarcinomas. Similar findings exist in hematologic malignancies. In chronic myeloid leukemia (CML) the fusion oncogene

BCR-ABL initially revealed the clonal nature of the disease, and subsequent studies demonstrated that cells harboring BCR-ABL constituted many different phenotypic myeloid blood cell types (10).

In addition to phenotypic heterogeneity, variability in the functional attributes of tumor cells has been described in many diseases. Early studies examining animal models of sarcoma, lymphoma, and multiple myeloma demonstrated that tumors could be transferred from one affected animal to another (11–13). However, this process was relatively inefficient and required large cell numbers in most cases. These results suggested that tumorigenic potential was not a feature shared by all cells within a tumor but rather was restricted to a minority. These animal studies were later extended to human cancers by Hamburger and Salmon, who examined the growth potential of individual tumor cells by developing methods that allowed the *in vitro* propagation of primary clinical specimens (14). These assays could determine the frequency capable of forming tumor cell colonies in semisolid media, and in both solid and hematologic malignancies, only a fraction of cells were found to be clonogenic. Thus, tumor cells may display both phenotypical and functional heterogeneity, despite their clonal origins.

The second factor that helped shape the cancer stem cell hypothesis is that normal tissues and organs are functionally organized in a hierarchical manner at the cellular level. Normal adult tissues are also composed of heterogeneous cell types. Furthermore, the different cell types are arranged in a functionally hierarchal manner with stem cells sitting at the apex and undergoing lineage-specific differentiation to generate the full complement of cell types required for specific effector functions within a particular tissue. A cardinal feature of stem cells is the ability to undergo self-renewal and retain proliferative capacity over a lifetime to both maintain tissue homeostasis and permit regeneration following injury. In most systems, normal stem cells are also phenotypically distinct from their mature progeny; thus, phenotype and function are firmly linked.

The cellular heterogeneity displayed by both cancers and normal tissues suggested that tumors are organized at the cellular level in a hierarchical manner similar to that of normal tissues, and that specific functional properties were limited to distinct tumor cell phenotypes. Parallels between the organization of normal and malignant tissues were initially described by reports in CML and acute myeloid leukemia (AML) (15,16). In these studies, phenotypically distinct cell populations consisting of leukemic blasts that lacked the hematopoietic stem cell surface antigen CD34 but expressed the differentiation antigen CD38 ($CD34^{neg}CD38^{+}$) and constituted the vast majority of tumor cells and displayed some degree of normal myeloid maturation and $CD34^{+}CD38^{neg}$ cells resembling normal hematopoietic stem cells were isolated and assessed prospectively for their ability to engraft immunodeficient NOD/SCID mice. Similar to normal hematopoiesis, bulk $CD34^{neg}CD38^{+}$ cells were unable to propagate disease *in vivo*, whereas immature $CD34^{+}CD38^{neg}$ cells could engraft and generate the full complement of cancer cell phenotypes and histological recapitulation of the original tumor. Moreover, tumors forming in these animals could continue to produce tumors following serial transplantation into secondary recipients. Therefore, cancer stem cells are capable of self-renewal similar to that of their normal counterparts.

Following these initial studies in leukemia, cancer stem cells have been identified in a wide number of human malignancies, including solid tumors. Initial findings were reported in breast cancer in which cancer stem cells were originally described as a rare cell population that expressed the surface phenotype $CD44^+CD24^{neg/low}ESA^+$ and could produce tumors following injection in NOD/SCID mice (17). Simultaneously in brain tumors, rare cells expressing CD133, a surface antigen expressed by normal central nervous system stem cells, were found to be highly tumorigenic and self-renewing (18). Recent work has extended these findings to many other solid tumors, including pancreatic, colorectal, bladder, squamous cell head and neck, and prostate carcinomas (19).

Tumor growth recapitulating the original clinical specimen and self-renewal are the defining features of cancer stem cells and have implicated a central role in disease initiation, maintenance, relapse, and progression. In many diseases, most notably chronic hematologic malignancies, relapses may occur despite the induction of complete remissions and complete eradication of macroscopic disease. Relapse and tumor regrowth in this setting suggests that cells capable of producing new cancer cells must be resistant to standard treatment approaches. Indeed, cancer stem cells have been found to be relatively resistant to several diseases. In plasma cell malignancy multiple myeloma, cancer stem cells appear phenotypically to resemble normal B cells rather than plasma cells (20). Furthermore, multiple myeloma cancer stem cells are relatively resistant to multiple standard therapies used routinely in the clinical setting compared to bulk plasma cells (21). These results may explain why, ultimately, these agents fail to cure the disease, despite impressive clinical activity and the ability to reduce disease burden substantially.

The mechanisms involved in the therapeutic resistance of cancer stem cells have begun to be elucidated, and cancer stem cells may express several intrinsic properties that promote the resistance of normal stem cells to toxic injury. In multiple myeloma, cancer stem cells express increased levels of membrane-bound drug transporters and intracellular detoxification enzymes that mediate drug efflux and metabolism (21). Studies in CML and mantle cell non-Hodgkin's lymphoma (NHL) have also demonstrated that cancer stem cells that are relatively quiescent may provide resistance to cytotoxic agents reliant on cell proliferation for activity (22,23). Moreover, cellular quiescence may also lead to lower expression or activity levels of cellular factors or pathways inhibited by targeted therapies. Another mechanism of resistance may be enhanced DNA repair mechanisms, as stem cells in glioblastoma and breast cancer have been found to be relatively radioresistant compared to the bulk tumor population (24,25). In glioblastoma, cancer stem cells display increased DNA damage checkpoint responses, whereas breast cancer stem cells have the ability to minimize DNA damage by enhanced abilities to handle ROSs. In addition to these laboratory findings, breast cancer stem cells have been found to be relatively resistant in the clinical setting, as the frequency of phenotypic cancer stem cells and the clonogenic growth potential of tumors were increased following treatment with conventional chemotherapy (26).

Cancer stem cells may also play a role in disease progression, especially the development of metastatic disease in solid tumors. In particular, a relationship has been identified between cancer stem cells and the epithelial-to-mesenchymal transition (EMT). In model systems of breast cancer, the induction of EMT through the forced expression of Twist-1 or silencing of E-cadherin results in cells phenotypically resembling breast cancer stem cells (27–29). Furthermore, these cells demonstrate increased tumorigenic potential both *in vitro* and *in vivo*. Given that EMT is thought to mediate tumor cell invasion and migration that is required for tumor dissemination, cancer stem cells may play a role in metastatic disease progression. Similar findings have been described in pancreatic cancer, and cancer cells marked by CD133 have been found to possess increased metastatic potential (30). In a separate study, pancreatic cancer stem cells identified by increased expression of the intracellular detoxification enzyme aldehyde dehydrogenase (ALDH) were found to express increased levels of Slug and lower levels of E-cadherin, suggestive of EMT, and were more invasive *in vitro* (31). Moreover, pancreatic cancer stem cells expressing ALDH activity (ALDH$^+$) are increased both at the invasive front of primary tumors as well as within metastatic lesions.

Although cancer stem cells have been identified in a wide range of diseases and their functional properties suggest a critical role in multiple aspects of disease pathogenesis, data regarding their clinical relevance are limited. Given the potential that cancer stem cells are responsible for disease relapse or progression, their frequency and/or functional characteristics should correlate with clinical outcomes. Several studies have investigated the potential for cancer stem cells as novel predictive biomarkers and examined the association between cancer stem cell frequency, cancer stem cell–specific gene expression signatures, and the functional capacity of cancer stem cells with clinical outcomes. In glioblastoma the frequency of CD133$^+$ cancer stem cells is associated with inferior progression free and overall survival rates (32). Similar studies in breast and pancreatic cancer have demonstrated that the frequency of ALDH$^+$ cancer stem cells within primary tumors is associated significantly with decreased rates of overall survival (31,33).

Studies examining the relationship between cancer stem cell–specific gene expression profiles and clinical outcomes have been carried out in breast cancer and AML. In breast cancer, gene expression profiles of CD44$^+$CD24$^{neg/low}$ breast cancer stem cells and normal breast epithelium were compared to generate a 186-gene "invasiveness" gene signature that correlated with decreased overall and metastasis-free survival (34). Moreover, this gene signature was also associated with a worse prognosis in patients with other diseases, including medulloblastoma, lung cancer, and prostate cancer, suggesting that similar regulatory pathways are utilized by cancer stem cells from disparate diseases. Similarly, a second study found that patients with primary breast tumors with gene expression patterns similar to that of isolated CD44$^+$CD24$^{neg/low}$ cells from normal and malignant breast tissues experienced poor clinical outcomes (35). In AML, gene expression signatures derived from normal hematopoietic and leukemic stem cells were similarly associated with decreased relapse-free and overall survival (36). Therefore, cancer stem cell–specific gene expression may be able to predict clinical outcomes either because the frequency of cancer stem cells is increased or the tumor cells in general acquire cancer stem cell characteristics.

Several studies have also examined the relationship between quantitative measurements of cancer stem cell function and disease outcomes. In AML, immunodeficient mouse engraftment of primary AML specimens is associated with poor overall survival (37). Similarly, *in vitro* tumor sphere formation and tumor formation in mice have been correlated with outcomes in glioblastoma patients (38). Therefore, a range of phenotypic, genetic, and functional assays based on cancer stem cells have been associated with clinical outcomes and strongly implicate an important role in clinical oncology.

CARCINOGENIC FACTORS ARISING FROM CHRONIC INFLAMMATION

Several potential factors arising during chronic inflammation may contribute to cancer formation and progression. The release of cytokines by immune effectors is a major feature of the chronic inflammatory response, and these soluble factors may play several roles in the development of cancer. Major inflammatory mediators include tumor necrosis factor-alpha (TNF-α), interleukin (IL)-1, IL-6, and transforming growth factor-beta (TGF-β), and each of these is associated strongly with human cancers (39,40). For example, increased levels of TNF-α have been documented to increase the risk of developing multiple myeloma and gastric cancer as well as correlating with a poorer prognosis (41,42). A major cellular response to TNF-α is increased generation of intracellular ROSs, which may result in DNA damage, mutagenesis, and tumor initiation or progression (43). IL-6 is also commonly increased in chronic inflammatory states, and in many malignancies, including multiple myeloma, non-Hodgkin's lymphoma, and hepatocellular carcinoma, IL-6 can directly stimulate tumor cell proliferation and enhance cell survival (44,45). TGF-β may both affect the tumor microenvironment and directly regulate cancer cells. In epithelial malignancies, TGF-β may induce EMT and enhance the invasive and migratory properties of cancer cells (46). In addition, increased levels of TGF-β in the microenvironment may promote tumor growth by suppressing the tumor surveillance activities of T cells, natural killer cells, neutrophils, monocytes, and macrophages (47). Therefore, cytokines and growth factors produced during chronic inflammation may have multiple effects on both the microenvironment and the tumor cells themselves.

Increased ROS levels are associated with inflammatory states as well as cancer. Major sources of increased ROSs (O_2^-, H_2O_2, ${}^\bullet OH$, HOCl) include immune effector cells, increased cytokine signaling, as well as altered tumor cell metabolism. In turn, elevated ROS levels may broadly enhance tumorigenesis, including the induction of genotoxic damage that leads to oncogenic mutations (39). In addition, increased intracellular ROS levels may lead to the activation of redox-sensitive transcription factors that enhance tumor formation (48,49). For example, the Forkhead box class O (FoxO) transcription factors are activated in response to increased levels of ROSs through the c-Jun N-terminal kinase–dependent signaling pathway and induce the expression of cellular factors that serve as ROS scavengers to attenuate cellular damage (50–52). FoxO transcription factors can also regulate a wide variety of additional cellular functions, such as proliferation, apoptosis, and differentiation, which may promote tumorigenesis and cancer progression (53–55). Increased intracellular ROS levels may also induce activation of Nrf2, a bZIP transcription factor that regulates the cellular response to increased oxidative states by inducing the expression of many cytoprotective genes, including those involved in the glutathione and thioredoxin pathways and xenobiotic metabolism (56). In developing tumors, Nrf2 activation appears to be required to protect cells against oncogene-induced increases in ROSs that would otherwise result in cellular senescence. In a mouse model of pancreatic cancer, the loss of Nrf2 signaling impaired efficient tumor formation (57). Furthermore, increased activation of Nrf2 in established tumors has been described in many diseases and may contribute to chemoresistance (58).

Most chronic inflammatory states are associated with anemia, which may result in relative hypoxia at the tissue level. Several cytokines released during inflammatory responses are well recognized to impair both hematopoiesis and erythropoiesis (59). For example, increased levels of IL-6 may decrease serum iron levels and the subsequent production of red blood cells (60). Despite the lack of direct evidence that anemia is a cause of cancer, a case–control study has found that a significantly increased proportion of patients with hematologic malignancies, including leukemia, lymphoma, and multiple myeloma, had documented anemia two to three years before diagnosis (61). Although this finding may be explained by impaired hematopoiesis as a consequence of early disruption of normal blood formation by a hematologic malignancy, a similar finding was found for patients with gastrointestinal cancer, suggesting that the anemia itself may contribute to clinical cancer development.

Chronic inflammation results in protracted tissue injury and subsequent tissue repair, which may also contribute to the development of cancer. In many organ systems, tissue repair and regeneration may involve specific pathways, such as Wnt, Notch, and Hedgehog (Hh), that regulate patterning events during normal embryonic development. These pathways are subsequently silenced in most tissues but are frequently reactivated to induce repair after injury. Interestingly, these pathways has been found to be activated aberrantly in a large number of human malignancies (62). Therefore, the early phases of tumorigenesis may represent deregulated attempts to repair tissues, and the signaling pathways associated with these processes may play a central role in this process.

CHRONIC INFLAMMATION AND NORMAL STEM CELL FUNCTION

Many factors that are generated during chronic inflammation may directly affect normal stem cell function. TNF-α has been found to induce NF-κB signaling that leads to neural stem cell proliferation as well as the inhibition of differentiation (63). Similarly, IL-6 can enhance the self-renewal of hematopoietic stem cells (64).

ROS may also be important regulators of normal stem cell function (50,65). Normal hematopoietic stem cells contain lower ROS levels than those of progenitors and other more differentiated blood cells, and the maintenance of low ROS levels is required to maintain self-renewal potential by preventing differentiation (66). Evidence that hematopoietic stem cell function is coupled to ROS levels has been provided by several studies involving transgenic mouse models. Here, conditional loss of specific genes, such as the FoxO3 transcription factors, ataxia telangiecta-sia mutated (ATM), p53, and p38 mitogen-activated protein (MAP) kinase, leads to increased ROS levels and impaired hematopoietic stem cell function (50,67–69). The mechanisms though which ROS levels regulate hematopoiesis and hematopoietic stem cells are not understood precisely, but in the case of the loss of FoxO3 tran-scription factors, hematopoietic stem cells undergo increased proliferation, and the loss of cellular quiescence is a well-recognized mechanism impairing self-renewal (70). Similarly, the loss of FoxO3 in the central nervous system results in a reduction in both the number and function of neural stem cells (71). Therefore, ROS levels may regulate normal stem cells in a wide range of organ systems by influencing both self-renewal and differentiation.

In addition to ROS levels, stem cell quiescence and self-renewal may be regu-lated by hypoxia (72,73). The cellular response to hypoxia is mediated by the activities of the hypoxia-inducible transcription factors (HIFs). In both normal neurogenesis and hematopoiesis, normal stem cells are located in areas of relatively low oxygen levels (74). These hypoxic conditions induce activation of HIF-1α transcriptional activity and the expression of its transcriptional targets, such as FoxO3, required for self-renewal and maintenance of the hematopoietic stem cell pool (75). Thus, hypoxia and HIF-1α may directly promote stem cell quiescence and maintain self-renewal potential.

Embryonic development pathways such as Hh, Notch, and Wnt that are acti-vated during tissue regeneration are also likely to regulate stem cells. During normal development these signaling pathways regulate critical stem cell fate decisions that allow proper patterning and organogenesis to be carried out. Following birth, these pathways have similarly been found to be required for normal stem cell self-renewal and differentiation in many adult tissues (9). Therefore, the tissue injury may lead to reactivation of developmental programs to institute repair, a process that requires stem cell activation.

INFLAMMATION AND THE DEVELOPMENT AND REGULATION OF CANCER STEM CELLS

Since both carcinogenesis and normal stem cell function may be affected by factors associated with chronic inflammatory states, it is possible that these activities directly influence the generation and/or regulation of cancer stem cells. The increased ROS levels produced by inflammatory cytokines and immune effectors within chronically inflamed tissues may affect cancer stem cells in multiple ways. Given the role of ROS levels in the maintenance of normal stem cells, it is possible that similar mechanisms influence cancer stem cells. In normal hematopoietic stem cells, the impact of ROS levels is mediated to some extent by the activity of the FoxO transcription factors (70). Similarly, in a mouse model of CML, FoxO3 has recently been found to be similarly required for the self-renewal of leukemic stem cells (76). Interestingly, cancer stem cells in AML have been found to higher levels of ROSs compared to their normal counterparts, but this relative increase may lead to novel targeting mechanisms that are specific for leukemic stem cells (77). Therefore, it is possible that increased ROS levels generated during chronic inflammatory states induce adaptive changes that lead to aberrant self-renewal through factors such as FoxO3. Furthermore, increased ROSs, levels within normal or cancer stem cells may result in DNA damage that may lead to initial cancer formation or additional mutational events that promote disease progression. In response to increased levels of ROS, several cellular responses are induced, and these may promote drug resistance (78–80). For example, the Nrf2 transcription factor is activated by increased levels of oxidative stress, and in lung cancer, mutational activation of its activity has been found to promote drug resistance, presumably by inducing the expression of a multitude of detoxification enzymes (81). Moreover, Nrf2 may play a role in the maintenance and self-renewal of cancer stem cells, as the loss of Nrf2 activity has been found to affect the function of normal hematopoietic stem cells (82). Therefore, ROS may regulate several aspects of cancer stem cell biology.

Increased local levels of cytokines and growth factors may also affect cancer stem cell function. IL-6 has been found to enhance the tumorigenic potential and self-renewal of glioblastoma cancer stem cells (83). In a similar fashion, TGF-β may affect CML cancer stem cells and induce their self-renewal through the induction of AKT signaling (76). In addition to its potential role in regulating cancer stem cells in hematopoietic malignancies, TGF-β may also influence cancer stem cells in solid tumors through its well-recognized effects on EMT (84). Emerging data suggest that EMT represents a transitional state for epithelial stem cells and differentiated cells, and several studies have demonstrated a direct relationship between EMT and cancer stem cells in solid tumors. For example, the induction of EMT through forced expression of the Snail family of transcriptional regulators, the loss of E-cadherin expression, or treatment with TGF-β in immortalized mammary epithelial or breast cancer cells can promote the formation of cells with phenotype identical to reported breast cancer stem cells as well as increased tumorigenic potential (27–29).

Hypoxia may also directly regulate cancer stem cells (85). Studies of brain tumors have demonstrated that HIF-1α is activated in cancer stem cells located within hypoxic regions of the tumor, and this activation inhibits differentiation and increases self-renewal potential (86,87). In normal hematopoiesis, HIF-1α is also expressed in hematopoietic stem cells that are located within hypoxic regions of the bone marrow (85). In an analogous situation, leukemia cells within the bone marrow are also found to express HIF-1α within hypoxic regions (88). Emerging data have suggested that relative tissue hypoxia and the induction of HIF-1α activity may also affect tumor metabolism. In AML, recurrent somatic mutations in isocitrate dehydrogenase (IDH) have been described, and in addition to epigenetic changes, the loss of IDH activity has been found to increase ROS levels and HIF-1α activity (89). Furthermore, similar mutations have been described during the conversion of myeloproliferative diseases to AML; thus, ROSs and HIF-1α may further affect cancer stem cell biology during disease progression (90,91). Hypoxia and the subsequent activation of HIF-1α may also affect cancer stem cells by promoting EMT through the induction of Twist-1 expression to enhance tumorigenicity and metastatic potential (92,93). Therefore, hypoxic stress may regulate cancer stem cells through a variety of mechanisms.

Finally, the factors mediating tissue repair and regeneration may also affect the formation and regulation of cancer stem cells. For example, the activation of developmental signaling pathways required for tissue regeneration may be maintained aberrantly during chronic injury and result in abnormal stem cell expansion or deregulated self-renewal. For example, mutations within components of the Hh signaling pathway that lead to constitutive pathway activity are frequently identified in skin cancers and brain tumors, suggesting that aberrant pathway activity promotes the formation of cancer stem cells (94,95). Increased Hh signaling pathway activity has also been reported in cancer stem cells in multiple myeloma, acute lymphoblastic leukemia, CML, glioblastoma, and pancreatic cancer (96–100). Furthermore, inhibition of Hh pathway activity can inhibit tumorigenicity, self-renewal, and metastatic potential (101). Aberrant activation of Wnt signaling has also been found to be associated with disease progression in CML by conferring self-renewal potential on typically self-limited progenitors (102). Therefore, the inhibition of developmental signaling pathways may serve as novel strategies to inhibit CSCs.

CONCLUDING REMARKS

The association between chronic inflammation and an increased risk of cancer is well recognized. Several factors involved in chronic inflammation, including inflammatory cytokines, increased ROS levels, hypoxia, and the activation of developmental signaling pathway, have been associated with both carcinogenesis and normal stem cell regulation. Furthermore, emerging evidence suggests that these factors also play a role in regulating cancer stem cells. It is likely that none of these factors act in an isolated manner, but interact to affect cancer stem cells properties, including self-renewal, drug resistance, and metastatic potential. Further studies better defining the precise ways in which these factors affect cancer stem cells may reveal novel strategies to prevent carcinogenesis and to target cancer stem cells.

REFERENCES

1. Jordan CT, Guzman ML, Noble M. Cancer stem cells. New Engl J Med 2006;355:1253–1261.
2. Clarke MF, Dick JE, Dirks PB, Eaves CJ, Jamieson CHM, Jones DL, et al. Cancer stem cells—perspectives on current status and future directions: AACR workshop on cancer stem cells. Cancer Res 2006;66:9339–9344.
3. Watson M, Saraiya M, Ahmed F, Cardinez CJ, Reichman ME, Weir HK, et al. Using population-based cancer registry data to assess the burden of human papillomavirus-associated cancers in the United States: overview of methods. Cancer 2008;113:2841–2854.
4. Thompson MP, Kurzrock R. Epstein–Barr virus and cancer. Clin Cancer Res 2004;10:803–821.
5. Martin D, Gutkind J. Human tumor-associated viruses and new insights into the molecular mechanisms of cancer. Oncogene 2008;27:S31–S42.
6. Kuper H, Adami HO, Trichopoulos D. Infections as a major preventable cause of human cancer. J Int Med 2000;248:171–183.
7. Eaden J, Abrams K, Mayberry J. The risk of colorectal cancer in ulcerative colitis: a meta-analysis. Gut 2001;48:526–535.
8. Shaheen N, Ransohoff DF. Gastroesophageal reflux, Barrett esophagus, and esophageal cancer. JAMA 2002;287:1972–1981.
9. Reya T, Morrison SJ, Clarke MF, Weissman IL. Stem cells, cancer, and cancer stem cells. Nature 2001;414:105–111.

10. Fialkow PJ, Jacobson RJ, Papayannopoulou T. Chronic myelocytic leukemia: clonal origin in a stem cell common to the granulocyte, erythrocyte, platelet and monocyte/macrophage. Am J Med 1977;63:125–130.

11. Hewitt H. Studies of the quantitative transplantation of mouse sarcoma. Br J Cancer 1953;7:367.

12. Bruce WR, Van Der Gaag H. A quantitative assay for the number of murine lymphoma cells capable of proliferation *in vivo*. Nature 1963;199:79–80.

13. Bergsagel DE, Valeriote FA. Growth characteristics of a mouse plasma cell tumor. Cancer Res 1968;28:2187–2196.

14. Hamburger AW, Salmon SE. Primary bioassay of human tumor stem cells. Science 1977;197:461–463.

15. Wang JC, Lapidot T, Cashman JD, Doedens M, Addy L, Sutherland DR, et al. High level engraftment of NOD/SCID mice by primitive normal and leukemic hematopoietic cells from patients with chronic myeloid leukemia in chronic phase. Blood 1998;91:2406–2414.

16. Lapidot T, Sirard C, Vormoor J, Murdoch B, Hoang T, Caceres-Cortes J, et al. A cell initiating human acute myeloid leukaemia after transplantation into SCID mice. Nature 1994;367:645–648.

17. Al-Hajj M, Wicha MS, Benito-Hernandez A, Morrison SJ, Clarke MF. Prospective identification of tumorigenic breast cancer cells. Proc Natl Acad Sci USA 2003;100:3983–3988.

18. Singh SK, Hawkins C, Clarke ID, Squire JA, Bayani J, Hide T, et al. Identification of human brain tumour initiating cells. Nature 2004;432:396–401.

19. Dalerba P, Cho RW, Clarke MF. Cancer stem cells: models and concepts. Ann Rev Med 2007;58:267–284.

20. Matsui W, Huff CA, Wang Q, Malehorn MT, Barber J, Tanhehco Y, et al. Characterization of clonogenic multiple myeloma cells. Blood 2004;103:2332–2336.

21. Matsui W, Wang Q, Barber JP, Brennan S, Smith BD, Borrello I, et al. Clonogenic multiple myeloma progenitors, stem cell properties, and drug resistance. Cancer Res 2008;68:190–197.

22. Graham SM, Jorgensen HG, Allan E, Pearson C, Alcorn MJ, Richmond L, et al. Primitive, quiescent, Philadelphia-positive stem cells from patients with chronic myeloid leukemia are insensitive to STI571 *in vitro*. Blood 2002;99:319–325.

23. Brennan SK, Meade B, Wang Q, Merchant AA, Kowalski J, Matsui W. Mantle cell lymphoma activation enhances bortezomib sensitivity. Blood 2010;116:4185–4191.

24. Bao S, Wu Q, Mclendon RE, Hao Y, Shi Q, Hjelmeland AB, et al. Glioma stem cells promote radioresistance by preferential activation of the DNA damage response. Nature 2006;444:756–760.

25. Diehn M, Cho RW, Lobo NA, Kalisky T, Dorie MJ, Kulp AN, et al. Association of reactive oxygen species levels and radioresistance in cancer stem cells. Nature 2009;458:780–783.

26. Creighton CJ, Li X, Landis M, Dixon JM, Neumeister VM, Sjolund A, et al. Residual breast cancers after conventional therapy display mesenchymal as well as tumor-initiating features. Proc Natl Acad Sci USA 2009;106:13820–13825.

27. Vesuna F, Lisok A, Kimble B, Raman V. Twist modulates breast cancer stem cells by transcriptional regulation of CD24 expression. Neoplasia 2009;11:1318–1328.

28. Gupta PB, Onder TT, Jiang G, Tao K, Kuperwasser C, Weinberg RA, et al. Identification of selective inhibitors of cancer stem cells by high-throughput screening. Cell 2009;138:645–659.

29. Mani SA, Guo W, Liao M-J, Eaton EN, Ayyanan A, Zhou AY, et al. The epithelial–mesenchymal transition generates cells with properties of stem cells. Cell 2008;133:704–715.

30. Hermann PC, Huber SL, Herrler T, Aicher A, Ellwart JW, Guba M, et al. Distinct populations of cancer stem cells determine tumor growth and metastatic activity in human pancreatic cancer. Cell Stem Cell 2007;1:313–323.

31. Rasheed ZA, Yang J, Wang Q, Kowalski J, Freed I, Murter C, et al. Prognostic significance of tumorigenic cells with mesenchymal features in pancreatic adenocarcinoma. J Natl Can Inst 2010;102:340–351.

32. Zeppernick F, Ahmadi R, Campos B, Dictus C, Helmke BM, Becker N, et al. Stem cell marker CD133 affects clinical outcome in glioma patients. Clin Cancer Res 2008;14:123–129.

33. Ginestier C, Hur MH, Charafe-Jauffret E, Monville F, Dutcher J, Brown M, et al. ALDH1 is a marker of normal and malignant human mammary stem cells and a predictor of poor clinical outcome. Cell Stem Cell 2007;1:555–567.

34. Liu R, Wang X, Chen GY, Dalerba P, Gurney A, Hoey T, et al. The prognostic role of a gene signature from tumorigenic breast-cancer cells. N Engl J Med 2007;356:217–226.

35. Shipitsin M, Campbell LL, Argani P, Weremowicz S, Bloushtain-Qimron N, Yao J, et al. Molecular definition of breast tumor heterogeneity. Cancer Cell 2007;11:259–273.

36. Eppert K, Takenaka K, Lechman ER, Waldron L, Nilsson B, van Galen P, et al. Stem cell gene expression programs influence clinical outcome in human leukemia. Nat Med 2011;17:1086–1093.

37. Pearce DJ, Taussig D, Zibara K, Smith L-L, Ridler CM, Preudhomme C, et al. AML engraftment in the NOD/SCID assay reflects the outcome of AML: implications for our understanding of the heterogeneity of AML. Blood 2006;107:1166–1173.

38. Laks DR, Masterman-Smith M, Visnyei K, Angenieux B, Orozco NM, Foran I, et al. Neurosphere formation is an independent predictor of clinical outcome in malignant glioma. Stem Cells 2009;27:980–987.

39. Perwez Hussain S, Harris CC. Inflammation and cancer: an ancient link with novel potentials. Int J Cancer 2007;121:2373–2380.

40. Vallabhapurapu S, Karin M. Regulation and function of NF-κB transcription factors in the immune system. Ann Rev Immun 2009;27:693–733.

41. Davies FE, Rollinson SJ, Rawstron AC, Roman E, Richards S, Drayson M, et al. High-producer haplotypes of tumor necrosis factor alpha and lymphotoxin alpha are associated with an increased risk of myeloma and have an improved progression-free survival after treatment. J Clin Oncol 2000;18:2843–2851.

42. Machado JC, Figueiredo C, Canedo P, Pharoah P, Carvalho R, Nabais S, et al. A proinflammatory genetic profile increases the risk for chronic atrophic gastritis and gastric carcinoma. Gastroenterology 2003;125:364–371.

43. Shoji Y, Uedono Y, Ishikura H, Takeyama N, Tanaka T. DNA damage induced by tumour necrosis factor-alpha in L929 cells is mediated by mitochondrial oxygen radical formation. Immunology 1995;84:543.

44. Aggarwal BB, Shishodia S, Sandur SK, Pandey MK, Sethi G. Inflammation and cancer: How hot is the link? Biochem Pharm 2006;72:1605–1621.

45. Grivennikov SI, Karin M. Dangerous liaisons: STAT3 and NF-κB collaboration and crosstalk in cancer. Cytokine Growth Factor Rev 2010;21:11–19.

46. Derynck R, Akhurst RJ, Balmain A. TGF-beta signaling in tumor suppression and cancer progression. Nat Genet 2001;29:117–130.

47. Bierie B, Moses HL. Tumour microenvironment: TGF-beta: the molecular Jekyll and Hyde of cancer. Nat Rev Cancer 2006;6:506–520.

48. Trachootham D, Lu W, Ogasawara MA, Valle NRD, Huang P. Redox regulation of cell survival. Antiox Redox Signaling 2008;10:1343–1374.

49. Giles GI. The redox regulation of thiol dependent signaling pathways in cancer. Curr Pharm Des 2006;12:4427–4443.

50. Tothova Z, Kollipara R, Huntly BJ, Lee BH, Castrillon DH, Cullen DE, et al. FoxOs are critical mediators of hematopoietic stem cell resistance to physiologic oxidative stress. Cell 2007;128:325–339.

51. Naka K, Muraguchi T, Hoshii T, Hirao A. Regulation of reactive oxygen species and genomic stability in hematopoietic stem cells. Antiox Redox Signal 2008;10:1883–1894.

52. Essers MAG, Weijzen S, de Vries-Smits AMM, Saarloos I, De Ruiter ND, Bos JL, et al. FoxO transcription factor activation by oxidative stress mediated by the small GTPase Ral and JNK. EMBOJ 2004;23:4802–4812.

53. Ho K, Myatt S, Lam EWF. Many forks in the path: cycling with FoxO. Oncogene 2008;27:2300–2311.

54. Gomes AR, Brosens JJ, Lam EWF. Resist or die. Cell Cycle 2008;7:3133–3136.

55. Burgering BMT. A brief introduction to FoxOlogy. Oncogene 2008;27:2258–2262.

56. Kensler TW, Wakabayashi N, Biswal S. Cell survival responses to environmental stresses via the Keap1–Nrf2–ARE pathway. Annu Rev Pharmacol Toxical 2007;47:89–116.

57. DeNicola GM, Karreth FA, Humpton TJ, Gopinathan A, Wei C, Frese K, et al. Oncogene-induced Nrf2 transcription promotes ROS detoxification and tumorigenesis. Nature 2011;475:106–109.

58. Lau A, Villeneuve NF, Sun Z, Wong PK, Zhang DD. Dual roles of Nrf2 in cancer. Pharm Res (NY) 2008;58:262–270.

59. Huber C, Batchelor J, Fuchs D, Hausen A, Lang A, Niederwieser D, et al. Immune response–associated production of neopterin: release from macrophages primarily under control of interferon-gamma. J Exp Med 1984;160:310.

60. Ganz T, Nemeth E. *Iron sequestration and anemia of inflammation*. NIH Publ Access; 2009; p. 387.

61. Edgren G, Bagnardi V, Bellocco R, Hjalgrim H, Rostgaard K, Melbye M, et al. Pattern of declining hemoglobin concentration before cancer diagnosis. Int J Cancer 2010;127:1429–1436.

62. Beachy PA, Karhadkar SS, Berman DM. Tissue repair and stem cell renewal in carcinogenesis. Nature 2004;432:324–331.

63. Widera D, Mikenberg I, Elvers M, Kaltschmidt C, Kaltschmidt B. Tumor necrosis factor alpha triggers proliferation of adult neural stem cells via IKK/NF-κB signaling. BMC Neurosci. 2006;7:64.

64. Audet J, Miller CL, Rose-John S, Piret JM, Eaves CJ. Distinct role of gp130 activation in promoting self-renewal divisions by mitogenically stimulated murine hematopoietic stem cells. Proc Natl Acad Sci USA 2001;98:1757.

65. Naka K, Muraguchi T, Hoshii T, Hirao A. Regulation of reactive oxygen species and genomic stability in hematopoietic stem cells. Antiox Redox Signal 2008;10:1883–1894.

66. Jang Y-Y, Sharkis SJ. A low level of reactive oxygen species selects for primitive hematopoietic stem cells that may reside in the low-oxygenic niche. Blood 2007;110:3056–3063.

67. Ito K, Takubo K, Arai F, Satoh H, Matsuoka S, Ohmura M, et al. Regulation of reactive oxygen species by ATM is essential for proper response to DNA double-strand breaks in lymphocytes. J Immunol 2007;178:103–110.

68. Ito K, Hirao A, Arai F, Takubo K, Matsuoka S, Miyamoto K, et al. Reactive oxygen species act through p38 MAPK to limit the lifespan of hematopoietic stem cells. Nat Med 2006;12:446–451.

69. Liu Y, Elf SE, Miyata Y, Sashida G, Huang G, Di Giandomenico S, et al. p53 regulates hematopoietic stem cell quiescence. Cell Stem Cell 2009;4:37–48.

70. Miyamoto K, Araki KY, Naka K, Arai F, Takubo K, Yamazaki S, et al. FoxO3a is essential for maintenance of the hematopoietic stem cell pool. Cell Stem Cell 2007;1:101–112.

71. Renault VM, Rafalski VA, Morgan AA, Salih DAM, Brett JO, Webb AE, et al. FoxO3 regulates neural stem cell homeostasis. Cell Stem Cell 2009;5:527–539.

72. Adelman DM, Maltepe E, Simon MC. Multilineage embryonic hematopoiesis requires hypoxic ARNT activity. Genes Dev 1999;13:2478–2483.

73. Gilbertson RJ, Rich JN. Making a tumour's bed: glioblastoma stem cells and the vascular niche. Nat Rev Cancer 2007;7:733–736.

74. Parmar K, Mauch P, Vergilio JA, Sackstein R, Down JD. Distribution of hematopoietic stem cells in the bone marrow according to regional hypoxia. Proc Natl Acad Sci USA 2007;104:5431.

75. Eliasson P, Jonsson JI. The hematopoietic stem cell niche: low in oxygen but a nice place to be. J Cell Physiol 2010;222:17–22.

76. Naka K, Hoshii T, Muraguchi T, Tadokoro Y, Ooshio T, Kondo Y, et al. TGF-beta-FoxO signalling maintains leukaemia-initiating cells in chronic myeloid leukaemia. Nature 2010;463:676–680.

77. Guzman ML, Rossi RM, Karnischky L, Li X, Peterson DR, Howard DS, Jordan CT. The sesquiter-pene lactone parthenolide induces apoptosis of human acute myelogenous leukemia stem and pro-genitor cells. Blood 2005;105:4163–4169.

78. Sullivan R, Graham CH. Chemosensitization of cancer by nitric oxide. Curr Pharm Des 2008;14:1113–1123.

79. Tiligada E. Chemotherapy: induction of stress responses. Endo Rel Cancer 2006;13:S115–S124.

80. Pervaiz S, Clement MV. Tumor intracellular redox status and drug resistance: Serendipity or a causal relationship? Curr Pharm Design 2004;10:1969–1977.

81. Solis LM, Behrens C, Dong W, Suraokar M, Ozburn NC, Moran CA, et al. Nrf2 and keap1 abnormal-ities in non-small cell lung carcinoma and association with clinicopathologic features. Clin Cancer Res 2010;16:3743–3753.

82. Merchant AA, Singh A, Matsui W, Biswal S. The redox-sensitive transcription factor Nrf2 reg-ulates murine hematopoietic stem cell survival independently of ROS levels. Blood 2011;118:6572–6579.

83. Nilsson CL, Dillon R, Devakumar A, Shi SDH, Greig M, Rogers JC, et al. Quantitative phos-phoproteomic analysis of the STAT3/IL-6/HIF1alpha signaling network: an initial study in GSC11 glioblastoma stem cells. J Protozool Res 2009;9:430–443.

84. Zavadil J, Bottinger EP. TGF-beta and epithelial-to-mesenchymal transitions. Oncogene 2005;24:5764–5774.

85. Mohyeldin A, Garzón-Muvdi T, Quiñones-Hinojosa A. Oxygen in stem cell biology: a critical component of the stem cell niche. Cell Stem Cell 2010;7:150–161.

86. Jogi A, Ora I, Nilsson H, Lindeheim O, Makino Y, Poellinger L, et al. Hypoxia alters gene expression in human neuroblastoma cells toward an immature and neural crest-like phenotype. Proc Natl Acad Sci USA 2002;99:7021.

87. Li Z, Bao S, Wu Q, Wang H, Eyler C, Sathornsumetee S, et al. Hypoxia-inducible factors regulate tumorigenic capacity of glioma stem cells. Cancer Cell 2009;15:501–513.

88. Wellmann S, Guschmann M, Griethe W, Eckert C, Stackelberg A, Lottaz C, et al. Activation of the HIF pathway in childhood ALL, prognostic implications of VEGF. Leukemia 2004;18:926–933.

89. Gross S, Cairns RA, Minden MD, Driggers EM, Bittinger MA, Jang HG, et al. Cancer-associated metabolite 2-hydroxyglutarate accumulates in acute myelogenous leukemia with isocitrate dehydrogenase 1 and 2 mutations. J Exp Med 2010;207:339–344.

90. Green A, Beer P. Somatic mutations of IDH1 and IDH2 in the leukemic transformation of myeloproliferative neoplasms. N Engl J Med 2010;362:369–370.

91. Ward PS, Patel J, Wise DR, Abdel-Wahab O, Bennett BD, Coller HA, et al. The common feature of leukemia-associated IDH1 and IDH2 mutations is a neomorphic enzyme activity converting [alpha]-ketoglutarate to 2-hydroxyglutarate. Cancer Cell 2010;17:225–234.

92. Nguyen QD, De Wever O, Bruyneel E, Hendrix A, Xie WZ, Lombet A, et al. Commutators of PAR-1 signaling in cancer cell invasion reveal an essential role of the Rho–Rho kinase axis and tumor microenvironment. Oncogene 2005;24:8240–8251.

93. Yang MH, Wu MZ, Chiou SH, Chen PM, Chang SY, Liu CJ, et al. Direct regulation of TWIST by HIF-alpha promotes metastasis. Nat Cell Biol 2008;10:295–305.

94. Johnson RL, Rothman AL, Xie J, Goodrich LV, Bare JW, Bonifas JM, et al. Human homolog of patched, a candidate gene for the basal cell nevus syndrome. Science 1996;272:1668–1671.

95. Goodrich LV, Milenkovic L, Higgins KM, Scott MP. Altered neural cell fates and medulloblastoma in mouse patched mutants. Science 1997;277:1109–1113.

96. Peacock CD, Wang Q, Gesell GS, Corcoran-Schwartz IM, Jones E, Kim J, et al. Hedgehog signaling maintains a tumor stem cell compartment in multiple myeloma. Proc Natl Acad Sci USA 2007;104:4048–4053.

97. Lin TL, Wang QH, Brown P, Peacock C, Merchant AA, Brennan S, et al. Self-renewal of acute lymphocytic leukemia cells is limited by the Hedgehog pathway inhibitors cyclopamine and IPI-926. PloS One 2010;5:e15262.

98. Zhao C, Chen A, Jamieson CH, Fereshteh M, Abrahamsson A, Blum J, et al. Hedgehog signalling is essential for maintenance of cancer stem cells in myeloid leukaemia. Nature 2009;458:776–779.

99. Bar EE, Chaudhry A, Lin A, Fan X, Schreck K, Matsui W, et al. Cyclopamine-mediated Hedgehog pathway inhibition depletes stem-like cancer cells in glioblastoma. Stem Cells 2007;25:2524–2533.

100. Feldmann G, Dhara S, Fendrich V, Bedja D, Beaty R, Mullendore M, et al. Blockade of Hedgehog signaling inhibits pancreatic cancer invasion and metastases: a new paradigm for combination therapy in solid cancers. Cancer Res 2007;67:2187–2196.

101. Merchant AA, Matsui W. Targeting Hedgehog—a cancer stem cell pathway. Clin Cancer Res 2010;16:3130–3140.

102. Jamieson CHM, Ailles LE, Dylla SJ, Muijtjens M, Jones C, Zehnder JL, et al. Granulocyte–macrophage progenitors as candidate leukemic stem cells in blast-crisis CML. N Engl J Med 2004;351:657–667.

CHAPTER 3

EPITHELIAL–MESENCHYMAL TRANSITION: A LINK BETWEEN CANCER AND INFLAMMATION

Jonas Fuxe and Mikael C. I. Karlsson

Inflammation has emerged as a key factor promoting cancer metastasis, the predominant cause of death in cancer (1,2). Different types of immune cells, including tumor-associated macrophages (TAMs), neutrophils, myeloid-derived suppressor cells (MDSCs), dendritic cells (DCs), and T and B cells, are found within and adjacent to tumor tissues, indicating that cancer-associated inflammation involves both innate and adaptive immune responses. Immune cells are rich sources of growth factors, cytokines, and proteases that upon secretion act in a paracrine fashion on multiple cell types and matrix components of the tumor microenvironment. Signaling events activated by these factors may lead to tissue remodeling through reactivation of certain latent developmental processes. For example, TAM-derived vascular endothelial growth factor-A (VEGF-A) is a major driving force of angiogenesis in experimental tumor models (2–4). Similarly, VEGF-C and VEGF-D produced by TAMs promote peritumoral lymphangiogenesis in cervical tumors (5). Another cytokine, transforming growth factor $\beta 1$ (TGF-$\beta 1$), is produced by various types of immune cells in the tumor microenvironment, including TAMs, DCs, and regulatory T cells, and is a potent inducer of epithelial–mesenchymal transition (EMT). EMT is a latent developmental process which provides epithelial cells with mesenchymal properties, allowing them to migrate away from their cellular context to establish new cell layers at other sites (6–8). Similarly, tumor cells of epithelial origin undergoing EMT acquire mesenchymal properties, allowing them to detach and migrate away from the primary tumor. Cancer-related EMT is also associated with stem cell properties, suggesting that cancer stem cells may at least partly originate from more differentiated tumor cells undergoing EMT (9).

In the following section, we highlight our current knowledge regarding the role of EMT as a link between cancer and inflammation. Specifically, we focus on how EMT is induced in tumors and how pro-EMT signals in the tumor microenvironment

Cancer and Inflammation Mechanisms: Chemical, Biological, and Clinical Aspects, First Edition.
Edited by Yusuke Hiraku, Shosuke Kawanishi, and Hiroshi Ohshima.
© 2014 John Wiley & Sons, Inc. Published 2014 by John Wiley & Sons, Inc.

might be targeted as a novel type of therapy. These include both tumor-intrinsic mechanisms and extrinsic factors imposed by the immune system.

EMT: A LINK BETWEEN CANCER AND INFLAMMATION

Preservation of the integrity of the epithelial barrier is a vital task in maintaining tissue homeostasis. In wound healing, epithelial cell proliferation is triggered to secure regeneration of the epithelial barrier (10). In more persistent diseases, such as chronic inflammation and cancer, repair mechanisms may be insufficient to maintain the barrier, with devastating consequences. For example, loss of the intestinal epithelial barrier, resulting in increased permeability, is associated with and suggested to play a role in chronic inflammatory bowel disease (IBD) (10). In cancer, invasion and migration of carcinoma cells into the tumor stroma can be triggered through induction of EMT and may be regarded as a collapse of the epithelial barrier.

EMT is a trans-differentiation process that involves transcriptional reprogramming of cells (11,12). Activation of EMT-promoting transcriptional repressors such as Snail and Zeb proteins leads to repression of epithelial genes, including junction proteins such as E-cadherin, coxsackie and adenovirus receptor (CAR), claudins, and occludin (13). As these junction proteins are lost, cells may detach from each other. In parallel, EMT-promoting transcriptional activators, such as AP-1, SP1, β-catenin, and Lef/TCF factors activate genes encoding mesenchymal proteins, such as vimentin, fibronectin, N-cadherin, and integrins (14). As a result, cells acquire increased capacity to migrate and interact with matrix components.

Members of the TGF-β/BMP family are important regulators of EMT at several stages of development (15). TGF-β1-induced EMT involves activation of Smad transcription factors that interact and form complexes with EMT-promoting regulators, such as Snail, Zeb, and AP-1 proteins (16,17). Reactivation of EMT in adult tissues is also associated with TGF-β expression and has been described in chronic inflammatory diseases, where EMT may contribute to fibrosis (11,18,19). This indicates that the capacity of tumor epithelial cells to undergo EMT (i.e., cellular sensitivity to EMT-inducing signals) is enhanced but is not dependent on intrinsic oncogenic mutations. They also suggest that EMT may not simply be regarded as a late event in a stepwise process whereby tumors progress from a low to a high malignancy grade. As an alternative view, EMT may be regarded as a process that can be triggered if two criteria are fulfilled: (1) cells are capable of undergoing EMT (*EMT-competent cells*), and (2) EMT-promoting signals are present (an *EMT-permissive microenvironment*) (Figure 3.1).

In terms of the first criterion it is known that not all epithelial cells, or tumor cells of epithelial origin, are equally EMT-competent. In one study, 18 different human and mouse epithelial cell lines, both normal and tumor cells, were tested for their capacity to undergo EMT in response to TGF-β (20). Although most cells responded to TGF-β treatment, which was evident by increased Smad phosphorylation, induction of EMT was a rare event. In conclusion, cellular sensitivity to TGF-β-induced EMT appears to vary between different epithelial cells and tumor cells and to be regulated downstream of signaling from receptors to Smads. It has been shown that Wnt signaling, which is

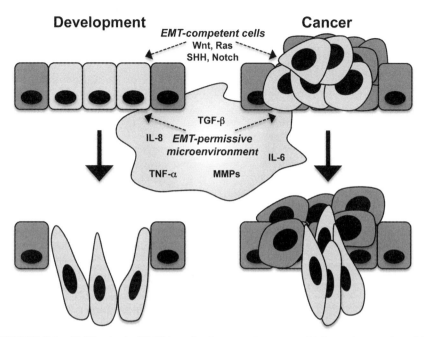

FIGURE 3.1   EMT criteria. EMT is a developmental process which can be reactivated in cancer and promote tumor cell invasion and metastasis. Induction of EMT requires two criteria to be fulfilled: (1) cells are competent to undergo EMT (EMT-competent cells); and (2) EMT-promoting signals are present (an EMT-permissive microenvironment). Wnt signaling, which promotes cellular stemness, renders epithelial cells EMT-competent, and members of the TGF-β/BMP family provide an EMT-permissive microenvironment during development. Hyperactivation of stem cell pathways, such as Wnt, Ras, sonic Hedgehog (SHH), and Notch signaling, is frequently observed in cancer tissues and renders tumor cells EMT-competent. Immune cells infiltrating tumor tissues promote an EMT-permissive tumor microenvironment by secreting EMT-promoting factors such as TGF-β1, IL-8, IL-6, MMPs, and TNF-α.

important in promoting and maintaining the undifferentiated state of stem cells (21), can prepare cells for TGF-β-induced EMT during development (12). Similar to this, hyperactivated Wnt signaling in tumor cells, which is frequently observed, sensitizes cells to TGF-β-induced EMT (22). A mechanism of this is inhibition of glycogen synthase kinase 3 beta (GSK-3β), which normally phosphorylates Snail, Zeb, and β-catenin and keeps these factors in an inactive state (23–25). Other oncogenic pathways, such as Ras, Notch, and Hedgehog, also promote cellular stemness and cooperate with TGF-β to induce EMT (14,26–28). Together, these data link *status of cell differentiation* to EMT competence. In support of this, bone morphogenetic protein 7 (BMP7), a promoter of epithelial cell differentiation in the mammary gland and kidney, antagonizes TGF-β-induced EMT in mammary ductal epithelial cells and in renal tubular epithelial cells (29). BMP7 counteracts TGF-β-induced nuclear translocation of activated Smad3/4 complexes and induces expression of E-cadherin (30). Systemic treatment with BMP7 was found to inhibit metastatic spread of human breast and prostate cancer cells.

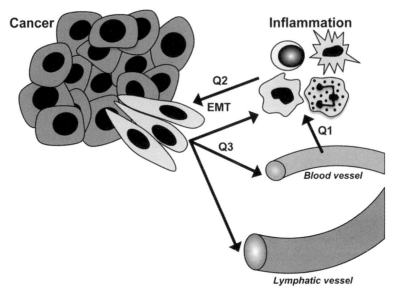

FIGURE 3.2 EMT: a link between cancer and inflammation. The inflammatory component in tumors provides EMT-inducing signals that promote tumor cell invasion and metastasis. Here we highlight research questions (Q) that might be important to answer to be able to understand and target these signals as a novel type of anticancer therapy. Q1: How are EMT-permissive, inflammatory tumor microenvironments created? Q2: Which EMT-promoting signals are provided by inflammatory tumor microenvironments? Q3: How do EMT-promoting signals influence the migratory behavior of tumor cells?

About the second criterion, EMT-permissive microenvironments may be created in different ways. For example, BMP4 secreted from neural folds promotes the conversion of competent neural plate cells into migratory neural crest cells during EMT in neural crest development (12). In tumors, the term *reactive stroma* has been used to describe a tumor microenvironment characterized by inflammation accompanied by tissue and matrix remodeling, which resembles wound healing (31,32). Similar to its broad role in wound healing, TGF-β plays a role in the generation of a reactive stroma in tumors (33).

Below we highlight three questions that from our point of view are important to answer to gain a better understanding of how EMT is induced in tumors and how it affects metastatic dissemination of tumor cells. Q1: How are EMT-permissive microenvironments created in tumors? Q2: Which EMT-promoting signals are present in such microenvironments? Q3: How do EMT-permissive microenvironments influence the migratory behavior of tumor cells (Figure 3.2)?

Q1: HOW ARE EMT-PERMISSIVE, INFLAMMATORY TUMOR MICROENVIRONMENTS CREATED?

Different myeloid cell populations and cellular components of the adaptive immune system, including B cells, T cells, DCs, and natural killer (NK) cells, are found in tumors. Previously, it was believed that tumor-associated immune cells predominantly are involved in antitumor immunity. However, extensive clinical and experimental evidence shows that infiltrating immune cells can play a major role in supporting tumor progression by promoting tumor cells. This includes invasion and migration, angiogenesis and lymphangiogenesis, and suppression of antitumor immune responses that could be induced by other immune cells. Thus, the inflammatory milieu within the tumor, which is regulated by different immune cell populations, constitutes an EMT-permissive microenvironment.

Role of Macrophages

TAMs represent a major component of immune cell infiltrates in tumors and have served as a prototype for the pro-tumor activity of inflammatory cells (2,34). Macrophages are recruited to tumors by cytokines, including CSF-1, granulocyte macrophage (GM)-CSF, IL-3, and chemokines such as CCL2, which also regulate macrophage growth and differentiation (35). Overexpression of CSF-1 and CCL2 is associated with poor prognosis in multiple types of cancer (34). Genetic deletion of CSF-1 in the polyoma middle T (PyMT) transgenic mouse model of breast cancer results in reduced macrophage density in tumors, slows tumor progression, and inhibits lung metastasis (36).

Macrophages are a heterogeneous cell population and are polarized onsite within the tumor microenvironment into M1 or M2 macrophages (37). M1 macrophages are induced by IFN-γ and other proinflammatory cytokines and are associated with CD4$^+$ Th1-polarized immune responses. M1 macrophages are believed to exist in more benign tumor lesions, where they are involved in antitumor immune responses. M2 macrophages are induced by the Th2-associated cytokines IL-10, IL-4, and IL-13 and tend to accumulate in tumors as the tumor-associated immune response becomes more Th2-polarized during tumor progression toward malignancy. M2 macrophages can sometimes be found as clusters within the tumor microenvironment, and these are thought to have primarily pro-tumor progression activities.

Recent data suggest that TAMs promote EMT in solid tumors (38). Carcinoma cells subjected to macrophage supernatants undergo EMT with characteristic loss of E-cadherin expression. On a signaling level this has been shown to occur through a nuclear factor-κB (NF-κB) pathway, which blocks degradation of Snail (39). TAMs also promote angiogenesis and lymphangiogenesis and suppress adaptive immunity (2). Thus, different macrophage populations represent a good example of how the activation state of a cell can balance inflammation in the tumor.

Role of Myeloid-Derived Suppressor Cells and Neutrophils

In addition to macrophages, at least two other classes of cells originating from the myeloid lineage have been defined in tumors: MDSCs and tumor-associated neutrophils (TANs). MDSCs are immature cells defined by their expression of CD11b and Gr-1, and most important by their ability to suppress T-cell activation (40,41). By this relatively broad definition, MDSCs include cells that have neutrophil, DC, monocyte, and macrophage characteristics. The presence of MDSCs in the tumor stroma is associated with poor prognosis (37,42). MDSCs influence other cells in the tumor stroma, including endothelial cells, and increase vascularization in part through production of factors such as MMP9 (43). MDSCs have also been shown to influence TAMs by decreasing their production of IL-12, which dampens antitumor T-cell responses (44).

Compared to TAMs, the role of TANs in tumor progression has been much less well studied. An increased number of TANs are linked to poorer outcome in patients with several types of cancer, including bronchioalveolar carcinoma (BAC) (45). In BAC, neutrophils and lymphocytes represent major immune cell populations within the tumor stroma. Tumor cell–produced IL-8 correlated with recruitment of neutrophils to the tumor microenvironment. It was recently discovered that TANs could be polarized into either N1 or N2 phenotypes (46). Similar to M1 macrophages, N1 neutrophils are antitumorigenic, due to their expression of immunoactivating cytokines and chemokines and their capacity to kill tumor cells. In contrast, and similar to M2 macrophages, N2 neutrophils are associated with tumor invasion and display pro-angiogenic and pro-metastatic properties (47).

Role of T Cells

Impressively, T-cell infiltrates have been reported to constitute as much as 45% of the total cellular mass in invasive breast cancer (53). However, the presence of T cells may also be a good prognostic indicator in benign breast tumors, which correlates with lymph node negativity and smaller tumor size (54). These conflicting data may reflect the presence of a different composition of T-cell populations in benign versus malignant breast cancers. T cells are good producers of cytokines and take on specialized phenotypes as effector cytokine producers, including Th1, Th2, and Th17 subsets. They can also differentiate into *regulatory subsets* (Tregs) that secrete IL-10. In addition, the $CD8^+$ T-cell subset and natural killer T (NKT) cells secrete cytokines (IFN-γ and IL-4), depending on their state of activation, and display cytotoxic activity. Thus, the presence of different T-cell populations may affect tumor progression in opposite directions. An indicator of this is that tumors with a good prognostic outcome are associated with the presence of $CD4^+$ Th1 cells that promote antitumor immune surveillance (55). In contrast, increasing amounts of $CD4^+$ Th2 and Tregs are found in invasive breast cancer, and the percentage of these cells correlates with lymph node metastasis (53,56,57).

Role of B Cells

B cells are recognized primarily for their ability to produce antibodies that play a role in eradicating tumors through antibody-dependent cell-mediated cytotoxicity. However, in addition to acting as antibody-producing cells, B cells are important antigen-presenting cells (APCs) and effector cells, thus having multiple functions in cancer progression (48). B cells are often observed within the tumor stroma and are found in Approximately 70% of all solid tumors (49). As an example, they can be found in about 20% of invasive breast cancers, where they may constitute up to 60% of the cancer-associated lymphocytes (50).

Experimentally, the role for B cells in tumor immunity and progression toward metastasis has been addressed in B-cell-deficient mice (51). In a model using melanoma and sarcoma cells expressing the murine leukemia virus gag, wild-type mice were not able to regulate tumor growth, whereas B-cell-deficient mice were cleared from tumors (52). Similar to this, T-cell responses against mammary adenocarcinoma were increased in B-cell-deficient mice, and the growth of melanoma and colon carcinoma cells decreased (53,54). These data suggest that B cells exert growth-inhibitory effects when they are present systemically but do not address their function locally in the tumor stroma. Results from other experiments in which B cells were depleted in the B16 melanoma model using anti-CD20 antibodies were contradictory since they showed decreased T cell activation and subsequently increased tumor growth (55). On the other hand, depletion of B cells in a syngeneic lymphoma model resulted in increased survival and decreased tumor growth (56). These discrepancies may or may not be due to differences in the immunogenicity of tumors as well as the relative contribution of B cells being activated to produce suppressive IL-10. In processes related directly to EMT, activated B cells can produce cytokines such as tumor necrosis factor (TNF), and are able to affect TAM polarization in the B16 melanoma model (57,58). The mechanisms was linked to B-cell-derived IL-10, which down-regulated TNF-α and IL-1β but up-regulated IL-10 production by TAMs. Yet another mechanism was elegantly discovered by de Visser et al., who showed that B cells sustained chronic inflammation needed for de novo carcinogenesis in a model of epithelial cancer (59). B- and T-cell-deficient mice had limited cancer progression, but transfer of B cells alone was enough to support the inflammatory status needed for tumors to arise.

Role of Dendritic Cells

DCs are positioned within epithelial cell layers in our bodies, where they sample the microenvironment. During mounting of an immune response, DCs become activated and migrate through the lymphatic system to peripheral lymph nodes, where they initiate primary and secondary immune responses by acting as professional APCs. The capacity of DCs to home to lymph nodes is dependent on the induction C-C chemokine receptors, such as CCR7 (60). This enables them to sense and migrate toward CCL21, which is expressed in lymphatic but not in vascular endothelial cells. Strikingly, tumor cell dissemination through the lymphatic system resembles how activated DCs migrate during inflammation. Metastatic tumor cells have also been reported to express CCRs (61). In particular, the expression of CCR7 is associated with poor prognosis and lymph node metastasis in breast cancer suggesting a role for CCR7-mediated chemotactic migration of tumor cells toward lymphatic vessels. Infiltration of DCs in colon cancer tumor stroma correlates with tumor progression and prognosis, where lower numbers are seen in patients with distant metastasis (62). Interestingly, melanoma cells induced to undergo EMT through overexpression of Snail were shown to support regulatory T-cell activation via induction of immunosuppressive DCs (63). Thus, EMT may alter the activation status of DCs and pave the way for subsequent metastasis to occur. This block or shift of DC function can also be mediated by infiltrating immune cells, producing TGF-β, IL-10, and PGE$_2$ (64).

Furthermore, infiltration of DCs in colon cancer tumor stroma correlates with tumor progression and prognosis, and lower numbers are seen in patients with distant metastasis (62). Immunosuppressive DCs were shown to mediate Treg activation in melanomas induced to undergo EMT by stable overexpression of Snail (63). This suggests that DCs may play a role in promoting pro-invasive and pro-metastatic effects downstream of the induction of EMT. Altered function of DCs in tumors can be mediated by other infiltrating immune cells producing TGF-β, IL-10, and PGE$_2$ (64).

Q2: WHICH EMT-PROMOTING SIGNALS ARE PROVIDED BY INFLAMMATORY TUMOR MICROENVIRONMENTS?

EMT-promoting signals generated by tumor-infiltrating immune cells are discussed below and summarized in Table 3.1.

Transforming Growth Factor Beta

Various cells within the tumor microenvironment produce TGF-β1, including tumor cells, TAMs, MDSCs, and Tregs. Elevated levels of TGF-β1 are found in plasma in cancer patients and at invasive fronts in human cancer tissues and are associated with metastasis and poor prognosis (8,65,66). TGF-β1 produced by MDSCs has been shown to have a profound impact on tumor progression and metastasis (67). In addition to its role as a potent inducer of EMT, TGF-β1 contributes to an EMT-permissive tumor microenvironment by regulating immune cell recruitment, polarization, and secretion of other EMT mediators (Figure 3.3). TGF-β1 promotes recruitment of macrophages into tumors and has been implicated in switching macrophage differentiation from an M1 to an M2 phenotype (40,68,69). TGF-β1 is also a potent chemotactic factor for neutrophils, promoting their migration but not degranulation,

TABLE 3.1   EMT-Promoting Factors Derived from Tumor-Infiltrating Immune Cells

Factor	Immune cell source	References supporting a role in EMT
TGF-β1	TAMs, MDSCs, Tregs	Reviewed in (8,10,16,17)
EGF	TAMs, MDSCs	(72–74)
IGF-1, IGF-2	TAMs	(76–78)
MMP9	TAMs, TANs, MDSCs, DCs	(80,81,84)
MMP2	TAMs, DCs	(80,81,83,85,86)
MMP3	DCs, lymphocytes	(81,87–90)
u-Pa	TAMs	(92–95)
TNF-α	TAMs	(96,100,101)
IL-6	TAMs, T cells	(104–106)
IL-8	TAMs	(107–109)

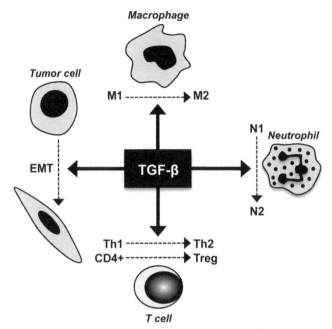

FIGURE 3.3   TGF-β1 is a master regulator of a pro-invasive, immunosuppressive tumor microenvironment. TGF-β1 plays a central role in promoting a pro-invasive and pro-metastatic tumor microenvironment by controlling the fate and function of both tumor cells and various subpopulations of immune cells. Tumor cells are induced to undergo EMT in response to TGF-β1. Macrophages are implicated to shift from an M1–M2 phenotype, and neutrophils from an N1–N2 phenotype, in response to TGF-β1. Compared to M1 macrophages and N1 neutrophils, which have antitumor effects, M2 macrophages and N2 neutrophils secrete cytokines and proteases that promote matrix remodeling, angiogenesis, lymphangiogenesis, and EMT. TGF-β1 also acts on CD4+ T cells from a Th1–Th2 phenotype and promotes differentiation of regulatory T cells (Tregs).

or activation (68). TGF-β1 controls the pro-tumor activities of neutrophils by regulating their N1–N2 polarization (46). Such N1–N2 polarization of neutrophils might mirror M1–M2 polarization of macrophages (70). Moreover, TGF-β1 inhibits Th1-mediated antitumor immune responses by shifting infiltrating T cells toward a Th2 cell phenotype (71).

Other Cytokines

TAMs express other cytokines mediating EMT responses, such as epidermal growth factor (EGF), which cooperates with TGF-β1 to induce EMT in breast cancer cells (72,73). MDSCs produce TGF-β and EGF to support EMT in cancer cells (41,74). TAMs also express insulin-like growth factors (IGFs), which are critical mediators of organogenesis and associated with increased risk for the development of several common carcinomas, such as breast, prostate, lung, and colon carcinomas. IGF-1 is expressed predominantly in stromal cells in breast cancer (75) and is induced by TGF-β1 in prostate cancer cells (76). IGF-2 is overexpressed in invasive breast cancer (77) and induces an EMT response in MDCK cells with concomitant intracellular sequestration and degradation of E-cadherin and nuclear translocation of β-catenin (78).

Proteases

Tumor-infiltrating immune cells express various proteases, such as matrix metalloproteinases, cathepsins, and urokinase (uPa), which contribute to the EMT-permissive tumor microenvironment (79). Overexpression of MMPs, which are produced by both macrophages and neutrophils, is correlated with poor prognosis and the induction of EMT in cancer (80,81). MMP9 produced by MDSCs promote tumor progression and metastasis by modulating tumor vascularization and tumor cell invasion (43). Knockout of MMP-9 in the mouse MMTV-polyoma virus middle T-antigen (PyMT) model of multistage cancer in the mammary gland reduces metastatic spread to the lungs by 80% (82). MMP-9 and MMP-2 are induced upon the induction of EMT in MCF-10A cells by TGF-β (83). MMP-9 has been shown to mediate EMT in cultured murine renal tubular cells (84). MMP-2, MMP-12, and MMP-13 are among the most up-regulated transcripts during TGF-β-induced EMT in Ras-transformed mouse mammary epithelial cells (85). MMP-2 plays an essential role for developmental EMT in the avian embryo (86). Transgenic overexpression of MMP-3 in the mammary gland results in the formation of tumors with mesenchymal characteristics (87). Subsequent studies have shown that MMP-3 can directly activate an EMT program in mammary epithelial cells by inducing Rac1b, which triggers EMT by increasing the levels of reactive oxygen species (88). Furthermore, MMP-3 can cleave E-cadherin, which leads to loss of cell–cell interactions and increased cell motility due to the release of a bioactive fragment of E-cadherin (89). Treatment of mouse mammary SCp2 epithelial cells with MMP-3 results in the induction of EMT features, including increased expression of Snail and vimentin, down-regulation of epithelial cytokeratins, and increased cell motility (90). MMP-3 is a key mediator of EMT during secondary branching of ducts in the mammary gland during pubertal mammary branching morphogenesis (91).

The serine protease uPa is expressed by TAMs and contributes to degradation of the extracellular matrix in tissue remodeling processes such as wound healing, mammary gland development, and invasion of tumor cells (79,92). TGF-β1 induces increased expression of uPa in TAMs (93). UPa mediates hypoxia-induced EMT in breast cancer cells through phosphorylation and inactivation of GSK-3β and increased expression of Snail1 (94). Overexpression of the uPa receptor (uPAR) in human MDA-MB-468 cells leads to EMT and the induction of cancer stem cell properties (95).

Inflammatory Cytokines and Chemokines

TAMs produce inflammatory cytokines, such as tumor necrosis factor-alpha (TNF-α), which significantly accelerates TGF-β-induced EMT through a mechanism involving p38 MAPK activation (96). NF-κB is a key transducer of inflammatory signals activated by TNF-α and has emerged as a promoter of tumor progression (2,97,98).

NF-κB stabilizes Snail-1 in response to TNF-α and is activated by TGF-β signaling through activation of TAK1 (39,99). TNF-α-induced EMT in breast cancer cells requires NF-κB-mediated transcriptional up-regulation of the transcriptional EMT inducer Twist-1 (100). Similarly, NF-κB was shown to be an important downstream mediator of both TGF-β- and TNF-α-induced EMT in pancreatic carcinoma cells (101). NF-κB was identified as an essential mediator of EMT in a model of EMT in breast cancer cells induced by cooperative action of TGF-β and Ha-Ras (102). NF-κB was also shown to play an important role in mediating collagen type I–induced EMT in pancreatic and colon carcinoma cells by up-regulating Snail1 and LEF1 transcription factors (99). NF-κB mediates upregulation of ZEB1 and ZEB2 in MCF-10A cells after treatment with TNF-α or IL-1α (103). In addition, overexpression of the p65 subunit results in increased levels of Zeb1 and Zeb2 in MCF-10A cells and is sufficient to activate the Zeb1 promoter in reporter assays.

TGF-β1 induces IL-6 expression in TAMs and T cells, and overexpression of IL-6 correlates with tumor grade and poor survival in breast cancer (104,105). IL-6 was shown to induce an EMT phenotype in MCF-7 breast adenocarcinoma cells with repression of E-cadherin and induction of vimentin, N-cadherin, Snail, and Twist (106). IL-8 is overexpressed in many carcinomas and is produced by TAMs (107). Recently, it was shown that IL-8 plays a critical role in mediating EMT in carcinoma cells (108). IL-8 is also induced during TGF-β-induced EMT in colon carcinoma cells in a NF-κB-dependent manner and in EMT induced by overexpression of the transcription factor Brachyury (108,109).

Q3: HOW DO EMT-PROMOTING SIGNALS INFLUENCE THE MIGRATORY BEHAVIOR OF TUMOR CELLS?

While the overall concept that EMT provides tumor cells with migratory properties is established, it remains to be determined more specifically how EMT cells migrate. How do tumor cells undergoing EMT make use of their novel migratory properties? Do they specifically migrate toward blood and/or lymphatic vessels (Figure 3.2)? Do they migrate toward areas with tumor-infiltrating immune cells? Parallels may be drawn between how tumor cells and immune cells migrate in our bodies, and one possibility is if they follow each other traveling through tissues. Metastasizing tumor cells excluded, immune cells are the only cells that can traffic through adult tissues and in and out of blood and lymphatic vessels as they are recruited to distant sites. This type of migration of immune cells is mediated by chemotaxis. Activated immune cells are equipped with chemokine receptors that can sense and respond to chemokine gradients present in the microenvironment. For example, DCs, which are positioned within epithelial cell layers in our bodies, become activated during an immune response and migrate through the lymphatic system to peripheral lymph nodes, where they initiate immune responses by acting as professional APCs (60). This type of targeted migration through the lymphatic system is at least in part mediated by CCR7, which is expressed on the surface of activated DCs and which allows them to sense and migrate toward gradients of CCL21 from lymphatic endothelial cells. Immune cells are also equipped with adhesion molecules, allowing them to interact and transmigrate through endothelial cell layers. The expression of chemokine receptors in immune cells and their ligands in other cell types is regulated by a delicate interplay between various cytokines produced and released as immune responses are mounted during inflammation.

CONCLUSIONS AND FUTURE PROSPECTS

Strikingly, tumor cell dissemination through the lymphatic system resembles how activated DCs migrate during inflammation. Metastatic tumor cells have been reported to express various CCRs (61). In particular, the expression of CCR7 is associated with poor prognosis and lymph node metastasis in breast cancer, suggesting a role for CCR7-mediated chemotactic migration of tumor cells toward lymphatic vessels. Recent studies have shown that tumor cells undergoing EMT may co-migrate with macrophages (73). Thus, there is a possibility that tumor cells undergoing EMT acquire immune cell properties, allowing them to sense and migrate toward chemokine gradients and interact with endothelial cells similar to activated immune cells. In this way, tumor cells may piggyback, using immune cell functions to move to distant sites. The tumor microenvironment is rich in various chemokines that potentially could act as chemoattractants for EMT cells. We expect this area of research to be important for further understanding of the functionality of EMT cells.

Acknowledgments

This work was supported by grants from the Swedish Research Council, the Swedish Cancer Society, the Swedish Childhood Cancer Foundation, and the Strategic Research Programme (StratCan) at Karolinska Institutet.

REFERENCES

1. Hanahan D, Weinberg RA. Hallmarks of cancer: the next generation. Cell 2011;144:646–674.
2. Mantovani A, Allavena P, Sica A, Balkwill F. Cancer-related inflammation. Nature 2008;454: 436–444.
3. Lin EY, Li JF, Gnatovskiy L, Deng Y, Zhu L, Grzesik DA, et al. Macrophages regulate the angiogenic switch in a mouse model of breast cancer. Cancer Res 2006;66:11238–11246.
4. Nozawa H, Chiu C, Hanahan D. Infiltrating neutrophils mediate the initial angiogenic switch in a mouse model of multistage carcinogenesis. Proc Natl Acad Sci USA 2006;103:12493–12498.
5. Schoppmann SF, Birner P, Stockl J, Kalt R, Ullrich R, Caucig C, et al. Tumor-associated macrophages express lymphatic endothelial growth factors and are related to peritumoral lymphangiogenesis. Am J Pathol 2002;161:947–956.
6. Dumitriu IE, Dunbar DR, Howie SE, Sethi T, Gregory CD. Human dendritic cells produce TGF-beta 1 under the influence of lung carcinoma cells and prime the differentiation of CD4+CD25+Foxp3 +regulatory T cells. J Immunol 2009;182:2795–2807.
7. Ghiringhelli F, Puig PE, Roux S, Parcellier A, Schmitt E, Solary E, et al. Tumor cells convert immature myeloid dendritic cells into TGF-beta-secreting cells inducing CD4+CD25+ regulatory T cell proliferation. J Exp Med 2005;202:919–929.
8. Massague J. TGFbeta in cancer. Cell 2008;134:215–230.

9. Mani SA, Guo W, Liao MJ, Eaton EN, Ayyanan A, Zhou AY, et al. The epithelial–mesenchymal transition generates cells with properties of stem cells. Cell 2008;133:704–715.

10. Sturm A, Dignass AU. Epithelial restitution and wound healing in inflammatory bowel disease. World J Gastroenterol 2008;14:348–353.

11. Kalluri R, Weinberg RA. The basics of epithelial–mesenchymal transition. J Clin Invest 2009;119:1420–1428.

12. Thiery JP, Acloque H, Huang RY, Nieto MA. Epithelial–mesenchymal transitions in development and disease. Cell 2009;139:871–890.

13. Peinado H, Olmeda D, Cano A. Snail, Zeb and bHLH factors in tumour progression: an alliance against the epithelial phenotype? Nat Rev Cancer 2007;7:415–428.

14. Fuxe J, Vincent T, Garcia de Herreros A. Transcriptional crosstalk between TGF-beta and stem cell pathways in tumor cell invasion: role of EMT promoting Smad complexes. Cell Cycle 2010;9:2363–2374.

15. Acloque H, Adams MS, Fishwick K, Bronner-Fraser M, Nieto MA. Epithelial–mesenchymal transitions: the importance of changing cell state in development and disease. J Clin Invest 2009;119:1438–1449.

16. Heldin CH, Landstrom M, Moustakas A. Mechanism of TGF-beta signaling to growth arrest, apoptosis, and epithelial–mesenchymal transition. Curr Opin Cell Biol 2009;21:166–176.

17. Xu J, Lamouille S, Derynck R. TGF-beta-induced epithelial to mesenchymal transition. Cell Res 2009;19:156–172.

18. Johnson JR, Roos A, Berg T, Nord M, Fuxe J. Chronic respiratory aeroallergen exposure in mice induces epithelial–mesenchymal transition in the large airways. PLoS One 2011;6:e16175.

19. Bataille F, Rohrmeier C, Bates R, Weber A, Rieder F, Brenmoehl J, et al. Evidence for a role of epithelial–mesenchymal transition during pathogenesis of fistulae in Crohn's disease. Inflamm Bowel Dis 2008;14:1514–1527.

20. Brown KA, Aakre ME, Gorska AE, Price JO, Eltom SE, Pietenpol JA, et al. Induction by transforming growth factor-beta1 of epithelial to mesenchymal transition is a rare event *in vitro*. Breast Cancer Res 2004;6:R215–R231.

21. Nusse R. Wnt signaling and stem cell control. Cell Res 2008;18:523–527.

22. Scheel C, Eaton EN, Li SH, Chaffer CL, Reinhardt F, Kah KJ, et al. Paracrine and autocrine signals induce and maintain mesenchymal and stem cell states in the breast. Cell 2011;145:926–940.

23. Zhou BP, Deng J, Xia W, Xu J, Li YM, Gunduz M, et al. Dual regulation of Snail by GSK-3beta-mediated phosphorylation in control of epithelial–mesenchymal transition. Nat Cell Biol 2004;6:931–940.

24. Schlessinger K, Hall A. GSK-3beta sets Snail's pace. Nat Cell Biol 2004;6:913–915.

25. Wu D, Pan W. GSK3: a multifaceted kinase in Wnt signaling. Trends Biochem Sci 2010;35:161–168.

26. Maitah MY, Ali S, Ahmad A, Gadgeel S, Sarkar FH. Up-regulation of sonic Hedgehog contributes to TGF-beta1-induced epithelial to mesenchymal transition in NSCLC cells. PLoS One 2011;6:e16068.

27. Oft M, Peli J, Rudaz C, Schwarz H, Beug H, Reichmann E. TGF-beta1 and Ha-Ras collaborate in modulating the phenotypic plasticity and invasiveness of epithelial tumor cells. Genes Dev 1996;10:2462–2477.

28. Zavadil J, Cermak L, Soto-Nieves N, Bottinger EP. Integration of TGF-beta/Smad and Jagged1/Notch signalling in epithelial-to-mesenchymal transition. EMBO J 2004;23:1155–1165.

29. Zeisberg M, Hanai J, Sugimoto H, Mammoto T, Charytan D, Strutz F, et al. BMP-7 counteracts TGF-beta1-induced epithelial-to-mesenchymal transition and reverses chronic renal injury. Nat Med 2003;9:964–968.

30. Buijs JT, Henriquez NV, van Overveld PG, van der Horst G, ten Dijke P, van der Pluijm G. TGF-beta and BMP7 interactions in tumour progression and bone metastasis. Clin Exp Metastasis 2007;24:609–617.

31. Dvorak HF. Tumors: wounds that do not heal. Similarities between tumor stroma generation and wound healing. N Engl J Med 1986;315:1650–1659.

32. Tuxhorn JA, Ayala GE, Rowley DR. Reactive stroma in prostate cancer progression. J Urol 2001;166:2472–2483.

33. Sieweke MH, Bissell MJ. The tumor-promoting effect of wounding: a possible role for TGF-beta-induced stromal alterations. Crit Rev Oncog 1994;5:297–311.

34. Qian BZ, Pollard JW. Macrophage diversity enhances tumor progression and metastasis. Cell 2010;141:39–51.
35. Pollard JW. Trophic macrophages in development and disease. Nat Rev Immunol 2009;9: 259–270.
36. Lin EY, Nguyen AV, Russell RG, Pollard JW. Colony-stimulating factor 1 promotes progression of mammary tumors to malignancy. J Exp Med 2001;193:727–740.
37. Mantovani A, Sica A, Allavena P, Garlanda C, Locati M. Tumor-associated macrophages and the related myeloid-derived suppressor cells as a paradigm of the diversity of macrophage activation. Hum Immunol 2009;70:325–330.
38. Bonde AK, Tischler V, Kumar S, Soltermann A, Schwendener R. Intratumoral macrophages contribute to epithelial–mesenchymal transition in solid tumors. BMC Cancer 2012;12:35.
39. Wu Y, Deng J, Rychahou PG, Qiu S, Evers BM, Zhou BP. Stabilization of Snail by NF-kappaB is required for inflammation-induced cell migration and invasion. Cancer Cell 2009;15: 416–428.
40. Mantovani A, Sozzani S, Locati M, Allavena P, Sica A. Macrophage polarization: tumor-associated macrophages as a paradigm for polarized M2 mononuclear phagocytes. Trends Immunol 2002;23:549–555.
41. Peranzoni E, Zilio S, Marigo I, Dolcetti L, Zanovello P, Mandruzzato S, et al. Myeloid-derived suppressor cell heterogeneity and subset definition. Curr Opin Immunol 2010;22:238–244.
42. Mantovani A, Schioppa T, Porta C, Allavena P, Sica A. Role of tumor-associated macrophages in tumor progression and invasion. Cancer Metastasis Rev 2006;25:315–322.
43. Yang L, DeBusk LM, Fukuda K, Fingleton B, Green-Jarvis B, Shyr Y, et al. Expansion of myeloid immune suppressor Gr+CD11b+ cells in tumor-bearing host directly promotes tumor angiogenesis. Cancer Cell 2004;6:409–421.
44. Sinha P, Clements VK, Bunt SK, Albelda SM, Ostrand-Rosenberg S. Cross-talk between myeloid-derived suppressor cells and macrophages subverts tumor immunity toward a type 2 response. J Immunol 2007;179:977–983.
45. Bellocq A, Antoine M, Flahault A, Philippe C, Crestani B, Bernaudin JF, et al. Neutrophil alveolitis in bronchioloalveolar carcinoma: induction by tumor-derived interleukin-8 and relation to clinical outcome. Am J Pathol 1998;152:83–92.
46. Fridlender ZG, Sun J, Kim S, Kapoor V, Cheng G, Ling L, et al. Polarization of tumor-associated neutrophil phenotype by TGF-beta: "N1" versus "N2" TAN. Cancer Cell 2009;16:183–194.
47. Coussens LM, Pollard JW. Leukocytes in mammary development and cancer. Cold Spring Harbor Perspect Biol 2011;3:1–22.
48. DeNardo DG, Coussens LM. Inflammation and breast cancer. Balancing immune response: crosstalk between adaptive and innate immune cells during breast cancer progression. Breast Cancer Res Treat 2007;9:212.
49. Punt CJ, Barbuto JA, Zhang H, Grimes WJ, Hatch KD, Hersh EM. Anti-tumor antibody produced by human tumor-infiltrating and peripheral blood B lymphocytes. Cancer Immunol Immunother 1994;38:225–232.
50. Coronella JA, Telleman P, Kingsbury GA, Truong TD, Hays S, Junghans RP. Evidence for an antigen-driven humoral immune response in medullary ductal breast cancer. Cancer Res 2001;61: 7889–7899.
51. DiLillo DJ, Matsushita T, Tedder TF. B10 cells and regulatory B cells balance immune responses during inflammation, autoimmunity, and cancer. Ann NY Acad Sci 2010;1183:38–57.
52. Inoue S, Leitner WW, Golding B, Scott D. Inhibitory effects of B cells on antitumor immunity. Cancer Res 2006;66:7741–7747.
53. Qin Z, Richter G, Schuler T, Ibe S, Cao X, Blankenstein T. B cells inhibit induction of T cell–dependent tumor immunity. Nat Med 1998;4:627–630.
54. Shah S, Divekar AA, Hilchey SP, Cho HM, Newman CL, Shin SU, et al. Increased rejection of primary tumors in mice lacking B cells: inhibition of anti-tumor CTL and TH1 cytokine responses by B cells. Int J Cancer 2005;117:574–586.
55. DiLillo DJ, Yanaba K, Tedder TF. B cells are required for optimal CD4+ and CD8 +T cell tumor immunity: therapeutic B cell depletion enhances B16 melanoma growth in mice. J Immunol 2010;184:4006–4016.

56. Minard-Colin V, Xiu Y, Poe JC, Horikawa M, Magro CM, Hamaguchi Y, et al. Lymphoma depletion during CD20 immunotherapy in mice is mediated by macrophage FcgammaRI, FcgammaRIII, and FcgammaRIV. Blood 2008;112:1205–1213.

57. Lund FE. Cytokine-producing B lymphocytes: key regulators of immunity. Curr Opin Immunol 2008;20:332–338.

58. Wong SC, Puaux AL, Chittezhath M, Shalova I, Kajiji TS, Wang X, et al. Macrophage polarization to a unique phenotype driven by B cells. Eur J Immunol 2010;40:2296–2307.

59. de Visser KE, Korets LV, Coussens LM. De novo carcinogenesis promoted by chronic inflammation is B lymphocyte dependent. Cancer Cell 2005;7:411–423.

60. Randolph GJ, Angeli V, Swartz MA. Dendritic-cell trafficking to lymph nodes through lymphatic vessels. Nat Rev Immunol 2005;5:617–628.

61. Zlotnik A. Involvement of chemokine receptors in organ-specific metastasis. Contrib Microbiol 2006;13:191–199.

62. Gulubova MV, Ananiev JR, Vlaykova TI, Yovchev Y, Tsoneva V, Manolova IM. Role of dendritic cells in progression and clinical outcome of colon cancer. Int J Colorectal Dis 2012;27:159–169.

63. Kudo-Saito C, Shirako H, Takeuchi T, Kawakami Y. Cancer metastasis is accelerated through immunosuppression during Snail-induced EMT of cancer cells. Cancer Cell 2009;15: 195–206.

64. Vicari AP, Caux C, Trinchieri G. Tumour escape from immune surveillance through dendritic cell inactivation. Semin Cancer Biol 2002;12:33–42.

65. Chod J, Zavadova E, Halaska MJ, Strnad P, Fucikova T, Rob L. Preoperative transforming growth factor-beta 1 (TGF-beta 1) plasma levels in operable breast cancer patients. Eur J Gynaecol Oncol 2008;29:613–616.

66. Dalal BI, Keown PA, Greenberg AH. Immunocytochemical localization of secreted transforming growth factor-beta 1 to the advancing edges of primary tumors and to lymph node metastases of human mammary carcinoma. Am J Pathol 1993;143:381–389.

67. Yang L, Huang J, Ren X, Gorska AE, Chytil A, Aakre M, et al. Abrogation of TGF beta signaling in mammary carcinomas recruits Gr-1+CD11b+ myeloid cells that promote metastasis. Cancer Cell 2008;13:23–35.

68. Flavell RA, Sanjabi S, Wrzesinski SH, Licona-Limon P. The polarization of immune cells in the tumour environment by TGFbeta. Nat Rev Immunol 2010;10:554–567.

69. Yang L, Pang Y, Moses HL. TGF-beta and immune cells: an important regulatory axis in the tumor microenvironment and progression. Trends Immunol 2010;31:220–227.

70. Mantovani A. The yin-yang of tumor-associated neutrophils. Cancer Cell 2009;16:173–174.

71. Maeda H, Shiraishi A. TGF-beta contributes to the shift toward Th2-type responses through direct and IL-10-mediated pathways in tumor-bearing mice. J Immunol 1996;156.73–78.

72. O'Sullivan C, Lewis CE, Harris AL, McGee JO. Secretion of epidermal growth factor by macrophages associated with breast carcinoma. Lancet 1993;342:148–149.

73. Wyckoff J, Wang W, Lin EY, Wang Y, Pixley F, Stanley ER, et al. A paracrine loop between tumor cells and macrophages is required for tumor cell migration in mammary tumors. Cancer Res 2004;64:7022–7029.

74. Toh B, Wang X, Keeble J, Sim WJ, Khoo K, Wong WC, et al. Mesenchymal transition and dissemination of cancer cells is driven by myeloid-derived suppressor cells infiltrating the primary tumor. PLoS Biol 2011;9:e1001162.

75. Yee D, Paik S, Lebovic GS, Marcus RR, Favoni RE, Cullen KJ, et al. Analysis of insulin-like growth factor I gene expression in malignancy: evidence for a paracrine role in human breast cancer. Mol Endocrinol 1989;3:509–517.

76. Kawada M, Inoue H, Arakawa M, Ikeda D. Transforming growth factor-beta1 modulates tumor–stromal cell interactions of prostate cancer through insulin-like growth factor-I. Anticancer Res 2008;28:721–730.

77. Cullen KJ, Yee D, Sly WS, Perdue J, Hampton B, Lippman ME, et al. Insulin-like growth factor receptor expression and function in human breast cancer. Cancer Res 1990;50:48–53.

78. Morali OG, Delmas V, Moore R, Jeanney C, Thiery JP, Larue L. IGF-II induces rapid beta-catenin relocation to the nucleus during epithelium to mesenchyme transition. Oncogene 2001;20: 4942–4950.

79. Mason SD, Joyce JA. Proteolytic networks in cancer. Trends Cell Biol 2011;21:228–237.
80. Radisky ES, Radisky DC. Matrix metalloproteinase-induced epithelial–mesenchymal transition in breast cancer. J Mammary Gland Biol Neoplasia 2010;15:201–212.
81. Kessenbrock K, Plaks V, Werb Z. Matrix metalloproteinases: regulators of the tumor microenvironment. Cell 2010;141:52–67.
82. Martin MD, Carter KJ, Jean-Philippe SR, Chang M, Mobashery S, Thiolloy S, et al. Effect of ablation or inhibition of stromal matrix metalloproteinase-9 on lung metastasis in a breast cancer model is dependent on genetic background. Cancer Res 2008;68:6251–6259.
83. Kim ES, Sohn YW, Moon A. TGF-beta-induced transcriptional activation of MMP-2 is mediated by activating transcription factor (ATF)2 in human breast epithelial cells. Cancer Lett 2007;252:147–156.
84. Tan TK, Zheng G, Hsu TT, Wang Y, Lee VW, Tian X, et al. Macrophage matrix metalloproteinase-9 mediates epithelial–mesenchymal transition *in vitro* in murine renal tubular cells. Am J Pathol 2010;176:1256–1270.
85. Jechlinger M, Grunert S, Tamir IH, Janda E, Ludemann S, Waerner T, et al. Expression profiling of epithelial plasticity in tumor progression. Oncogene 2003;22:7155–7169.
86. Duong TD, Erickson CA. MMP-2 plays an essential role in producing epithelial–mesenchymal transformations in the avian embryo. Dev Dyn 2004;229:42–53.
87. Sternlicht MD, Lochter A, Sympson CJ, Huey B, Rougier JP, Gray JW, et al. The stromal proteinase MMP3/stromelysin-1 promotes mammary carcinogenesis. Cell 1999;98:137–146.
88. Radisky DC, Levy DD, Littlepage LE, Liu H, Nelson CM, Fata JE, et al. Rac1b and reactive oxygen species mediate MMP-3-induced EMT and genomic instability. Nature 2005;436: 123–127.
89. Lochter A, Galosy S, Muschler J, Freedman N, Werb Z, Bissell MJ. Matrix metalloproteinase stromelysin-1 triggers a cascade of molecular alterations that leads to stable epithelial-to-mesenchymal conversion and a premalignant phenotype in mammary epithelial cells. J Cell Biol 1997;139:1861–1872.
90. Przybylo JA, Radisky DC. Matrix metalloproteinase-induced epithelial–mesenchymal transition: tumor progression at Snail's pace. Int J Biochem Cell Biol 2007;39:1082–1088.
91. Wiseman BS, Sternlicht MD, Lund LR, Alexander CM, Mott J, Bissell MJ, et al. Site-specific inductive and inhibitory activities of MMP-2 and MMP-3 orchestrate mammary gland branching morphogenesis. J Cell Biol 2003;162:1123–1133.
92. Nielsen BS, Sehested M, Duun S, Rank F, Timshel S, Rygaard J, et al. Urokinase plasminogen activator is localized in stromal cells in ductal breast cancer. Lab Invest 2001;81:1485–1501.
93. Hildenbrand R, Jansen C, Wolf G, Bohme B, Berger S, von Minckwitz G, et al. Transforming growth factor-beta stimulates urokinase expression in tumor-associated macrophages of the breast. Lab Invest 1998;78:59–71.
94. Lester RD, Jo M, Montel V, Takimoto S, Gonias SL. uPAR induces epithelial–mesenchymal transition in hypoxic breast cancer cells. J Cell Biol 2007;178:425–436.
95. Jo M, Eastman BM, Webb DL, Stoletov K, Klemke R, Gonias SL. Cell signaling by urokinase-type plasminogen activator receptor induces stem cell-like properties in breast cancer cells. Cancer Res 2010;70:8948–8958.
96. Bates RC, Mercurio AM. Tumor necrosis factor-alpha stimulates the epithelial-to-mesenchymal transition of human colonic organoids. Mol Biol Cell 2003;14:1790–1800.
97. Karin M. Nuclear factor-kappaB in cancer development and progression. Nature 2006;441:431–436.
98. Lopez-Novoa JM, Nieto MA. Inflammation and EMT: an alliance towards organ fibrosis and cancer progression. EMBO Mol Med 2009;1:303–314.
99. Medici D, Nawshad A. Type I collagen promotes epithelial-mesenchymal transition through ILK-dependent activation of NF-kappaB and LEF-1. Matrix Biol 2010;29:161–165.
100. Li CW, Xia W, Huo L, Lim SO, Wu Y, Hsu JL, et al. Epithelial–mesenchymal transition induced by TNF-alpha requires NF-kappaB-mediated transcriptional upregulation of Twist1. Cancer Res 2012;72:1290–1300.
101. Maier HJ, Schmidt-Strassburger U, Huber MA, Wiedemann EM, Beug H, Wirth T. NF-kappaB promotes epithelial–mesenchymal transition, migration and invasion of pancreatic carcinoma cells. Cancer Lett 2010;295:214–228.

102. Huber MA, Azoitei N, Baumann B, Grunert S, Sommer A, Pehamberger H, et al. NF-kappaB is essential for epithelial–mesenchymal transition and metastasis in a model of breast cancer progression. J Clin Invest 2004;114:569–581.

103. Chua HL, Bhat-Nakshatri P, Clare SE, Morimiya A, Badve S, Nakshatri H. NF-kappaB represses E-cadherin expression and enhances epithelial to mesenchymal transition of mammary epithelial cells: potential involvement of ZEB-1 and ZEB-2. Oncogene 2007;26:711–724.

104. Chavey C, Bibeau F, Gourgou-Bourgade S, Burlinchon S, Boissiere F, Laune D, et al. Oestrogen receptor negative breast cancers exhibit high cytokine content. Breast Cancer Res Treat 2007;9:R15.

105. Salgado R, Junius S, Benoy I, Van Dam P, Vermeulen P, Van Marck E, et al. Circulating interleukin-6 predicts survival in patients with metastatic breast cancer. Int J Cancer 2003;103:642–646.

106. Sullivan NJ, Sasser AK, Axel AE, Vesuna F, Raman V, Ramirez N, et al. Interleukin-6 induces an epithelial–mesenchymal transition phenotype in human breast cancer cells. Oncogene 2009;28:2940–2947.

107. Waugh DJ, Wilson C. The interleukin-8 pathway in cancer. Clin Cancer Res 2008;14:6735–6741.

108. Fernando RI, Castillo MD, Litzinger M, Hamilton DH, Palena C. IL-8 signaling plays a critical role in the epithelial–mesenchymal transition of human carcinoma cells. Cancer Res 2011;71:5296–5306.

109. Bates RC, DeLeo MJ, III, Mercurio AM. The epithelial–mesenchymal transition of colon carcinoma involves expression of IL-8 and CXCR-1-mediated chemotaxis. Exp Cell Res 2004;299:315–324.

CHAPTER 4

ROLE OF NITRATIVE DNA DAMAGE IN INFLAMMATION-RELATED CARCINOGENESIS

Yusuke Hiraku and Shosuke Kawanishi

Chronic inflammation is considered to be one of the most important environmental factors involved in human carcinogenesis. In the nineteenth century, Rudolf Virchow noted leucocytes in neoplastic tissues and suggested that the "lymphoreticular infiltrate" reflected the origin of cancer at sites of chronic inflammation (1). Actually, many malignancies arise from areas of infection and inflammation (1,2). Epidemiological and experimental studies have provided evidence indicating that a substantial part of environmental carcinogenesis involves chronic infection and inflammatory conditions (2,3). It has been estimated that chronic inflammation accounts for approximately 25% of human cancers (4). The International Agency for Research on Cancer (IARC) has estimated that infectious diseases contribute to approximately 18% of cancer cases worldwide, which are largely attributed to oncogenic viruses and, to a lesser extent, bacteria and parasites (3). In addition, inflammatory diseases independent of infection, including inflammatory bowel diseases and Barrett's esophagus, and physicochemical factors, such as exposure to asbestos and ultraviolet (UV) light, contribute to carcinogenesis (2). However, the precise molecular mechanism of inflammation-related carcinogenesis remains to be clarified.

DNA DAMAGE DURING CHRONIC INFLAMMATION

DNA damage is a key molecular event causing genetic instability involved in human carcinogenesis. Under inflammatory conditions, reactive oxygen species (ROSs) and reactive nitrogen species (RNSs), including nitric oxide (NO), are generated from inflammatory and epithelial cells. These highly reactive species are capable of causing oxidative and nitrative DNA damage, which may contribute to carcinogenesis (5,6) (Figure 4.1). ROSs induce the formation of various oxidative DNA lesions, including mutagenic 8-oxo-7,8-dihydro-2′-deoxyguanosine (8-oxodG) (7,8). If 8-oxodG is

Cancer and Inflammation Mechanisms: Chemical, Biological, and Clinical Aspects, First Edition.
Edited by Yusuke Hiraku, Shosuke Kawanishi, and Hiroshi Ohshima.
© 2014 John Wiley & Sons, Inc. Published 2014 by John Wiley & Sons, Inc.

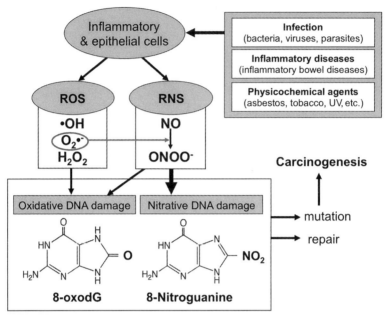

FIGURE 4.1   Formation of oxidative and nitrative DNA lesions under inflammatory conditions.

not repaired, adenine is incorporated preferentially opposite this lesion during DNA synthesis, leading to G → T transversion (9,10). ROSs are generated from multiple sources, including not only inflammatory cells but also carcinogenic chemicals and electron transport chain in mitochondria. Accumulation of 8-oxodG in the human body has been demonstrated in patients with cancer and cancer-prone diseases. A significant increase in urinary 8-oxodG levels has been observed in patients with various types of cancer compared with those in control subjects (11–13). We have demonstrated that oxidative DNA damage occurs in the tissues of animals and clinical specimens in association with inflammation-related carcinogenesis (14,15). In addition, oxidative DNA damage is mediated by ROSs generated from a variety of carcinogenic chemicals (7,16).

In contrast to ROSs, RNSs are generated primarily under inflammatory conditions (14,15). Therefore, RNS-mediated DNA damage may play a role in inflammation-related carcinogenesis and is expected as a potential biomarker to evaluate the cancer risk. 8-Nitroguanine is a nitrative DNA lesion, formed under inflammatory conditions. NO and superoxide anion $(O_2 \cdot^-)$ are generated from inflammatory and epithelial cells, and these species react with each other to form peroxynitrite $(ONOO^-)$, a highly reactive species capable of causing nitrative and oxidative DNA damage (Figure 4.1). *In vitro* experiments revealed that 8-nitroguanine was formed by the interaction of guanine with $ONOO^-$ (17) as well as 8-oxodG (18). In experimental animals, 8-nitroguanine was formed via inflammation in lung tissues of mice with viral pneumonia (19).

8-Nitroguanine formed in DNA is chemically unstable and can be released spontaneously, resulting in the formation of an apurinic site (20). During DNA synthesis, adenine is incorporated preferentially opposite an apurinic site, leading to G → T transversion (21). Cells deficient in subunits of DNA polymerase ζ were hypersensitive to NO, and translesion DNA synthesis past an apurinic site mediated by this polymerase might contribute to extensive point mutations (22). Adenine is also incorporated preferentially opposite 8-nitroguanine during DNA synthesis mediated by DNA polymerase, leading to G → T transversion (23,24). Therefore, 8-nitroguanine is a potentially mutagenic DNA lesion, which may contribute to inflammation-related carcinogenesis. In the ONOO⁻-treated supF shuttle vector plasmid, which was replicated in *Escherichia coli*, the majority of mutations occurred at G:C base pairs, predominantly involving G → T transversions (25,26). This type of mutation also occurred *in vivo* in the *ras* gene (27) and the *p53* tumor suppressor gene in lung and liver cancer (28,29). These findings imply that DNA damage mediated by inflammatory reactions participates in carcinogenesis via activation of protooncogenes and inactivation of tumor suppressor genes.

To investigate the formation of 8-nitroguanine in relation to inflammation-related carcinogenesis, we produced a specific antibody against this DNA lesion (30,31) and carried out immunohistochemistry using various tissues obtained from experimental animals and patients with cancer-prone inflammatory diseases. Our studies on 8-nitroguanine formation are summarized in Table 4.1. We first demonstrated that this DNA lesion was formed in intrahepatic bile duct epithelium of hamsters infected with the liver fluke *Opisthorchis viverrini* (OV), causing cholangiocarcinoma (32,33). In addition, we reported that 8-nitroguanine was formed in human tissues, such as gastric gland epithelium of gastritis patients with *Helicobacter pylori* infection (34), hepatocytes of patients with chronic hepatitis C (35), cervical epithelium of patients infected with human papillomavirus (HPV) (36), nasopharyngeal epithelium of patients infected with Epstein–Barr virus (EBV) (37), and urinary bladder epithelium of patients infected with *Schistosoma haematobium* (38). 8-Nitroguanine formation was also observed under inflammatory conditions independent of infection. This DNA lesion was seen in colon epithelial cells of a mouse model of inflammatory bowel disease (39), oral epithelial cells of patients with oral premalignant lesions (40,41), and esophageal epithelium of patients with Barrett's esophagus (42). In asbestos-exposed mice, 8-nitroguanine formation was observed in bronchial epithelial cells (43). An *in vitro* study revealed that 8-nitroguanine was formed in lung epithelial cell line treated with carbon nanotube (CNT) (44). These findings raise a possibility that 8-nitroguanine participates in inflammation-related carcinogenesis caused by a wide variety of environmental factors. In this chapter we discuss the role of 8-nitroguanine in inflammation-related carcinogenesis caused by representative carcinogenic infectious agents such as parasites (e.g., OV) and viruses (e.g., HPV and EBV), and physicochemical factors (e.g., asbestos and CNT). In addition, 8-nitroguanine was formed in lung adenocarcinoma tissues of conditional transgenic mice bearing mutation in the K-*ras* gene (45). The staining intensity of 8-nitroguanine in soft tumor tissue was associated with poor patient prognosis (46). We also discuss the role of this DNA lesion in tumor progression.

TABLE 4.1 8-Nitroguanine Formation Associated with Inflammation-Related
Carcinogenesis in Human and Animal Tissues

Cause	Site of DNA damage	Species	Ref.
Infectious Diseases			
Helicobacter pyroli	Gastric epithelium	Human	(34)
HPV	Cervical epithelium	Human	(36)
Hepatitis C virus	Hepatocyte	Human	(35)
EBV	Nasopharyngeal epithelium	Human	(37)
Liver fluke	Bile duct epithelium	Hamster	(30,32,33,51)
(*Opisthorchis viverrini*)	Tumor tissue	Human	(52)
Schistosoma haematobium	Urinary bladder epithelium	Human	(38)
Inflammatory Conditions			
Inflammatory bowel disease	Colon epithelium	Mouse	(39)
Oral licken planus	Oral epithelium	Human	(40)
Oral leukoplakia	Oral epithelium	Human	(41)
Barrett's esophagus	Esophageal epithelium	Human	(42)
Soft tissue tumor (malignant histiocytoma)	Tumor tissue	Human	(46,120)
Asbestos	Bronchial epithelium	Mouse	(43)
	Alveolar and bronchial epithelium	Human	Unpublished

LIVER FLUKE–INDUCED CHOLANGIOCARCINOMA

The liver fluke OV is endemic in Southeast Asia, including Thailand, Lao People's Democratic Republic, Cambodia, and central Vietnam, and is a major risk factor for cholangiocarcinoma (47,48). Actually, in northeastern Thailand, the incidence of this disease is much higher than that in other regions (49). Approximately 70% of OV-induced cholangiocarcinoma occurs in the intrahepatic bile ducts (50), while the incidence of intrahepatic cholangiocarcinoma is generally low. OV metacecariae is contained in the muscle of cyprinoid fish, and repeated ingestion of raw or under-cooked freshwater fish is the major cause of infection (48).

We investigated DNA damage in the liver of OV-infected hamsters as a model of inflammation-related carcinogenesis. To detect 8-nitroguanine in tissues, we produced a specific antibody, which was used for immunohistochemical analysis (30,31). We have first demonstrated that 8-nitroguanine was formed in bile duct epithelial cells of OV-infected hamsters (32). Double immunofluorescence staining revealed that 8-nitroguanine and 8-oxodG were formed prominently in inflammatory cells in the acute phase (21 to 30 days post-infection), and these DNA lesions remained in bile duct epithelial cells in the chronic phase (180 days post-infection) (30,33). Repeated OV infection augmented the formation of these DNA lesions in bile duct epithelium compared with a single infection (33). Moreover, the treatment of OV-infected hamsters with the antiparasitic drug praziquantel reduced DNA damage dramatically (51). On the basis of these findings, we proposed that 8-nitroganine could be used as a biomarker to evaluate the risk of inflammation-related carcinogenesis and the efficacy of drug treatment. In addition, 8-nitroguanine and 8-oxodG were formed in tumor and adjacent tissues of patients with inhrahepatic cholongiocarcinoma, and significantly associated with tumor invasion (52), suggesting that these DNA lesions participate in tumor progression.

In parasitic infection, local parasite-specific inflammatory response may participate in carcinogenesis. OV antigen was distributed in bile duct epithelium of the liver of infected hamsters and associated with inflammatory cell infiltration (53). Our *in vitro* study using a macrophage cell line and *in vivo* study using OV-infected hamsters revealed that OV antigen induced Toll-like receptor (TLR)-2-mediated pathway leading to inflammatory responses, including the expression of iNOS and cyclooxygenase (COX)-2 via activation of nuclear factor-κB (NF-κB) (51,54). TLRs activate homologous signal transduction pathways, leading to nuclear localization of NF-κB/Rel-type transcription factors (55) and participate in inflammation-related carcinogenesis (56). NF-κB is a transcriptional factor that plays a key role in the expression of genes involved in inflammatory responses, including iNOS (57,58), and participates in tumor promotion and progression (59,60). Therefore, molecules involved in this process could be potential therapeutic targets for inflammation-related cancer.

Recently, we have demonstrated that administration of an antioxidant curcumin, derived from turmeric (*Curcuma longa*), reduced the incidence of cholangiocarcinoma significantly in OV-induced hamsters via suppression of oxidative and nitrative DNA damage (61). Curcumin also reduced the expression of NF-κB-related gene products involved in cell survival (bcl-2 and bcl-xL), proliferation (cyclin D1 and c-myc), tumor invasion [matrix metalloproteinase (MMP)-9 and intracellular adhesion molecule (ICAM)-1], and angiogenesis [vascular endothelial growth factor (VEGF)] (61). Therefore, curcumin exhibits an anticarcinogenic effect via suppression of molecular events involved in multiple steps of carcinogenesis, and may serve as a safe and promising nutraceutical agent to prevent inflammation-related carcinogenesis.

HPV-INDUCED CERVICAL CANCER

Cervical cancer is the third most common female cancer, ranking after breast and colorectal cancer in 2008 (62). This disease is the leading cause of cancer-related death among women in developing countries in eastern, western, and middle Africa, central America, south-central Asia, and Melanesia (62). Virtually all cases of cervical cancer are mediated by HPV. Actually, HPV DNA can be identified in almost all specimens of patients with invasive cervical cancer (63).The IARC has classified high-risk HPVs, including HPV-16 and HPV-18, to be carcinogenic to humans (group 1) (64). Molecular epidemiological studies have demonstrated that specific subtypes of HPV are closely associated with cervical cancer, although the risk of cervical cancer varies with HPV types (63,64). The molecular mechanisms of HPV-induced carcinogenesis have been investigated extensively in terms of HPV oncoproteins E6 and E7 (65). E6 forms a complex with E6-associated protein (E6AP), and this complex promotes the degradation of p53 protein, a tumor suppressor gene product, and induces the transcription of the catalytic subunit of human telomerase reverse transcriptase, leading to cell immortalization (66). E7 binds and degrades retinoblastoma protein (RB), a major negative regulator of cell cycle, and inactivates cyclin-dependent kinase (CDK) inhibitors p21 and p27 (67). E7-induced RB degradation leads to release of the transcription factor E2F from the RB/E2F complex (68), which mediates the expression of cyclin E and cyclin A and aberrant CDK2 activity (67).

In addition to HPV oncoproteins, epidemiological studies have revealed that chronic inflammation is associated with cervical carcinogenesis, although no evidence suggesting that HPV infection alone induces inflammatory conditions has been provided. An epidemiological study showed that cervical inflammation was associated with high-grade lesions in HPV-infected women (69). Several epidemiological studies have suggested that other pathogens, including *Chlamydia trachomatis* and herpes simplex virus-2, act in conjunction with HPV infection to increase the risk of cervical cancer (70–72). In a recent study using cervical smears, the relative risk of high-grade squamous intraepithelial lesions was significantly higher in patients infected with *Gardinerella vaginalis, Chlamydia*, and leucocytosis than with normal subjects (73).

To clarify the role of inflammation-mediated DNA damage in cervical carcinogenesis, we obtained biopsy specimens from patients with cervical intraepithelial neoplasia (CIN) and performed immunohistochemistry to examine the formation of 8-nitroguanine and 8-oxodG. In CIN patients, 8-nitroguanine formation was observed in the nuclei of atypical cells, whereas no or weak 8-nitroguanine formation occurred in patients with condyloma acuminatum, benign cervical warts caused by low-risk HPV (Figure 4.2A). Immunoreactivity of 8-nitroguanine in cervical epithelium was increased, significantly in the order condyloma acuminatum < CIN1 < CIN2–3 (36). The formation of nitrative DNA lesion during cervical carcinogenesis is supported by a recent study showing that NO exposure induced DNA damage and increased mutation rates in HPV-positive cervical epithelial cell lines established from CIN patients (74). 8-OxodG showed a staining pattern similar to that of 8-nitroguanine, but there was no significant difference in staining intensity in different CIN grades, although it has been reported that 8-oxodG content in cervical cells increased significantly with the grade of squamous intraepithelial lesion (75).

Several reports have shown that the cyclin-dependent kinase inhibitor p16 is overexpressed in cervical neoplasia (76,77) and proposed as a biomarker of cervical carcinogenesis. It has been reported that E2F accumulation mediated by E7 leads to the expression of p16-related transcript (78). We demonstrated that strong p16 expression and 8-nitroguanine formation were observed in cervical epithelial cells of CIN patients, but in patients with condyloma acuminatum and HPV-positive cervicitis,

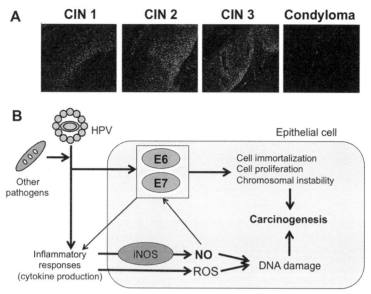

FIGURE 4.2 DNA damage and carcinogenesis induced by HPV infection. (A) 8-Nitroguanine formation in cervical epithelium. 8-Nitroguanine formation in cervical biopsy specimens was assessed by immunohistochemical analysis. Tissue sections were incubated with rabbit polyclonal anti-8-nitroguanine antibody and then with Alexa 594-labeled goat anti-rabbit IgG antibody. 8-Nitroguanine formation was observed in the nuclei of atypical epithelial cells. In patients with condyloma acuminatum, no or weak 8-nitroguanine formation occurred (36). (B) Possible mechanisms of HPV-induced cervical carcinogenesis involving inflammatory reactions.

only p16 expression was observed (36). This result suggests that 8-nitroguanine is a more suitable marker than p16 to discriminate high- and low-risk cervical lesions.

There are studies suggesting that HPV oncoproteins crosstalk with signaling pathways of inflammatory responses. NO exposure increased the expression of E6 and E7 genes, resulting in decreased p53 and RB protein levels (74). On the other hand, HPV oncoproteins mediate inflammatory responses: E6 and E7 derived from a high-risk HPV enhanced the release of interleukin (IL)-1α from cultures of normal cervical keratinocytes (79). E7 expression increased the promoter activity of COX-2 and the downstream molecule IL-32 in HPV-positive cervical cancer cell lines (80). IL-32 expression in cervical tissues of patients with squamous cell carcinoma increased with the tumor stage (80). These findings raise the possibility that inflammatory reactions reciprocally mediate the expression of HPV oncoproteins, resulting in the persistence of nitrative DNA damage and dysregulated cell proliferation, contributing to tumor progression. In addition, HPV oncoprotein-expressing cells have an impaired ability to respond to DNA damage (81,82). Possible mechanisms of HPV-induced cervical carcinogenesis involving inflammatory reactions are shown in Figure 4.2B. The evidence for the participation of inflammation in cervical carcinogenesis has been accumulating, but further studies are required to clarify the precise molecular mechanism.

EBV-INDUCED NASOPHARYNGEAL CARCINOMA

Nasopharyngeal carcinoma (NPC) and Burkitt lymphoma are strongly associated with EBV infection (83) and account for approximately 1% of world cancer cases (3). NPC is an epithelial tumor with a high prevalence in southern China, where the incidence rate is approximately 100-fold higher than that in the Western world (84,85). Latent EBV infection is detected in cancer cells of virtually all cases of undifferentiated NPC in endemic regions (3,86). Among viral genes, latent membrane protein 1 (LMP1) may play the most important role in EBV-mediated NPC. In addition to EBV infection, environmental and dietary factors are proposed to contribute to the etiology of NPC. In southern China, the traditional foods such as salted fish and other preserved foods contain volatile nitrosoamines, considered to be important carcinogenic factors of NPC (87). Chinese-style salted fish contains substances capable of causing activation of latently infected EBV, and has recently been classified by the IARC as a group 1 carcinogen (88). An epidemiological study has revealed that herbal medicine use increases the risk of NPC, probably through reactivation of EBV or a direct promoting effect on EBV-transformed cells (89). Phorbol diester is an EBV-activating substance, which was identified in the soil collected from under *Sapium sebiferum* (Chinese tallow) (90).

We obtained biopsy specimens of nasopharyngeal tissues from patients with nasopharyngitis and NPC in southern China, and examined DNA damage. 8-Nitroguanine formation and iNOS expression were observed in epithelial cells of EBV-positive patients with chronic nasopharyngitis, and interestingly, their intensities were significantly stronger in cancer cells of NPC patients than in those of nasopharyngitis patients (37). In EBV-negative subjects, no or little DNA damage was observed. We have examined the mechanism by which EBV infection causes 8-nirtroguanine formation in these tissues. Lo et al. have demonstrated that epidermal growth factor receptor (EGFR) is translocated to the nucleus and then interacts with the transcriptional factor signal transducer and activator of transcription 3 (STAT3) to mediate iNOS expression (91). In our study, phosphorylated STAT3 (active form) and EGFR were co-localized in the nucleus of cancer cells of NPC patients. IL-6, which is known to activate STAT3, was expressed mainly in macrophages in nasopharyngeal tissues of EBV-infected patients (37). These findings suggest that nuclear accumulation of EGFR and STAT3 activation by IL-6 plays the key role in iNOS expression and resulting DNA damage, leading to EBV-mediated carcinogenesis. This mechanism of DNA damage has been confirmed by stimulating LMP1-transfected cells, in which EGFR is consistently expressed, by IL-6 (37). A recent study has demonstrated that excess $ONOO^-$ mediates the activation of JAK/STAT signaling pathway in experimental animals (92). Therefore, a positive loop between STATs and iNOS may exist and contribute to tumor development. Moreover, inflammation is known to be associated with epigenetic changes, including DNA methylation (93). Recently, it has been reported that Ras-related associated with a diabetes (RRAD) gene was aberrantly methylated and down-regulated in the tumor tissues of most NPC patients (94). Therefore, such epigenetic changes may involve inflammatory responses and contribute to EBV-mediated carcinogenesis.

DNA DAMAGE INDUCED BY ASBESTOS EXPOSURE

Particulate and fibrous materials are known to cause inflammation in the lung tissues, leading to carcinogenesis. Asbestos is a naturally occurring fiber causing lung cancer and malignant mesothelioma of the pleura and peritoneum. The IARC has evaluated asbestos to be carcinogenic to humans (group 1) (95). Crocidolite (blue asbestos) and amosite (brown asbestos) contain iron (approximately 30%) and are much more potent carcinogens than chrysotile (white asbestos), which contains only a trace amount of iron as a contaminant. Although possible mechanisms for asbestos-induced carcinogenesis have been proposed (96), the precise mechanism remains to be clarified. The roles of inflammatory responses and nitrative stress in asbestos-related diseases have been examined. iNOS expression and RNS generation occurred in cultured human alveolar macrophages (97), lung epithelial cells (98), and mesothelial cells treated with asbestos (99), and in bronchiolar epithelium of asbestos-exposed animals (100,101). In a clinical study, alveolar NO concentration was increased in asbestosis patients compared with healthy controls (102). These findings indicate that nitrative stress plays a role in the development of asbestos-induced diseases, and use of nitratively modified biomolecules, such as 8-nitroguanine, may be of clinical relevance.

We performed immunohistochemical analysis to examine DNA damage in the lung tissues of mice intratracheally administered asbestos. 8-Nitroguanine and 8-oxodG formation was detected particularly in the nucleus of bronchial epithelial cells. Interestingly, quantitative image analysis revealed that crocidolite induced 8-nitroguanine formation to a significantly greater extent than chrysotile, consistent with their carcinogenic potentials (43). We have recently demonstrated that the extent of 8-nitroguanine formation is correlated with asbestos contents in human lung tissues (unpublished data). G → T transversion was most frequently detected in the omentum of crocidolite-exposed rats (103). As mentioned above, this type of mutation is likely to be induced via 8-nitroguanine formation. The association of asbestos exposure with this DNA lesion is supported by these observations. Therefore, 8-nitroguanine could be a biomarker to evaluate the risk of carcinogenesis in persons exposed to fibrous and particulate matters.

DNA DAMAGE INDUCED BY CARBON NANOTUBE

Nanomaterials are defined as substances with at least one dimension smaller than 100 nm in size. Recently, various nanomaterials have been used in various industries. Carbon nanotube (CNT) is one of the most promising nanomaterials used for many industrial and medical purposes, because of its unique physicochemical properties, such as high electrical conductivity and excellent strength (104,105). However, there is a growing concern that CNT may exert toxic and carcinogenic effects on humans, because its fibrous-like shape and durability appear to be analogous to those of asbestos (106,107). Actually, animal studies demonstrated that intraperitoneal application of CNT caused mesothelioma in mice (107–109), and its intrascrotal administration induced masothelioma in rats (110). However, the mechanism of CNT-induced carcinogenesis has not been well understood. Several studies have demonstrated that administration of CNT induces inflammatory responses in the lung tissues of animals (111–113). *In vitro* studies have shown that CNT induced inflammatory reactions such as cytokine production in lung epithelial cell lines (114). These findings raise the possibility that CNT induces inflammatory reactions directly within the target cells, which may contribute to carcinogenesis.

We examined 8-nitroguanine formation in human lung alveolar epithelial cells treated with multiwalled CNT (MWCNT) by immunocytochemistry. MWCNT induced 8-nitroguanine formation in the nucleus in a dose-dependent manner. MWCNT significantly increased 8-nitroguanine formation at a lower concentration (44) than in previous studies, in which the genotoxicity of CNT was evaluated by comet assay and micronucleus test (115–117). This result implies that nitrative DNA lesion is more likely to occur than other types of DNA damage. MWCNT-induced 8-nitroguanine formation was largely suppressed by inhibitors of iNOS, NF-κB, and caveola- and clathrin-mediated endocytosis (Figure 4.3A). Electron microscopy

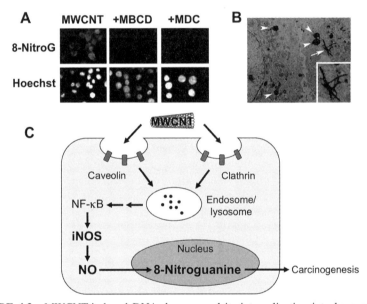

FIGURE 4.3 MWCNT-induced DNA damage and its internalization into lung epithelial cells. (A) 8-Nitroguanine formation induced by MWCNT. A549 cells were treated with 10 µg/mL of MWCNT for 8 h at 37°C. 8-Nitroguanine formation was assessed by immunocytochemical analysis. Cells were incubated with rabbit polyclonal anti-8-nitroguanine antibody and then with Alexa 594-labeled goat anti-rabbit IgG antibody, followed by the treatment with Hoechst 33258 (Hoechst) to stain the nuclei. 8-Nitroguanine (8-NitroG) formation was observed in the nuclei and suppressed by the addition of inhibitors of caveolae- (2 mM methyl-β-cyclodextrin, MBCD) and clathrin-mediated endocytosis (50 µM monodansylcadaverine, MDC) (44). (B) Cellular internalization of MWCNT. A549 cells were treated with 10 µg/mL MWCNT for 4 h at 37°C, and then examined by transmission electron microscopy. MWCNT is largely localized in vesicular structures in the cytoplasm (arrowheads). Bundles of MWCNT (arrow) are also seen in the cell, and an enlarged image is shown in an inset. (C) Proposed mechanism of MWCNT-induced 8-nitroguanine formation.

revealed that MWCNT was located in vesicular structures in the cytoplasm (Figure 4.3B), and its cellular internalization was reduced by endocytosis inhibitors. These results suggest that MWCNT is internalized into cells via endocytosis, leading to inflammatory reactions including iNOS expression and resulting nitrative DNA damage, which may contribute to carcinogenesis (44) (Figure 4.3C). It is speculated that cytokines released from CNT-exposed cells may mediate additional inflammatory reactions and DNA damage in adjacent cells. Since CNT is expected as a promising material that will be used in various industrial fields in the future, the risk assessment for its harmful effects on humans is of extreme importance.

SOFT TISSUE TUMORS AND PROGNOSIS OF TUMOR PATIENTS

Malignant fibrous histiocytoma (MFH) is one of the most common soft tissue sarcomas occurring in adults. MFH has been proposed to be accompanied by inflammatory responses. The expression of cytokines in MFH may account for the local inflammatory infiltrate and aggressive nature observed in malignant cells (118). In experimentally induced rat sarcoma, an inflammatory reaction characterized by infiltration of lymphocytes, monocytes, and macrophages occurred (119). These findings raise the possibility that the inflammatory responses contribute to the etiology of MFH.

We examined DNA damage and the expression of inflammation-related molecules in surgical specimens from MFH patients by immunohistochemistry. 8-Nitroguanine was formed in tumor tissues but not in adjacent nontumor tissues. iNOS, NF-κB, COX-2, and hypoxia-inducible factor (HIF)-1α were co-localized with 8-nitroguanine in tumor tissues (46,120). HIF-1α is an oxygen-sensing transcription factor which is up-regulated in a hypoxic environment during tumor growth. Statistical analysis using the Kaplan–Meier method revealed that strong 8-nitroguanine staining was associated with a poor prognosis (46). Tumor cells that adapt to hypoxia by increasing the synthesis of HIF-1α mediate the transcription of various genes, including iNOS (121). On the other hand, an increase in NO production induces the accumulation and activation of HIF-1α (122,123). Therefore, HIF-1α and iNOS positively regulate each other during tumor growth, leading to persistent DNA damage, which may contribute to poor prognosis of cancer patients. IκB kinase (IKK)-β, involved in NF-κB activation, is also required for HIF-1α accumulation under hypoxia in cultured cells and animals (124), whereas NF-κB is regulated under hypoxia in an HIF-1α-dependent manner (125). Recent studies have demonstrated that HIF-1α mediates the expression of receptors of alarmins or damage-associated molecular patterns (DAMPs), which are released from necrotic cells under hypoxic conditions, and these receptors strongly activate NF-κB (126). Thus, reciprocal activation of HIF-1α and NF-κB may also be involved in iNOS-dependent DNA damage and tumor development. Therefore, 8-nitroguanine may participate in tumor progression and could be used as a biomarker to predict the prognosis of cancer patients.

CONCLUSIONS AND FUTURE PROSPECTS

As described above, we have demonstrated that 8-nitroguanine was formed in various animal models and patients with cancer-prone inflammatory diseases. On the basis of our studies, the possible mechanism of inflammation-related carcinogenesis and tumor development via DNA damage is shown in Figure 4.4. Various infectious agents, inflammatory diseases, and physicochemical factors induce inflammatory responses and the production of ROSs and RNSs from inflammatory and epithelial cells. Inflammatory cytokines mediate the activation of transcription factors, including NF-κB and STATs. NF-κB regulates the expression of a wide variety of inflammation-related molecules, including iNOS (58,60). RNSs can induce the activation of NF-κB under certain circumstances (127). Therefore, reciprocal and positive regulation between RNSs and NF-κB may lead to persistent inflammatory reactions. Collectively, these molecular events converge to nitrative stress, and resulting DNA damage contributes to an accumulation of genetic changes in tissues throughout the carcinogenic process. In particular, 8-nitroguanine accumulates during the development of chronic inflammation to cancer as a common mechanism and raises the possibility that this DNA lesion can be used as a promising biomarker to assess the risk of inflammation-mediated carcinogenesis at the precancerous stage and predict the prognosis of cancer patients.

Quantitative analysis of 8-nitroguanine in biological samples, such as blood and urine, would be useful for evaluation of the risk of inflammation-related carcinogenesis. Since 8-nitroguanine formed in DNA is chemically unstable and likely to be released from DNA, an attempt to utilize free 8-nitroguanine in urine has been made for analysis using high-performance liquid chromatography coupled with

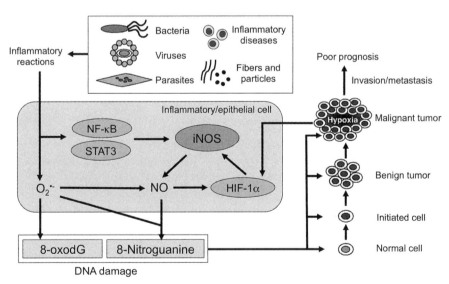

FIGURE 4.4 Proposed mechanisms of inflammation-related carcinogenesis via nitrative DNA damage. (*See insert for color representation of the figure.*)

electrochemical detection and immunoaffinity purification (128). 8-Nitroguanine has also been measured by liquid chromatography with mass spectrometry and glyoxal derivatization (129). Recently, urinary 8-nitroguanine has been analyzed by capillary electrophoresis with amperometric detection (130).

Moreover, the development of therapeutics targeting inflammation-related molecules may contribute to cancer prevention and improvement of prognosis of cancer patients. Several animal experiments demonstrated that iNOS inhibitors, such as ONO-1714 and 1400 W, effectively reduced inflammation-related carcinogenesis. ONO-1714 significantly decreased the degree of cholangitis and reduced the incidence of intrahepatic biliary tumors in bilioenterostomized hamsters (131). 1400 W reduced tumorigenesis in the mammary glands of γ-irradiated mice treated with diethylstilbestrol (132). To develop a strategy for cancer prevention, further studies are needed to clarify the precise molecular mechanisms of inflammation-related carcinogenesis.

Acknowledgments

The author is grateful to Ning Ma (Suzuka University of Medical Sciences, Japan) for technical advice. The author is also grateful to Mariko Murata (Mie University Graduate School of Medicine) and all collaborators for encouragement throughout. This work was supported by Grants-in-Aid for Scientific Research from the Ministry of Education, Culture, Sports, Science and Technology and the National Cancer Center Research and Development Fund in Japan.

REFERENCES

1. Balkwill F, Mantovani A. Inflammation and cancer: Back to Virchow? Lancet 2001;357: 539–545.
2. Coussens LM, Werb Z. Inflammation and cancer. Nature 2002;420:860–867.
3. International Agency for Research on Cancer. World Cancer Report. Lyon, France: IARC Press; 2003.
4. Hussain SP, Harris CC. Inflammation and cancer: an ancient link with novel potentials. Int J Cancer 2007;121:2373–2380.
5. Hussain SP, Hofseth LJ, Harris CC. Radical causes of cancer. Nat Rev Cancer 2003;3:276–285.
6. Ohshima H, Tatemichi M, Sawa T. Chemical basis of inflammation-induced carcinogenesis. Arch Biochem Biophys 2003;417:3–11.
7. Kawanishi S, Hiraku Y, Murata M, Oikawa S. The role of metals in site-specific DNA damage with reference to carcinogenesis. Free Radic Biol Med 2002;32:822–832.
8. Dizdaroglu M, Jaruga P. Mechanisms of free radical-induced damage to DNA. Free Radic Res 2012;46:382–419.
9. Shibutani S, Takeshita M, Grollman AP. Insertion of specific bases during DNA synthesis past the oxidation-damaged base 8-oxodG. Nature 1991;349:431–434.
10. David SS, O'Shea VL, Kundu S. Base-excision repair of oxidative DNA damage. Nature 2007; 447:941–950.
11. Thanan R, Murata M, Pinlaor S, Sithithaworn P, Khuntikeo N, Tangkanakul W, et al. Urinary 8-oxo-7,8-dihydro-2′-deoxyguanosine in patients with parasite infection and effect of antiparasitic drug in relation to cholangiocarcinogenesis. Cancer Epidemiol Biomark Prev 2008;17:518–524.

12. Erhola M, Toyokuni S, Okada K, Tanaka T, Hiai H, Ochi H, et al. Biomarker evidence of DNA oxidation in lung cancer patients: association of urinary 8-hydroxy-2′-deoxyguanosine excretion with radiotherapy, chemotherapy, and response to treatment. FEBS Lett 1997;409:287–291.

13. Tagesson C, Kallberg M, Klintenberg C, Starkhammar H. Determination of urinary 8-hydroxydeoxyguanosine by automated coupled-column high performance liquid chromatography: a powerful technique for assaying *in vivo* oxidative DNA damage in cancer patients. Eur J Cancer 1995;31A:934–940.

14. Kawanishi S, Hiraku Y, Pinlaor S, Ma N. Oxidative and nitrative DNA damage in animals and patients with inflammatory diseases in relation to inflammation-related carcinogenesis. Biol Chem 2006;387:365–372.

15. Hiraku Y. Formation of 8-nitroguanine, a nitrative DNA lesion, in inflammation-related carcinogenesis and its significance. Environ Health Prev Med 2010;15:63–72.

16. Kawanishi S, Hiraku Y. Oxidative and nitrative DNA damage as biomarker for carcinogenesis with special reference to inflammation. Antioxid Redox Signal 2006;8:1047–1058.

17. Yermilov V, Rubio J, Becchi M, Friesen MD, Pignatelli B, Ohshima H. Formation of 8-nitroguanine by the reaction of guanine with peroxynitrite *in vitro*. Carcinogenesis 1995;16:2045–2050.

18. Inoue S, Kawanishi S. Oxidative DNA damage induced by simultaneous generation of nitric oxide and superoxide. FEBS Lett 1995;371:86–88.

19. Akaike T, Okamoto S, Sawa T, Yoshitake J, Tamura F, Ichimori K, et al. 8-nitroguanosine formation in viral pneumonia and its implication for pathogenesis. Proc Natl Acad Sci USA 2003;100: 685–690.

20. Yermilov V, Rubio J, Ohshima H. Formation of 8-nitroguanine in DNA treated with peroxynitrite *in vitro* and its rapid removal from DNA by depurination. FEBS Lett 1995;376:207–210.

21. Loeb LA, Preston BD. Mutagenesis by apurinic/apyrimidinic sites. Annu Rev Genet 1986;20:201–230.

22. Wu X, Takenaka K, Sonoda E, Hochegger H, Kawanishi S, Kawamoto T, et al. Critical roles for polymerase ζ in cellular tolerance to nitric oxide-induced DNA damage. Cancer Res 2006;66:748–754.

23. Suzuki N, Yasui M, Geacintov NE, Shafirovich V, Shibutani S. Miscoding events during DNA synthesis past the nitration-damaged base 8-nitroguanine. Biochemistry 2005;44:9238–9245.

24. Bhamra I, Compagnone-Post P, O'Neil IA, Iwanejko LA, Bates AD, Cosstick R. Base-pairing preferences, physicochemical properties and mutational behaviour of the DNA lesion 8-nitroguanine. Nucleic Acids Res, 2012;40:11126–11138.

25. Juedes MJ, Wogan GN. Peroxynitrite-induced mutation spectra of pSP189 following replication in bacteria and in human cells. Mutat Res 1996;349:51–61.

26. Kim MY, Dong M, Dedon PC, Wogan GN. Effects of peroxynitrite dose and dose rate on DNA damage and mutation in the *supF* shuttle vector. Chem Res Toxicol 2005;18:76–86.

27. Bos JL. The ras gene family and human carcinogenesis. Mutat Res 1988;195:255–271.

28. Takahashi T, Nau MM, Chiba I, Birrer MJ, Rosenberg RK, Vinocour M, et al. p53: a frequent target for genetic abnormalities in lung cancer. Science 1989;246:491–494.

29. Hsu IC, Metcalf RA, Sun T, Welsh JA, Wang NJ, Harris CC. Mutational hotspot in the *p53* gene in human hepatocellular carcinomas. Nature 1991;350:427–428.

30. Pinlaor S, Hiraku Y, Ma N, Yongvanit P, Semba R, Oikawa S, et al. Mechanism of NO-mediated oxidative and nitrative DNA damage in hamsters infected with *Opisthorchis viverrini*: a model of inflammation-mediated carcinogenesis. Nitric Oxide 2004;11:175–183.

31. Hiraku Y, Kawanishi S. Immunohistochemical analysis of 8-nitroguanine, a nitrative DNA lesion, in relation to inflammation-associated carcinogenesis. Methods Mol Biol 2009;512:3–13.

32. Pinlaor S, Yongvanit P, Hiraku Y, Ma N, Semba R, Oikawa S, et al. 8-Nitroguanine formation in the liver of hamsters infected with *Opisthorchis viverrini*. Biochem Biophys Res Commun 2003;309:567–571.

33. Pinlaor S, Ma N, Hiraku Y, Yongvanit P, Semba R, Oikawa S, et al. Repeated infection with *Opisthorchis viverrini* induces accumulation of 8-nitroguanine and 8-oxo-7,8-dihydro-2′-deoxyguanine in the bile duct of hamsters via inducible nitric oxide synthase. Carcinogenesis 2004;25:1535–1542.

34. Ma N, Adachi Y, Hiraku Y, Horiki N, Horiike S, Imoto I, et al. Accumulation of 8-nitroguanine in human gastric epithelium induced by *Helicobacter pylori* infection. Biochem Biophys Res Commun 2004;319:506–510.

35. Horiike S, Kawanishi S, Kaito M, Ma N, Tanaka H, Fujita N, et al. Accumulation of 8-nitroguanine in the liver of patients with chronic hepatitis C. J Hepatol 2005;43:403–410.

36. Hiraku Y, Tabata T, Ma N, Murata M, Ding X, Kawanishi S. Nitrative and oxidative DNA damage in cervical intraepithelial neoplasia associated with human papilloma virus infection. Cancer Sci 2007;98:964–972.

37. Ma N, Kawanishi M, Hiraku Y, Murata M, Huang GW, Huang Y, et al. Reactive nitrogen species–dependent DNA damage in EBV-associated nasopharyngeal carcinoma: the relation to STAT3 activation and EGFR expression. Int J Cancer 2008;122:2517–2525.

38. Ma N, Thanan R, Kobayashi H, Hammam O, Wishahi M, Leithy TE, et al. Nitrative DNA damage and Oct3/4 expression in urinary bladder cancer with *Schistosoma haematobium* infection. Biochem Biophys Res Commun 2011;414:344–349.

39. Ding X, Hiraku Y, Ma N, Kato T, Saito K, Nagahama M, et al. Inducible nitric oxide synthase–dependent DNA damage in mouse model of inflammatory bowel disease. Cancer Sci 2005;96:157–163.

40. Chaiyarit P, Ma N, Hiraku Y, Pinlaor S, Yongvanit P, Jintakanon D, et al. Nitrative and oxidative DNA damage in oral lichen planus in relation to human oral carcinogenesis. Cancer Sci 2005;96: 553–559.

41. Ma N, Tagawa T, Hiraku Y, Murata M, Ding X, Kawanishi S. 8-Nitroguanine formation in oral leukoplakia, a premalignant lesion. Nitric Oxide 2006;14:137–143.

42. Thanan R, Ma N, Iijima K, Abe Y, Koike T, Shimosegawa T, et al. Proton pump inhibitors suppress iNOS-dependent DNA damage in Barrett's esophagus by increasing Mn-SOD expression. Biochem Biophys Res Commun 2012;421:280–285.

43. Hiraku Y, Kawanishi S, Ichinose T, Murata M. The role of iNOS-mediated DNA damage in infection- and asbestos-induced carcinogenesis. Ann NY Acad Sci 2010;1203:15–22.

44. Guo F, Ma N, Horibe Y, Kawanishi S, Murata M, Hiraku Y. Nitrative DNA damage induced by multi-walled carbon nanotube via endocytosis in human lung epithelial cells. Toxicol Appl Pharmacol 2012;260:183–192.

45. Ohnishi S, Saito H, Suzuki N, Ma N, Hiraku Y, Murata M, Kawanishi S. Nitrative and oxidative DNA damage caused by K-ras mutation in mice. Biochem Biophys Res Commun 2011;413:236–240.

46. Hoki Y, Hiraku Y, Ma N, Murata M, Matsumine A, Nagahama M, et al. iNOS-dependent DNA damage in patients with malignant fibrous histiocytoma in relation to prognosis. Cancer Sci 2007;98:163–168.

47. International Agency for Research on Cancer. *Opisthorchis viverrini* and *Clonorchis sinensis*. IARC Monographs on the Evaluation of Carcinogenic Risks to Humans, vol. 100, A Review of Human Carcinogens, Part B: Biological Agents. Lyon, France: IARC Press; 2012, pp. 341–370.

48. Sripa B, Kaewkes S, Intapan PM, Maleewong W, Brindley PJ. Food-borne trematodiases in Southeast Asia: epidemiology, pathology, clinical manifestation and control. Adv Parasitol 2010;72: 305–350.

49. Vatanasapt V, Sriamporn S, Vatanasapt P. Cancer control in Thailand. Jpn J Clin Oncol 2002;32 Suppl:S82–S91.

50. Uttaravichen T, Buddhiswasdi V, Pairojkul C. Bile duct cancer and the liver fluke: pathology, presentation and surgical management. Asian J Surg 1996;19:267–270.

51. Pinlaor S, Hiraku Y, Yongvanit P, Tada-Oikawa S, Ma N, Pinlaor P, et al. iNOS-dependent DNA damage via NF-κB expression in hamsters infected with *Opisthorchis viverrini* and its suppression by the antihelminthic drug praziquantel. Int J Cancer 2006;119:1067–1072.

52. Pinlaor S, Sripa B, Ma N, Hiraku Y, Yongvanit P, Wongkham S, et al. Nitrative and oxidative DNA damage in intrahepatic cholangiocarcinoma patients in relation to tumor invasion. World J Gastroenterol 2005;11:4644–4649.

53. Sripa B, Kaewkes S. Localisation of parasite antigens and inflammatory responses in experimental opisthorchiasis. Int J Parasitol 2000;30:735–740.

54. Pinlaor S, Tada-Oikawa S, Hiraku Y, Pinlaor P, Ma N, Sithithaworn P, et al. *Opisthorchis viverrini* antigen induces the expression of Toll-like receptor 2 in macrophage RAW cell line. Int J Parasitol 2005;35:591–596.

55. Takeuchi O, Akira S. Pattern recognition receptors and inflammation. Cell 2010;140:805–820.

56. Rakoff-Nahoum S, Medzhitov R. Toll-like receptors and cancer. Nat Rev Cancer 2009;9:57–63.

57. Karin M, Greten FR. NF-κB: linking inflammation and immunity to cancer development and progression. Nat Rev Immunol 2005;5:749–759.

58. Kundu JK, Surh YJ. Inflammation: gearing the journey to cancer. Mutat Res 2008;659:15–30.

59. Pikarsky E, Porat RM, Stein I, Abramovitch R, Amit S, Kasem S, et al. NF-κB functions as a tumour promoter in inflammation-associated cancer. Nature 2004;431:461–466.

60. Karin M. Nuclear factor-κB in cancer development and progression. Nature 2006;441:431–436.

61. Prakobwong S, Khoontawad J, Yongvanit P, Pairojkul C, Hiraku Y, Sithithaworn P, et al. Curcumin decreases cholangiocarcinogenesis in hamsters by suppressing inflammation-mediated molecular events related to multistep carcinogenesis. Int J Cancer 2011;129:88–100.

62. Arbyn M, Castellsague X, de Sanjose S, Bruni L, Saraiya M, Bray F, et al. Worldwide burden of cervical cancer in 2008. Ann Oncol 2011;22:2675–2686.

63. Munoz N, Bosch FX, de Sanjose S, Herrero R, Castellsague X, Shah KV, et al. Epidemiologic classification of human papillomavirus types associated with cervical cancer. N Engl J Med 2003;348:518–527.

64. International Agency for Research on Cancer. Human Papillomaviruses. IARC Monographs on the Evaluation of Carcinogenic Risks to Humans, vol 100, A Review of Human Carcinogens, Part B: Biological Agents. Lyon, France: IARC Press; 2012, pp. 261–319.

65. Yugawa T, Kiyono T. Molecular mechanisms of cervical carcinogenesis by high-risk human papillomaviruses: novel functions of E6 and E7 oncoproteins. Rev Med Virol 2009;19:97–113.

66. Xu M, Luo W, Elzi DJ, Grandori C, Galloway DA. NFX1 interacts with mSin3A/histone deacetylase to repress hTERT transcription in keratinocytes. Mol Cell Biol 2008;28:4819–4828.

67. Duensing S, Munger K. Mechanisms of genomic instability in human cancer: insights from studies with human papillomavirus oncoproteins. Int J Cancer 2004;109:157–162.

68. von Knebel Doeberitz M. New markers for cervical dysplasia to visualise the genomic chaos created by aberrant oncogenic papillomavirus infections. Eur J Cancer 2002;38:2229–2242.

69. Castle PE, Hillier SL, Rabe LK, Hildesheim A, Herrero R, Bratti MC, et al. An association of cervical inflammation with high-grade cervical neoplasia in women infected with oncogenic human papillomavirus (HPV). Cancer Epidemiol Biomark Prev 2001;10:1021–1027.

70. Munoz N, Franceschi S, Bosetti C, Moreno V, Herrero R, Smith JS, et al. Role of parity and human papillomavirus in cervical cancer: the IARC multicentric case–control study. Lancet 2002;359:1093–1101.

71. Smith JS, Herrero R, Bosetti C, Munoz N, Bosch FX, Eluf-Neto J, et al. Herpes simplex virus-2 as a human papillomavirus cofactor in the etiology of invasive cervical cancer. J Natl Cancer Inst 2002;94:1604–1613.

72. Smith JS, Munoz N, Herrero R, Eluf-Neto J, Ngelangel C, Franceschi S, et al. Evidence for *Chlamydia trachomatis* as a human papillomavirus cofactor in the etiology of invasive cervical cancer in Brazil and the Philippines. J Infect Dis 2002;185:324–331.

73. Roeters AM, Boon ME, van Haaften M, Vernooij F, Bontekoe TR, Heintz AP. Inflammatory events as detected in cervical smears and squamous intraepithelial lesions. Diagn Cytopathol 2010;38: 85–93.

74. Wei L, Gravitt PE, Song H, Maldonado AM, Ozbun MA. Nitric oxide induces early viral transcription coincident with increased DNA damage and mutation rates in human papillomavirus-infected cells. Cancer Res 2009;69:4878–4884.

75. Romano G, Sgambato A, Mancini R, Capelli G, Giovagnoli MR, Flamini G, et al. 8-Hydroxy-2′-deoxyguanosine in cervical cells: correlation with grade of dysplasia and human papillomavirus infection. Carcinogenesis 2000;21:1143–1147.

76. Klaes R, Friedrich T, Spitkovsky D, Ridder R, Rudy W, Petry U, et al. Overexpression of p16(INK4A) as a specific marker for dysplastic and neoplastic epithelial cells of the cervix uteri. Int J Cancer 2001;92:276–284.

77. Wang JL, Zheng BY, Li XD, Angstrom T, Lindstrom MS, Wallin KL. Predictive significance of the alterations of p16INK4A, p14ARF, p53, and proliferating cell nuclear antigen expression in the progression of cervical cancer. Clin Cancer Res 2004;10:2407–2414.

78. Khleif SN, DeGregori J, Yee CL, Otterson GA, Kaye FJ, Nevins JR, et al. Inhibition of cyclin D-CDK4/CDK6 activity is associated with an E2F-mediated induction of cyclin kinase inhibitor activity. Proc Natl Acad Sci USA 1996;93:4350–4354.

79. Iglesias M, Yen K, Gaiotti D, Hildesheim A, Stoler MH, Woodworth CD. Human papillomavirus type 16 E7 protein sensitizes cervical keratinocytes to apoptosis and release of interleukin-1α. Oncogene 1998;17:1195–1205.

80. Lee S, Kim JH, Kim H, Kang JW, Kim SH, Yang Y, et al. Activation of the interleukin-32 pro-inflammatory pathway in response to human papillomavirus infection and over-expression of interleukin-32 controls the expression of the human papillomavirus oncogene. Immunology 2011;132:410–420.

81. Kessis TD, Slebos RJ, Nelson WG, Kastan MB, Plunkett BS, Han SM, et al. Human papillomavirus 16 E6 expression disrupts the p53-mediated cellular response to DNA damage. Proc Natl Acad Sci USA 1993;90:3988–3992.

82. Song S, Gulliver GA, Lambert PF. Human papillomavirus type 16 *E6* and *E7* oncogenes abrogate radiation-induced DNA damage responses *in vivo* through p53-dependent and p53-independent pathways. Proc Natl Acad Sci USA 1998;95:2290–2295.

83. International Agency for Research on Cancer. Epstein–Barr virus. IARC Monographs on the Evaluation of Carcinogenic Risks to Humans, vol. 100, A Review of Human Carcinogens, Part B: Biological Agents. Lyon, France: IARC Press; 2012, pp. 49–92.

84. Jeannel D, Bouvier G, Huber A. Nasopharyngeal carcinoma: an epidemiological approach to carcinogenesis. Cancer Surv 1999;33:125–155.

85. McDermott AL, Dutt SN, Watkinson JC. The aetiology of nasopharyngeal carcinoma. Clin Otolaryngol Allied Sci 2001;26:82–92.

86. Tsai ST, Jin YT, Mann RB, Ambinder RF. Epstein–Barr virus detection in nasopharyngeal tissues of patients with suspected nasopharyngeal carcinoma. Cancer 1998;82:1449–1453.

87. Lo KW, To KF, Huang DP. Focus on nasopharyngeal carcinoma. Cancer Cell 2004;5:423–428.

88. International Agency for Research on Cancer. Chinese–style Salted Fish. IARC Monographs on the Evaluation of Carcinogenic Risks to Humans, vol. 100, A Review of Human Carcinogens, Part E: Personal Habits and Indoor Combustions. Lyon, France: IARC Press; 2012, pp. 505–518.

89. Hildesheim A, West S, DeVeyra E, De Guzman MF, Jurado A, Jones C, et al. Herbal medicine use, Epstein–Barr virus, and risk of nasopharyngeal carcinoma. Cancer Res 1992;52:3048–3051.

90. Takeda N, Ohigashi H, Hirai N, Koshimizu K, Suzuki M, Tatematsu A, et al. Mass spectrometric identification of a phorbol diester 12-*O*-hexadecanoylphorbol-13-acetate, an Epstein–Barr virus-activating substance, in the soil collected from under *Sapium sebiferum*. Cancer Lett 1991;59:153–158.

91. Lo HW, Hsu SC, Ali-Seyed M, Gunduz M, Xia W, Wei Y, et al. Nuclear interaction of EGFR and STAT3 in the activation of the iNOS/NO pathway. Cancer Cell 2005;7:575–589.

92. Wang H, Li Y, Liu H, Liu S, Liu Q, Wang XM, et al. Peroxynitrite mediates glomerular lesion of diabetic rat via JAK/STAT signaling pathway. J Endocrinol Invest 2009;32:844–851.

93. Schetter AJ, Heegaard NH, Harris CC. Inflammation and cancer: interweaving microRNA, free radical, cytokine and p53 pathways. Carcinogenesis 2010;31:37–49.

94. Mo Y, Midorikawa K, Zhang Z, Zhou X, Ma N, Huang G, et al. Promoter hypermethylation of Ras-related GTPase gene RRAD inactivates a tumor suppressor function in nasopharyngeal carcinoma. Cancer Lett 2012;323:147–154.

95. International Agency for Research on Cancer. Asbestos (chrysotile, amosite, crocidolite, tremolite, actinolite, and anthophyllite). IARC Monographs on the Evaluation of Carcinogenic Risks to Humans, vol. 100, A Review of Human Carcinogens, Part C: Arsenic, Metals, Fibres, and Dusts. Lyon, France: IARC Press, 2012, pp. 219–309.

96. Robinson BW, Lake RA. Advances in malignant mesothelioma. N Engl J Med 2005;353:1591–1603.

97. Quinlan TR, BeruBe KA, Hacker MP, Taatjes DJ, Timblin CR, Goldberg J, et al. Mechanisms of asbestos-induced nitric oxide production by rat alveolar macrophages in inhalation and *in vitro* models. Free Radic Biol Med 1998;24:778–788.

98. Chao CC, Park SH, Aust AE. Participation of nitric oxide and iron in the oxidation of DNA in asbestos-treated human lung epithelial cells. Arch Biochem Biophys 1996;326:152–157.

99. Riganti C, Orecchia S, Silvagno F, Pescarmona G, Betta PG, Gazzano E, et al. Asbestos induces nitric oxide synthesis in mesothelioma cells via Rho signaling inhibition. Am J Respir Cell Mol Biol 2007;36:746–756.

100. Dorger M, Allmeling AM, Kiefmann R, Munzing S, Messmer K, Krombach F. Early inflammatory response to asbestos exposure in rat and hamster lungs: role of inducible nitric oxide synthase. Toxicol Appl Pharmacol 2002;181:93–105.

101. Tanaka S, Choe N, Hemenway DR, Zhu S, Matalon S, Kagan E. Asbestos inhalation induces reactive nitrogen species and nitrotyrosine formation in the lungs and pleura of the rat. J Clin Invest 1998;102:445–454.

102. Lehtonen H, Oksa P, Lehtimaki L, Sepponen A, Nieminen R, Kankaanranta H, et al. Increased alveolar nitric oxide concentration and high levels of leukotriene B_4 and 8-isoprostane in exhaled breath condensate in patients with asbestosis. Thorax 2007;62:602–607.

103. Unfried K, Schurkes C, Abel J. Distinct spectrum of mutations induced by crocidolite asbestos: clue for 8-hydroxydeoxyguanosine-dependent mutagenesis *in vivo*. Cancer Res 2002;62:99–104.

104. Pacurari M, Castranova V, Vallyathan V. Single- and multi-wall carbon nanotubes versus asbestos: Are the carbon nanotubes a new health risk to humans? J Toxicol Environ Health A 2010;73: 378–395.

105. Medina C, Santos-Martinez MJ, Radomski A, Corrigan OI, Radomski MW. Nanoparticles: pharmacological and toxicological significance. Br J Pharmacol 2007;150:552–558.

106. Oberdörster G. Safety assessment for nanotechnology and nanomedicine: concepts of nanotoxicology. J Intern Med 2010;267:89–105.

107. Poland CA, Duffin R, Kinloch I, Maynard A, Wallace WA, Seaton A, et al. Carbon nanotubes introduced into the abdominal cavity of mice show asbestos-like pathogenicity in a pilot study. Nat Nanotechnol 2008;3:423–428.

108. Takagi A, Hirose A, Nishimura T, Fukumori N, Ogata A, Ohashi N, et al. Induction of mesothelioma in p53$^{+/-}$ mouse by intraperitoneal application of multi-wall carbon nanotube. J Toxicol Sci 2008;33:105–116.

109. Nagai H, Okazaki Y, Chew SH, Misawa N, Yamashita Y, Akatsuka S, et al. Diameter and rigidity of multiwalled carbon nanotubes are critical factors in mesothelial injury and carcinogenesis. Proc Natl Acad Sci USA 2011;108:E1330–E1338.

110. Sakamoto Y, Nakae D, Fukumori N, Tayama K, Maekawa A, Imai K, et al. Induction of mesothelioma by a single intrascrotal administration of multi-wall carbon nanotube in intact male Fischer 344 rats. J Toxicol Sci 2009;34:65–76.

111. Ma-Hock L, Treumann S, Strauss V, Brill S, Luizi F, Mertler M, et al. Inhalation toxicity of multiwall carbon nanotubes in rats exposed for 3 months. Toxicol Sci 2009;112:468–481.

112. Shvedova AA, Kisin ER, Porter D, Schulte P, Kagan VE, Fadeel B, et al. Mechanisms of pulmonary toxicity and medical applications of carbon nanotubes: Two faces of Janus? Pharmacol Ther 2009;121:192–204.

113. Porter DW, Hubbs AF, Mercer RR, Wu N, Wolfarth MG, Sriram K, et al. Mouse pulmonary dose- and time course-responses induced by exposure to multi-walled carbon nanotubes. Toxicology 2010;269:136–147.

114. Ye SF, Wu YH, Hou ZQ, Zhang QQ. ROS and NF-κB are involved in upregulation of IL-8 in A549 cells exposed to multi-walled carbon nanotubes. Biochem Biophys Res Commun 2009;379: 643–648.

115. Pacurari M, Yin XJ, Zhao J, Ding M, Leonard SS, Schwegler-Berry D, et al. Raw single-wall carbon nanotubes induce oxidative stress and activate MAPKs, AP-1, NF-κB, and Akt in normal and malignant human mesothelial cells. Environ Health Perspect 2008;116:1211–1217.

116. Lindberg HK, Falck GC, Suhonen S, Vippola M, Vanhala E, Catalan J, et al. Genotoxicity of nanomaterials: DNA damage and micronuclei induced by carbon nanotubes and graphite nanofibres in human bronchial epithelial cells *in vitro*. Toxicol Lett 2009;186:166–173.

117. Karlsson HL, Gustafsson J, Cronholm P, Moller L. Size-dependent toxicity of metal oxide particles: a comparison between nano- and micrometer size. Toxicol Lett 2009;188:112–118.

118. Melhem MF, Meisler AI, Saito R, Finley GG, Hockman HR, Koski RA. Cytokines in inflammatory malignant fibrous histiocytoma presenting with leukemoid reaction. Blood 1993;82:2038–2044.

119. Richter KK, Parham DM, Scheele J, Hinze R, Rath FW. Presarcomatous lesions of experimentally induced sarcomas in rats: morphologic, histochemical, and immunohistochemical features. *In Vivo* 1999;13:349–355.

120. Hoki Y, Murata M, Hiraku Y, Ma N, Matsumine A, Uchida A, et al. 8-Nitroguanine as a potential biomarker for progression of malignant fibrous histiocytoma, a model of inflammation-related cancer. Oncol Rep 2007;18:1165–1169.

121. Harris AL. Hypoxia: a key regulatory factor in tumour growth. Nat Rev Cancer 2002;2:38–47.

122. Mateo J, Garcia-Lecea M, Cadenas S, Hernandez C, Moncada S. Regulation of hypoxia-inducible factor-1α by nitric oxide through mitochondria-dependent and -independent pathways. Biochem J 2003;376:537–544.

123. Thomas DD, Espey MG, Ridnour LA, Hofseth LJ, Mancardi D, Harris CC, et al. Hypoxic inducible factor 1α, extracellular signal-regulated kinase, and p53 are regulated by distinct threshold concentrations of nitric oxide. Proc Natl Acad Sci USA 2004;101:8894–8899.

124. Rius J, Guma M, Schachtrup C, Akassoglou K, Zinkernagel AS, Nizet V, et al. NF-κB links innate immunity to the hypoxic response through transcriptional regulation of HIF-1α. Nature 2008;453:807–811.

125. Walmsley SR, Print C, Farahi N, Peyssonnaux C, Johnson RS, Cramer T, et al. Hypoxia-induced neutrophil survival is mediated by HIF-1α-dependent NF-κB activity. J Exp Med 2005;201:105–115.

126. Tafani M, Pucci B, Russo A, Schito L, Pellegrini L, Perrone GA, et al. Modulators of HIF1α and NFκB in cancer treatment: Is it a rational approach for controlling malignant progression? Front Pharmacol 2013;4:13.

127. Janssen-Heininger YM, Poynter ME, Baeuerle PA. Recent advances towards understanding redox mechanisms in the activation of nuclear factor κB. Free Radic Biol Med 2000;28:1317–1327.

128. Sawa T, Tatemichi M, Akaike T, Barbin A, Ohshima H. Analysis of urinary 8-nitroguanine, a marker of nitrative nucleic acid damage, by high-performance liquid chromatography–electrochemical detection coupled with immunoaffinity purification: association with cigarette smoking. Free Radic Biol Med 2006;40:711–720.

129. Ishii Y, Ogara A, Okamura T, Umemura T, Nishikawa A, Iwasaki Y, et al. Development of quantitative analysis of 8-nitroguanine concomitant with 8-hydroxydeoxyguanosine formation by liquid chromatography with mass spectrometry and glyoxal derivatization. J Pharm Biomed Anal 2007;43:1737–1743.

130. Li MJ, Zhang JB, Li WL, Chu QC, Ye JN. Capillary electrophoretic determination of DNA damage markers: content of 8-hydroxy-2′-deoxyguanosine and 8-nitroguanine in urine. J Chromatogr B Analyt Technol Biomed Life Sci 2011;879:3818–3822.

131. Mishima T, Tajima Y, Kuroki T, Kosaka T, Adachi T, Kitasato A, et al. Chemopreventative effect of an inducible nitric oxide synthase inhibitor, ONO-1714, on inflammation-associated biliary carcinogenesis in hamsters. Carcinogenesis 2009;30:1763–1767.

132. Inano H, Onoda M. Nitric oxide produced by inducible nitric oxide synthase is associated with mammary tumorigenesis in irradiated rats. Nitric Oxide 2005;12:15–20.

CHAPTER 5

LIPID PEROXIDATION–DERIVED DNA ADDUCTS AND THE ROLE IN INFLAMMATION-RELATED CARCINOGENESIS

Helmut Bartsch and Urmila Jagadeesan Nair

Chronic infection and persistent inflammation are now recognized as important risk factors in many human cancers. Emerging evidence suggests that cancer-related inflammatory processes cause tissue damage and genetic instability involved in the initiation, promotion, and progression of carcinogenesis (1–5). Early and persistent inflammatory responses observed in or around developing preneoplastic lesions regulate many aspects of tumor development by controlling growth and survival factors implicated in maintaining tissue homeostasis. Increased levels of pro-inflammatory cytokines and stress response enzymes such as cyclooxygenase-2 (COX-2) and inducible nitric oxide synthase (iNOS), encoded by genes activated by the transcription factor NF-κB, are also involved in inflammatory processes (2,6). Reactive oxygen (ROS) and nitrogen (RNS) species and lipid peroxidation (LPO)–mediated tissue damage play a major role in inflammation-related malignancies (5–9).

An important mechanistic link between inflammation and cancer involves the generation of nitric oxide (NO), superoxide anion ($O_2 \bullet^-$) and other ROSs and RNSs by macrophages and neutrophils that infiltrate the sites of inflammation (10). Pathologically high levels of these reactive species can cause damage to cellular nucleic acids, lipids, carbohydrates and proteins. Oxidative and nitrative DNA damage measured as 8-oxo-7,8-dihydro-2′-deoxyguanine (8-oxodG) and 8-nitroguanine are recognized prevalent inflammation-related lesions (3,11).

LPO-DERIVED REACTIVE PRODUCTS CONTRIBUTE TO DNA DAMAGE

Oxidation of lipids by excess amounts of ROSs and RNSs generates numerous endogenous reactive intermediates (12,13) to include 4-hydroxy-2(E)-nonenal (HNE)

Cancer and Inflammation Mechanisms: Chemical, Biological, and Clinical Aspects, First Edition.
Edited by Yusuke Hiraku, Shosuke Kawanishi, and Hiroshi Ohshima.

and malondialdehyde (MDA) as major LPO products, which can react with DNA bases to form a variety of exocyclic DNA adducts *in vivo, inter alia* etheno-modified DNA bases, reviewed in (14). These miscoding lesions (15), now detected in many inflamed, cancer-prone human tissues (16), are thought to contribute to the onset of carcinogenesis through the induction of point mutations and genetic instability.

Since Virchow's early postulate on inflammation-related tumor growth in 1863 (17), the self-perpetuating accumulation of DNA and protein damage by inflammatory processes over time is now recognized as a novel target for diagnostic, therapeutic, and preventive strategies. Thus, LPO-derived DNA adducts, in combination with oxidative and nitrative damage markers, could be explored as early risk indicators of malignancies in afflicted human cancer-prone organs. LPO-related DNA damage also appears to be involved in the pathogenesis of other chronic degenerative diseases (18,19).

In this chapter we update information on methods for quantifying some representative LPO-derived DNA adducts in human biomonitoring studies. We also summarize results from biomarker applications that have provided insights into mechanisms of cancer causation and possibilities of preventive measures in human at- risk subjects. Due to space limitation, the literature citations are not exhaustive and review articles are often cited.

RESULTS AND DISCUSSION

LPO-induced DNA Damage: Adduct Types and Levels in Humans

Inflammatory processes induce persistent oxidative/nitrative stress and excess LPO, which cause massive DNA damage, now considered to be among the pathogenic pathways leading to malignancy. Reactive endogenous agents produced by LPO, such as HNE, MDA, acrolein, and crotonaldehyde, or by nucleobase oxidation via propenals (9), react with DNA either directly or through bifunctional intermediates to form various promutagenic exocyclic DNA adducts. Some major types are shown in Figure 5.1.

LPO products derived from γ-linoleic acid, including HNE and its electrophilic epoxy-, hydroperoxy-, and oxo-enal intermediates (20), react with the DNA bases A, C, and G to yield *inter alia* the etheno-DNA adducts $1,N^6$-etheno-2′-deoxyadenosine (εdA), $3,N^4$-etheno-2′-deoxycytidine (εdC), $1,N^2$–etheno-2′-deoxyguanosine, and $N^2,3$-etheno-2′-deoxyguanosine (Figure 5.1) all of which have been detected in human specimens (14).

LPO-derived reactive products and their macromolecular interactions have been characterized primarily by *in vitro* studies. The complex pharmacokinetics and DNA chemistry prevailing in inflamed cells makes it difficult, however, to pinpoint the main precursors and pathways involved in the generation of cancer-relevant DNA damage in humans *in vivo*. One important challenge in the biomarker field (21) is to find metabolically stable DNA damage indicators characteristic for inflammatory conditions and related malignancies. One promising approach consists of measuring DNA adduct patterns in inflamed tissues, now amenable by sensitive

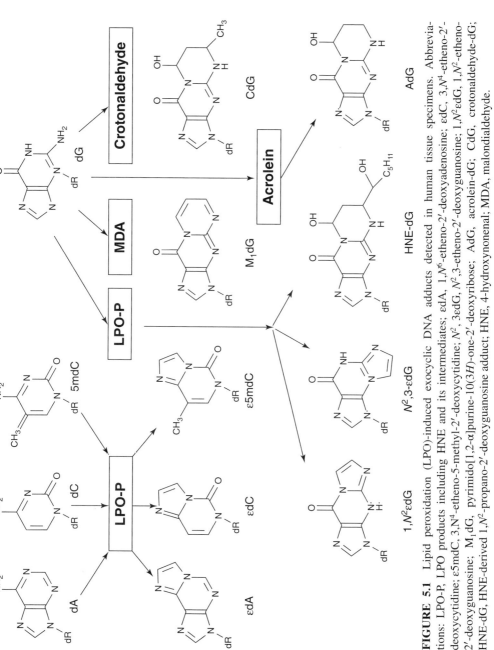

FIGURE 5.1 Lipid peroxidation (LPO)-induced exocyclic DNA adducts detected in human tissue specimens. Abbreviations: LPO-P, LPO products including HNE and its intermediates; εdA, 1,N^6-etheno-2'-deoxyadenosine; εdC, 3,N^4-etheno-2'-deoxycytidine; ε5mdC, 3,N^4-etheno-5-methyl-2'-deoxycytidine; N^2,3εdG, N^2,3-etheno-2'-deoxyguanosine; 1,N^2εdG, 1,N^2-etheno-2'-deoxyguanosine; M₁dG, pyrimido[1,2-α]purine-10(3H)-one-2'-deoxyribose; AdG, acrolein-dG; CdG, crotonaldehyde-dG; HNE-dG, HNE-derived 1,N^2-propano-2'-deoxyguanosine adduct; HNE, 4-hydroxynonenal; MDA, malondialdehyde.

63

mass spectrometry (MS) techniques and immunoassays, so as to establish disease-related damage profiles. As examplified in a mouse model of inflammation (22), analyses of the prevalent DNA damage by liquid chromatography (LC)–MS/MS and immunoblot techniques revealed a modest increases for nucleobase deamination products (10 to 30%), 8-oxodG (10%), and pyrimido[1,2-α]purine-10(3H)-one-2′-deoxyribose (M_1dG) (50%), while there were large three- to four-fold increases in the level of εdA and 1,N^2-etheno-2′-deoxyguanosine. The feasibility demonstrated for such analyses, as well as in human saliva and white blood cell (WBC) specimens (23), holds great promise of establishing inflammatory-related and, hopefully, tumor-specific DNA fingerprints as potential diagnostic and prognostic markers.

In a previous review (14) we discussed the reliability and validation status of several detection methods for the human sample analyses available at that time. These included immunoaffinity–^{32}P-postlabeling, high-performance liquid chromatography (HPLC)–electrochemical detection, gas chromatography–MS, liquid chromatography–tandem-MS, immunoslotblot, and immunohistochemical assays. Their application aimed to quantify exocyclic DNA adduct levels in human tissues, white blood cells (WBCs), and urine, and to compare adduct types and levels in human cancer-prone inflammatory and asymptomatic tissues as well as in rodent tumor models. As a notable limitation, only some of the methods noted above were able to detect a background of adducts in DNA arising from physiological LPO *in vivo*, a prerequisite to reliable monitoring of disease-related increase in human samples. Background adduct levels in asymptomatic human tissues ranged from not detectable to about 1 adduct per 10^8 DNA bases, while in organs affected by inflammatory cancer-prone conditions, levels increased by one to two orders of magnitude. Meanwhile, progress in the analytical field of adduct detection has been updated (9,21,23,24).

Quantifying Etheno-DNA Adducts in Human Tissues, White Blood Cells, and Urine by Ultrasensitive ^{32}P-Postlabeling and Immunohistochemical Methods

Detection methods for εdA and εdC residues in nuclear (25) and mitochondrial DNA (26) of human and rodent specimens have been developed in our laboratory: (1) combined immunopurification by monoclonal antibodies and subsequent ^{32}P-postlabeling and HPLC or thin-layer chromatography (TLC) analysis requiring only 5 to 10 μg of DNA per sample (with a detection limit as low as about 5 adducts per 10^{10} parent DNA nucleotides, it is one of the most sensitive and specific analytical techniques for etheno-adducts available todate); (2) immunohistochemical detection procedures for εdA and εdC in nuclear DNA of fine-needle tissue biopsies; and (3) quantification of εdA and εdC, excreted in urine as ethenodeoxyribonucleosides, achieved by a combination of immunoprecipitation–HPLC–fluorescence detection (for εdA) and ^{32}P-postlabeling –TLC (for εdC).

3,N^4-Etheno-5-methyl-2′-deoxycytidine (ε5mdC), an endogenous, hitherto unknown LPO-derived adduct (Figure 5.1), was recently identified in the DNA of human tissue and WBC specimens (27). Methylation of cytidine in dCpdG sequences regulates gene expression and patterns which are altered in chronic inflammatory diseases (28). Thus, etheno-adduction of 5-methyl-2'-deoxycytidine (5mdC) is of considerable interest, as it could play a role in epigenetic mechanisms of carcinogenesis. LPO products react with 2'-deoxycytidine to form $3,N^4$-etheno-2'-deoxycytidine (εdC). Because of an increased nucleophilicity at the N3 position, 5mdC residues in DNA were investigated as better targets for etheno-adduction. Known etheno-adduct forming agents, chloroacetaldehyde, and LPO products generated from arachidonic acid *in situ* produced the corresponding etheno-5mdC adducts (ε5mdC) upon reaction with DNA *in vitro*. A specific ^{32}P-postlabeling method developed unequivocally quantified both background and surprisingly high levels of ε5mdC lesions in the DNA of several human samples, which had been preanalyzed for εdC. The number of ε5mdC adducts per 10^8 parent nucleosides (in parentheses) were in asymptomatic human livers (ca. 7), in WBCs from women on a high γ-linoleic acid diet (ca. 700) (29), and in nontumorous lung tissue specimens from lung cancer patients (ca. 400). ε5mdC levels were found to be up to 10 times higher than those observed for etheno-adducts of the nonmethylated cytidine (εdC).

Based on our findings and method development, several possible pathogenic mechanisms should now be explored: Do ε5mdC lesions perturb the DNA methylation status, a hallmark observed in many chronic degenerative diseases (28), leading to hypo- and hyper-DNA methylation? If so, does etheno-adduction of 5mdC in dCpdG sequences play a role in epigenetic mechanisms of inflammation-driven carcinogenesis?

To conclude, the utility of the immunopurification/^{32}P-postlabeling detection methods for etheno-adducts described above in human biomonitoring studies is based on (1) their high sensitivity and specificity, (2) low amounts of DNA sample required, (3) capability to detect low background levels of adducts in biopsies, WBCs, and urine samples of healthy subjects; and (4) reliable measurements of disease-related increase of adduct levels in human at-risk subjects as well as a decrease after preventive and therapeutic interventions.

Detailed method protocols recently published (25) should facilitate the application of adduct monitoring in human studies, aimed at a better understanding of causative and preventive mechanisms in chronic degenerative diseases. However, these methods are labor intensive and require handling of ^{32}P. Therefore, the development of MS methods for easier application in clinical and epidemiologic studies should be pursued with priority. The simultaneous quantification of multiple exocyclic DNA adducts recently achieved in human saliva and WBC samples by nanoflow liquid chromatography–nanospray ionization–tandem MS (23) augurs progress in this analytical field.

Etheno-DNA Adducts as Potential Risk Markers for Inflammation-Driven Malignancies

Results from studies in at-risk patients (Table 5.1) provided strong evidence that oxidative stress-related DNA damage is induced by known inherited and acquired cancer risk factors, including inflammatory processes, persistent infections, and metal storage diseases (16). Sensitive and specific detection methods proved unequivocally

TABLE 5.1 Biomonitoring Studies in Humans with Cancer-prone Conditions or Exposed to Known Inherited or Acquired Cancer Risk Factors[a]

Predisposing condition	Organ	Relative risk for cancer site	Fold increase in εdA or εdC levels in cases and respective controls[b]	Method[c]	Refs.
Familial adenomatous polyposis	Colon (polyps)	100% Adenocarcinoma if not treated	2	A	(32)
Crohn's disease	Colon	3	2–20	A	(52)
Ulcerative colitis	Colon	6	4 (εdC only)	A	(52)
Chronic pancreatitis	Pancreas	2–3	3–28	A	(52)
Chronic viral hepatitis B or cirrhosis	Liver	90	90 (εdA urine)	C	(40)
OV infection	Bile duct, liver	5 (Cholangiocarcinoma)	3–4 (urine)	D	(34,36)
			3–5 (WBCs)	A	
Helicobacter pylori (HP) infection and high salt intake	Stomach	10	εdA (urine) increased with high salt excretion in HP-infected subjects	C	(53)
Wilson's disease	Liver	Increased if not treated by copper-chelating therapy	3	A	(51)
Primary hemochromatosis	Liver	220	2	A	(51)
Thalassemia	Liver	5–6	9 (εdA urine)	D	(49,50)
			13 (εdC urine)		
Alcoholic fatty liver	Liver	5–6	5 (εdA)	B	(42)
Alcoholic fibrosis (ALD patients with high alcohol intake, ca. 100 g/day)	Liver	5–6	16 (εdA)	B	(42)
			2–3 (high *vs.* low HNE-modified proteins in liver biopsies)	B	(44)
Esophagitis (in smokers + drinkers)	Esophagus	12–83	ca. 30 (εdA), ca. 80 (εdC) (increase vs. nonsmokers + nondrinkers)	B	(47)

[a] LPO-derived etheno-DNA adducts were measured in cancer-prone tissue specimens, WBCs and urine, where adducts are excreted as modified nucleosides.

[b] Statistically significant.

[c] Methods: A, Nair et al. (54); B, Frank et al. (42); C, Nair (55); D, Hanaoka et al., Sun et al. (53,56).

that pro-mutagenic LPO-derived DNA adducts are increased significantly in affected organs of cancer-prone subjects suffering from chronic pancreatitis, ulcerative colitis, Crohn's disease, and alcohol- and viral-related chronic hepatitis. In these studies, two etheno-adducts, εdA and εdC, were quantified as damage markers and shown to accumulate in target organs over time, paralleling progression to malignancy. Probably as a result of DNA repair and cellular apoptotic processes, excretion of ethenodeoxyribonucleosides in urine was recognized as a new noninvasive approach for *in vivo* monitoring of LPO-related pathogenic processes.

The mechanisms of etheno-adduct formation in inflamed organs appear to involve up-regulation or overexpression of stress response enzymes such as iNOS, lipoxygenases (LOXs), and COX-2. Inflammatory conditions are frequently associated with elevated enzyme activities in affected organs and in preneoplastic lesions where tumors develop later. As a consequence, self-perpetuating overproduction of ROSs, RNSs, and LPO-derived DNA damage is set in motion. This paradigm was supported by results in two rodent tumor models and in the pre-neoplastic tissue of the human colon.

1. In SJL (Swiss Jim Lambert) mice exhibiting inflammation-related NO overproduction by iNOS, etheno-DNA adduct formation was strongly interrelated (30) and showed the highest increase among other damaged DNA bases (22,30).

2. In a multistage mouse skin carcinogenesis model, etheno-adduct levels in skin papilloma correlated with LOX-catalyzed tumor-associated arachidonic acid metabolites (31).

3. In familial adenomatous polyposis (FAP)-patients, colonic polyps display a high COX-2 activity which was paralleled by increased etheno-DNA adduct levels (32). This is significant in view of the fact that if not treated, pre-neoplastic polyps in the FAP colon develop to 100% adenocarcinoma.

In studies extended to other at-risk patients, increased DNA damage was also detected in patients with *Opisthorchis viverrini* infection, thalassemia-related iron storage, and in the liver and esophagus of alcohol abusers. The mechanistic findings were supported by studies in rodent models that mimic human diseases (see below).

Etheno-DNA Adducts in Liver Fluke–Infected Patients: Damage Protection by Praziquantel and α-Tocopherol

Chronic infection by *Opisthorchis viverrini* (OV) or liver fluke is a strong risk factor for developing cholangiocarcinoma (CCA) in Southeast Asia. Etiology and pathogenesis have been reviewed (33). To clarify the involvement of LPO-derived DNA damage, the excretion of etheno-DNA adducts as a consequence of DNA repair and apoptosis in internal organs was measured in urine samples collected from healthy and OV-infected Thai subjects (34). εdA and εdC levels were quantified by immunoprecipitation/HPLC using fluorescence detection and ^{32}P-postlabeling. Mean urinary etheno-adduct levels were three to four times higher in OV-infected patients. Metabolic indicators of inflammatory conditions, MDA, and nitrate/nitrite levels in urine and plasma alkaline phosphatase (ALP) activity were increased up to two-fold. MDA and ALP activity were positively related to εdA excretion. Two months after a single dose of the antiparasitic drug Praziquantel, εdA and εdC concentrations in OV-infected subjects were significantly decreased and inflammatory indicators concomitantly lowered. Development of Praziquantel as an antischistosomal drug by pharmaceutical companies was encouraged by initial results that proved the absence of adverse genotoxic effects of this drug, now in use worldwide (35).

To support the mechanism that chronic OV infection also increases LPO-derived DNA damage *in vivo*, εdA and εdC in DNA were measured in WBC samples from OV-infected Thai patients and noninfected volunteers (36). Etheno-adduct levels were three to five times higher ($p < 0.001$) in patients when inflammatory indicators in plasma were increased two or three times. Mean plasma α-tocopherol concentration was two times lower in OV patients than in healthy controls ($p < 0.001$). A single dose of the antiparasitic drug Praziquantel given to infected patients decreased εdA and εdC adducts in WBC–DNA to control levels, paralleled by lowered values of inflammatory indicators. Plasma 8-isoprostane, MDA, nitrate/nitrite concentrations, and ALP were each positively correlated with εdA and εdC levels in WBCs. In contrast, plasma α-tocopherol and adduct levels showed inverse correlations. In a hamster model for liver fluke–induced cholangiocarcinogenesis, it was demonstrated that bile duct epithelial cells, where CCA develops later, were targets for LPO–DNA damage. In inflamed areas of the hamster liver, etheno-adducts were immunohistochemically localized in nuclei of inflammatory cells and in the epithelial lining of the bile duct (37).

Conclusions: Chronic *O. viverrini* infection via inflammation-related oxidative/nitrative stress led to increased urinary excretion of the ethenonucleosides εdA and εdC, reflecting that heavy LPO-derived DNA damage is also likely to occur in internal organs. As an accessible surrogate tissue, WBC samples were analyzed and accumulation of etheno-DNA adducts was detected in OV-infected subjects. Such damage generated in target bile duct epithelial cells was clearly shown in a hamster model for liver fluke–induced cholangiocarcinogenesis (36). LPO-related damage in OV-infected patients was reduced by a single dose of the effective antiparasitic drug Praziquantel. A lowered α-tocopherol plasma concentration found in OV patients was associated with an apparent diminished protection against LPO damage in WBC–DNA. Therefore, etheno-adducts in WBCs and urine should be explored: (1) as noninvasive risk markers for developing opisthorchiasis-related cholangiocarcinoma, and (2) to assess the efficacy of preventive measures (e.g., by dietary α-tocopherol) and interventions by therapeutic and antiparasitic drugs in infected at-risk patients.

High Urinary Excretion of LPO-Derived DNA Damage Markers in Patients with Cancer-prone Liver Diseases

Patients with chronic HBV and HCV infections of the liver are at very high risk for developing hepatocellular carcinoma (HCC) that proceeds via chronic hepatitis and liver cirrhosis (38). Oxidative and nitrative stress imposed in the virus-infected liver (39) can trigger LPO-related DNA damage, similar to that observed after liver fluke infection. Etheno-DNA adducts once formed *in vivo* from reactive α,β-unsaturated aldehydes (enals) can be monitored in urine following their elimination from internal organs as etheno-deoxyribonucleosides. A sensitive immunoprecipitation/HPLC–fluorescence detection method (25) allowed the quantification of εdA levels in urine samples of Thai and European patients with cancer-prone liver diseases and comparison with asymptomatic controls (40). When expressed as fmol εdA/μmol creatinine, levels in the controls ranged from 3 to 6 units, while patients with HBV-related inflammatory conditions excreted massively increased but highly variable εdA concentrations. When compared with asymptomatic HBsAg carriers, Thai patients with chronic hepatitis, liver cirrhosis, and HCC, had 20, 73, and 39 times higher urinary εdA levels, respectively. Of interest for potential diagnostic applications is the observed peak adduct level in the premalignant cirrhotic liver, followed by a decline once the tumor (HCC) had developed. A similar time course was also observed in rodent carcinogenesis models and in other human organ tumors (41). In European patients with HBV-, HCV-, and alcohol-related liver disease, urinary εdA levels were not as drastically elevated, but still seven- to 10- fold increased in over two-thirds of the patients (40). Thus, etheno-adducts in urine could be explored in clinical trials as markers in patients with chronic inflammatory diseases and premalignant conditions. Thereby, the preventive efficacy of prophylactive vaccination, dietary antioxidants, inhibitors of iNOS, LOX, and COX-2 functions, and scavengers of peroxynitrite and of N- and O-centered free radicals could be verified noninvasively.

Based on the exploratory studies noted above, we conclude that (1) high urinary εdA levels, reflecting massive LPO-derived hepatic DNA damage *in vivo*, probably contribute to the pathogenesis of HCC; (2) noninvasive monitoring of etheno-adducts in urine and in fine-needle liver biopsies (42) should be explored as a predictive marker to follow malignant progression of inflammatory liver disease; and (3) once validated, etheno-adducts may serve as clinical markers to assess the efficacy of preventive and therapeutic interventions.

Alcohol-Induced DNA Damage in Human Liver and Esophageal Tissue

Oxidative stress is thought to play a major role in the pathogenesis of HCC, a frequent complication of alcoholic liver disease (ALD) (43). In hepatocytes of ALD patients, immunohistochemical localization had previously revealed increased levels of εdA adducts in nuclei (42). In a follow-up study in fine-needle liver biopsies from ALD patients, etheno-DNA adducts (εdA, εdC) and protein-bound HNE in nuclei stained positively in immunoassays. As shown for the first time (44), both macromolecular damage markers were strongly correlated with CYP2E1 expression ($r = 0.9, p < 0.01$).

This ethanol-mediated simultaneous formation of protein and DNA damage by HNE was confirmed in rat models and human cells. In hepatocytes of alcohol fed lean and obese Zucker rats, increased levels of etheno-DNA adducts were detected, which correlated with Cyp2e1 expression. In HepG2 cells, stably transfected with human CYP2E1, ethanol exposure increased etheno-DNA adducts in nuclei in a concentration- and time-dependent manner. Chlormethiazole, a specific CYP2E1 inhibitor, blocked etheno-adducts by about 80%. These results further incriminate HNE as a bioactive molecule of pathophysiological importance in carcinogenesis and other chronic degenerative diseases (45).

Chronic alcohol consumption is a major risk factor for esophageal cancer (46), acting through various mechanisms, including oxidative stress-mediated damage. CYP2E1 induction and the generation of miscoding etheno-DNA adducts were analyzed in nontumorous esophageal biopsies of patients with upper aerodigestive tract cancer and high alcohol consumption (mean ca. 100 g/day). Comparisons were made with adduct levels in biopsies from tumor-free subjects that had no or low ethanol intake (47). CYP2E1 expression, etheno-DNA adducts, and Ki67 as a marker for cell proliferation were determined immunohistochemically. Chronic alcohol ingestion resulted in a significant induction of CYP2E1, which correlated with the amount of alcohol consumed ($r = 0.6, p < 0.001$). Analyses of esophageal mucosa specimens revealed a significant correlation ($r = 0.9, p < 0.001$) between CYP2E1 induction and the generation of miscoding etheno-DNA adducts (εdA, εdC) as seen in the liver. Etheno-DNA adducts correlated significantly with the cell proliferation rate, adducts being 30 to 80 times higher in patients who both drank and smoked. Nonsmokers and nondrinkers had the lowest values for cell proliferation, CYP2E1 expression, and DNA lesions.

Conclusions: In ALD patients, etheno-adducts mediated the induction of hepatic CYP2E1 and generated both HNE-derived miscoding DNA lesions and HNE-modified proteins in the liver. The same HNE-related damage pathway was shown to occur in nontumorous esophageal biopsies of patients with upper aerodigestive tract cancer and high chronic alcohol consumption. Cell proliferation rate, dose-dependent CYP2E1 induction by ethanol, and etheno-DNA adducts were strongly interrelated. Therefore, LPO-derived DNA lesions are, among other mechanisms, also implicated in esophageal tumorigenesis. Etheno-adducts in target organs or urine could be explored as markers to follow malignant progression of alcohol-related upper digestive tract cancers.

DNA Damage in Thalassemia-Related Iron Storage Disease Thalassemic diseases, including homozygous β-thalassemia and β-thalassemia/Hb E (β-Thal/Hb E), are prevalent in Southeast Asia and predispose patients to about a five-fold higher relative risk of developing HCC (48). Iron overload, a common complication in β-thalassemia patients, induces intracellular oxidative stress and stimulates LPO. To verify whether LPO products generate miscoding etheno-DNA adducts *in vivo*, εdA and εdC formation was monitored noninvasively in urine (49), where they are excreted as ethenodeoxyribonucleosides following their elimination from the DNA of internal organs. εdA and εdC in urine samples from β-Thal/Hb E patients and controls were assayed by immunoprecipitation–HPLC–fluorescence detection or ^{32}P-postlabeling–TLC, respectively (25). When expressed as fmol εdA/μmol creatinine, levels in patients ranged from 5 to 120 (mean 34) and were about nine times higher then levels in asymptomatic controls. εdC levels ranged from 0.2 to 33 (mean 5) and were increased about 13 times over controls. Adducts were correlated positively with non-transferin-bound iron, confirming that in these patients an iron overload stimulates LPO, thereby generating miscoding DNA adducts.

Three exocyclic DNA adducts—εdA, εdC, and the MDA-dG adduct (M_1dG)— were also quantified in lymphocyte DNA of β-Thal/Hb E patients and healthy controls by ^{32}P-postlabeling–TLC or HPLC (50). While etheno-adducts were twofold but not increased significantly, M_1dG levels in patients were four times higher than in controls.

In thalassemic mice strains, etheno-adducted DNA accumulated in the liver when iron concentration was above a threshhold of 2.7 mg Fe/g dry weight tissue. εdA and εdC adduct levels and iron were highly correlated. DNA damage was randomly distributed in all liver cell types (50).

To conclude, the highly increased levels of etheno-adducts excreted in the urine of β-thalassemia/HbE patients and their presence in the liver of thalassemic mice implicate that this massive type of hepatic DNA damage also occurs in thalassemia patients. Promutagenic LPO-derived DNA lesions are thus likely to contribute to the onset of HCC, as found previously in the iron-overloaded liver of hemochromatosis patients at extremely high risk for HCC (51).

The efficacy of preventive treatments using antioxidants and/or iron-chelating agents could be tested in thalassemia mouse models and the findings later applied to patients. Adduct measurements in patients' urine and in lymphocytes (M_1dG) could assess the protective effect of these interventions.

CONCLUSIONS AND FUTURE PROSPECTS

Data compiled in this volume provide overwhelming evidence that persistent oxidative/nitrative stress and excess LPO are induced by chronic inflammatory processes and infections, causing massive DNA damage from endogenous sources. Thus, continuous and self-perpetuating production of excess ROSs, RNSs, and LPO products leads to exocyclic DNA adduct formation and to other multiple damage in DNA, lipids, and proteins through oxidation, nitration, nitrosation, and halogenation reactions. As a consequence, altered cellular functions, and genetic and epigenetic changes accumulate over time and drive cells in inflamed tissues to malignancy.

Using ultrasensitive and specific detection methods for one type of the many LPO lesions generated, miscoding etheno-DNA adducts were unequivocally identified in human samples. Through collaborative efforts with clinicians, tissue samples were collected from patients affected by chronic inflammatory processes, persistent viral and parasitic infections, and iron storage- and alcohol-related diseases. Adduct levels increased progressively in the cancer-prone organs so far investigated, including liver, bile duct, esophagus, colon, and pancreas.

Consistent results were also observed in rodent tumor models and incriminate LPO-derived adducts as putative cancer-causing lesions. Etheno-adducts could thus be explored as markers to follow malignant progression of inflammatory diseases. Once validated, they may serve as predictive indicators to assess the efficacy of preventive and therapeutic interventions. The novel findings that ε5mdC adducts are easily generated by LPO products and showed large increases in several human organ DNA samples warrant investigation as to whether these lesions in dCpG sequences play a role in epigenetic mechanisms of inflammation-driven carcinogenesis.

To facilitate clinical and epidemiological applications, we have developed (1) noninvasive urine assays to quantify excreted ethenodeoxyribonucleosides as a dosimeter of DNA damage in internal organs, (2) methods for simultaneous determination of etheno- and M_1dG-adducts in human WBC-DNA, and (3) immunoassays with monoclonal antibodies for detecting etheno-adducts in the nuclei of needle biopsies.

Because some methods are labor intensive and require handling radioactive ^{32}P, the development of mass spectrometric methods for adduct analyses of small human DNA samples is urgently needed. Measurement of multiple LPO-derived DNA adducts in a single run has now become feasible by nanospray ionization–tandem mass spectrometry. Such novel analytical approaches could assess the largely unknown burden and diversity of DNA lesions generated from endogenous sources, to establish inflammation-related DNA fingerprints as potential diagnostic and prognostic tools. The use of early cancer predictive markers within prevention strategies is warranted in view of the enormous global cancer burden linked to infectious and chronic inflammatory diseases.

Acknowledgments

The authors gratefully acknowledge the contributions of and collaborative efforts by numerous investigators, clinicians, and technical staff (as cited, in part in the references). The chapter is dedicated to the late J. Nair, without whose perseverance our achievements would not have been possible. (See the obituary of Jagadeesan Nair, Senior Scientist at the German Cancer Research Center, in Carcinogenesis 2008;29,887–888.)

REFERENCES

1. Coussens LM, Werb Z. Inflammation and cancer. Nature 2002;420:860–867.
2. Hussain SP, Hofseth LJ, Harris CC. Radical causes of cancer. Nat Rev Cancer 2003;3:276–285.
3. Kawanishi S, Hiraku Y. Oxidative and nitrative DNA damage as biomarker for carcinogenesis with special reference to inflammation. Antioxid Redox Signal 2006;8:1047–1058.
4. Mantovani A, Allavena P, Sica A, Balkwill F. Cancer-related inflammation. Nature 2008;454: 436–444.
5. Ohshima H, Bartsch H. Chronic infections and inflammatory processes as cancer risk factors: possible role of nitric oxide in carcinogenesis. Mutat Res 1994;305:253–264.
6. Karin M, Greten FR. NF-kappaB: linking inflammation and immunity to cancer development and progression. Nat Rev Immunol 2005;5:749–759.
7. Inoue S, Kawanishi S. Oxidative DNA damage induced by simultaneous generation of nitric oxide and superoxide. FEBS Lett 1995;371:86–88.
8. Hussain SP, Harris CC. Inflammation and cancer: an ancient link with novel potentials. Int J Cancer 2007;121:2373–2380.
9. Lonkar P, Dedon PC. Reactive species and DNA damage in chronic inflammation: reconciling chemical mechanisms and biological fates. Int J Cancer 2011;128:1999–2009.
10. Karin M, Lawrence T, Nizet V. Innate immunity gone awry: linking microbial infections to chronic inflammation and cancer. Cell 2006;124:823–835.
11. Ohshima H, Sawa T, Akaike T. 8-Nitroguanine, a product of nitrative DNA damage caused by reactive nitrogen species: formation, occurrence, and implications in inflammation and carcinogenesis. Antioxid Redox Signal 2006;8:1033–1045.
12. Marnett LJ. Oxy radicals, lipid peroxidation and DNA damage. Toxicology 2002;181-182:219–222.
13. West JD, Marnett LJ. Endogenous reactive intermediates as modulators of cell signaling and cell death. Chem Res Toxicol 2006;19:173–194.

14. Nair U, Bartsch H, Nair J. Lipid peroxidation-induced DNA damage in cancer-prone inflammatory diseases: a review of published adduct types and levels in humans. Free Radic Biol Med 2007;43:1109–1120.

15. Barbin A. Etheno-adduct-forming chemicals: from mutagenicity testing to tumor mutation spectra. Mutat Res 2000;462:55–69.

16. Bartsch H, Nair J. Chronic inflammation and oxidative stress in the genesis and perpetuation of cancer: role of lipid peroxidation, DNA damage, and repair. Langenbecks Arch Surg 2006;391:499–510.

17. Virchow R. Die krankhaften Geschwülste: Dreissig Vorlesungen gehalten während des Wintersemesters 1862–1863 an der Universität zu Berlin. Berlin: Verlag August Hirschwald; 1863.

18. Nair J, De FS, Izzotti A, Bartsch H. Lipid peroxidation-derived etheno-DNA adducts in human atherosclerotic lesions. Mutat Res 2007;621:95–105.

19. DeFlora S., Izzotti A, Randerath K, Randerath E, Bartsch H, Nair J, et al. DNA adducts and chronic degenerative disease: pathogenetic relevance and implications in preventive medicine. Mutat Res 1996;366:197–238.

20. Blair IA. DNA adducts with lipid peroxidation products. J Biol Chem 2008;283:15545–15549.

21. Dedon PC, DeMott MS, Elmquist CE, Prestwich EG, McFaline JL, Pang B. Challenges in developing DNA and RNA biomarkers of inflammation. Biomark Med 2007;1:293–312.

22. Pang B, Zhou X, Yu H, Dong M, Taghizadeh K, Wishnok JS, et al. Lipid peroxidation dominates the chemistry of DNA adduct formation in a mouse model of inflammation. Carcinogenesis 2007;28:1807–1813.

23. Chen HJ, Lin WP. Quantitative analysis of multiple exocyclic DNA adducts in human salivary DNA by stable isotope dilution nanoflow liquid chromatography–nanospray ionization tandem mass spectrometry. Anal Chem 2011;83:8543–8551.

24. Medeiros MH. Exocyclic DNA adducts as biomarkers of lipid oxidation and predictors of disease: challenges in developing sensitive and specific methods for clinical studies. Chem Res Toxicol 2009;22:419–425.

25. Nair J, Nair UJ, Sun X, Wang Y, Arab K, Bartsch H. Quantifying etheno-DNA adducts in human tissues, white blood cells, and urine by ultrasensitive (32)P-postlabeling and immunohistochemistry. Methods Mol Biol 2011;682:189–205.

26. Nair J, Strand S, Frank N, Knauft J, Wesch H, Galle PR, et al. Apoptosis and age-dependent induction of nuclear and mitochondrial etheno-DNA adducts in Long–Evans Cinnamon (LEC) rats: enhanced DNA damage by dietary curcumin upon copper accumulation. Carcinogenesis 2005;26:1307–1315.

27. Nair J, Godschalk RW, Nair U, Owen RW, Hull WE, Bartsch H. Identification of 3,N(4)-etheno-5-methyl-2′-deoxycytidine in human DNA: a new modified nucleoside which may perturb genome methylation. Chem Res Toxicol 2012;25:162–169.

28. Robertson KD. DNA methylation and human disease. Nat Rev Genet 2005;6:597–610.

29. Nair J, Vaca CE, Velic I, Mutanen M, Valsta LM, Bartsch H. High dietary omega-6 polyunsaturated fatty acids drastically increase the formation of etheno-DNA base adducts in white blood cells of female subjects. Cancer Epidemiol Biomark Prev 1997;6:597–601.

30. Nair J, Gal A, Tamir S, Tannenbaum SR, Wogan GN, Bartsch H. Etheno adducts in spleen DNA of SJL mice stimulated to overproduce nitric oxide. Carcinogenesis 1998;19:2081–2084.

31. Nair J, Furstenberger G, Burger F, Marks F, Bartsch H. Promutagenic etheno-DNA adducts in multistage mouse skin carcinogenesis: correlation with lipoxygenase-catalyzed arachidonic acid metabolism. Chem Res Toxicol 2000;13:703–709.

32. Schmid K, Nair J, Winde G, Velic I, Bartsch H. Increased levels of promutagenic etheno-DNA adducts in colonic polyps of FAP patients. Int J Cancer 2000;87:1–4.

33. Yongvanit P, Pinlaor S, Bartsch H. Oxidative and nitrative DNA damage: key events in opisthorchiasis-induced carcinogenesis. Parasitol Int 2012;61:130–135.

34. Dechakhamphu S, Yongvanit P, Nair J, Pinlaor S, Sitthithaworn P, Bartsch H. High excretion of etheno adducts in liver fluke–infected patients: protection by praziquantel against DNA damage. Cancer Epidemiol Biomark Prev 2008;17:1658–1664.

35. Bartsch H, Kuroki T, Malaveille C, Loprieno N, Barale R, Abbondandolo A, et al. Absence of mutagenicity of praziquantel, a new, effective, anti-schistosomal drug, in bacteria, yeasts, insects and mammalian cells. Mutat Res 1978;58:133–142.

36. Dechakhamphu S, Pinlaor S, Sitthithaworn P, Nair J, Bartsch H, Yongvanit P. Lipid peroxidation and etheno DNA adducts in white blood cells of liver fluke-infected patients: protection by plasma alpha-tocopherol and praziquantel. Cancer Epidemiol Biomark Prev 2010;19:310–318.

37. Dechakhamphu S, Pinlaor S, Sitthithaworn P, Bartsch H, Yongvanit P. Accumulation of miscoding etheno-DNA adducts and highly expressed DNA repair during liver fluke–induced cholangiocarcinogenesis in hamsters. Mutat Res 2010;691:9–16.

38. Brechot C. Pathogenesis of hepatitis B virus–related hepatocellular carcinoma: old and new paradigms. Gastroenterology 2004;127:S56–S61.

39. Loguercio C, Federico A. Oxidative stress in viral and alcoholic hepatitis. Free Radic Biol Med 2003;34:1–10.

40. Nair J, Srivatanakul P, Haas C, Jedpiyawongse A, Khuhaprema T, Seitz HK, et al. High urinary excretion of lipid peroxidation-derived DNA damage in patients with cancer-prone liver diseases. Mutat Res 2010;683:23–28.

41. Bartsch H, Nair J. Accumulation of lipid peroxidation-derived DNA lesions: potential lead markers for chemoprevention of inflammation-driven malignancies. Mutat Res 2005;591:34–44.

42. Frank A, Seitz HK, Bartsch H, Frank N, Nair J. Immunohistochemical detection of 1,N6-ethenodeoxyadenosine in nuclei of human liver affected by diseases predisposing to hepatocarcinogenesis. Carcinogenesis 2004;25:1027–1031.

43. Seitz HK, Stickel F. Molecular mechanisms of alcohol-mediated carcinogenesis. Nat Rev Cancer 2007;7:599–612.

44. Wang Y, Millonig G, Nair J, Patsenker E, Stickel F, Mueller S, et al. Ethanol-induced cytochrome P4502E1 causes carcinogenic etheno-DNA lesions in alcoholic liver disease. Hepatology 2009;50:453–461.

45. Zarkovic N. 4-Hydroxynonenal as a bioactive marker of pathophysiological processes. Mol Aspects Med 2003;24:281–291.

46. Secretan B, Straif K, Baan R, Grosse Y, El GF, Bouvard V, et al. A review of human carcinogens—Part E: tobacco, areca nut, alcohol, coal smoke, and salted fish. Lancet Oncol 2009;10:1033–1034.

47. Millonig G, Wang Y, Homann N, Bernhardt F, Qin H, Mueller S, et al. Ethanol-mediated carcinogenesis in the human esophagus implicates CYP2E1 induction and the generation of carcinogenic DNA-lesions. Int J Cancer 2011;128:533–540.

48. Borgna-Pignatti C, Vergine G, Lombardo T, Cappellini MD, Cianciulli P, Maggio A, et al. Hepatocellular carcinoma in the thalassaemia syndromes. Br J Haematol 2004;124:114–117.

49. Meerang M, Nair J, Sirankapracha P, Thephinlap C, Srichairatanakool S, Fucharoen S, et al. Increased urinary 1,N6-ethenodeoxyadenosine and 3,N4-ethenodeoxycytidine excretion in thalassemia patients: markers for lipid peroxidation-induced DNA damage. Free Radic Biol Med 2008;44: 1863–1868.

50. Meerang M, Nair J, Sirankapracha P, Thephinlap C, Srichairatanakool S, Arab K, et al. Accumulation of lipid peroxidation-derived DNA lesions in iron-overloaded thalassemic mouse livers: comparison with levels in the lymphocytes of thalassemia patients. Int J Cancer 2009;125:759–766.

51. Nair J, Carmichael PL, Fernando RC, Phillips DH, Strain AJ, Bartsch H. Lipid peroxidation-induced etheno-DNA adducts in the liver of patients with the genetic metal storage disorders Wilson's disease and primary hemochromatosis. Cancer Epidemiol Biomark Prev 1998;7:435–440.

52. Nair J, Gansauge F, Beger H, Dolara P, Winde G, Bartsch H. Increased etheno-DNA adducts in affected tissues of patients suffering from Crohn's disease, ulcerative colitis, and chronic pancreatitis. Antioxid Redox Signal 2006;8:1003–1010.

53. Hanaoka T, Nair J, Takahashi Y, Sasaki S, Bartsch H, Tsugane S. Urinary level of 1,N(6)-ethenodeoxyadenosine, a marker of oxidative stress, is associated with salt excretion and omega 6-polyunsaturated fatty acid intake in postmenopausal Japanese women. Int J Cancer 2002;100:71–75.

54. Nair J, Barbin A, Guichard Y, Bartsch H. 1,N6-ethenodeoxyadenosine and 3,N4-ethenodeoxycytine in liver DNA from humans and untreated rodents detected by immunoaffinity/32P-postlabeling. Carcinogenesis 1995;16:613–617.

55. Nair J. Lipid peroxidation-induced etheno-DNA adducts in humans. IARC Sci Publ 1999;55–61.

56. Sun X, Karlsson A, Bartsch H, Nair J. New ultrasensitive 32P-postlabelling method for the analysis of 3,N4-etheno-2'-deoxycytidine in human urine. Biomarkers 2006;11:329–340.

CHAPTER 6

LEVEL OF INFLAMMATION-RELATED DNA ADDUCTS IN HUMAN TISSUES

Tomonari Matsuda, Pei-Hsin Chou, and Haruhiko Sugimura

Formation of DNA adducts is thought to be an initial event of carcinogenesis. Improvement in the performance of liquid chromatography–tandem mass spectrometry (LC/MS-MS) enables us to detect trace amounts of DNA adducts in tissue samples. Recently, we carried out a comprehensive analysis of DNA adducts by using LC/MS-MS and found that DNA adducts derived from 4-oxo-2(*E*)-nonenal (4-ONE) and 4-oxo-2(*E*)-hexenal (4-OHE) were one of the major DNA adduct types presented in various human tissues (1,2). In this chapter we describe the level of 4-ONE- and 4-OHE-derived DNA adducts in human tissues, the genotoxicity of the adducts, and their possible relation to inflammation.

LEVEL OF 4-ONE- AND 4-OHE-DERIVED DNA ADDUCTS IN HUMAN TISSUES

Fatty acids are the major components of the various lipids in living organisms. Saturated fatty acids are carboxylic acids with long-chain alkyl groups. The representative saturated fatty acids are palmitic acid and stearic acid. Unsaturated fatty acids are fatty acids that contain a double bond in their side chain. Two important classes of polyunsaturated fatty acids (PUFAs) are n-6 (or ω-6) and n-3 (or ω-3) fatty acids. In n-6 PUFAs, the first double-bonded carbon atom appears in the sixth carbon atom from the methyl terminal (ω) end of the chain, and in n-3 PUFAs, the first double-bonded carbon atom appears in the third position. The major n-6 PUFAs in animal tissues are linoleic acid and arachidonic acid, and the major n-3 PUFAs are α-linolenic acid, 5,8,11,14,17-eicosapentaenoic acid (EPA), and 4,7,10,13,16,19-docosahexaenoic acid (DHA) (Figure 6.1). In a rat study, the amount of n-6 PUFAs was several times higher than n-3 PUFAs in many organs. However, the brain, in which the amount of n-3 PUFAs was almost as same as n-6 PUFAs, was an exception (3).

Cancer and Inflammation Mechanisms: Chemical, Biological, and Clinical Aspects, First Edition.
Edited by Yusuke Hiraku, Shosuke Kawanishi, and Hiroshi Ohshima.

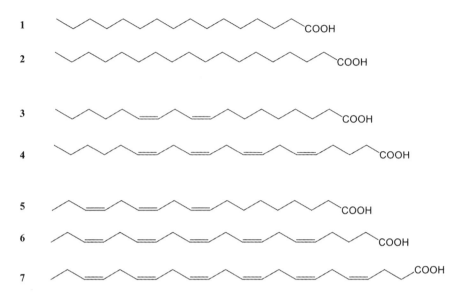

FIGURE 6.1 Structure of ubiquitous fatty acids in cells.

No	Symbol	Common Name	Systematic Name	Structure
Saturated fatty acids				
1	16:0	Palmitic acid	Hexadecanoic acid	$CH_3(CH_2)_{14}COOH$
2	18:0	Stearic acid	Octadecanoic acid	$CH_3(CH_2)_{16}COOH$
n-6 PUFA				
3	18:2n-6	Linoleic acid	9,12-Octadecadienoic acid	$CH_3(CH_2)_4(CH=CHCH_2)_2(CH_2)_6COOH$
4	20:4n-6	Arachidonic acid	5,8,11,14-Eicosatetraenoic acid	$CH_3(CH_2)_4(CH=CHCH_2)_4(CH_2)_2COOH$
n-3 PUFA				
5	18:3n-3	α-Linolenic acid	9,12,15-Octadecatrienoic acid	$CH_3CH_2(CH=CHCH_2)_3(CH_2)_6COOH$
6	20:5n-3	EPA	5,8,11,14,17-Eicosapentaenoic acid	$CH_3CH_2(CH=CHCH_2)_5(CH_2)_2COOH$
7	22:6n-3	DHA	4,7,10,13,16,19-Docosahexaenoic acid	$CH_3CH_2(CH=CHCH_2)_6CH_2COOH$

It is well known that purified PUFAs are easily autooxidized and produce many types of toxic aldehydes, such as acrolein, crotonaldehyde, and malondialdehyde, and these aldehydes attack DNA bases and make various DNA adducts. In addition to these toxic aldehydes, 4-ONE and 4-OHE are recognized as important degradation products of PUFAs in terms of DNA damaging potential. 4-ONE, a decomposition product of n-6 PUFAs, has been shown to induce the formation of etheno-DNA adducts carrying aliphatic side chains, including heptanone-etheno-2′-deoxycytidine (HεdC), heptanone-etheno-2′-deoxyguanosine (HεdG), and heptanone-etheno-2′-deoxyadenosine (HεdA) (4–6). On the other hand, 4-OHE, an n-3 PUFA-peroxidation product, was recently reported to be able to produce etheno-DNA adducts as well, such as butanone-etheno-2′-deoxycytidine (BεdC), butanone-etheno-2′-deoxyguanosine (BεdG), and butanone-etheno-2′-deoxyadenosine (BεdA) (7–9) (Figure 6.2).

We have found these DNA adducts as major peaks of DNA adductome analysis. This DNA adductome approach utilized LC/MS-MS analysis methods designed to

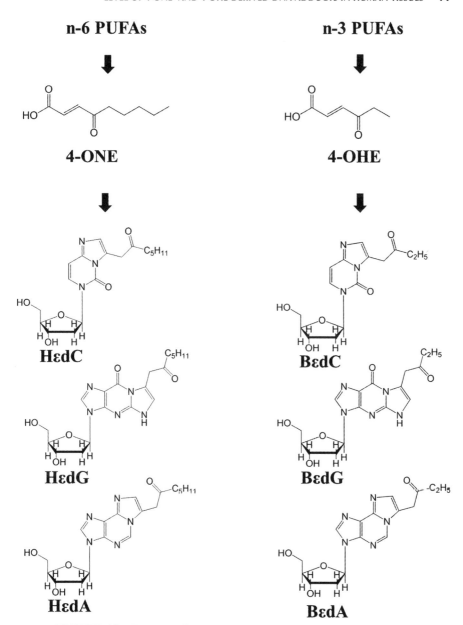

FIGURE 6.2   Structure of DNA adducts derived from 4-ONE and 4-OHE.

detect the neutral loss of 2′-deoxyribose from positively ionized 2′-deoxynucleoside adducts transmitting the $[M+H]^+ > [M+H-116]^+$ transition over several hundred transitions. In this setting, -116 corresponds to the neutral loss of 2′-deoxyribose. Although this method is semiquantitative, it helps us to grasp a complete picture of a DNA adducts in test samples. Figure 6.3 is an example of a DNA adductome map.

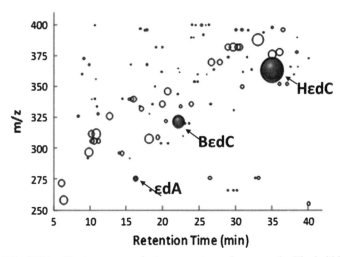

FIGURE 6.3 DNA adductome map of a human autopsy lung sample. The bubble size represents the peak area of the LC/MS-MS chromatogram of putative adducted deoxynucleosides detected by monitoring the $[M + H]^+ > [M + H - 116]^+$ transition. The bubbles that correspond to known DNA adducts are highlighted and annotated.

DNA was extracted from a lung autopsy sample and digested to $2'$-deoxynucleoside, followed by DNA adductome analysis. In this adductome map, the horizontal axis is the retention time of LC/MS-MS, the vertical axis is the mass-to-charge ratio (m/z) of the parental ion, and the size of the bubbles indicates the relative peak area of the DNA adducts. Although many DNA adducts were detected in this sample, we found that several major peaks were identical with those of 4-ONE- and 4-OHE-derived DNA adducts (1). Next, we measured the levels of these 4-ONE- and 4-OHE-derived DNA adducts in various human autopsy tissues (1) and in several specimens from gastric cancer patients (nontumor tissue) (2) using LC/MS-MS. The 4-ONE-derived DNA adducts—HɛdC, HɛdG, and HɛdA—were detected ubiquitously in human tissues and 4-OHE-derived BɛdC was detected at a relatively high rate, but the other adducts, BɛdG and BɛdA, were rarely detected. Table 6.1 summarizes the level of the representative 4-OHE- and 4-ONE-derived DNA adducts BɛdC and HɛdC, along with ubiquitous DNA adducts etheno-$2'$-deoxyadenosine (ɛdA) in tissues of Japanese patients. The level of HɛdC was higher than that of ɛdA in various tissues; the approximate median value in liver, lung, and pancreas was 6 to 18 adducts per 10^8 bases. The level in the stomach was one order of magnitude higher than in the other tissues. Considering that the approximate level of 8-oxo-7,8-dihydro-$2'$-deoxyguanosine (8-oxodG) is around 100 adducts per 10^8 bases, the level of the 4-ONE-derived DNA adduct was very high and it may be an important source of somatic mutations and could contribute significantly to cancer formation in humans. The level of 4-OHE-derived BɛdC was one order of magnitude smaller than that of HɛdC. However, several samples showed very high levels of this adduct.

TABLE 6.1   Levels of 4-OHE- and 4-ONE-Derived DNA Adducts

	DNA adduct level (per 10^8 bases)			
	Autopsy			Specimen (nontumor tissue)
	Liver	Lung	Pancreas	Stomach
εdA				
Maximum	76	246	39	276
Median	4	9	2	54
Minimum	n.d.[a]	4	n.d.	26
Detection rate[b]	17/19	12/12	7/9	10/10
BεdC				
Maximum	6	1190	60	762
Median	0.4	0.4	n.d.	84
Minimum	n.d.	n.d.	n.d.	9
Detection rate	10/19	6/12	3/9	10/10
HεdC				
Maximum	58	8200	583	2930
Median	6	18	10	134
Minimum	n.d.	7	3	20
Detection rate	17/19	12/12	9/9	10/10

Source: References 1 and 2.

[a] n.d.; not detected.

[b] Detection rate: no. of detections/no. of samples.

BIOLOGICAL PROPERTIES OF 4-ONE- AND 4-OHE-DERIVED DNA ADDUCTS

Several lines of studies using mammalian cell lines and *Escherichia coli* demonstrated that HεdC is a mutatgenic DNA adduct that induces C → A or C → T base substitutions (10,11). Pollack et al. (10) reported that in human cell lines, HεdC blocked DNA synthesis and also miscoded markedly during replication of a shuttle vector site-specifically modified with HεdC. The miscoding frequency was higher than 90%. However, information about the potential mutagenic properties of the other 4-ONE- and 4-OHE-derived DNA adducts and about the DNA repair pathways of these DNA adducts is still unavailable.

INFLAMMATION AND FORMATION OF 4-ONE

4-ONE is produced from n-6 PUFAs not only by lipid peroxidation but also by some enzymes related to inflammation. Such pathways have been well studied by Blair's group (12–14), and the enzymes involved in the pathways are cyclooxygenase (COX) and lipoxygenase (LO) enzymes (Figure 6.4). Among those enzymes, COX-2

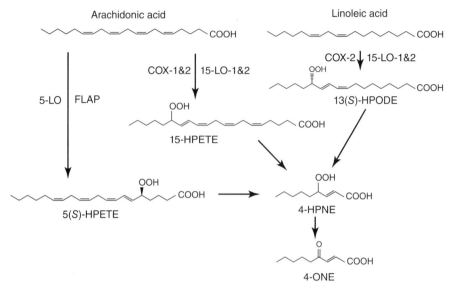

FIGURE 6.4 Cyclooxygenase- and lipoxygenase-mediated pathways for 4-ONE production. This figure was made by modifying schemes in references 13 and 14.

is responsible for the formation of prostaglandin H2, and 5-LO is responsible for the formation of leulotriene A_4, both of which are important mediators of inflammation. These enzymes are thought to be activated during inflammation.

Other than the prostaglandin biosynthesis pathway, COX-2 can enzymatically convert arachidonic acid into 15-hydroperoxy-5,8,11,13-(Z,Z,Z,E)-eicosatetraenoic acid (15-HPETE), and linoleic acid into 13(S)-hydroperoxy-9,11-(Z,E)-octadecadienoic acid [13(S)-HPODE]. The other enzymes—COX-1, 15-LO-1, and 15-LO-2—also carry out these reactions. 15-HPETE and 13(S)-HPODE are decomposed in the presence of vitamin C to form DNA-reactive bifunctional electrophiles: 4-hydroperoxy-2(E)-nonenal (4-HPNE), 4-hydroxy-2(E)-nonenal (4-HNE), 4-ONE, and 4,5-epoxy-2(E)-nonenal (EDE). The C57BL/6J APC^{min} mouse is a colorectal cancer mouse model in which COX-2 is up-regulated. The DNA adducts derived from the 4-ONE, HɛdC, and HɛdG levels in the small intestine was several times higher in this COX-2 up-regulated mouse than in control C57BL/6J mice. On the other hand, 5-LO can enzymatically convert arachidonic acid, with the assistance of 5-lipoxygenase-activating protein (FLAP), into 5(S)-hydroperoxy-6,8,11,14-(E,Z,Z,Z)-eicosatetraenoic acid, which is also decomposed to 4-HPNE and 4-ONE, and this pathway is also assumed to contribute to the enzymatic production of 4-ONE under inflammation.

Although the pathways for 4-ONE production were well studied, the mechanisms involved in the production of 4-OHE by n-3 PUFAs under inflammation have not yet been well elucidated. Further studies are needed to understand this point.

CONCLUDING REMARKS

In conclusion, inflammation-related DNA adducts produced by 4-ONE and 4-OHE are present ubiquitously in human tissues and are even predominant in some cases. This suggests a possible link between inflammation and cancer. Further studies are anticipated to clarify the relationship between these DNA adducts and inflammation status.

REFERENCES

1. Chou PH, Kageyama S, Matsuda S, Kanemoto K, Sasada Y, Oka M, et al. Detection of lipid peroxidation-induced DNA adducts caused by 4-oxo-2(*E*)-nonenal and 4-oxo-2(*E*)-hexenal in human autopsy tissues. Chem Res Toxicol 2010;23:1442–1448.
2. Matsuda T, Tao H, Goto M, Yamada H, Suzuki M, Wu Y, et al. Lipid peroxidation induced DNA adducts in human gastric mucosa. Carcinogenesis 2012;34:121–127.
3. Engler MM, Engler MB, Kroetz DL, Boswell KDB, Neeley E, Krassner SM. The effects of a diet rich in docosahexaenoic acid on organ and vascular fatty acid composition in spontaneously hypertensive rats. Prostaglandins Leukot Essent Fatty Acids 1999;61:289–295.
4. Rindgen D, Nakajima M, Wehrli S, Xu K, Blair IA. Covalent modifications to 2′-deoxyguanosine by 4-oxo-2-nonenal, a novel product of lipid peroxidation. Chem Res Toxicol 1999;12:1195–1204.
5. Rindgen D, Lee SH, Nakajima M, Blair IA. Formation of a substituted 1,*N*(6)-etheno-2′-deoxyadenosine adduct by lipid hydroperoxide-mediated generation of 4-oxo-2-nonenal. Chem Res Toxicol 2000;13:846–852.
6. Pollack M, Oe T, Lee SH, Silva Elipe MV, Arison BH, Blair IA. Characterization of 2′-deoxycytidine adducts derived from 4-oxo-2-nonenal, a novel lipid peroxidation product. Chem Res Toxicol 2003;16:893–900.
7. Kasai H, Maekawa M, Kawai K, Hachisuka K, Takahashi Y, Nakamura H, et al. 4-Oxo-2-hexenal, a mutagen formed by omega-3 fat peroxidation, causes DNA adduct formation in mouse organs. Ind Health 2005;43:699–701.
8. Maekawa M, Kawai K, Takahashi Y, Nakamura H, Watanabe T, Sawa R, et al. Identification of 4-oxo-2-hexenal and other direct mutagens formed in model lipid peroxidation reactions as dGuo adducts. Chem Res Toxicol 2006;19:130–138.
9. Kasai H, Kawai K. 4-Oxo-2-hexenal, a mutagen formed by omega-3 fat peroxidation: occurrence, detection and adduct formation. Mutat Res 2008;659:56–59.
10. Pollack M, Yang IY, Kim HY, Blair IA, Moriya M. Translesion DNA synthesis across the heptanone–etheno-2′-deoxycytidine adduct in cells. Chem Res Toxicol 2006;19:1074–1079.
11. Yang IY, Hashimoto K, de Wind N, Blair IA, Moriya M. Two distinct translesion synthesis pathways across a lipid peroxidation-derived DNA adduct in mammalian cells. J Biol Chem 2009;284:191–198.
12. Lee SH, Williams MV, Dubois RN, Blair IA. Cyclooxygenase-2-mediated DNA damage. J Biol Chem 2005;280:28337–28346.
13. Williams MV, Lee SH, Pollack M, Blair IA. Endogenous lipid hydroperoxide-mediated DNA-adduct formation in min mice. J Biol Chem 2006;281:10127–10133.
14. Jian W, Lee SH, Williams MV, Blair IA. 5-Lipoxygenase-mediated endogenous DNA damage. J Biol Chem 2009;284:16799–16807.

CHAPTER 7

TOLL-LIKE RECEPTORS: ROLE IN INFLAMMATION AND CANCER

Sarang Tartey and Osamu Takeuchi

The Toll receptor and its involvement in innate immunity was first described in *Drosophila* (1,2). A year after the discovery of the role of *Drosophila* Toll in host defense against fungal infection, Toll-like receptors (TLRs) were identified as a mammalian homolog of *Drosophila* Toll and are so far the best-characterized family of pattern recognition receptors (PRRs), which sense foreign material, called *pathogen-associated molecular patterns* (PAMPs), derived from microbial origin (3,4). Recent studies revealed that TLRs can recognize endogenous molecules released in the course of tissue damage or during cell death. These molecules are called *damage-associated molecular patterns* (DAMPs). TLR family members are characterized structurally by the presence of a leucine-rich repeat (LRR) domain in their extracellular domain, a trans-membrane domain, and a conserved Toll/IL-1R homology (TIR) domain in the cytosolic region, which activates common signaling pathways (Figure 7.1), most notably those leading to activation of the transcription factor nuclear factor-κB (NF-κB) and stress-activated protein kinases (5), transactivating the expression of genes involved in inflammation, such as pro-inflammatory cytokines, chemokines, and type I interferons. TLRs are expressed on various innate immune cells, including macrophages, neutrophils, and dendritic cells (DCs) and are essential for evoking inflammation in response to infection and cellular stimuli. In addition to TLRs, recent studies identified three PRR families, including RIG-I-like receptors (RLRs) sensing viral RNAs in the cytoplasm, nucleotide-binding oligomerization domain-like receptors (NLRs) activating inflammasomes, and C-type lectin receptors (CLRs) recognizing fungal components on the cell surface.

The relationship between cancer and inflammation has been well documented. On one hand, inflammation caused by innate immunity is beneficial for eliminating cancer cells by activating natural killer cells and cytotoxic T cells via production of pro-inflammatory cytokines. On the other hand, it has been suggested that proliferation of cells alone does not cause cancer but, rather, sustained proliferation in an environment rich in inflammatory cells, growth factors, activated stroma, and DNA damage-promoting agents certainly potentiates and/or promotes neoplastic risks (6,7).

Cancer and Inflammation Mechanisms: Chemical, Biological, and Clinical Aspects, First Edition.
Edited by Yusuke Hiraku, Shosuke Kawanishi, and Hiroshi Ohshima.

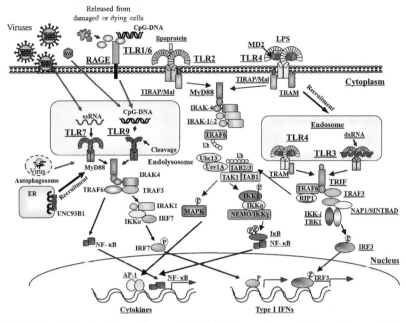

FIGURE 7.1 TLR signaling pathways. TLR4–MD2 complex and a heterodimer of TLR1/6 and TLR2 recognize the LPS and the lipoproteins on the cell surface, respectively, thereby recruiting MyD88 and TIR domains containing adaptor protein (TIRAP) to the TLR, and a complex of IRAKs and TRAF6 is subsequently formed. TRAF6 acts as an E3 ubiquitin ligase and catalyzes formation of a K63-linked polyubiquitin chain on TRAF6 itself and generation of an unconjugated polyubiquitin chain with an E2 ubiquitin ligase complex of Ubc13 and Uev1A. Ubiquitination activates a complex of TAK1, TAB1, and TAB2/3, resulting in the phosphorylation of NF-κB essential modulator (NEMO) and activation of an IKK complex. Degradation of the phosphorylated IκB by the ubiquitin proteasome system allows the translocation of NF-κB to the nucleus, where it crusades the expression of cytokine genes. Simultaneously, TAK1 activates MAP kinase cascades, leading to the activation of activator protein 1 (AP-1), which is also decisive for the induction of cytokine genes. LPS induces translocation of TLR4 to the endosome together with TRIF-related adaptor molecule (TRAM). TLR3 is present in the endosome and recognizes dsRNA. TLR3 and TLR4 activate TRIF-dependent signaling, which activates NF-κB and IRF3, resulting in the induction of pro-inflammatory cytokine genes and type I IFNs. TRAF6 and receptor-interacting protein 1 (RIP1) activate NF-κB, whereas TRAF3 is responsible for the phosphorylation of IRF3 by TBK1/IKK-i. Nucleosome assembly protein 1 (NAP1) and (similar to NAP1) TBK1 adaptor (SINTBAD) are required for the activation of TBK1/IKK-i. Phosphorylated IRF3 translocates into the nucleus to induce expression of type I IFN genes. Viral ssRNA and CpG DNA are recognized by TLR7 and TLR9, respectively. Viruses that have entered the cytoplasm are engulfed by autophagosomes and deliver viral nucleic acids to the endolysosome. An HMGB1-DNA complex released from damaged cells is captured by a receptor of advanced glycation end products (RAGE). Autoantibodies recognizing self-DNA or -RNA bind to FcγRIIa. LL37, an antimicrobial peptide, associates with endogenous DNA. These proteins are responsible for the delivery of endogenous nucleic acids to endolysosomes. Stimulation with ligands or infection by viruses induces trafficking of TLR7 and TLR9 from the ER to the endolysosome via UNC93B1. TLR9 undergoes cleavage by proteases present in the endolysosome. A complex of MyD88, IRAK-4, TRAF6, TRAF3, IRAK-1, IKK-α, and IRF7 is recruited to the TLR. Phosphorylated IRF7 translocates into the nucleus and up-regulates the expression of type I IFN genes. (*See insert for color representation of the figure.*)

An inflammatory cytokine network influences survival, growth, mutation, proliferation, differentiation, and movement of both tumor and stromal cells. Moreover, these cytokines can regulate communication between tumor and stromal cells and tumor interactions with the extracellular matrix (6). In this chapter we discuss the complex relationship between TLRs and cancer progression and elimination.

TLRs AND THEIR LIGANDS

There are 13 different TLRs in mammals: 10 in humans and 12 in mice. Different TLRs recognize the different molecular patterns of microorganisms and endogenous components (Table 7.1). TLRs 1, 2, 4, 5, and 6 are reported to be expressed on the plasma membrane, whereas TLRs 3, 7, 8, 9, 10, 11, and 13 are found almost exclusively within the endosomal compartments (8). TLR2 senses various components from bacteria, mycoplasma, fungi, and viruses. Although most TLRs act alone, TLR2 forms heterodimers with TLR1 or TLR6, allowing it to recognize a wide range of exogenous or endogenous molecules from pathogen and host, respectively (9,10). TLR4 recognizes lipopolysaccharide (LPS) together with myeloid differentiation factor 2 (MD2) on the cell surface (11,12). TLR4 is also involved in the recognition of viruses by binding to viral envelope proteins. TLR5 is highly expressed by DCs of lamina propria (LPDCs) in the small intestine, where it recognizes flagellin from flagellated bacteria (13,14). TLR11, present in mouse only, recognizes uropathogenic bacteria and a profilin-like molecule derived from the intracellular protozoan *Toxoplasma gondii* (15). A group of TLRs, comprising TLR3, TLR7, TLR8, TLR9, and TLR13 recognize nucleic acids derived from viruses and bacteria as well as endogenous nucleic acids in pathogenic contexts (3). Activation of these TLRs leads to the production of type I interferons (IFNs) in addition to pro-inflammatory cytokines. TLR3 detects viral double-stranded RNA (dsRNA) in the endolysosome. Mouse TLR7 and human TLR7/8 recognize single-stranded RNAs (ssRNA) from RNA viruses as well as small purine analog compounds (imidazoquinolines). TLR7 also detects RNAs from bacteria such as group B *Streptococcus* in endolysosomes in conventional DCs (cDCs) (16). In addition to cytosine–phosphate–guanosine (CpG) DNA (17), TLR9 also recognizes hemozoin, which is a crystalline metabolite of hemoglobin produced by the malaria parasite (18). Recent studies have demonstrated that TLR13 acts as a bacterial ribosomal RNA sensor (19,20), even though it was earlier identified to be linked with the recognition of vesicular stomatitis virus (21). New data have unscrambled an unforeseen link between the antibiotic resistance and equivocation from TLR13 recognition, which may explain why TLR13 expression has been abandoned in certain mammalian species, including human. The exact mechanism of TLR13 action and whether or not it is involved in cancer-related inflammation are yet to be established, but this study may provide a cue for further action. The ligands for TLR 10 and 12 are yet to be identified. TLR8 is not functional in mice, whereas TLRs 11, 12, and 13 are not functional in humans.

 In addition to detection of the pathogen-derived ligands, TLRs are also reported to interact with a diverse variety of host molecules, including antimicrobial molecules

TABLE 7.1 TLRs, Ligands, and Their Expression in Various Tumor Cells and Tissues

TLR	Localization	Origin of ligands	Ligands	Expressing cancer cells and tissues
TLR1	Plasma membrane	Bacteria and mycobacteria	Triacyl lipopeptide	Colon cancer (78)
		Neisseria meningitides	Soluble factors	
TLR2	Plasma membrane	Gram-positive bacteria	Peptidoglycan	Colon cancer (79)
		A variety of pathogens	Lipoprotein/lipopeptides	Gastric cancer (80)
		Gram-positive bacteria	Lipoteichoic acid	Hepatocellular carcinoma (81)
		Mycobacteria	Lipoarabinomannan	
		Staphylococcus epidermidis	Phenol-soluble modulin	
		Trypanosoma cruzi	Glycoinositolphospholipids	
		Treponema maltophilum	Glycolipids	
		Neisseria	Porins	
		Fungi	Zymosan	
		Leptospira interrogans	Atypical LPS	
		Porphyromonas gingivalis	Atypical LPS	
		Host	Hsp70	
		Host	EDN	
TLR3	Endolysosome	Virus	Double-stranded RNA	Breast cancer (82)
				Colon cancer (78)
				Melanoma (83)
				Hepatocellular carcinoma (81)
TLR4	Plasma membrane	Gram-negative bacteria	Lipopolysaccharides	Breast cancer (84)
		Plant	Taxol	Colon cancer (78)
		RSV	Fusion protein	Melanoma (85)
		MMTV	Envelope proteins	Gastric cancer (86)
		Chlamydia pneumoniae	Hsp60	Lung cancer (87)
		Host	Hsp60, Hsp70	Hepatocellular carcinoma (81)

TLR	Localization	Source	Ligand	Associated cancers
	Host		Type III repeat extra domain A of fibronectin	Ovarian cancer (88)
	Host		Oligosaccharides of hyaluronic acid	
	Host		Polysaccharide fragments of heparan sulfate	
	Host		Fibrinogen	
TLR5	Plasma membrane	Bacteria	Flagellin	Gastric cancer (86); Cervical squamous cell carcinomas (66); Breast cancer (56)
TLR6	Plasma membrane	Mycoplasma	Diacyl lipopeptides	Hepatocellular carcinoma (81)
TLR7	Endolysosome	Synthetic compounds	Imidazoquinoline	Chronic lymphocytic leukemia (89)
		Synthetic compounds	Loxoribine	
		Synthetic compounds	Bropirimine	
		Virus	Single-stranded RNA	
hTLR8	Endolysosome	Virus	Single-stranded RNA (93)	Lung cancer (69)
TLR9	Endolysosome	Bacteria and viruses	CpG ODN	Breast cancer (90); Gastric cancer (86); Hepatocellular carcinoma (81); Cervical squamous cell carcinomas (91); Glioma (92); Prostate cancer (68)
		Malaria	Pigment hemozoin	
hTLR10	Endolysosome	ND		
TLR11	Endolysosome (94)	Toxoplasma gondii	Profilin-like protein	
TLR12				
TLR13	Endolysosome	Bacteria	23S ribosomal RNA (19)	

Source: References 4 and 77.

such as murin defensins 2 (mDF2) (22), reactive oxygen species (ROSs) (23), proteins released from dead or dying cells such as high-mobility group box protein-1 (HMGB1) (24), surfactant protein A (25), fibrinogen (26), breakdown products of extracellular matrix such as fibronectin (27), hyaluronic acid oligosaccharides (28), and eosinophil-derived neurotoxin (EDN) (29). Heat shock proteins (Hsps) such as Hsp60, Hsp70, and Hsp90 (gp90) have also been reported to induce pro-inflammatory cytokine production by monocyte–macrophages and the maturation of DCs through TLRs. The data need to be interpreted carefully since recombinant proteins with possible contamination of bacterial components are used in many cases.

TLR SIGNALING

Once TLRs sense the presence of their ligands, they initiate intracellular signaling pathways leading to transcriptional activation of pro-inflammatory molecules. TLR signaling has been investigated extensively in the last several years (8,9,30,31). TLR2/1 heterodimers respond to triacyl lipopeptide, while TLR2/6 heterodimers recognize diacyl lipopeptide. CD14 and MD-2 are accessory proteins required for LPS/TLR4 ligation. Following ligation with their respective ligands, TLRs recruit downstream adaptor molecules, myeloid differentiation primary response gene 88 (MyD88), and Toll/IL-1R domain-containing adapter, inducing IFN-β (TRIF) to activate interleukin-1 receptor-associated kinase 4 (IRAK4), TNF receptor-associated factor 6 (TRAF6), and inhibitor of nuclear factor-κB kinase ε/TANK-binding kinase 1 (IKKε/TBK1). These sequentially activate the signaling pathway of NF-κB, mitogen-activated protein kinase (p38 MAPK), and c-Jun N-terminal kinase (JNK), leading to a series of specific cellular responses related to cell survival, proliferation, and inflammation. There are two major TLR signaling pathways depending on the use of adaptor proteins, MyD88 and TRIF.

MYD88-DEPENDENT SIGNALING PATHWAYS

Stimulation of macrophages and DCs with TLR PAMPs initiates MyD88-dependent signaling via recruitment of IRAK-family signaling proteins (4,8). First, IRAK4 is activated and recruited by MyD88. Then, IRAK1 and IRAK2 are sequentially activated and recruited to form an active signaling complex which interacts with TRAF6 [an E3 ligase required for Lys63 (K63)-linked poly-ubiquitination]. Recently, the crystal structure of the MyD88–IRAK4–IRAK2 complex was reported and shows a left-handed helical oligomeric signaling complex, which consists of six, four, and four molecules of MyD88, IRAK4, and the IRAK2 death domain, respectively (32). MyD88 then recruits IRAK4, and this complex recruits IRAK4 substrate IRAK2 or IRAK1. Formation of this signaling complex facilitates the phosphorylation of all these kinases, resulting in the activation of downstream signaling molecules. TRAF6, along with E2 ubiquitin-conjugating enzymes such as Ubc13 and Uev1A, ubiquitinates both itself and IRAK1 to activate TAK1 (33). Activated TAK1 then activates NF-κB and MAP kinases to initiate the transcription and translation of various pro-inflammatory cytokines, chemokines, interferons, and other TLR-inducible genes. TLR4-mediated signaling also recruits TRAF3, which inhibits MAP kinases and cytokine production, to the MyD88 multiprotein complex, where it undergoes K48-linked ubiquitination and degradation. This results in TAK1 activation and subsequent induction of inflammatory cytokines (34).

TRIF-DEPENDENT SIGNALING PATHWAYS

Macrophages and DCs stimulated with TLR3 and TLR4 PAMPs trigger TRIF-dependent signaling, which leads to the production of pro-inflammatory cytokines and type I interferons via activation of NF-κB, MAP kinases, and IRF3 (4,8). TRIF-dependent signaling is initiated through the recruitment of noncanonical IKKs such as TBK1 and IKKi (IKKε), via TRAF3, to phosphorylate IRF3. Phosphorylated IRF3 translocates to the nucleus and initiates the transcription of type I interferons. TRAF3 plays a critical role in the regulation of both MyD88- and TRIF-dependent signaling because TRAF3 degradation via K48-linked ubiquitination activates MyD88-dependent signaling and suppresses TRIF-dependent signaling (and vice versa). In addition, TRAF6 and RIP1 are also recruited to the distinct domain of TRIF. The interaction of TRAF6 with TRIF (N-terminal TRAF-binding domain) activates TAK1 via mechanisms similar to those in the MyD88-dependent pathway. However, TRIF interacts with RIP1 and undergoes K63-linked polyubiquitination. RIP1 also interacts with TNFR1 associated death domain protein (TRADD), and this multiprotein complex is required for NF-κB activation. These signals are essential not only for the innate inflammatory responses, but also for instructing adaptive immune responses by the action of cytokines, enhancing the surface expression of co-stimulatory molecules on innate immune cells.

INNATE IMMUNE CELLS IN CANCER TISSUES

In response to tissue injury, a multifactorial network of chemical signals initiates and maintains a host response designed to "heal" the stricken tissue, which involves activation and maneuvered migration of leukocytes (neutrophils, monocytes, and eosinophils) from the venous system to sites of damage, and tissue mast cells also have a significant role. For neutrophils, a four-step mechanism is believed to coordinate recruitment of these inflammatory cells to sites of tissue injury and to the provisional extracellular matrix (ECM) that forms a scaffolding upon which fibroblast and endothelial cells proliferate and migrate, thus providing a nidus for reconstitution of the normal microenvironment (35). These neutrophils, monocytes, and myeloid-derived suppressor cells (MDSCs) invade the tumor microenvironment in response to diverse tumor-derived chemoattractants, including chemokines, cytokines, and growth factors. Myeloid cells may differentiate into tumor-associated macrophages (TAMs) or tumor-associated neutrophils (TANs), which express proangiogenic and immunosuppressive factors, thereby promoting tumor growth (36–38). TAMs, being a key component of the tumor microenvironment, can stimulate tumor growth (7,39,40), and their compactness is associated with contrary outcomes and shorter survival in several types of cancer, including breast cancer, Hodgkin's lymphoma, and lung adenocarcinoma (41–43).

TLRs AND TUMOR MICROENVIRONMENT

Various cancer cells have been shown to express TLRs (Table 7.1). This TLR-mediated recognition of PAMPs and DAMPs triggers the expression of several adapter proteins and downstream kinases, leading to the induction of key pro-inflammatory mediators, which results in the activation of both the innate immune response (enhanced expression of anti-apoptotic proteins, pro-inflammatory cytokines, and antibacterial proteins) and the adaptive immune response (maturation of the DCs, antigen presentation, etc.), thereby facilitating a direct or indirect tumor-stimulating effect.

Immune cell infiltration is another noteworthy phenomenon in tumor ontogenesis. Formerly, these immune infiltrates were thought to assist the host against the developing tumor, however, subsequent studies have demonstrated that instead of combating the tumor, the infiltrating immune cells contribute to cancer growth and metastasis as well as immunosuppression (6). Indeed, it was reported that the infiltration of macrophages stimulated cancer progression and metastasis in humans (44). Studies on colony-stimulated factor-1 (CSF-1, a cytokine that regulates macrophage differentiation and function) knockout mice has further explicated that knocking down CSF-1 in mouse strains inhibited breast cancer growth and metastasis. A study on oral epithelial squamous cell carcinoma also divulged that the level of immune cell infiltration was positively correlated with the level of morphological and pathological transformation from normal to malignant phenotypes (45).

As discussed above, the inflammatory process is triggered not only by infection but also by dying cells, and in case of an infection, the products of dying cells need to be recognized by immune cells, or tumor cells, to elicit an inflammatory response. During both cancer development and the tissue repair process, the immune infiltrate is characterized by the presence of a large number of macrophages. Further characterization of these cells led to the conclusion of the existence of two subclasses of macrophages (M1 and M2 macrophages) based on their cytokine profile. M2 macrophages are the dominant subtype recognized during tissue repair and in the pro-tumor inflammatory microenvironment. M2 macrophages promote tissue repair and remodeling and are present in established tumors and may promote tumor growth (46). Unlike M1 macrophages, which are known to mediate response against intracellular parasites and tumors by producing IL-12, IL-23, IFN-γ, IL-18, and TNF-α, M2 macrophages have high levels of scavenger, mannose, and galactose-type receptors; produce vascular endothelial growth factor (VEGF), IL-6, IL-10, prostaglandin (PG), inducible nitric oxide synthase (iNOS), and indoleamine di-oxygenase metabolites (IDO); and have immunoregulatory and proliferation-stimulating functions.

For all these stages, the expression of TLRs by cancer cells is a major factor. Through TLRs, cancer cells can recognize either microbial pathogens (PAMPs) or cellular debris (DAMPs) and promote the expression and secretion of chemokines

TABLE 7.2 TLRs: Tumor Progression and Suppression

Tumor promoting	TLRs	Refs.	Antitumor	TLRs	Refs.
Proangiogenic	2,9	(95)	Antiangiogenic	7,9	(96–98)
Proliferation	3,4	(99–102)	Apoptosis	3,4,7,9	(103–107)
Chemoresistance	4	(88,108)	Chemosensitivity	2,4,7	(48,109–111)
Treg activation	4,5	(112,113)	Treg inhibition	4,5,7	(113–119)
			Antigen presentation	8,9	
			Cytotoxicity	9	(120–124)

and cytokines which would work as mediators for the regulation of immune cell migration, differentiation, and function.

TOLL-LIKE RECEPTORS IN CANCER AND TUMOR THERAPY

Various aspects of TLRs have been accounted for to date. From one point of view, the inflammation caused by treatment with TLR-stimulating PAMPs can suppress tumor growth; on the other hand, PAMPs can participate directly or indirectly in tumor progression and metastasis (47) (Table 7.2). The role of different TLRs as negative and/or positive regulators of cancer needs to be evaluated in order to step up their potential in therapeutics.

ANTITUMOR ACTIVITY OF TLR LIGANDS

In the eighteenth century, Deidier observed a positive correlation between infection and the remission of malignant disease, making the first inference that microbes could have anticarcinogenic properties (48). At the end of the nineteenth century, William Coley observed that repeated injections of a mixture of bacterial toxins from the gram-positive bacterium *Streptococcus pneumonia* and the gram-negative bacterium *Serratia marcescens* served as efficient antitumor therapeutic agents, providing evidence that microbial products, rather than infection itself, may mediate an antitumor effect (49). It was later discovered by Shear and Turner that LPS was the "haemorrhage producing fraction" of Coley's toxin that accounted for its antitumor effect (48). As LPS stimulates TLR4, these results suggest that the antitumor effect of Coley's toxin was a result of TLR activation.

As described above, TLR3 is activated by viral or synthetic dsRNA, such as poly I:C in normal cells or cancer cells. Although described initially as expressed specifically by dendritic cells, TLR3 has recently been reported to be expressed by a broad array of epithelial cells, including pulmonary cells and hepatocytes. Interestingly, tumor cells such as breast cancers or melanoma can also express TLR3, which triggering may lead to apoptosis and/or to chemoattraction of cytotoxic lymphocytes to tumor beds (50). Our group reported previously that both the RLR–IFN–promoter stimulator-1 (IPS-1) and TLR3–TRIF-dependent pathways contributed to suppression of implanted B16 tumor growth in response to poly I:C administration via natural killer (NK) cell activation in CD8α^+ conventional dendritic cells (cDCs) (51) and in myeloid dendritic cells (mDCs) (52). Salaun et al. showed both the predictive value of immunoreactive TLR3 expression by tumor cells for the efficacy of dsRNA treatment in breast cancer patients and the functional relevance of TLR3 expression by cancer cells for the antitumor effects mediated by dsRNA. Their findings establish a rationale for the assessment of TLR3 expression by tumor cells as a biomarker for the efficacy of dsRNA treatment in selected human cancers (50).

Tumor cell lysates or purified tumor-associated antigens for vaccines have been used for therapeutic or prophylactic cancer vaccine. TLR4 is particularly important for the development of a strong adaptive immune response by stimulation of the antibody class switching, affinity maturation, and formation of memory cells (53). MPLA is a monophosphorylated lipid A derivative and was reported to preferentially activate a TRIF-dependent pathway. It is the only TLR4 agonist that has been tested clinically as an adjuvant for cancer vaccines (54,55). Further research of the role and mechanisms of TLR4 activation in cancer may provide novel antitumor vaccine adjuvants as well as TLR4 inhibitors that could prevent inflammation-induced carcinogenesis.

Cai et al. investigated the role of TLRs in breast cancer epithelial cells and found that TLR5 was highly expressed in breast carcinomas and that TLR5 signaling pathway is overly responsive in breast cancer cells (56). Interestingly, flagellin/TLR5 signaling in breast cancer cells inhibits cell proliferation and an anchorage-independent growth which is a hallmark of tumorigenic transformation. These findings indicate that TLR5 activation by flagellin mediates innate immune response to elicit potent antitumor activity in breast cancer cells themselves, which may serve as a novel therapeutic target for human breast cancer therapy. Recently, one group designed and tested a series of engineered flagellin derivatives for NF-κB activation *in vitro* (57). They showed that the most potent NF-κB activator, designated CBLB502, has a radioprotective effect in a mouse model.

Another study highlights the importance of TLR5 in the intestinal epithelium, which could be crucial in controlling the innate immune response against tumors. It reveals that TLR5/MyD88-dependent signaling elicits innate immune responses that are likely to mediate antitumor activity in a mouse xenograft model of human colon cancer, indicating that regulating TLR5/MyD88-mediated signaling could be an immunotherapeutic approach against tumors (58). TLR9 agonists, commonly referred to as CpG oligodeoxyneuclotides (ODNs), are another set of agonists being considered in the armory of anticancer drugs as monotherapy or in combination with chemotherapy, radiotherapy, or other known immunotherapeutic approaches, as they increase the antigen presentation by activation of dendritic cells and boost antitumor B- and T-cell responses, thereby bridging the innate and adaptive immunity (59).

One of the most appreciated functions of TLRs in cancer therapy is stimulation of the adaptive immune system. In these studies, tolerance to tumor self-antigens is broken, presumably by TLR-mediated up-regulation of co-stimulatory signals to the adaptive immune response, a property known as *adjuvanticity*. This has been used in cancer vaccines, as targets of gene therapy, and in raising antitumor antigen-specific T cells *in vitro* for adoptive transfers. A recent study by Shatz et al. suggested that a profile of TLR gene expression patterns in specific tumors in response to p53 and DNA-damaging agents combined with knowledge of p53 expression and mutation status in these tumors can be an important tool in cancer diagnosis and in strategies that target TLR pathway for cancer therapy (60).

Classically, the ability of TLR signaling to activate the adaptive immune system has led to attempts to harness this response against cancer cells through the use of exogenous administration of TLR ligands. More research is needed to determine the role of microbial and endogenous TLR ligands in inhibiting tumorigenesis in both infectious and noninfectious situations.

ROLES OF TLRs IN CANCER PROGRESSION

Although we discussed potential use of TLR ligands as antitumor reagents, TLRs can play a totally opposite role in tumorigenesis and tumor metastasis. This is in part because of chronic inflammations caused by innate immune cells to stimulate the development of a malignant neoplasm. The role of TLRs in the development and progression of a wide variety of tumors is recently becoming a subject of study of great importance.

The interrelationship between TLR2 and tumor progression has been confirmed in a study by Karin et al. (61), who have proved this receptor's key role in lung cancer metastasis formation. Furthermore, it was shown that TLR2 signaling by the endogenous extracellular matrix-derived proteoglycan versican can promote tumor metastasis (62). Versican strongly enhances Lewis lung carcinoma (LLC) metastatic growth by activating TLR2:TLR6 complexes and inducing TNF-α secretion by myeloid cells. Another example of endogenous TLR ligand potentially involved in tumor progression is HMGB-1. Although HMGB-1 is a nuclear protein interacting with nucleic acids in resting cells, it is secreted from damaged cells and acts as a DAMP that activates TLR2 and TLR4 on the surface of tumor and immune cells, thereby inducing tumor progression and metastasis (63).

TLR4 signaling in cancer is considered to be a double-edged sword. Activated TLR4 on immune cells can enhance antitumor immunity. On the other hand, chronic inflammation is a major risk factor in cancer development (7). Increased translocation of intestinal bacteria is a hallmark of chronic liver disease, which contributes to hepatic inflammation and fibrosis (64). TLR4 is expressed by a variety of tumor cell lines, in both mouse and humans (Table 7.3). Many of them are not limited to a single TLR but, like immune cells, utilize an assortment of TLRs.

The role of TLR5 in cancer progression has also been studied extensively. TLR5 expression in gastric cancer cells has suggested the crucial role of this receptor in tumor progression via the regulation of cell proliferation (65). Elevated expression of TLR5 and TLR9 on the surface of cervical epithelial cells has been observed which can be colligated with cervical cancer progression (66). In lung cancer, tumor cell lines (67), and human prostate cancer cells (68), stimulation of TLR9

TABLE 7.3 Tumor Cell Lines Expressing TLR4

Human		Mice	
Tumor type	Tumor cell line	Tumor type	Tumor cell line
Bladder cancer	T24 (125)	Breast cancer	4T1 (126)
Breast cancer	MDA-MB-231 (127)	Colon cancer	MC26 (126)
Colon cancer	SW480, HT29, KM20 (128,129)	Glioma	GL261 (130)
Laryngeal and oral cancer	PCI-1, PCI-30 (131)	Lung cancer	LLC1 (126)
Melanoma	SkMEL-28, BN1, 9923M, ME5, ME16, ME17 (75,85)	Melanoma	B16 (126)
Neuroblastoma	NB-1 (132)	Prostate cancer	RM1 (126)
Ovarian cancer	SKOV3, AD10, A2780, CP70 (88,133,134)		

by specific agonists resulted in increased production of tumor-associated cytokines as well.

Small-molecule agonists at TLR7 and TLR8 have sparked a vivid interest in cancer research, owing to their profound activity for innate immunity in viral infections and cancer. Cherfila-Vicini et al. have demonstrated the expression of TLR7 and TLR8 by tumor cells in human lung cancer *in situ* and also in human lung tumor cell lines, suggesting that stimulation with TLR7 and TLR8 agonists led to the activation of NF-κB, up-regulated expression of the anti-apoptotic protein Bcl-2, increased tumor cell survival, and chemoresistance (69). Furthermore, microRNAs (miRNAs), which are small noncoding RNAs 19 to 24 nucleotides in length with gene expression regulatory functions (70,71), are expressed aberrantly in most types of cancers (72,73) and can act as paracrine agonists of murine TLR7 and human TLR8, thereby triggering a TLR-mediated pro-metastatic inflammatory response that may ultimately lead to tumor growth and metastasis (74).

CURRENT AND FUTURE DEVELOPMENTS

Although TLRs are involved in the induction of innate immune response against microbial insult from the surroundings, such as bacteria and viruses, their relation to tumor cells is complicated, as described above. The TLR ligands show antitumor effects in some conditions, although they are also known to facilitate tumorigenesis and cancer metastasis.

When TLRs expressed on tumor cells are activated by their ligands, the malignant cells are prone to survive and proliferate due to NF-κB activation. Similarly, in carcinogenesis and tumor progression, TLRs play an active role in the tumor microenvironment. During chronic inflammation, abnormal activation of TLRs in normal fibroblasts and epithelial cells might facilitate neoplastic transformation and carcinogenesis. Cancer cells activated by TLR signals can release pro-inflammatory cytokines (TNF-α, G-CSF, IL-1α, and IL-6), pro-inflammatory chemokines (CCL2 and CXCL2), immunosuppressive cytokines (IL-10), and an inflammatory factor (COX-2) (75) that recruit and optimize immune cells to release further cytokines and chemokines and also induce regulatory T cells (Tregs) in the tumor microenvironment (76). The result is an aberrant cytokine profile associated with immune tolerance, cancer progression, and propagation of the tumor microenvironment (Figure 7.2). In addition, overexpression of TLRs has been found in many tumor cells themselves. Furthermore, it has been demonstrated that TLR-mediated signaling promotes tumor growth, and escape of tumor cells from a host immune system has been observed in some cancer models.

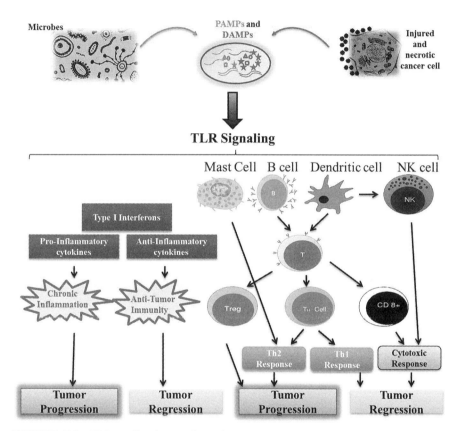

FIGURE 7.2   TLR-mediated tumorigenesis. PAMPs derived from microorganisms and DAMPs derived from injured and necrotic cancer cells might activate TLRs expressed on immune cells and on cancer cells. These activated cells release cytokines and chemokines, the aberrant molecular pattern of which might significantly affect the tumorigenesis. The activation of antigen-presenting cells (APCs) such as DCs, mast cells, and B cells via TLR signaling can lead to differential adaptive immunity. The induction of a Th1 and cytotoxic T-cell response leads to tumor regression, different than a Th2 response and T-reg-mediated immunity, which could facilitate tumor progression. The induction of pro-inflammatory as well as anti-inflammatory cytokines may lead to chronic inflammation and tumor-mediated immunity, which could result in tumor progression and regression, respectively.

In contrast, treatment with TLR-stimulating reagents can lead to totally opposite outcome in cancer progression. Using TLR agonist in combination with antigen isolated from tumor, we can anticipate an increasing effect of vaccination and evoking specific innate immunity against cancer. These studies and observations strongly suggest that biological outcomes induced by high levels of the protein are determined by both an inter- and an intracellular environment. Although recent issues have claimed continuously that specific TLR or signaling molecules in TLR pathway are involved in the pathogenesis of cancers, choices of the use of either agonist or antagonist for TLR in the context of therapy could not be decided immediately. This mechanism is complex and thus far not well understood; however, it is clear that carcinogenesis, cancer progression, and site-specific metastasis are related to interactions between cancer cells, immune cells, DAMPs, and PAMPs through TLR signals in the tumor microenvironment. Better understanding of these signals and pathways will lead to the development of novel therapeutic approaches to a wide variety of cancers.

Acknowledgments

We thank all the members of the Takeuchi Lab for active discussions. The work was supported in part by the Japan Society for the Promotion of Science (JSPS) through a Funding Program for World-Leading Innovative R&D on Science and Technology (FIRST Program). The work was also supported by a JSPS Grant-in-Aid for Young Scientists and a grant from the Ministry of Health, Labour and Welfare in Japan.

REFERENCES

1. Lemaitre B, Nicolas E, Michaut L, Reichhart JM, Hoffmann JA. The dorsoventral regulatory gene cassette spatzle/*Toll*/cactus controls the potent antifungal response in *Drosophila* adults. Cell 1996;86:973–983.
2. Hashimoto C, Hudson KL, Anderson KV. The *Toll* gene of *Drosophila*, required for dorsal–ventral embryonic polarity, appears to encode a transmembrane protein. Cell 1988;52:269–279.
3. Akira S, Uematsu S, Takeuchi O. Pathogen recognition and innate immunity. Cell 2006;124:783–801.
4. Takeuchi O, Akira S. Pattern recognition receptors and inflammation. Cell 2010;140:805–820.
5. Beutler B. Inferences, questions and possibilities in Toll-like receptor signalling. Nature 2004;430:257–263.
6. Balkwill F, Mantovani A. Inflammation and cancer: Back to Virchow? Lancet 2001;357:539–545.
7. Mantovani A, Allavena P, Sica A, Balkwill F. Cancer-related inflammation. Nature 2008;454:436–444.
8. Kawai T, Akira S. The role of pattern-recognition receptors in innate immunity: update on Toll-like receptors. Nat Immunol 2010;11:373–384.
9. Takeda K, Kaisho T, Akira S. Toll-like receptors. Annu Rev Immunol 2003;21:335–376.
10. Liew FY, Xu D, Brint EK, O'Neill LA. Negative regulation of Toll-like receptor-mediated immune responses. Nat Rev Immunol 2005;5:446–458.
11. Park BS, Song DH, Kim HM, Choi BS, Lee H, Lee JO. The structural basis of lipopolysaccharide recognition by the TLR4-MD-2 complex. Nature 2009;458:1191–1195.
12. Takeuchi O, Hoshino K, Kawai T, Sanjo H, Takada H, Ogawa T, et al. Differential roles of TLR2 and TLR4 in recognition of gram-negative and gram-positive bacterial cell wall components. Immunity 1999;11:443–451.

13. Uematsu S, Fujimoto K, Jang MH, Yang BG, Jung YJ, Nishiyama M, et al. Regulation of humoral and cellular gut immunity by lamina propria dendritic cells expressing Toll-like receptor 5. Nat Immunol 2008;9:769–776.

14. Uematsu S, Jang MH, Chevrier N, Guo Z, Kumagai Y, Yamamoto M, et al. Detection of pathogenic intestinal bacteria by Toll-like receptor 5 on intestinal CD11c+ lamina propria cells. Nat Immunol 2006;7:868–874.

15. Yarovinsky F, Zhang D, Andersen JF, Bannenberg GL, Serhan CN, Hayden MS, et al. TLR11 activation of dendritic cells by a protozoan profilin-like protein. Science 2005;308:1626–1629.

16. Mancuso G, Gambuzza M, Midiri A, Biondo C, Papasergi S, Akira S, et al. Bacterial recognition by TLR7 in the lysosomes of conventional dendritic cells. Nat Immunol 2009;10:587–594.

17. Hemmi H, Takeuchi O, Kawai T, Kaisho T, Sato S, Sanjo H, et al. A Toll-like receptor recognizes bacterial DNA. Nature 2000;408:740–745.

18. Coban C, Ishii KJ, Kawai T, Hemmi H, Sato S, Uematsu S, et al. Toll-like receptor 9 mediates innate immune activation by the malaria pigment hemozoin. J Exp Med 2005;201:19–25.

19. Oldenburg M, Kruger A, Ferstl R, Kaufmann A, Nees G, Sigmund A, et al. TLR13 recognizes bacterial 23S rRNA devoid of erythromycin resistance-forming modification. Science 2012.

20. Hidmark A, von Saint Paul A, Dalpke AH. Cutting edge: TLR13 is a receptor for bacterial RNA. J Immunol 2012;189:2717–2721.

21. Shi Z, Cai Z, Sanchez A, Zhang T, Wen S, Wang J, et al. A novel Toll-like receptor that recognizes vesicular stomatitis virus. J Biol Chem 2011;286:4517–4524.

22. Biragyn A, Ruffini PA, Leifer CA, Klyushnenkova E, Shakhov A, Chertov O, et al. Toll-like receptor 4-dependent activation of dendritic cells by beta-defensin 2. Science 2002;298:1025–1029.

23. Frantz S, Kelly RA, Bourcier T. Role of TLR-2 in the activation of nuclear factor kappaB by oxidative stress in cardiac myocytes. J Biol Chem 2001;276:5197–5203.

24. Park JS, Svetkauskaite D, He Q, Kim JY, Strassheim D, Ishizaka A, et al. Involvement of Toll-like receptors 2 and 4 in cellular activation by high mobility group box 1 protein. J Biol Chem 2004;279:7370–7377.

25. Guillot L, Balloy V, McCormack FX, Golenbock DT, Chignard M, Si-Tahar M. Cutting edge: the immunostimulatory activity of the lung surfactant protein-A involves Toll-like receptor 4. J Immunol 2002;168:5989–5992.

26. Smiley ST, King JA, Hancock WW. Fibrinogen stimulates macrophage chemokine secretion through Toll-like receptor 4. J Immunol 2001;167:2887–2894.

27. Okamura Y, Watari M, Jerud ES, Young DW, Ishizaka ST, Rose J, et al. The extra domain A of fibronectin activates Toll-like receptor 4. J Biol Chem 2001;276:10229–10233.

28. Termeer C, Benedix F, Sleeman J, Fieber C, Voith U, Ahrens T, et al. Oligosaccharides of hyaluronan activate dendritic cells via Toll-like receptor 4. J Exp Med 2002;195:99–111.

29. Oppenheim JJ, Yang D. Alarmins: chemotactic activators of immune responses. Curr Opin Immunol 2005;17:359–365.

30. Takeda K, Akira S. Toll-like receptors in innate immunity. Int Immunol 2005;17:1–14.

31. Takeda K, Akira S. TLR signaling pathways. Semin Immunol 2004;16:3–9.

32. Lin SC, Lo YC, Wu H. Helical assembly in the MyD88–IRAK4–IRAK2 complex in TLR/IL-1R signalling. Nature 2010;465:885–890.

33. Yamamoto M, Okamoto T, Takeda K, Sato S, Sanjo H, Uematsu S, et al. Key function for the Ubc13 E2 ubiquitin-conjugating enzyme in immune receptor signaling. Nat Immunol 2006;7:962–970.

34. Tseng PH, Matsuzawa A, Zhang W, Mino T, Vignali DA, Karin M. Different modes of ubiquitination of the adaptor TRAF3 selectively activate the expression of type I interferons and proinflammatory cytokines. Nat Immunol 2010;11:70–75.

35. Coussens LM, Werb Z. Inflammation and cancer. Nature 2002;420:860–867.

36. Biswas SK, Mantovani A. Macrophage plasticity and interaction with lymphocyte subsets: cancer as a paradigm. Nat Immunol 2010;11:889–896.

37. Lazennec G, Richmond A. Chemokines and chemokine receptors: new insights into cancer-related inflammation. Trends Mol Med 2010;16:133–144.

38. Yang L, Pang Y, Moses HL. TGF-beta and immune cells: an important regulatory axis in the tumor microenvironment and progression. Trends Immunol 2010;31:220–227.

39. Grivennikov SI, Greten FR, Karin M. Immunity, inflammation, and cancer. Cell 2010;140:883–899.

40. Fridlender ZG, Sun J, Kim S, Kapoor V, Cheng G, Ling L, et al. Polarization of tumor-associated neutrophil phenotype by TGF-beta: "N1" versus "N2" TAN. Cancer Cell 2009;16:183–194.
41. Steidl C, Lee T, Shah SP, Farinha P, Han G, Nayar T, et al. Tumor-associated macrophages and survival in classic Hodgkin's lymphoma. N Engl J Med 2010;362:875–885.
42. Qian BZ, Pollard JW. Macrophage diversity enhances tumor progression and metastasis. Cell 2010;141:39–51.
43. Cortez-Retamozo V, Etzrodt M, Newton A, Rauch PJ, Chudnovskiy A, Berger C, et al. Origins of tumor-associated macrophages and neutrophils. Proc Natl Acad Sci USA 2012;109:2491–2496.
44. Pollard JW. Tumour-educated macrophages promote tumour progression and metastasis. Nat Rev Cancer 2004;4:71–78.
45. Gannot G, Gannot I, Vered H, Buchner A, Keisari Y. Increase in immune cell infiltration with progression of oral epithelium from hyperkeratosis to dysplasia and carcinoma. Br J Cancer 2002;86:1444–1448.
46. Lewis CE, Pollard JW. Distinct role of macrophages in different tumor microenvironments. Cancer Res 2006;66:605–612.
47. Shcheblyakov DV, Logunov DY, Tukhvatulin AI, Shmarov MM, Naroditsky BS, Gintsburg AL. Toll-like receptors (TLRs): The role in tumor progression. Acta Naturae 2010;2:21–29.
48. Garay RP, Viens P, Bauer J, Normier G, Bardou M, Jeannin JF, et al. Cancer relapse under chemotherapy: why TLR2/4 receptor agonists can help. Eur J Pharmacol 2007;563:1–17.
49. Coley WB. The treatment of malignant tumors by repeated inoculations of erysipelas. With a report of ten original cases. 1893. Clin Orthop Relat Res 1991:3–11.
50. Salaun B, Zitvogel L, Asselin-Paturel C, Morel Y, Chemin K, Dubois C, et al. TLR3 as a biomarker for the therapeutic efficacy of double-stranded RNA in breast cancer. Cancer Res 2011;71:1607–1614.
51. Miyake T, Kumagai Y, Kato H, Guo Z, Matsushita K, Satoh T, et al. Poly I:C-induced activation of NK cells by CD8 alpha+ dendritic cells via the IPS-1 and TRIF-dependent pathways. J Immunol 2009;183:2522–2528.
52. Akazawa T, Ebihara T, Okuno M, Okuda Y, Shingai M, Tsujimura K, et al. Antitumor NK activation induced by the Toll-like receptor 3-TICAM-1 (TRIF) pathway in myeloid dendritic cells. Proc Natl Acad Sci USA 2007;104:252–257.
53. Kasturi SP, Skountzou I, Albrecht RA, Koutsonanos D, Hua T, Nakaya HI, et al. Programming the magnitude and persistence of antibody responses with innate immunity. Nature 2011;470:543–547.
54. Cluff CW. Monophosphoryl lipid A (MPL) as an adjuvant for anti-cancer vaccines: clinical results. Adv Exp Med Biol 2009;667:111–123.
55. Mata-Haro V, Cekic C, Martin M, Chilton PM, Casella CR, Mitchell TC. The vaccine adjuvant monophosphoryl lipid A as a TRIF-biased agonist of TLR4. Science 2007;316:1628–1632.
56. Cai Z, Sanchez A, Shi Z, Zhang T, Liu M, Zhang D. Activation of Toll-like receptor 5 on breast cancer cells by flagellin suppresses cell proliferation and tumor growth. Cancer Res 2011;71:2466–2475.
57. Burdelya LG, Krivokrysenko VI, Tallant TC, Strom E, Gleiberman AS, Gupta D, et al. An agonist of Toll-like receptor 5 has radioprotective activity in mouse and primate models. Science 2008;320:226–230.
58. Rhee SH, Im E, Pothoulakis C. Toll-like receptor 5 engagement modulates tumor development and growth in a mouse xenograft model of human colon cancer. Gastroenterology 2008;135:518–528.
59. Krieg AM. Therapeutic potential of Toll-like receptor 9 activation. Nat Rev Drug Discov 2006;5:471–484.
60. Shatz M, Menendez D, Resnick MA. The human TLR innate immune gene family is differentially influenced by DNA stress and p53 status in cancer cells. Cancer Res 2012;72:3948–3957.
61. Karin M, Yamamoto Y, Wang QM. The IKK NF-kappa B system: a treasure trove for drug development. Nat Rev Drug Discov 2004;3:17–26.
62. Kim S, Takahashi H, Lin WW, Descargues P, Grivennikov S, Kim Y, et al. Carcinoma-produced factors activate myeloid cells through TLR2 to stimulate metastasis. Nature 2009;457:102–106.
63. Sims GP, Rowe DC, Rietdijk ST, Herbst R, Coyle AJ. HMGB1 and RAGE in inflammation and cancer. Annu Rev Immunol 2010;28:367–388.
64. Dapito DH, Mencin A, Gwak GY, Pradere JP, Jang MK, Mederacke I, et al. Promotion of hepatocellular carcinoma by the intestinal microbiota and TLR4. Cancer Cell 2012;21:504–516.

65. Song EJ, Kang MJ, Kim YS, Kim SM, Lee SE, Kim CH, et al. Flagellin promotes the proliferation of gastric cancer cells via the Toll-like receptor 5. Int J Mol Med 2011;28:115–119.

66. Kim WY, Lee JW, Choi JJ, Choi CH, Kim TJ, Kim BG, et al. Increased expression of Toll-like receptor 5 during progression of cervical neoplasia. Int J Gynecol Cancer 2008;18:300–305.

67. Droemann D, Albrecht D, Gerdes J, Ulmer AJ, Branscheid D, Vollmer E, et al. Human lung cancer cells express functionally active Toll-like receptor 9. Respir Res 2005;6:1.

68. Ilvesaro JM, Merrell MA, Swain TM, Davidson J, Zayzafoon M, Harris KW, et al. Toll like receptor-9 agonists stimulate prostate cancer invasion *in vitro*. Prostate 2007;67:774–781.

69. Cherfils-Vicini J, Platonova S, Gillard M, Laurans L, Validire P, Caliandro R, et al. Triggering of TLR7 and TLR8 expressed by human lung cancer cells induces cell survival and chemoresistance. J Clin Invest 2010;120:1285–1297.

70. Bartel DP. MicroRNAs: target recognition and regulatory functions. Cell 2009;136:215–233.

71. Ambros V. MicroRNAs: tiny regulators with great potential. Cell 2001;107:823–826.

72. Fabbri M, Croce CM. Role of microRNAs in lymphoid biology and disease. Curr Opin Hematol 2011;18:266–272.

73. Croce CM. Causes and consequences of microRNA dysregulation in cancer. Nat Rev Genet 2009;10:704–714.

74. Fabbri M, Paone A, Calore F, Galli R, Gaudio E, Santhanam R, et al. MicroRNAs bind to Toll-like receptors to induce prometastatic inflammatory response. Proc Natl Acad Sci USA 2012;109:E2110–2116.

75. Goto Y, Arigami T, Kitago M, Nguyen SL, Narita N, Ferrone S, et al. Activation of Toll-like receptors 2, 3, and 4 on human melanoma cells induces inflammatory factors. Mol Cancer Ther 2008;7:3642–3653.

76. Byrne WL, Mills KH, Lederer JA, O'Sullivan GC. Targeting regulatory T cells in cancer. Cancer Res 2011;71:6915–6920.

77. Akira S, Takeda K. Toll-like receptor signalling. Nat Rev Immunol 2004;4:499–511.

78. Furrie E, Macfarlane S, Thomson G, Macfarlane GT. Toll-like receptors-2, -3 and -4 expression patterns on human colon and their regulation by mucosal-associated bacteria. Immunology 2005;115:565–574.

79. Yoshioka T, Morimoto Y, Iwagaki H, Itoh H, Saito S, Kobayashi N, et al. Bacterial lipopolysaccharide induces transforming growth factor beta and hepatocyte growth factor through Toll-like receptor 2 in cultured human colon cancer cells. J Int Med Res 2001;29:409–420.

80. Chang YJ, Wu MS, Lin JT, Sheu BS, Muta T, Inoue H, et al. Induction of cyclooxygenase-2 overexpression in human gastric epithelial cells by *Helicobacter pylori* involves TLR2/TLR9 and c-Src-dependent nuclear factor-kappaB activation. Mol Pharmacol 2004;66:1465–1477.

81. Nishimura M, Naito S. Tissue-specific mRNA expression profiles of human Toll-like receptors and related genes. Biol Pharm Bull 2005;28:886–892.

82. Salaun B, Coste I, Rissoan MC, Lebecque SJ, Renno T. TLR3 can directly trigger apoptosis in human cancer cells. J Immunol 2006;176:4894–4901.

83. Salaun B, Lebecque S, Matikainen S, Rimoldi D, Romero P. Toll-like receptor 3 expressed by melanoma cells as a target for therapy? Clin Cancer Res 2007;13:4565–4574.

84. Harmey JH, Bucana CD, Lu W, Byrne AM, McDonnell S, Lynch C, et al. Lipopolysaccharide-induced metastatic growth is associated with increased angiogenesis, vascular permeability and tumor cell invasion. Int J Cancer 2002;101:415–422.

85. Molteni M, Marabella D, Orlandi C, Rossetti C. Melanoma cell lines are responsive *in vitro* to lipopolysaccharide and express TLR-4. Cancer Lett 2006;235:75–83.

86. Schmausser B, Andrulis M, Endrich S, Muller-Hermelink HK, Eck M. Toll-like receptors TLR4, TLR5 and TLR9 on gastric carcinoma cells: an implication for interaction with *Helicobacter pylori*. Int J Med Microbiol 2005;295:179–185.

87. He W, Liu Q, Wang L, Chen W, Li N, Cao X. TLR4 signaling promotes immune escape of human lung cancer cells by inducing immunosuppressive cytokines and apoptosis resistance. Mol Immunol 2007;44:2850–2859.

88. Kelly MG, Alvero AB, Chen R, Silasi DA, Abrahams VM, Chan S, et al. TLR-4 signaling promotes tumor growth and paclitaxel chemoresistance in ovarian cancer. Cancer Res 2006;66:3859–3868.

89. Koski GK, Czerniecki BJ. Combining innate immunity with radiation therapy for cancer treatment. Clin Cancer Res 2005;11:7–11.

90. Merrell MA, Ilvesaro JM, Lehtonen N, Sorsa T, Gehrs B, Rosenthal E, et al. Toll-like receptor 9 agonists promote cellular invasion by increasing matrix metalloproteinase activity. Mol Cancer Res 2006;4:437–447.

91. Lee JW, Choi JJ, Seo ES, Kim MJ, Kim WY, Choi CH, et al. Increased toll-like receptor 9 expression in cervical neoplasia. Mol Carcinog 2007;46:941–947.

92. El Andaloussi A, Sonabend AM, Han Y, Lesniak MS. Stimulation of TLR9 with CpG ODN enhances apoptosis of glioma and prolongs the survival of mice with experimental brain tumors. Glia 2006;54:526–535.

93. Heil F, Hemmi H, Hochrein H, Ampenberger F, Kirschning C, Akira S, et al. Species-specific recognition of single-stranded RNA via Toll-like receptor 7 and 8. Science 2004;303:1526–1529.

94. Pifer R, Benson A, Sturge CR, Yarovinsky F. UNC93B1 is essential for TLR11 activation and IL-12-dependent host resistance to *Toxoplasma gondii*. J Biol Chem 2011;286:3307–3314.

95. Chang YJ, Wu MS, Lin JT, Chen CC. *Helicobacter pylori*–induced invasion and angiogenesis of gastric cells is mediated by cyclooxygenase-2 induction through TLR2/TLR9 and promoter regulation. J Immunol 2005;175:8242–8252.

96. Li VW, Li WW, Talcott KE, Zhai AW. Imiquimod as an antiangiogenic agent. J Drugs Dermatol 2005;4:708–717.

97. Majewski S, Marczak M, Mlynarczyk B, Benninghoff B, Jablonska S. Imiquimod is a strong inhibitor of tumor cell-induced angiogenesis. Int J Dermatol 2005;44:14–19.

98. Damiano V, Caputo R, Bianco R, D'Armiento FP, Leonardi A, De Placido S, et al. Novel Toll-like receptor 9 agonist induces epidermal growth factor receptor (EGFR) inhibition and synergistic antitumor activity with EGFR inhibitors. Clin Cancer Res 2006;12:577–583.

99. Kundu SD, Lee C, Billips BK, Habermacher GM, Zhang Q, Liu V, et al. The Toll-like receptor pathway: a novel mechanism of infection-induced carcinogenesis of prostate epithelial cells. Prostate 2008;68:223–229.

100. Pries R, Hogrefe L, Xie L, Frenzel H, Brocks C, Ditz C, et al. Induction of c-Myc-dependent cell proliferation through Toll-like receptor 3 in head and neck cancer. Int J Mol Med 2008;21:209–215.

101. Jego G, Bataille R, Geffroy-Luseau A, Descamps G, Pellat-Deceunynck C. Pathogen-associated molecular patterns are growth and survival factors for human myeloma cells through Toll-like receptors. Leukemia 2006;20:1130–1137.

102. Chochi K, Ichikura T, Kinoshita M, Majima T, Shinomiya N, Tsujimoto H, et al. *Helicobacter pylori* augments growth of gastric cancers via the lipopolysaccharide-Toll-like receptor 4 pathway whereas its lipopolysaccharide attenuates antitumor activities of human mononuclear cells. Clin Cancer Res 2008;14:2909–2917.

103. Paone A, Starace D, Galli R, Padula F, De Cesaris P, Filippini A, et al. Toll-like receptor 3 triggers apoptosis of human prostate cancer cells through a PKC-alpha-dependent mechanism. Carcinogenesis 2008;29:1334–1342.

104. Jahrsdorfer B, Wooldridge JE, Blackwell SE, Taylor CM, Griffith TS, Link BK, et al. Immunostimulatory oligodeoxynucleotides induce apoptosis of B cell chronic lymphocytic leukemia cells. J Leukoc Biol 2005;77:378–387.

105. Jahrsdorfer B, Jox R, Muhlenhoff L, Tschoep K, Krug A, Rothenfusser S, et al. Modulation of malignant B cell activation and apoptosis by bcl-2 antisense ODN and immunostimulatory CpG ODN. J Leukoc Biol 2002;72:83–92.

106. Smits EL, Ponsaerts P, Van de Velde AL, Van Driessche A, Cools N, Lenjou M, et al. Proinflammatory response of human leukemic cells to dsRNA transfection linked to activation of dendritic cells. Leukemia 2007;21:1691–1699.

107. Lehner M, Bailo M, Stachel D, Roesler W, Parolini O, Holter W. Caspase-8 dependent apoptosis induction in malignant myeloid cells by TLR stimulation in the presence of IFN-alpha. Leuk Res 2007;31:1729–1735.

108. Bottero V, Busuttil V, Loubat A, Magne N, Fischel JL, Milano G, et al. Activation of nuclear factor kappaB through the IKK complex by the topoisomerase poisons SN38 and doxorubicin: a brake to apoptosis in HeLa human carcinoma cells. Cancer Res 2001;61:7785–7791.

109. Tosi P, Zinzani PL, Pellacani A, Ottaviani E, Magagnoli M, Tura S. Loxoribine affects fludarabine activity on freshly isolated B-chronic lymphocytic leukemia cells. Leuk Lymphoma 1997;26:343–348.

110. Pellacani A, Tosi P, Zinzani PL, Ottaviani E, Albertini P, Magagnoli M, et al. Cytotoxic combination of loxoribine with fludarabine and mafosfamide on freshly isolated B-chronic lymphocytic leukemia cells. Leuk Lymphoma 1999;33:147–153.

111. Shi Y, White D, He L, Miller RL, Spaner DE. Toll-like receptor-7 tolerizes malignant B cells and enhances killing by cytotoxic agents. Cancer Res 2007;67:1823–1831.

112. Caramalho I, Lopes-Carvalho T, Ostler D, Zelenay S, Haury M, Demengeot J. Regulatory T cells selectively express toll-like receptors and are activated by lipopolysaccharide. J Exp Med 2003;197:403–411.

113. Sfondrini L, Rossini A, Besusso D, Merlo A, Tagliabue E, Menard S, et al. Antitumor activity of the TLR-5 ligand flagellin in mouse models of cancer. J Immunol 2006;176:6624–6630.

114. Peng G, Guo Z, Kiniwa Y, Voo KS, Peng W, Fu T, et al. Toll-like receptor 8-mediated reversal of CD4+ regulatory T cell function. Science 2005;309:1380–1384.

115. Pasare C, Medzhitov R. Toll pathway-dependent blockade of CD4+CD25+ T cell–mediated suppression by dendritic cells. Science 2003;299:1033–1036.

116. Spaner DE, Shi Y, White D, Mena J, Hammond C, Tomic J, et al. Immunomodulatory effects of Toll-like receptor-7 activation on chronic lymphocytic leukemia cells. Leukemia 2006;20:286–295.

117. Tomic J, White D, Shi Y, Mena J, Hammond C, He L, et al. Sensitization of IL-2 signaling through TLR-7 enhances B lymphoma cell immunogenicity. J Immunol 2006;176:3830–3839.

118. Decker T, Schneller F, Sparwasser T, Tretter T, Lipford GB, Wagner H, et al. Immunostimulatory CpG-oligonucleotides cause proliferation, cytokine production, and an immunogenic phenotype in chronic lymphocytic leukemia B cells. Blood 2000;95:999–1006.

119. Decker T, Schneller F, Kronschnabl M, Dechow T, Lipford GB, Wagner H, et al. Immunostimulatory CpG-oligonucleotides induce functional high affinity IL-2 receptors on B-CLL cells: costimulation with IL-2 results in a highly immunogenic phenotype. Exp Hematol 2000;28:558–568.

120. Decker T, Hipp S, Kreitman RJ, Pastan I, Peschel C, Licht T. Sensitization of B-cell chronic lymphocytic leukemia cells to recombinant immunotoxin by immunostimulatory phosphorothioate oligodeoxynucleotides. Blood 2002;99:1320–1326.

121. Wysocka M, Benoit BM, Newton S, Azzoni L, Montaner LJ, Rook AH. Enhancement of the host immune responses in cutaneous T-cell lymphoma by CpG oligodeoxynucleotides and IL-15. Blood 2004;104:4142–4149.

122. Mangsbo SM, Ninalga C, Essand M, Loskog A, Totterman TH. CpG therapy is superior to BCG in an orthotopic bladder cancer model and generates CD4+ T-cell immunity. J Immunother 2008;31:34–42.

123. Ren T, Wen ZK, Liu ZM, Qian C, Liang YJ, Jin ML, et al. Targeting Toll-like receptor 9 with CpG oligodeoxynucleotides enhances anti-tumor responses of peripheral blood mononuclear cells from human lung cancer patients. Cancer Invest 2008;26:448–455.

124. Roda JM, Parihar R, Carson WE, 3rd. CpG-containing oligodeoxynucleotides act through TLR9 to enhance the NK cell cytokine response to antibody-coated tumor cells. J Immunol 2005;175:1619–1627.

125. Qian Y, Deng J, Xie H, Geng L, Zhou L, Wang Y, et al. Regulation of TLR4-induced IL-6 response in bladder cancer cells by opposing actions of MAPK and PI3K signaling. J Cancer Res Clin Oncol 2009;135:379–386.

126. Huang B, Zhao J, Li H, He KL, Chen Y, Chen SH, et al. Toll-like receptors on tumor cells facilitate evasion of immune surveillance. Cancer Res 2005;65:5009–5014.

127. Yang H, Zhou H, Feng P, Zhou X, Wen H, Xie X, et al. Reduced expression of Toll-like receptor 4 inhibits human breast cancer cells proliferation and inflammatory cytokines secretion. J Exp Clin Cancer Res 2010;29:92.

128. Tang XY, Zhu YQ, Wei B, Wang H. Expression and functional research of TLR4 in human colon carcinoma. Am J Med Sci 2010;339:319–326.

129. Doan HQ, Bowen KA, Jackson LA, Evers BM. Toll-like receptor 4 activation increases Akt phosphorylation in colon cancer cells. Anticancer Res 2009;29:2473–2478.

130. Grauer OM, Molling JW, Bennink E, Toonen LW, Sutmuller RP, Nierkens S, et al. TLR ligands in the local treatment of established intracerebral murine gliomas. J Immunol 2008;181:6720–6729.

131. Szczepanski M, Stelmachowska M, Stryczynski L, Golusinski W, Samara H, Mozer-Lisewska I, et al. Assessment of expression of Toll-like receptors 2, 3 and 4 in laryngeal carcinoma. Eur Arch Otorhinolaryngol 2007;264:525–530.

132. Hassan F, Islam S, Tumurkhuu G, Naiki Y, Koide N, Mori I, et al. Intracellular expression of Toll-like receptor 4 in neuroblastoma cells and their unresponsiveness to lipopolysaccharide. BMC Cancer 2006;6:281.

133. Szajnik M, Szczepanski MJ, Czystowska M, Elishaev E, Mandapathil M, Nowak–Markwitz E, et al. TLR4 signaling induced by lipopolysaccharide or paclitaxel regulates tumor survival and chemoresistance in ovarian cancer. Oncogene 2009;28:4353–4363.

134. Zhou M, McFarland-Mancini MM, Funk HM, Husseinzadeh N, Mounajjed T, Drew AF. Toll-like receptor expression in normal ovary and ovarian tumors. Cancer Immunol Immunother 2009;58:1375–1385.

CHAPTER 8

INFLAMMASOMES AND INFLAMMATION

Kaiwen W. Chen, Ayanthi A. Richards, Alina Zamoshnikova, and Kate Schroder

The innate immune system provides the first line of defense against infection and coordinates tissue repair during injury. Cells of the innate immune system use germ-line-encoded receptors called *pattern recognition receptors* (PRRs) to recognize conserved microbial structures called *pathogen-associated molecular patterns*, or host-derived molecules released during cell or tissue damage (known as *danger-associated molecular patterns*) (1). PRRs are expressed primarily on cells at the front line of defense against infection, including macrophages, dendritic cells, neutrophils, and epithelial cells. Engagement of PRRs initiates signal transduction pathways that usually culminate in the elimination of pathogens and/or the initiation of tissue repair processes. A major pro-inflammatory pathway mediating host defense against infection is triggered by inflammasomes, multiprotein complexes that regulate the function of the pro-inflammatory caspase, caspase-1 (2). Caspase-1, in turn, controls the maturation and secretion of pro-inflammatory cytokines such as interleukin (IL)-1β and IL-18 and the induction of inflammatory cell death (pyroptosis). Inflammasome assembly is mediated by members of the NOD-like receptor (NLR) or pyrin and HIN domain–containing (PYHIN) families of "danger" sensor proteins.

NOD-LIKE RECEPTORS AND THE INFLAMMASOMES

NLRs are intracellular PRRs that play key roles in detecting cellular damage and cytoplasmic microbial infection. The NLR gene family is comprised of 22 human genes and 34 mouse genes, several of which assemble into inflammasome complexes (3,4). The NLR family is defined by the presence of a common central nucleotide-binding and oligomerization (NACHT) domain. Inflammasome-forming NLRs also contain a leucine-rich repeat (LRR) domain and a caspase recruitment (CARD) and/or pyrin (PYD) effector domain (Figure 8.1) (4). The LRR domain is thought to be involved in sensing microbial or endogenous stimuli, whereas the CARD or

Cancer and Inflammation Mechanisms: Chemical, Biological, and Clinical Aspects, First Edition.
Edited by Yusuke Hiraku, Shosuke Kawanishi, and Hiroshi Ohshima.

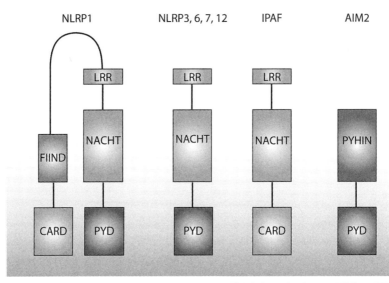

FIGURE 8.1 Protein domains of inflammasome scaffolds from the human NLR and PYHIN families. Domains are classified according to the NCBI domain annotation tool (3). Abbreviations for domain names: PYD, pyrin domain; CARD, caspase activation and recruitment domain; NACHT, nucleotide-binding and oligomerization domain; LRR, leucine-rich repeat; FIIND, domain with function to find; PYHIN, pyrin- and HIN-domain-containing domain. Abbreviations for protein names: NLRP, NOD-like receptor protein; IPAF, ICE protease-activating factor; AIM2, absent in melanoma 2.

PYD domain mediates coupling to downstream effectors via homotypic interactions between protein–protein binding domains.

Upon activation by specific signals, NLRs oligomerize through their NACHT domain in an ATP-dependent manner (5). In the case of many inflammasome-forming NLR proteins, oligomerization of NLRs facilitates the recruitment of an apoptosis-associated speck-like protein containing a caspase recruitment domain (ASC; domain structure PYD-CARD) through homotypic PYD interactions. ASC serves as an adaptor to recruit the CARD domain of pro-caspase-1 via CARD–CARD interactions. For CARD-containing inflammasome scaffolds (i.e., NLRP1 and IPAF), pro-caspase-1 can be coupled directly to the inflammasome complex through CARD–CARD interactions, bypassing the requirement for the ASC adaptor. Upon formation of the inflammasome complex, pro-caspase-1 undergoes proximity-induced autoproteolytic cleavage to generate three caspase-1 cleavage fragments: CARD, p20, and p10. Two molecules of p20 heterodimerize with two molecules of p10 to form the active caspase-1 enzyme (Figure 8.2).

Major cellular targets of caspase-1-dependent proteolysis are the pro-inflammatory cytokines IL-1β and IL-18 (4). The action of these cytokines is critical for protective inflammatory responses during host defense; however, they can also trigger tissue pathology. Therefore, the function of these cytokines is tightly regulated (Figure 8.2). Cells express these cytokines as inactive precursor proteins that

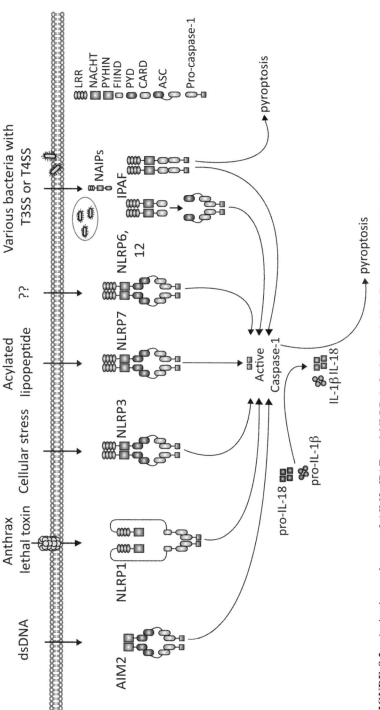

FIGURE 8.2 Activation pathways of AIM2, IPAF, and NLRP 1, 3, 6, 7, and 12 inflammasomes. The AIM2, IPAF, and NLRP 1, 3, 6, 7 and 12 inflammasomes are activated by distinct microbial triggers or cellular pathways. Activation of the IPAF inflammasome is mediated by NAIP proteins of the NLR family. Inflammasome assembly culminates in the activation of caspase-1, the caspase-1-directed maturation and secretion of IL-1β and IL-18, and pyroptotic cell death. (*See insert for color representation of the figure.*)

105

accumulate intracellularly until the inflammasome pathway is engaged. Upon inflammasome triggering, caspase-1 cleaves these cytokine precursors into their mature forms and regulates their release via an unconventional protein secretion pathway that is poorly characterized (6). This unusual cytokine secretion pathway also regulates the release of the related cytokine, IL-1α, which is not directly cleaved by caspase-1 (7).

In addition to regulating the maturation of IL-1β and IL-18, activation of caspase-1 is also associated with the initiation of pyroptosis, a form of inflammatory cell death that is distinct from apoptosis and necrosis (8). Pyroptosis is dependent on caspase-1 activation and does not involve apoptotic caspases. It is characterized by rapid pore formation, swelling, rupture, and release of cellular contents to the extracellular space (9).

THE NLRP3 INFLAMMASOME

The NLRP3 inflammasome is currently the most studied inflammasome. It is composed of the NLRP3 scaffold, the adaptor protein ASC, and pro-caspase-1 (Figure 8.2). A diverse range of signals from a microbial, endogenous, and environmental origin can activate the NLRP3 inflammasome. Several pathogenic microbes, such as *Candida albicans*, *Neisseria gonorrhoeae*, and influenza viruses, activate the NLRP3 inflammasome (10–12). In addition, the NLRP3 inflammasome is triggered by the presence of endogenous molecules that are released as a result of cellular injury or dysfunction, such as ATP, monosodium urate, and β-amyloid (13–15).

Despite the growing list of NLRP3 agonists, the exact molecular mechanisms governing the assembly and activation of the NLRP3 inflammasome are still unclear. The chemical diversity of NLRP3 agonists suggests that these molecules activate a common stress pathway sensed by NLRP3, but the mechanism by which NLRP3 senses cellular stress remains unclear. Three major models are supported in the literature (4). The first model suggests that NLRP3 is a sensor for pathways triggering the production of reactive oxygen species (ROSs), an ancient danger signal. The second model pertains to particulate or crystalline NLRP3 agonists; these are endocytosed by phagocytes and lead to lysosomal rupture. Lysosomal rupture is proposed to cause leakage of lysosomal proteases (e.g., cathepsin B) into the cytoplasm, to enable NLRP3 activation by an unknown mechanism. The third model posits NLRP3 as a sensor for potassium efflux, as a number of NLRP3 agonists trigger intracellular ionic flux (e.g., bacterial pore-forming toxins, and extracellular ATP, which engages the P2X7 potassium efflux pump), and suppression of potassium flux (e.g., by culturing cells in high extracellular potassium) can inhibit NLRP3-dependent caspase-1 activation (16,17).

THE IPAF INFLAMMASOME

The IPAF (also known as NLRC4) inflammasome plays an important role in host defense against pathogenic bacteria such as *Salmonella enterica* serovar *typhimurium*

(*S. typhimurium*), *Legionella pneumophila*, *Shigella flexneria*, and *Pseudomonas aeruginosa* (18–21). Recently, IPAF was demonstrated to cooperate with NLRP3 to provide antifungal defense against *C. albicans* during mucosal challenge *in vivo* (22).

IPAF appears to be responsive to a narrower spectrum of agonists than does NLRP3, and IPAF inflammasome activation by pathogenic bacteria depends on the presence of a functional bacterial secretion system (23). These bacterial secretion systems, such as the type 3 and 4 secretion systems (T3SS, T4SS), are molecular needle-like structures that inject bacterial effector proteins into the host cytosol and mediate bacterial virulence (24,25). IPAF does not interact directly with its microbial agonists, but senses these molecules via co-receptors also encoded by the NLR family, neuronal apoptosis inhibitor proteins (NAIPs) (26,27). The mouse NLR family contains seven NAIP proteins, many of which are uncharacterized. NAIP5 and 6 sense bacterial flagellin, whereas NAIP2 recognizes bacterial PrgJ-like proteins (26,27). Interestingly, the human genome encodes only a single NAIP protein that does not interact with either flagellin or PrgJ-like proteins (27), suggesting that some level of species specificity exists in IPAF activation mechanisms.

The IPAF protein contains a CARD domain and so is able to interact directly with pro-caspase-1 without an absolute requirement for the adaptor protein ASC (Figure 8.2). Indeed, ASC was dispensable for IPAF-dependent pyroptosis (18,21) and was not absolutely required for IPAF-dependent IL-1β maturation and secretion (19,21,28).

THE NLRP1 INFLAMMASOME

The NLRP1 inflammasome was the first to be identified (2). Humans encode a single NLRP1 gene, whereas three orthologous genes are present in mice (*Nlrp1a–c*) (29). The *Nlrp1* gene is highly polymorphic between inbred mice, and strain variation in the mouse *Nlrp1b* locus confers susceptibility to anthrax lethal toxin (Figure 8.2) (29). Similar to IPAF, human NLRP1 contains a CARD domain that can interact directly with pro-caspase-1, bypassing an absolute requirement for ASC, although ASC can still bind to the N-terminal PYD domain of NLRP1 (Figure 8.2) (30). Unlike human NLRP1, murine NLRP1 proteins lack functional PYD domains and are unable to interact with ASC. Indeed, ASC is not required for caspase1 activation and IL-1β secretion in mouse macrophages (31). The molecular mechanisms regulating NLRP1 activation are still uncharacterized, but potassium efflux and cleavage of the NLRP1 protein have been implicated (32,33).

THE NLRP12 INFLAMMASOME

NLRP12 (also known as Nalp12, Monarch-1, and Pypaf-7) is the closest phylogenetic relative of NLRP3 (4). Human mutations in NLRP12 are associated with a hereditary periodic fever syndrome that resembles strongly that caused by gain-of-function NLRP3 mutation (34). These conditions are driven by excess IL-1 activity, as IL-1 receptor antagonists block disease (35).

The precise biochemical function of NLRP12 remains controversial. Although various studies suggest that NLRP12 forms an ASC-containing caspase-1-activating platform analogous to NLRP3 (Figure 8.2), concrete biochemical evidence for such a complex is still lacking. However, such a role is supported by a study in which overexpression of NLRP12 and ASC triggered caspase-1 activation and IL-1β secretion (36), and another report demonstrating that a NLRP12/IL-18 axis was critical for host defense against *Yersinia pestis in vivo* (37). Other reports suggest inflammasome-independent roles for NLRP12 in leukocyte migration (38), MHC class I expression (39), NF-κB inhibition (40), and colonic inflammation and tumorigenesis (41).

THE NLRP7 INFLAMMASOME

Microbial acylated lipopeptide (ac-LP) was recently demonstrated to promote caspase-1 activation by a NLRP7 inflammasome in human cells, leading to IL-1β and IL-18 maturation (Figure 8.2) (42). The exact mechanism by which ac-LP activates the NLRP7 inflammasome remains unclear, but it appears to follow a pathway distinct from that of NLRP3, as treatment with known inhibitors of the NLRP3 pathway (e.g., ROS scavengers and a cathepsin B inhibitor) did not suppress ac-LP-dependent caspase-1 activity (42). Interestingly, the NLRP7 inflammasome was a poor inducer of pyroptosis (42).

THE NLRP6 INFLAMMASOME

Limited studies on the NLRP6 inflammasome in colonic epithelial cells suggest an important function for this pathway in the regulation of gut microbiota, colitis, and colorectal cancer. NLRP6 was demonstrated to enhance tissue repair and regeneration in colonic epithelial cells (43), as well as maintaining a healthy gut microflora composition via IL-18 (44). Thus, it is presumed that NLRP6 forms a caspase-1 activation complex similar to that of NLRP3, to regulate IL-18 production (Figure 8.2). NLRP6-deficient mice displayed altered gut flora (dysbiosis) (44) and deregulated epithelial proliferation upon experimental colitis (43), predisposing these mice to colitis-related tumorigenesis. Controversially, a recent study described NLRP6 as a negative regulator of inflammation (45). In this report, *Nlrp6* deficiency rendered mice more resistant to infection with *Listeria monocytogenes, S. typhimurium*, and *Escherichia coli*. Their resistance was ascribed to enhanced activation of MAPK and the canonical NF-κB pathway, leading to increased cytokine and chemokine production and heightened granulocyte recruitment (45).

THE AIM2 INFLAMMASOME

The cytosolic protein AIM2 was the first non-NLR family member identified to nucleate an inflammasome (46–48). The AIM2 inflammasome is activated by cytosolic double-stranded DNA, which indicates the presence of intracellular pathogens such as

L. monocytogenes, Francisella tularensis, cytomegalovirus, and vaccinia virus (49). The AIM2 inflammasome is comprised of AIM2, ASC, and caspase-1 (Figure 8.2). As for other inflammasomes, such as the NLRP3 inflammasome, AIM2 activation triggers the recruitment of ASC and pro-caspase-1 through PYD–PYD (AIM2-ASC) and CARD–CARD (ASC-procaspase-1) interactions, to enable cytokine maturation and secretion, and lytic cell death.

CYTOKINE TARGETS OF THE INFLAMMASOME

Interleukin-1

IL-1α and IL-1β (collectively referred to as IL-1) are potent cytokines that mediate pro-inflammatory responses. Both IL-1α and IL-1β signal through a common receptor, IL-1 receptor 1 (IL-1R1). Effective signaling requires the IL-1R accessory protein (IL-1R1AcP). Ligation of the IL-1 signaling complex triggers downstream signaling, driving inflammatory responses via mitogen-activated protein kinase (MAPK) and the transcription factor NF-κB (50,51). IL-1 causes both local and systemic responses. IL-1 induces fever, recruits leukocytes to a site of infection or injury, and induces their activation (51). Whereas the expression of pro-IL-1β is largely restricted to cells of the myeloid lineage, pro-IL-1α can be expressed by most cell types (52). Expression of intracellular pro-IL-1α/β is induced upon NF-κB activation by TLRs (e.g., TLR4) or pro-inflammatory cytokines (e.g., TNF, IL-1). Maturation and secretion of IL-1β requires engagement of the inflammasome pathway. The majority of studies investigating IL-1α biology describe its passive release during cell death, where it functions as an alarmin to promote inflammation (52,53), because unlike IL-1β, both the pro- and mature forms of IL-1α can trigger IL-1R1 signaling. Under some circumstances, caspase-1 can regulate IL-1α secretion (54), and inflammasome activators induce IL-1α secretion in murine bone marrow dendritic cells; however, IL-1α secretion does not always depend on NLRs (7). Soluble or infectious inflammasome agonists such as ATP, imiquimod, *C. albicans*, and *S. typhimurium* triggered IL-1α secretion in an NLRP3-dependent manner, whereas IL-1α release upon treatment with particulate agonists such as alum, silica, and monosodium urate did not require NLRP3 (7).

Interleukin-18

Similar to IL-1β, IL-18 is synthesized as an inactive cytokine and requires proteolytic maturation by caspase-1 (55). IL-18 is involved in both innate and adaptive immune responses and is best known for inducing interferon-γ (IFN-γ) production by T cells and natural killer (NK) cells. IFN-γ in turn potentiates macrophage antimicrobial activities such as the production of reactive oxygen and nitrogen species (56). In addition, IFN-γ skews the adaptive immune system toward a Th1 type of response, which allows for effective microbial clearance. The importance of IL-18 in host defense was demonstrated in several experimental models whereby pharmacological or genetic ablation of IL-18 increased the susceptibility of mice to infection by *Yersinia* sp. (57), *Mycobacterium leprae* (58), *Leishmania*, and *Staphylococcus* (59).

IL-18 also provides tumor surveillance by inducing tumor cell apoptosis via Fas/FasL, by inducing tumor cell FasL expression (60), or by increasing the antitumor action of NK cells (61).

INFLAMMASOME DYSREGULATION IN DISEASE

Numerous and diverse human disorders of wide-ranging etiology arise from, or are exacerbated by, the dysregulation of inflammasome activity. They include genetic disorders, autoimmune conditions,and metabolic diseases. Heightened IL-1 production has been identified as a key mechanism in the majority of these conditions, and several IL-1-antagonizing therapies have shown stunning efficacy in clinical trials [reviewed in (62)]. Some of these therapies (e.g., the IL-1 receptor antagonist, anakinra) are currently in routine clinical use.

Hereditary Fever Syndromes

The role of inflammasomes in human disease states was first appreciated in the context of rare hereditary periodic fever syndromes, where inherited mutations in inflammasome proteins result in pathway hyperactivation in the absence of overt infection. Autosomal dominant mutations of *NLRP3* (also called "cryopyrin") give rise to a range of diseases known collectively as *cryopyrin-associated periodic syndromes* (CAPSs) [reviewed in (63)]. Mutations in *NLRP12* also give rise to NLRP12-associated autoinflammatory disorders, which show a clinical presentation similar to that of CAPSs (34). Both of these diseases are now managed routinely with inhibitors of IL-1 or its receptor. Autoinflammatory diseases resulting from dysregulated inflammasome activity also arise from mutations in noninflammasome proteins that regulate inflammasome activity and/or IL-1 production. Among these are familial mediterranean fever, hyperimmunoglobulin D syndrome, pyogenic arthritis, pyoderma gangrenosum and acne, TNF receptor–associated periodic syndrome, and Blau syndrome (63).

Neurodegenerative Diseases

Local and systemic increases in IL-1 levels have been described in multiple neurodegenerative diseases, including Alzheimer's disease, stroke, and Parkinson's disease [reviewed in (64)]. Of these, the mechanism leading to inflammasome activation is most apparent for Alzheimer's disease, an increasingly prevalent form of dementia characterized by abnormal accumulation of oligomeric and fibrillar amyloid-β peptide. Fibrillar amyloid-β peptide was shown to be a particulate agonist of the NLRP3 inflammasome (13).

Pulmonary Diseases

Other particulate agonists of the NLRP3 inflammasome include the environmental irritants silica and asbestos, which are the causative agents of the pulmonary inflammatory disorders silicosis and asbestosis, respectively (65–67).

Metabolic Diseases

The NLRP3 inflammasome has in recent years also become implicated in obesity-related diseases, where dyslipidemia and low-grade chronic inflammation are causative and exacerbating factors. Obesity and overnutrition are major risk factors for diseases of the metabolic syndrome, including insulin-resistance and type 2 diabetes, nonalcoholic fatty liver disease, and cardiovascular disease. Palmitate is one of the most abundant saturated fatty acids in plasma and becomes elevated in response to a high-fat diet (68). Palmitate, but not the unsaturated fatty acid, oleate, induced NLRP3-dependent caspase-1 activation and cytokine secretion in mouse macrophages (69). Ceramide, an obesity-associated derivative of palmitate, also triggered NLRP3 inflammasome function in macrophages (70). Supporting a pathogenic role for inflammasomes in metabolic disease, mice deficient in the NLRP3 inflammasome pathway were resistant to high-fat-diet-induced obesity and resulting insulin resistance (70–72). Other lipid toxins shown to activate the NLRP3 inflammasome in macrophages include oxidized low-density lipoprotein and cholesterol crystals in models of high-fat-diet-induced atherosclerosis (73). Late stages of type 2 diabetes are characterized by destruction of the insulin-secreting beta cells of pancreatic islets. IL-1β, elicited by NLRP3 stimulation by high levels of circulating glucose, is thought to contribute to beta-cell death and dysfunction, as therapy with an IL-1 receptor antagonist protected pancreatic beta-cell function and improved insulin secretion [reviewed in (74)]. However, such treatment does not appear to ameliorate insulin resistance, the cause of elevated plasma glucose and islet failure (75). Therefore, the mechanism that caused inflammasome activity obesity-induced insulin-resistance and type 2 diabetes to progress may involve factors additional to IL-1. Finally, obesity and diseases of the metabolic syndrome are also associated with increased levels of monosodium urate (76), a crystalline agonist of the NLRP3 inflammasome responsible for gouty arthritis [reviewed in (77)].

Cancer

Inflammation is a common feature of tumor microenvironments and is often observed to promote tumorigenesis and drive cancer progression. Pro-inflammatory cytokines such as IL-1 are among the inflammatory mediators in tumor microenvironments and may be produced by infiltrating leukocytes or by the malignant cells themselves [reviewed in (78)]. The role of inflammasomes in tumorigenesis and cancer progression/metastasis is presently unclear, with evidence for both beneficial and detrimental effects of inflammasome activity on these processes.

Many tumors show constitutive production of IL-1β (79–81), allowing its use as a cancer biomarker (82), and IL-1 has been implicated in the progression of several types of cancer. In multiple myeloma, IL-1β released by myeloma cells induces the production of a myeloma growth factor, IL-6, from bone marrow stromal cells (83). IL-1 also contributes to the generation of CD4$^+$ T cells, which produce IL-17 (83), which in turn, promotes skin carcinogenesis in concert with IL-23 (83). Indeed, chemically induced subcutaneous carcinogenesis was retarded in mice deficient in IL-1F, the IL-1 receptor, or caspase-1, but was accelerated in *Il1rn*$^{-/-}$

mice, which lack the endogenous IL1-receptor antagonist, IL-1RA (84,85). Another study demonstrated that IL-1 signaling deficiency ($Il1r^{-/-}$) increased tumor infiltration of myeloid-derived suppressor cells (MDSC), and decreased tumor progression, while the opposite was observed in mice in which IL-1 signaling is enhanced ($Il1rn^{-/-}$) (86). Tumor-derived IL-18 can also suppress NK-cell immunosurveillance and antimetastatic functions (87). The IL-1 receptor antagonist anakinra has been used successfully to treat patients with multiple myeloma (79).

Inflammasomes can also mediate protective anticancer functions. For example, IL-18-dependent IFN-γ can elicit immunostimulatory and anticancer effects (56). Deficiency in NLRP3, NLRP6, NLRP12, ASC, and caspase-1 confers susceptibility to dextran sulfate sodium (DSS)-induced colitis and tumorigenesis in mice, suggesting that inflammasomes mediate protective functions in intestinal homeostasis (43,88–92), despite some controversy [reviewed in (93)]. Mice lacking IL-1β and IL-18 were more susceptible than wild-type mice to chemically induced colitis (94,95), and IL-18 deficiency also rendered mice sensitive to azoxymethane (AOM)-DSS-induced tumorigenesis (94). In keeping with this, IL-18 administration successfully limited murine colorectal carcinogenesis (96). Finally, inflammasome activity triggered by radiotherapy/chemotherapy-induced cell death (and resulting ATP release) may stimulate the adaptive antitumor immune responses important for the success of these treatments (97).

CONCLUDING REMARKS

A decade has passed since the initial description of the inflammasome. During that time, significance advances in the field have clarified the physiological role of inflammasomes in protecting against infection and mediating wound-healing responses during injury, and have defined the pathogenic role of inflammasomes in human genetic and acquired inflammatory disease. Many such human diseases of dysregulated inflammasome function are now treated successfully with antagonists of IL-1 or its receptor. A key outstanding question is how the inflammasome pathway can drive some forms of cancer while protecting against others. Whereas inflammasomes are reported to protect against colorectal cancer in mice, they appear to contribute to the progression of multiple myeloma and skin cancer. Such contrasting roles may depend on the nature of the inflammasome elicited, its microbial or endogenous trigger(s), or its anatomical location. Further research is required to unravel the protective and deleterious roles of inflammasomes in cancer.

REFERENCES

1. Medzhitov R. Approaching the asymptote: 20 years later. Immunity 2009;30:766–775.
2. Martinon F, Burns K, Tschopp J. The inflammasome: a molecular platform triggering activation of inflammatory caspases and processing of proIL-beta. Mol Cell 2002;10:417–426.
3. Rathinam VA, Vanaja SK, Fitzgerald KA. Regulation of inflammasome signaling. Nat Immunol 2012;13:332–333.

4. Schroder K, Tschopp J. The inflammasomes. Cell 2010;140:821–832.
5. Koonin EV, Aravind L. The NACHT family: a new group of predicted NTPases implicated in apoptosis and MHC transcription activation. Trends Biochem Sci 2000;25:223–224.
6. Keller M, Ruegg A, Werner S, Beer HD. Active caspase-1 is a regulator of unconventional protein secretion. Cell 2008;132:818–831.
7. Groß O, Yazdi Amir S, Thomas Christina J, Masin M, Heinz Leonhard X, Guarda G, et al. Inflammasome activators induce interleukin-1α secretion via distinct pathways with differential requirement for the protease function of caspase-1. Immunity 2012;36:388–400.
8. Fink SL, Cookson BT. Caspase-1-dependent pore formation during pyroptosis leads to osmotic lysis of infected host macrophages. Cell Microbiol 2006;8:1812–1825.
9. Bergsbaken T, Fink SL, Cookson BT. Pyroptosis: host cell death and inflammation. Nat Rev Microbiol 2009;7:99–109.
10. Allen IC, Scull MA, Moore CB, Holl EK, McElvania-TeKippe E, Taxman DJ, et al. The NLRP3 inflammasome mediates *in vivo* innate immunity to influenza A virus through recognition of viral RNA. Immunity 2009;30:556–565.
11. Duncan JA, Gao X, Huang MT, O'Connor BP, Thomas CE, Willingham SB, et al. *Neisseria gonorrhoeae* activates the proteinase cathepsin B to mediate the signaling activities of the NLRP3 and ASC-containing inflammasome. J Immunol 2009;182:6460–6469.
12. Gross O, Poeck H, Bscheider M, Dostert C, Hannesschlager N, Endres S, et al. Syk kinase signalling couples to the Nlrp3 inflammasome for anti-fungal host defence. Nature 2009;459:433–436.
13. Halle A, Hornung V, Petzold GC, Stewart CR, Monks BG, Reinheckel T, et al. The NALP3 inflammasome is involved in the innate immune response to amyloid-beta. Nat Immunol 2008;9:857–865.
14. Mariathasan S, Weiss DS, Newton K, McBride J, O'Rourke K, Roose-Girma M, et al. Cryopyrin activates the inflammasome in response to toxins and ATP. Nature 2006;440:228–232.
15. Martinon F, Petrilli V, Mayor A, Tardivel A, Tschopp J. Gout-associated uric acid crystals activate the NALP3 inflammasome. Nature 2006;440:237–241.
16. Franchi L, Kanneganti TD, Dubyak GR, Nunez G. Differential requirement of P2×7 receptor and intracellular K+ for caspase-1 activation induced by intracellular and extracellular bacteria. J Biol Chem 2007;282:18810–18818.
17. Petrilli V, Papin S, Dostert C, Mayor A, Martinon F, Tschopp J. Activation of the NALP3 inflammasome is triggered by low intracellular potassium concentration. Cell Death Differ 2007;14:1583–1589.
18. Amer A, Franchi L, Kanneganti TD, Body-Malapel M, Ozoren N, Brady G, et al. Regulation of Legionella phagosome maturation and infection through flagellin and host Ipaf. J Biol Chem 2006;281:35217–35223.
19. Mariathasan S, Newton K, Monack DM, Vucic D, French DM, Lee WP, et al. Differential activation of the inflammasome by caspase-1 adaptors ASC and Ipaf. Nature 2004;430:213–218.
20. Miao EA, Ernst RK, Dors M, Mao DP, Aderem A. *Pseudomonas aeruginosa* activates caspase 1 through Ipaf. Proc Natl Acad Sci USA 2008;105:2562–2567.
21. Suzuki T, Franchi L, Toma C, Ashida H, Ogawa M, Yoshikawa Y, et al. Differential regulation of caspase-1 activation, pyroptosis, and autophagy via Ipaf and ASC in *Shigella*-infected macrophages. PLoS Pathog 2007;3:e111.
22. Tomalka J, Ganesan S, Azodi E, Patel K, Majmudar P, Hall BA, et al. A novel role for the NLRC4 inflammasome in mucosal defenses against the fungal pathogen *Candida albicans*. PLoS Pathog 2011;7:e1002379.
23. Miao EA, Mao DP, Yudkovsky N, Bonneau R, Lorang CG, Warren SE, et al. Innate immune detection of the type III secretion apparatus through the NLRC4 inflammasome. Proc Natl Acad Sci USA 2010;107:3076–3080.
24. Cascales E, Christie PJ. The versatile bacterial type IV secretion systems. Nat Rev Microbiol 2003;1:137–149.
25. Cornelis GR. The type III secretion injectisome. Nat Rev Microbiol 2006;4:811–825.
26. Kofoed EM, Vance RE. Innate immune recognition of bacterial ligands by NAIPs determines inflammasome specificity. Nature 2011;477:592–595.
27. Zhao Y, Yang J, Shi J, Gong YN, Lu Q, Xu H, et al. The NLRC4 inflammasome receptors for bacterial flagellin and type III secretion apparatus. Nature 2011;477:596–600.

28. Case CL, Roy CR. Asc modulates the function of NLRC4 in response to infection of macrophages by *Legionella pneumophila*. MBio 2011;2:e00117–00111.
29. Boyden ED, Dietrich WF. Nalp1b controls mouse macrophage susceptibility to anthrax lethal toxin. Nat Genet 2006;38:240–244.
30. Faustin B, Lartigue L, Bruey JM, Luciano F, Sergienko E, Bailly-Maitre B, et al. Reconstituted NALP1 inflammasome reveals two-step mechanism of caspase-1 activation. Mol Cell 2007;25:713–724.
31. Nour AM, Yeung YG, Santambrogio L, Boyden ED, Stanley ER, Brojatsch J. Anthrax lethal toxin triggers the formation of a membrane-associated inflammasome complex in murine macrophages. Infect Immun 2009;77:1262–1271.
32. Fink SL, Bergsbaken T, Cookson BT. Anthrax lethal toxin and *Salmonella* elicit the common cell death pathway of caspase-1-dependent pyroptosis via distinct mechanisms. Proc Natl Acad Sci USA 2008;105:4312–4317.
33. Levinsohn JL, Newman ZL, Hellmich KA, Fattah R, Getz MA, Liu S, et al. Anthrax lethal factor cleavage of Nlrp1 is required for activation of the inflammasome. PLoS Pathog 2012;8:e1002638.
34. Jeru I, Duquesnoy P, Fernandes-Alnemri T, Cochet E, Yu JW, Lackmy-Port-Lis M, et al. Mutations in NALP12 cause hereditary periodic fever syndromes. Proc Natl Acad Sci U S A 2008;105:1614–1619.
35. Jeru I, Hentgen V, Normand S, Duquesnoy P, Cochet E, Delwail A, et al. Role of interleukin-1beta in NLRP12-associated autoinflammatory disorders and resistance to anti-interleukin-1 therapy. Arthritis Rheum 2011;63:2142–2148.
36. Wang L, Manji GA, Grenier JM, Al-Garawi A, Merriam S, Lora JM, et al. PYPAF7, a novel PYRIN-containing Apaf1-like protein that regulates activation of NF-kappa B and caspase-1-dependent cytokine processing. J Biol Chem 2002;277:29874–29880.
37. Vladimer GI, Weng D, Paquette SW, Vanaja SK, Rathinam VA, Aune MH, et al. The NLRP12 inflammasome recognizes *Yersinia pestis*. Immunity 2012;37:96–107.
38. Arthur JC, Lich JD, Ye Z, Allen IC, Gris D, Wilson JE, et al. Cutting edge: NLRP12 controls dendritic and myeloid cell migration to affect contact hypersensitivity. J Immunol 2010;185:4515–4519.
39. Williams KL, Taxman DJ, Linhoff MW, Reed W, Ting JP. Cutting edge: Monarch-1: a pyrin/nucleotide-binding domain/leucine-rich repeat protein that controls classical and nonclassical MHC class I genes. J Immunol 2003;170:5354–5358.
40. Lich JD, Williams KL, Moore CB, Arthur JC, Davis BK, Taxman DJ, et al. Monarch-1 suppresses non-canonical NF-kappaB activation and p52-dependent chemokine expression in monocytes. J Immunol 2007;178:1256–1260.
41. Allen IC, Wilson JE, Schneider M, Lich JD, Roberts RA, Arthur JC, et al. NLRP12 suppresses colon inflammation and tumorigenesis through the negative regulation of noncanonical NF-kappaB signaling. Immunity 2012;36:742–754.
42. Khare S, Dorfleutner A, Bryan NB, Yun C, Radian AD, de Almeida L, et al. An NLRP7-containing inflammasome mediates recognition of microbial lipopeptides in human macrophages. Immunity 2012;36:464–476.
43. Normand S, Delanoye-Crespin A, Bressenot A, Huot L, Grandjean T, Peyrin-Biroulet L, et al. NOD-like receptor pyrin domain-containing protein 6 (NLRP6) controls epithelial self-renewal and colorectal carcinogenesis upon injury. Proc Natl Acad Sci U S A 2011;108:9601–9606.
44. Elinav E, Strowig T, Kau AL, Henao-Mejia J, Thaiss CA, Booth CJ, et al. NLRP6 inflammasome regulates colonic microbial ecology and risk for colitis. Cell 2011;145:745–757.
45. Anand PK, Malireddi RK, Lukens JR, Vogel P, Bertin J, Lamkanfi M, et al. NLRP6 negatively regulates innate immunity and host defence against bacterial pathogens. Nature 2012;488:389–393.
46. Fernandes-Alnemri T, Yu JW, Datta P, Wu J, Alnemri ES. AIM2 activates the inflammasome and cell death in response to cytoplasmic DNA. Nature 2009;458:509–513.
47. Hornung V, Ablasser A, Charrel-Dennis M, Bauernfeind F, Horvath G, Caffrey DR, Latz E, and Fitzgerald KA. AIM2 recognizes cytosolic dsDNA and forms a caspase-1-activating inflammasome with ASC. Nature 2009;458:514–518.
48. Roberts TL, Idris A, Dunn JA, Kelly GM, Burnton CM, Hodgson S, et al. HIN-200 proteins regulate caspase activation in response to foreign cytoplasmic DNA. Science 2009;323:1057–1060.

49. Rathinam VA, Jiang Z, Waggoner SN, Sharma S, Cole LE, Waggoner L, et al. The AIM2 inflammasome is essential for host defense against cytosolic bacteria and DNA viruses. Nat Immunol 2010;11:395–402.

50. Bagby GC, Jr. Interleukin-1 and hematopoiesis. Blood Rev 1989;3:152–161.

51. Dinarello CA. Interleukin-1 and interleukin-1 antagonism. Blood 1991;77:1627–1652.

52. Rider P, Carmi Y, Guttman O, Braiman A, Cohen I, Voronov E, et al. IL-1alpha and IL-1beta recruit different myeloid cells and promote different stages of sterile inflammation. J Immunol 2011;187:4835–4843.

53. Botelho FM, Bauer CM, Finch D, Nikota JK, Zavitz CC, Kelly A, et al. IL-1alpha/IL-1R1 expression in chronic obstructive pulmonary disease and mechanistic relevance to smoke-induced neutrophilia in mice. PLoS One 2011;6:e28457.

54. Kuida K, Lippke JA, Ku G, Harding MW, Livingston DJ, Su MS, et al. Altered cytokine export and apoptosis in mice deficient in interleukin-1 beta converting enzyme. Science 1995;267:2000–2003.

55. Gu Y, Kuida K, Tsutsui H, Ku G, Hsiao K, Fleming MA, et al. Activation of interferon-gamma inducing factor mediated by interleukin-1beta converting enzyme. Science 1997;275:206–209.

56. Schroder K, Hertzog PJ, Ravasi T, Hume DA. Interferon-gamma: an overview of signals, mechanisms and functions. J Leukoc Biol 2004;75:163–189.

57. Bohn E, Sing A, Zumbihl R, Bielfeldt C, Okamura H, Kurimoto M, et al. IL-18 (IFN-gamma-inducing factor) regulates early cytokine production in, and promotes resolution of, bacterial infection in mice. J Immunol 1998;160:299–307.

58. Kobayashi K, Kai M, Gidoh M, Nakata N, Endoh M, Singh RP, et al. The possible role of interleukin (IL)-12 and interferon-gamma-inducing factor/IL-18 in protection against experimental *Mycobacterium leprae* infection in mice. Clin Immunol Immunopathol 1998;88:226–231.

59. Wei XQ, Leung BP, Niedbala W, Piedrafita D, Feng GJ, Sweet M, et al. Altered immune responses and susceptibility to *Leishmania major* and *Staphylococcus* aureus infection in IL-18-deficient mice. J Immunol 1999;163:2821–2828.

60. Ohtsuki T, Micallef MJ, Kohno K, Tanimoto T, Ikeda M, Kurimoto M. Interleukin 18 enhances Fas ligand expression and induces apoptosis in Fas-expressing human myelomonocytic KG-1 cells. Anticancer Res 1997;17:3253–3258.

61. Hashimoto W, Osaki T, Okamura H, Robbins PD, Kurimoto M, Nagata S, et al. Differential antitumor effects of administration of recombinant IL-18 or recombinant IL-12 are mediated primarily by Fas-Fas ligand- and perforin-induced tumor apoptosis, respectively. J Immunol 1999;163: 583–589.

62. Dinarello CA, Simon A, van der Meer JW. Treating inflammation by blocking interleukin-1 in a broad spectrum of diseases. Nat Rev Drug Discov 2012;11:633–652.

63. Zeft AS, Spalding SJ. Autoinflammatory syndromes: fever is not always a sign of infection. Clevel Clin J Med 2012;79:569–581.

64. Lukens JR, Gross JM, Kanneganti TD. IL-1 family cytokines trigger sterile inflammatory disease. Front Immunol 2012;3:315.

65. Cassel SL, Eisenbarth SC, Iyer SS, Sadler JJ, Colegio OR, Tephly LA, et al. The Nalp3 inflammasome is essential for the development of silicosis. Proc Natl Acad Sci USA 2008;105:9035–9040.

66. Dostert C, Petrilli V, Van Bruggen R, Steele C, Mossman BT, Tschopp J. Innate immune activation through Nalp3 inflammasome sensing of asbestos and silica. Science 2008;320:674–677.

67. Hornung V, Bauernfeind F, Halle A, Samstad EO, Kono H, Rock KL, et al. Silica crystals and aluminum salts activate the NALP3 inflammasome through phagosomal destabilization. Nat Immunol 2008;9:847–856.

68. Boden G. Interaction between free fatty acids and glucose metabolism. Curr Opin Clin Nutr Metab Care 2002;5:545–549.

69. Wen H, Gris D, Lei Y, Jha S, Zhang L, Huang MT, et al. Fatty acid-induced NLRP3-ASC inflammasome activation interferes with insulin signaling. Nat Immunol 2011;12:408–415.

70. Vandanmagsar B, Youm YH, Ravussin A, Galgani JE, Stadler K, Mynatt RL, et al. The NLRP3 inflammasome instigates obesity-induced inflammation and insulin resistance. Nat Med 2011;17: 179–188.

71. Stienstra R, van Diepen JA, Tack CJ, Zaki MH, van de Veerdonk FL, Perera D, et al. Inflammasome is a central player in the induction of obesity and insulin resistance. Proc Natl Acad Sci USA 2011;108:15324–15329.

72. Zhou R, Tardivel A, Thorens B, Choi I, Tschopp J. Thioredoxin-interacting protein links oxidative stress to inflammasome activation. Nat Immunol 2010;11:136–140.

73. Duewell P, Kono H, Rayner KJ, Sirois CM, Vladimer G, Bauernfeind FG, et al. NLRP3 inflammasomes are required for atherogenesis and activated by cholesterol crystals. Nature 2010;464:1357–1361.

74. Schroder K, Zhou R, Tschopp J. The NLRP3 inflammasome: a sensor for metabolic danger? Science 2010;327:296–300.

75. Larsen CM, Faulenbach M, Vaag A, Volund A, Ehses JA, Seifert B, et al. Interleukin-1-receptor antagonist in type 2 diabetes mellitus. N Engl J Med 2007;356:1517–1526.

76. Ford ES, Li C, Cook S, Choi HK. Serum concentrations of uric acid and the metabolic syndrome among US children and adolescents. Circulation 2007;115:2526–2532.

77. Kingsbury SR, Conaghan PG, McDermott MF. The role of the NLRP3 inflammasome in gout. J Inflamm Res 2011;4:39–49.

78. Balkwill FR, Mantovani A. Cancer-related inflammation: common themes and therapeutic opportunities. Semin Cancer Biol 2012;22:33–40.

79. Lust JA, Lacy MQ, Zeldenrust SR, Dispenzieri A, Gertz MA, Witzig TE, et al. Induction of a chronic disease state in patients with smoldering or indolent multiple myeloma by targeting interleukin 1{beta}-induced interleukin 6 production and the myeloma proliferative component. Mayo Clin Proc 2009;84:114–122.

80. Okamoto M, Liu W, Luo Y, Tanaka A, Cai X, Norris DA, et al. Constitutively active inflammasome in human melanoma cells mediating autoinflammation via caspase-1 processing and secretion of interleukin-1beta. J Biol Chem 2010;285:6477–6488.

81. Qin Y, Ekmekcioglu S, Liu P, Duncan LM, Lizee G, Poindexter N, et al. Constitutive aberrant endogenous interleukin-1 facilitates inflammation and growth in human melanoma. Mol Cancer Res 2011;9:1537–1550.

82. Scheede-Bergdahl C, Watt HL, Trutschnigg B, Kilgour RD, Haggarty A, Lucar E, et al. Is IL-6 the best pro-inflammatory biomarker of clinical outcomes of cancer cachexia? Clin Nutr 2012;31: 85–88.

83. Langowski JL, Zhang X, Wu L, Mattson JD, Chen T, Smith K, et al. IL-23 promotes tumour incidence and growth. Nature 2006;442:461–465.

84. Drexler SK, Bonsignore L, Masin M, Tardivel A, Jackstadt R, Hermeking H, et al. Tissue-specific opposing functions of the inflammasome adaptor ASC in the regulation of epithelial skin carcinogenesis. Proc Natl Acad Sci USA 2012;109:18384–18389.

85. Krelin Y, Voronov E, Dotan S, Elkabets M, Reich E, Fogel M, et al. Interleukin-1beta-driven inflammation promotes the development and invasiveness of chemical carcinogen-induced tumors. Cancer Res 2007;67:1062–1071.

86. Bunt SK, Yang L, Sinha P, Clements VK, Leips J, Ostrand-Rosenberg S. Reduced inflammation in the tumor microenvironment delays the accumulation of myeloid-derived suppressor cells and limits tumor progression. Cancer Res 2007;67:10019–10026.

87. Terme M, Ullrich E, Aymeric L, Meinhardt K, Desbois M, Delahaye N, et al. IL-18 induces PD-1-dependent immunosuppression in cancer. Cancer Res 2011;71:5393–5399.

88. Allen IC, TeKippe EM, Woodford RM, Uronis JM, Holl EK, Rogers AB, et al. The NLRP3 inflammasome functions as a negative regulator of tumorigenesis during colitis-associated cancer. J Exp Med 2010;207:1045–1056.

89. Chen GY, Liu M, Wang F, Bertin J, Nunez G. A functional role for Nlrp6 in intestinal inflammation and tumorigenesis. J Immunol 2011;186:7187–7194.

90. Dupaul-Chicoine J, Yeretssian G, Doiron K, Bergstrom KS, McIntire CR, LeBlanc PM, et al. Control of intestinal homeostasis, colitis, and colitis-associated colorectal cancer by the inflammatory caspases. Immunity 2010;32:367–378.

91. Zaki MH, Boyd KL, Vogel P, Kastan MB, Lamkanfi M, Kanneganti TD. The NLRP3 inflammasome protects against loss of epithelial integrity and mortality during experimental colitis. Immunity 2010;32:379–391.

92. Zaki MH, Vogel P, Malireddi RK, Body-Malapel M, Anand PK, Bertin J, et al. The NOD-like receptor NLRP12 attenuates colon inflammation and tumorigenesis. Cancer Cell 2011;20:649–660.
93. Zitvogel L, Kepp O, Galluzzi L, Kroemer G. Inflammasomes in carcinogenesis and anticancer immune responses. Nat Immunol 2012;13:343–351.
94. Salcedo R, Worschech A, Cardone M, Jones Y, Gyulai Z, Dai RM, et al. MyD88-mediated signaling prevents development of adenocarcinomas of the colon: role of interleukin 18. J Exp Med 2010;207:1625–1636.
95. Takagi H, Kanai T, Okazawa A, Kishi Y, Sato T, Takaishi H, et al. Contrasting action of IL-12 and IL-18 in the development of dextran sodium sulphate colitis in mice. Scand J Gastroenterol 2003;38:837–844.
96. Zaki MH, Vogel P, Body-Malapel M, Lamkanfi M, Kanneganti TD. IL-18 production downstream of the Nlrp3 inflammasome confers protection against colorectal tumor formation. J Immunol 2010;185:4912–4920.
97. Ghiringhelli F, Apetoh L, Tesniere A, Aymeric L, Ma Y, Ortiz C, et al. Activation of the NLRP3 inflammasome in dendritic cells induces IL-1beta-dependent adaptive immunity against tumors. Nat Med 2009;15:1170–1178.

CHAPTER 9

ACTIVATION-INDUCED CYTIDINE DEAMINASE: AN INTRINSIC GENOME MODULATOR IN INFLAMMATION-ASSOCIATED CANCER DEVELOPMENT

Hiroyuki Marusawa and Tsutomu Chiba

Since Virchow's era it has been well recognized that inflammation plays important roles in cancer development. In particular, many cancers of digestive organs arise on the background of inflammation. These include *Helicobacter pylori*–induced gastric cancer, hepatitis C virus (HCV)- and hepatitis B virus (HBV)-related liver cancers, Barrett's adenocarcinoma, colitis-associated colorectal cancer, and cholangiocarcinoma accompanied by sclerosing cholangitis (PSC) (1,2). There appear many pathways that can lead to cancer development through inflammation. First, microorganisms such as *H. pylori*, HBV, and HCV may directly modulate cellular functions, giving the cells growth advantages and resistance to apoptosis (3,4). Inflammation also induces many mediators that appear to contribute to carcinogenesis, including various cytokines, chemokines, and growth factors (5,6). Moreover, cyclooxygenase 2 (COX-2) is produced under inflammatory conditions to promote tumorigenesis through various mechanisms, such as enhancement of angiogenesis (7). In addition, reactive oxygen species (ROSs) and reactive nitrogen species (RNSs) induced during inflammation have been shown to have mutagenic activity (8).

On the other hand, genetic changes such as mutation, deletion, and translocation are a hallmark of cancer. Indeed, cancers are derived from a clonal proliferation of the transformed cells caused by the accumulation of various genetic alterations in proto-oncogenes, tumor-suppressor genes, and other genes that control cell proliferation, regeneration, and apoptosis. Recent genome-wide analysis of human cancer tissues revealed that a single cancer cell generally possesses approximately 100 mutations

Cancer and Inflammation Mechanisms: Chemical, Biological, and Clinical Aspects, First Edition.
Edited by Yusuke Hiraku, Shosuke Kawanishi, and Hiroshi Ohshima.
© 2014 John Wiley & Sons, Inc. Published 2014 by John Wiley & Sons, Inc.

in coding regions, 10 to 20 of which, known as *driver genes*, contribute to cancer development (9,10). Because normal mutation rates cannot account for such multiple mutations in cancer cells, certain molecular mechanisms must be present to explain a large number of nucleotide alterations. One mechanism for the enhanced susceptibility to mutagenesis may be a defect in DNA repair systems. Indeed, dysfunction of the mismatch repair system results in familial colorectal cancer syndrome, and defects in the nucleotide excision repair system are also associated with colon cancer and skin cancer (11,12). However, the frequency of such defects in the DNA repair system is generally low in human cancers. Considering these facts, it is tempting to hypothesize that inflammation may enhance gene alterations during carcinogenesis.

Previous studies have identified many exogenous gene modulators and mutagens, and several possible intrinsic mutagens have also been proposed. Among these, ROS produced during inflammation are known to elicit mutations, particularly G-to-T transversions (12). However, a recent study has shown that G-to-T transversion accounts for only a minor proportion of the total mutations in human cancers; instead, C/G-to-T/A transition is the most prevalent mutation, especially in gastrointestinal malignancies (13). Thus, alternative mechanisms for mutagenesis during inflammation-associated carcinogenesis should be considered.

PHYSIOLOGICAL ROLE OF ACTIVATION-INDUCED CYTIDINE DEAMINASE

Generally, cells have several systems to prevent genetic alterations, including mutations. However, there is one type of cell in which somatic mutations and gene recombinations occur frequently under physiological conditions. That cell type is the B lymphocyte, in which immunoglobulin genes undergo somatic hypermutations and class-switch recombination to generate molecular diversity against many antigens. Muramatsu et al. first cloned the gene responsible for immunoglobulin class-switch recombination, in 1999 (14). They found that this gene is homologous with an apolipoprotein B RNA-editing enzyme catalytic polypeptide (APOBEC) family that inserts nucleotide alterations in target DNA or RNA through cytidine deamination, and named it activation-induced cytidine deaminase (AID). Interestingly, they found subsequently that this molecule is also responsible for somatic hypermutation of immunoglobulin genes (15). Notably, AID is the only enzyme known to induce DNA mutation in human genomes, although under normal conditions it is expressed only in B cells. Through its enzymatic activity, AID can deaminate C on the immunoglobulin gene to produce a uracil (U), and therefore turns a DNA C:G pair into a U:G mismatch. When DNA replication begins before recognition by the repair system, a U:G mismatch gives rise to C/G-to-T/A transition. Alternatively, recognition of a U:G mismatch by uracil–DNA–glycosylase (UNG) or mutS homolog 2 (MSH2)/mutS homolog 6 (MSH6) heterodimer induces mutations in the U:G mismatch or at the nearby A:T site (Figure 9.1). As a result, AID can induce any type of mutation (16). As described, however, AID basically induces C/G-to-T/A transitions by its cytidine deaminase activity. In this regard, it should be emphasized that a recent report on systemic sequencing of cancer genomes clearly demonstrated that the most prevalent

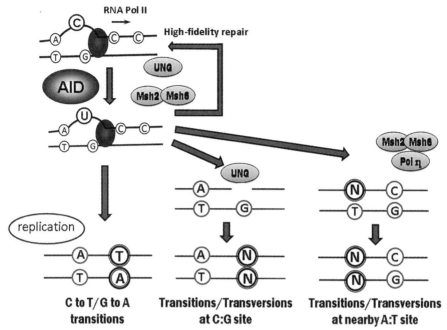

FIGURE 9.1   Mutation induction by AID. AID deaminates cytosine (C), resulting in the generation of a uracil (U) and therefore can transform a DNA C:G pair into a U:G mismatch. The AID-generated U:G mismatch can be recognized by UNG or MSH2/MSH6 heterodimer and repaired correctly. If DNA replication starts before recognition by the repair system, a U:G mismatch gives rise to C/G-to-T/A transition. Alternatively, generation of an abasic site by UNG or recognition of the U:G mismatch by the MSH2/MSH6 heterodimer induces any mutations in the AID-generated U:G mismatch or at a nearby A:T site, respectively, in an error-prone manner (indicated as N).

mutation pattern in human cancers is C/G-to-T/A transition, a pattern similar to that induced by AID (13).

After cloning of the *AID* gene, many investigators have found overexpression of *AID* in human lymphoid malignancies, suggesting involvement of AID in human lymphoma development (17,18). Indeed, AID has subsequently been shown to be responsible for the chromosomal breaks in *c-MYC* leading to a *c-MYC/immunoglobulin H* (IGH) translocation in B-cell lymphoma (19). Moreover, AID induces *breakpoint cluster region* (BCR)–*Abelson murine leukemia viral oncogene homolog 1* (ABL1) mutations leading to Imatinib resistance in chronic myeloid leukemia cells (20). Since the target of AID-mediated genotoxic effects was not restricted to immunoglobulin genes and a variety of other genes also received the AID-mediated mutations in B cells (16), it was not surprising that aberrant up-regulation of AID induced genetic alterations in various tumor-related genes, leading to the transformation of hematopoietic cells.

Honjo's group subsequently established AID transgenic mice, and as expected, they found that nearly 100% of these mice developed lymphomas (21). We then wanted to see whether these mice also develop cancers, because they expressed AID not only in lymphocytes but also in other cells, including epithelial cells. Interestingly, we found that in addition to lymphomas, these mice developed many types of cancers, including lung cancers, liver cancers, gastric cancers, and cholangiocarcinomas (21–23). These observations led us to speculate that AID may be involved in human carcinogenesis, although expression of AID was believed to be restricted to B cells.

ABERRANT EXPRESSION OF AID IN INFLAMMATORY TISSUES

Accordingly, we first examined whether AID is expressed in clinical specimens of liver tissues of patients with HCV infection, and surprisingly, we found strong expression of AID not only in liver cancers but also in chronic hepatitis tissues (24,25). AID expression was also found in both chronic gastritis mucosa and gastric cancer tissues of *H. pylori*–infected patients, and eradication of *H. pylori* reduced its expression (23,26). Later, aberrant AID expression was also demonstrated in the cholangioepithelium of patients with PSC (27), colonic mucosa and cancer tissues of patients with inflammatory bowel disease (28), and Barrett's epithelium and Barrett's cancers (29). All of these data suggested important roles of AID in inflammation-associated carcinogenesis in humans.

MECHANISMS FOR AID EXRESSION BY *H. PYLORI* AND HCV CORE PROTEIN

Then, to elucidate the mechanisms for AID expression in non-B cells, AID expression was examined using human hepatocytes and gastric cells *in vitro*. It was found that AID is induced by expression of HCV core protein or by *H. pylori* infection. Then, because AID expression in B cells was known to be dependent on NF-κB activation through CD40 ligation by T cells, and because both *H. pylori* and the core protein of HCV enhance NF-κB activation (4,30), the role of NF-κB in AID expression in epithelial cells was examined. As expected, introduction of the gene for the core protein of HCV into human hepatocytes induced AID expression via NF-κB activation (24). AID induction in human gastric cells by *H. pylori* infection was also dependent on NF-κB (23). Because *H. pylori* deficient for the Cag pathogenicity island (PAI) completely lost its ability to induce both NF-κB activation and AID expression (23), and *H. pylori* deficient for CagA reduced its ability by nearly 50%, it was considered that certain *H. pylori* factors other than CagA that are introduced into epithelial cells through the *H. pylori* type IV secretion machinery cause AID expression via NF-κB activation together with CagA. We observed further *in vitro* that *H. pylori* infection resulted in mutations in various genes, including *p53*, which could be inhibited by knockdown of endogenous AID using AID siRNA (23). Taken together, the following scenario may be illustrated: Both *H. pylori* and HCV infection generate gene mutations by inducing AID expression through NF-κB activation (Figures 9.2 and 9.3).

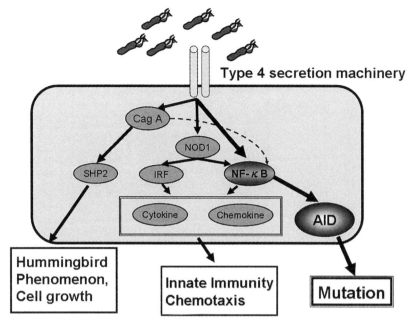

Type 4 secretion machinery

FIGURE 9.2 *H. pylori*–induced intracellular signaling events in gastric epithelial cells that are involved in gastric cancer development.

In addition to *H. pylori* and HCV core protein themselves, AID expression is also induced by IL-1β and TNF-α through NF-κB activation, by IL-4 and IL-13 through STAT6, and by TGF-β (25,28). Interestingly, these cytokines are known to enhance AID expression in B lymphocytes, facilitating class-switch recombination and somatic hypermutation of immunoglobulin genes. Thus, these cytokines appear to be involved in inflammation-associated cancer development by accelerating gene mutations. Another interesting point to be noted is that the mechanisms for AID expressions in B cells and epithelial cells have many similarities.

INDUCTION OF GENOMIC INSTABILITY BY AID

Recent studies have shown that in addition to mutation induction, AID is also responsible for chromosomal translocations through production of double-stranded DNA breaks in B-cell lymphomagenesis. Indeed, Nussenzweig's group (19,31) reported that AID is required for chromosomal breaks not only in *IgH* but also in *c-myc*, leading to *c-myc/IgH* translocations. In this connection, recent genome-wide studies have revealed that cancer cells possess a considerable number of genetic alterations, such as duplications, deletions, and translocations (32). Because these genetic alterations require DNA breaks and AID-induced uracil formation can cause DNA breaks through the creation of abasic sites by uracil–glycosylase, it appears reasonable to consider that ectopically expressed AID is involved not only in gene mutations but also in genetic alterations in cancer cells. Indeed, we recently found that AID

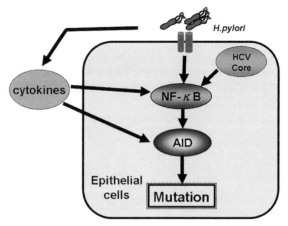

FIGURE 9.3 Induction of AID and gene mutations by *H. pylori* and HCV infection. Certain *H. pylori* factors that are introduced into epithelial cells through the *H. pylori* type IV secretion machinery, and HCV core protein could cause AID expression via NF-κB activation. *H. pylori* and HCV can also induce AID expression indirectly by enhancing production of various cytokines.

induces copy number changes as well as point mutation at various chromosomal loci, particularly in the tumor suppressor genes *CDKN2A* and *CDKN2B* in both human gastric mucosal cell lines and gastric mucosal cells of AID transgenic mice (33). Furthermore, we observed association of AID expression and frequent alteration of the *Cdkn2a–Cdkn2b* locus in both mouse and human *H. pylori*–infected gastric mucosa. These data suggest that AID is responsible for the accumulation of both somatic mutation and gene aberrations in *non-Ig* genes in non-B cells (33). Supporting such an idea, a recent study demonstrated involvement of AID in androgen receptor–dependent gene translocations in human prostate cancer cells (34).

LESSONS FOR CANCER RESEARCH IN THE STUDY OF AID

Roles of NF-κB in Carcinogenesis

Recent cancer studies have focused on the important roles of NF-κB in inflammation-associated carcinogenesis. Indeed, NF-κB in epithelial cells is activated not only by microorganisms through TLR or NLR stimulation but also by many cytokines, such as IL-1β and TNF-α. Moreover, NF-κB activation in blood cells accelerates inflammation through the production of various cytokines and chemokines. Thus, most investigators are interested in its growth-promoting and anti-apoptotic activity and its role in enhancing inflammation during cancer development. In addition, our studies demonstrated that NF-κB activation is involved in generating gene alterations by inducing AID expression in epithelial cells, suggesting a new role for NF-κB in inflammation-associated carcinogenesis (Figure 9.4).

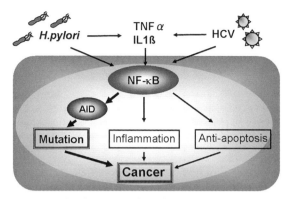

FIGURE 9.4 Important roles for NF-κB in inflammation-associated carcinogenesis. NF-κB is known to play roles in inflammation-associated carcinogenesis by further enhancing inflammation and also through its anti-apoptotic or growth-promoting action. Moreover, NF-κB contributes to inflammation-associated carcinogenesis by enhancing gene mutations through AID induction.

Different Spectra of Gene Mutations in Different Cancers

With the exception of *TP53*, the cancer genes that are frequently mutated differ for different types of cancers. For example, *k-RAS* mutations are found in the majority of pancreatic cancers, whereas few gastric cancers have *k-RAS* mutations (35). Moreover, although *APC* mutations are rarely found in hepatocellular carcinomas, many colon cancers possess this mutation. The reason for such distinct mutation patterns in different cancers is not clear. Because the importance of each gene as a tumor suppressor or as a tumor enhancer may be different in different cell types, such as pancreatic cells, gastric cells, and colonocytes, we may be able to find distinct spectra of mutated genes in cancer cells that have arisen from different tissues. In this regard, it is interesting to note that the genes targeted by AID are different in different cells (22). Indeed, induction of AID causes mutations at *TP53* and β-*catenin* but not at *c-MYC* in gastric cells, whereas it induces *c-MYC* mutations in hepatocytes and lymphocytes (23–25). In the case of immunoglobulin gene somatic hypermutation in B lymphocytes, AID recognizes a consensus sequence in the immunoglobulin gene to induce mutations. However, in addition to the immunoglobulin gene, *BCL6* is known to be an excellent target for AID, for reasons that are not clear at present. Unlike the immunoglobulin genes, we have so far been unable to find any consensus sequence in various cancer-related genes that is recognized by AID in epithelial cells. One mechanism may be AID targeting genes that are undergoing active transcription (36); however, we know that *BCL6* could be the target for AID not only in B cells but also in epithelial cells, in which it is not actively transcribed (33). Thus, other mechanisms by which AID recognizes specific target genes are also present. For example, we found that AID induction is associated with frequent deletion as well as mutation at the *Cdkn2b–Cdkn2a* locus, which codes p16, ARF, and p15 in both mice and humans (33). Interestingly, AID is required to introduce DNA single-stranded breaks into both rearranged *IgH* variable-region genes and the *CDKN2B* gene in leukemia cells, and frequent deletion of *CDKN2A* and *CDKN2B* was also reported in AID-positive lymphoid blast crisis leukemia cells (37). In support of these findings, we confirmed that the upstream sequences of the *Cdkn2b–Cdkn2a* locus carries E box motifs (CAGGTG), which are tightly associated with AID hypermutation activity at both *Ig* and *non-Ig* genes (20).

Relationship Between Mutation Induction and Repair Systems

As already mentioned, AID deaminates cytotsine to produce uracil, and after DNA replication the paired guanine is mutated to adenine, and after further DNA replication the original cytosine gets mutated to thymine; thus, eventually, a C/G-to-T/A transition develops (Figure 9.1). The importance of the induction of C/G-to-T/A mutation by AID may be supported by the fact that C/G-to-T/A transitions are the most prevalent mutation found in human gastrointestinal cancers (13). However, although C/G-to-T/A transitions are the most prevalent after AID induction, other mutations also develop (38). In agreement with these data, a range of mutations for which AID is entirely responsible do occur, even in immunoglobulin gene somatic hypermutation. The reason that AID induces different mutations is not completely clear at present. However, as described previously, it is possible that several repair systems, including mismatch repair and excision repair, become involved after cytosine deamination by AID, and it is important to note that these repair systems do not always provide high-fidelity repair (16,39). Thus, the ultimate mutation spectrum may be determined by a balance between the mutagenic activity of AID and the involvement of repair systems (16,40) (Figure 9.1).

Importance of Mutation Signature for Prediction of Mutagens

It is interesting to note that a mutation signature is different in different cancers, even in the same codon of the same gene, and also different in different ethnic groups, even in the same codon of the same gene of the same cancer. These differences in different cancers or in different ethnic groups suggest the involvement of different mutagens. For example, the mutation pattern of *p53* in hepatocellular carcinoma in China is almost exclusively G-to-T transversion, for which exposure to afflatoxin is believed to be responsible (41). However, C-to-T transition is most frequently observed in Japan, ruling out major involvement of afflatoxin in hepatocellular carcinogenesis in Japan. On the other hand, GGT-to-GTT transversion is most prevalent in *K-RAS* codon 12 mutation in lung cancer, probably due to cigarette smoking, whereas GGT-to-GAT transition is the major mutation pattern in pancreatic cancer (42), suggesting different mutagens between the two cancers. In this regard, as discussed above, one might take into account the fact that upon inflammation, AID induces mainly C/G-to-T/A transition. In any case, these differences in mutation signature might give us a hint as to what extent AID is involved in mutation induction of certain genes in the context of inflammation.

SUMMARY

The finding that AID expression is induced in nonlymphoid cells in various inflammatory conditions has not only proposed a new mechanism of inflammation-associated carcinogenesis but has also opened a new field of tumor biology. Recent studies have suggested the importance of both genetic and epigenetic changes in cancer development in the background of inflammation, and the relative importance of each is always a matter of discussion. Under such circumstances we believe that the study of AID is able to "reinforce" the importance of genetic changes in cancer development. Our recent finding that a deficiency of endogenous AID reduced the incidence of both accumulation of somatic mutations in the *Trp53* gene and the development of colitis-associated colorectal cancers further supports the critical role of AID in inflammation-associated cancer development via its ability to induce genetic alterations in tumor-related genes (43).

Finally, it is interesting to note that AID has recently been shown to be involved in active DNA demethylation during fetal development (44). In addition to aberrant DNA methylation of promoter CpG islands, cancer cells are characterized by global DNA hypomethylation as well as aberrant hypomethylation of oncogenes (45,46). Gastric mucosa infected by *H. pylori* displays global hypomethylation (47). Of note, when 5-methyl cytosine (5-mC) is deaminated by AID, the T yielded would subsequently be removed by either of the T:G mismatch-specific glycosylases: thymidine DNA glycosylase or methyl-CpG binding domain protein 4 (MBD4). The resulting abasic site would then be replaced by an unmethylated C via base excision repair processes, resulting in DNA demethylation. Notably, AID participates in active demethylation by 5-mC hydroxylase, 10,11-translocation 1 (TET1), and subsequent gene expression in the dentate gyrus of adult mouse brain (48). Thus, whether AID is involved in DNA demethylation during cancer development is an interesting topic for future studies (49).

REFERENCES

1. Chiba T, Marusawa H. A novel mechanism for inflammation-associated carcinogenesis; an important role of activation-induced cytidine deaminase (AID) in mutation induction. J Mol Med 2009;87:1023–1027.
2. Mantovani A, Allavena P, Sica A, Balkwill F. Cancer-related inflammation. Nature 2008;454: 436–444.
3. Higashi H, Nakaya A, Tsutsumi R, Yokoyama K, Fujii Y, Ishikawa S, et al. *Helicobacter pylori* CagA induces Ras-independent morphogenic response through SHP2 recruitment and activation. J Biol Chem 2004;279:17205–17216.
4. Marusawa H, Hijikata M, Chiba T, Shimotono K. Hepatitis C virus core protein inhibits Fas- and tumor necrosis facto alpha-mediated apoptosis via NF-κB activation. J Virol 1999;73:4713–4720.
5. Seno H, Satoh K, Tsuji S, Shiratsuchi T, Harada Y, Hamajima N, et al. Novel IL4 and IL1RN variations associated with non-cardia gastric cancer in Japan: comprehensive analysis of 207 polymorphisms of 11 cytokines. J Gastroenterol Hepatol 2007;22:729–737.
6. Tu S, Bhagat G, Cui G, Takaishi S, Kurt-Jones EA, Rickman B, et al. Overexpression of interleukin-1β induces gastric inflammation and cancer and mobilizes myeloid-derived suppressor cells in mice. Cancer Cell 2008;14:408–419.

7. Seno H, Oshima M, Ishikawa TO, Oshima H, Takaku K, Chiba T, et al. Cyclooxygenase 2 and prostaglandin E2 receptor EP2-dependent angiogenesis in Apc716mouse intestinal polyps. Cancer Res 2002;62:506–511.

8. Friedberg EC, McDaniel LD, Schultz RA. The role of endogenous and exogenous DNA damage and mutagenesis. Curr Opin Genet Dev 2004;14:5–10.

9. Sjöblom T, Jones S, Wood LD, Parsons DW, Lin J, Barber TD, et al. The consensus coding sequences of human breat and colorectal cancers. Science 2006;314:268–274.

10. Pleasance ED, Stephens PJ, O'Meara S, McBride DJ, Meynert A, Jones D, et al. A small-cell lung cancer genome with complex signatures of tobacco exposure. Nature 2010;463:184–190.

11. Hoeijmakers JH: Genome maintenance mechanisms for preventing cancer. Nature 2001;411:366–374.

12. Cleary SP, Cotterchio M, Jenkins MA, Kim H, Bristow R, Green R, et al. Germline MutY human homologue mutations and colorectal cancer: a multisite case–control study. Gastroenterology 2009;136:1251–1260.

13. Greenman C, Stephens P, Smith R, Dalgliesh GL, Hunter C, Bignell G, et al. Patterns of somatic mutation in human cancer genomes. Nature 2007;446:153–158.

14. Muramatsu M, Sankaranand VS, Anant S, Sugai M, Kinoshita K, Davidson NO, et al. Specific expression of activation-induced cytidine deaminase (AID), a novel member of the RNA-editing deaminase family in germinal center B cells. J Biol Chem 1999;274:18470–18476.

15. Muramatsu M, Sankaranand VS, Anant S, Sugai M, Kinoshita K, Davidson NO, et al. Class switch recombination and hypermutation require activation-induced cytidine deaminase (AID), a potential RNA editing enzyme. Cell 2000;102: 553–563.

16. Liu M, Duke JL, Richter DJ, Vinuesa CG, Goodnow CC, Kleinstein SH, et al. Two levels of protection for the B cell genome during somatic hypermutation. Nature 2008;451:841–845.

17. Greeve J, Philipsen A, Krause K, Klapper W, Heidorn K, Castle BE, et al. Expression of activation-induced cytidine deaminase in human B-cell non-Hodgkin lymphomas. Blood 2003;101:3574–3580.

18. Albesiano E, Messmer BT, Damle RN, Allen SL, Rai KR, Chiorazzi N. Activation-induced cytidine deaminase in chronic lymphocytic leukemia B cells: expression as multiple forms in a dynamic, variably sized fraction of the clone. Blood 2003;102:3333–3339.

19. Robbiani DF, Bothmer A, Callen E, Reina-San-Martin B, Dorsett Y, Difilippantonio S, et al. AID is required for the chromosomal breaks in *c-myc* that lead to *c-myc/IgH* translocations. Cell 2008;135:1028–1038.

20. Klemm L, Duy C, Iacobucci I, Kuchen S, von Levetzow G, Feldhahn N, et al. The B cell mutator AID promotes B lymphoid blast crisis and drug resistance in chronic myeloid leukemia. Cancer Cell 2009;16:232–245.

21. Okazaki IM, Hiai H, Kakazu N, Yamada S, Muramatsu M, Kinoshita K, et al. Constitutive expression of AID leads to tumorigenesis. J Exp Med 2003;197:1173–1181.

22. Morisawa T, Marusawa H, Ueda Y, Iwai A, Okazaki IM, Honjo T, et al. Organ-specific profiles of genetic changes in cancers caused by activation-induced cytidine deaminase expression. Int J Cancer 2008;123:2735–2740.

23. Matsumoto Y, Marusawa H, Kinoshita K, Endo Y, Kou T, Morisawa T, et al. *Helicobacter pylori* infection triggers aberrant expression of activation-induced cytidine deaminase in gastric epithelium. Nat Med 2007;13:470–476.

24. Endo Y, Marusawa H, Kinoshita K, Morisawa T, Sakurai T, Okazaki IM, et al. Expression of human activation-induced cytidine deaminase in human hepatocytes via NF-κB signaling. Oncogene 2007;26:5587–5595.

25. Kou T, Marusawa H, Kinoshita K, Endo Y, Okazaki IM, Ueda Y, et al. Expression of activation-induced cytidine deaminase in human hepatocytes during hepatocarcinogenesis. Int J Cancer 2007;120:469–476.

26. Nagata N, Akiyama J, Marusawa H, Shimbo T, Liu Y, Igari T, et al. Enhanced expression of activation-induced cytidine deaminase in human gastric mucosa infected by Helicobacter pylori and its decrease following eradication. J Gastroenterol 2013; (in press).

27. Komori J, Marusawa H, Machimoto T, Endo Y, Kinoshita K, Kou T, et al. Activation-induced cytidine deaminase links bile duct inflammation to human cholangiocarcinoma. Hepatology 2008;47:888–896.

28. Endo Y, Marusawa H, Kou T, Nakase H, Fujii S, Fujimori T, et al. Activation-induced cytidine deaminase links between inflammation to colitis-associated colorectal cancers. Gastroenterology 2008;135:889–898.

29. Morita S, Matsumoto Y, Okuyama S, Ono K, Kitamura Y, Tomori A, et al. Bile acid–induced expression of activation-induced cytidine deaminase during the development of Barrett's oesophageal adenocarcinoma. Carcinogenesis 2011;32:1706–1712,

30. Maeda S, Akanuma M, Mitsuno Y, Hirata Y, Ogura K, Yoshida H, et al. Distinct mechanism of *Helicobacter pylori*–mediated NF-κB activation between gastric cancer cells and monocytic cells. J Biol Chem 2001;276:44856–44864.

31. Dorsett Y, Robbiani DF, Jankovic M, Reina-San-Martin B, Eisenreich TR, Nussenzweig MC. A role for AID in chromosome translocations between c-myc and the IgH variable region. J Exp Med 2007;204:2225–2232.

32. Pleasance ED, Cheetham RK, Stephens PJ, McBride DJ, Humphray SJ, Greenman CD, et al. A comprehensive catalogue of somatic mutations from a human cancer genome. Nature 2010;463: 191–196.

33. Matsumoto Y, Marusawa H, Kinoshita K, Niwa Y, Sakai Y, Chiba T. Upregulation of activation-induced cytidine deaminase causes genetic aberrations at the *CDKN2b-CDKN2a* in gastric cancer. Gastroenterology 2010;139:1984–1994.

34. Lin C, Yang L, Tanasa B, Hutt K, Ju BG, Ohgi K, et al. Nuclear receptor-induced chromosomal proximity and DNA breaks underlie specific translocations in cancer. Cell 2009;139:1069–1083.

35. Bos JL. Ras oncogenes in human cancer: a review. Cancer Res 1989;49:4682–4689.

36. Shen HM, Poirier MG, Allen MJ, North J, Lal R, Widom J, et al. The activation-induced cytidine deaminase (AID) efficiently targets DNA in nucleosomes but only during transcription. J Exp Med 2009;206:1057–1071.

37. Feldhahn N, Henke N, Melchior K, Duy C, Soh BN, Klein F, et al. Activation-induced cytidine deaminase acts as a mutator in BCR-ABL1-transformed acute lymphoblastic leukemia cells. J Exp Med 2007;204;1157–1166.

38. Takai A, Toyoshima T, Uemura M, Kitawaki Y, Marusawa H, Hiai H, et al. A novel mouse model of hepatocarcinogenesis triggered by AID causing deleterious p53 mutations. Oncogene 2009;28: 469–478.

39. Wilson TM, Vaisman A, Martomo SA, Sullivan P, Lan L, Hanaoka F, et al. MSH2-MSH6 stimulates DNA polymerase η, suggesting a role for A:T mutations in antibody genes. J Exp Med 2005;201: 637–645.

40. Liu M, Schatz DG. Balancing AID and DNA repair during somatic hypermutation. Trends Immunol 2009;30:173–181.

41. Kensler ATW, Qian GS, Chen JG, Groopman JD. Translational strategies for cancer prevention in liver. Nat Rev Cancer 2003;3:321–329.

42. Tada M, Ohashi M, Shiratori Y, Okudaira T, Komatsu Y, Kawabe T, et al. Analysis of K-*ras* gene mutation in hyperplastic duct cells of the pancreas without pancreatic disease. Gastroenterology 1996;110:227–231.

43. Takai A, Marusawa H, Minaki Y, Watanabe T, Nakase H, Kinoshita K, et al. Targeting activation-induced cytidine deaminase prevents colon cancer development despite persistent colonic inflammation. Oncogene 2012;31:1733–1742.

44. Cortellino S, Xu J, Sannai M, Moore R, Caretti E, Cigliano A, et al. Thymine DNA glycosylase is essential for active DNA demethylation by linked deamination-base excision repair. Cell 2011; 146:67–79.

45. Feinberg AP, Tycko B. The history of cancer epigenetics. Nat Rev Cancer 2004;4:143–153.

46. Ehrlich M. DNA hypomethylation in cancer cells. Epigenomics 2009;1:239–259.

47. Ushijima T, Hattori N. Molecular pathways: Involvement of *Helicobacter pylori*–triggered inflammation in the formation of an epigenetic field defect, and its usefulness as cancer risk and exposure markers. Clin Cancer Res 2012;18:923–929.

48. Guo JU, Su Y, Zhong C, Ming GL, Song H. Hydroxylation of 5-methylcytosine by TET1 promotes active DNA demethylation in the adult brain. Cell 2011;145:423–434.

49. Fritz EL, Papavasiliou N. Cytidine deaminases: AIDing DNA demethylation? Genes Dev 2010;24:2107–2114.

CHAPTER 10

MicroRNA AND INFLAMMATION-RELATED CANCER

*Zhaojian Gong, Zhaoyang Zeng, Pranab Behari Mazumder,
Jian Ma, Ming Zhou, Xiayu Li, Xiaoling Li, Wei Xiong, Yong Li,
and Guiyuan Li*

Links between inflammation and cancers were first made in the nineteenth century, based on the observations that tumors often arose at sites of chronic inflammation and that inflammatory cells were present in tumors (1,2). Epidemiologic evidence suggests that approximately 25% of all human cancers worldwide may be caused by chronic inflammation, such as inflammatory bowel diseases for colon carcinoma, prostatitis for prostate cancer, *Helicobacter pylori* infection for gastric cancer (GC), hepatitis B or C virus (HBV/HCV) infection for hepatocellular carcinoma (HCC), Epstein–Barr virus (EBV) infection for nasopharyngeal carcinoma (NPC) and lymphomas, and so on (2–4). Conversely, inflammatory cells are found in the microenvironment of most, if not all, tumors, including those not causally related to an obvious inflammatory condition (such as breast cancer) (2,3,5). In the tumor microenvironment, inflammatory cells and molecules influence almost every aspect of cancer, including proliferation and survival of malignant cells, angiogenesis, metastasis, and response to chemotherapeutic agents (2,3,5). Thus, inflammation has been recognized as a hallmark of cancer (5).

MicroRNAs (miRNAs) are small (19 to 24 nucleotides (nt)) endogenous noncoding RNA molecules that control gene expression post-transcriptionally, either by degradation of target mRNAs or by inhibition of protein translation (6–12). Substantial evidence supports the finding that miRNAs not only play important roles in tumor initiation but also participate in the invasion and metastasis of numerous inflammation-related cancers (13–19). Investigation of the interrelation between miRNAs and inflammation-related cancers may provide novel preventive, diagnostic, and therapeutic strategies for cancers.

Cancer and Inflammation Mechanisms: Chemical, Biological, and Clinical Aspects, First Edition.
Edited by Yusuke Hiraku, Shosuke Kawanishi, and Hiroshi Ohshima.

BIOGENESIS AND FUNCTION OF miRNAs

Since first discovered in 1993 (20), miRNAs have attracted wide attention due to their unique functional significance and modes of action. To date, more than 1000 human miRNAs have been identified. miRNA genes are transcribed into long primary miRNAs (pri-miRNAs) by RNA polymerase II or, in some cases, by RNA polymerase III (21) (Figure 10.1). Pri-miRNAs are subsequently processed into smaller, stem-looped, hairpin-like miRNA precursors (pre-miRNAs) about 70 nt in length by the RNase III–type enzyme Drosha, which forms a microprocessor complex with the double-stranded RNA binding protein DGCR8. Afterward, pre-miRNAs are exported from the nucleus across the nuclear membrane into the cytoplasm through an Exportin-5/Ran complex. In the cytoplasm, pri-miRNAs are cleaved by the highly conserved RNase III–type enzyme Dicer to generate a 19- to 24-nt RNA duplex. Then one strand, the guide strand, is incorporated into the RNA-induced silencing complex (RISC), while the other strand, miRNA* (also known as the antiguide or passenger strand), is degraded. Although the biosynthesis of most miRNAs is through the Drosha pathway, a Drosha-independent pathway has also been found recently (22,23). miRNAs produced without Drosha-mediated cleavage are called *mirtrons* (Figure 10.1).

By complementary binding to the 3′ untranslated regions (UTRs) of their target mRNAs, miRNAs regulate the expression of many key genes that are involved in cell proliferation, differentiation, apoptosis, and development as well as other biological processes. It is estimated that over 30% of all genes and the majority of genetic pathways are regulated by miRNAs (24,25). Mounting evidence indicates that abnormal expression of specific miRNAs is related to a broad range of human diseases, especially cancers (10,19,26).

One of the earlier studies that established the role of miRNAs in cancers came with the observation that two miRNA genes, *miR-15a* and *miR-16a*, located at chromosome 13q14, were deleted in the majority of B-cell chronic lymphocytic leukemia (B-CLL) patients (27). Following this observation, many other miRNAs in various cancers have been found to be dysregulated, based primarily on differential miRNA profiling between cancer and normal tissue or from genetic screens with miRNA expression libraries (28). miRNAs have been proposed to contribute to oncogenesis because they can function either as tumor suppressors (as is the case for the *miR-15a*, *miR-16-1*, and *miR-34* family) (29,30) or oncogenes (as is the case for *miR-155*, *miR-372*, *miR-373* and members of the *miR-17-92* cluster) (31–33).

miRNAs AS MEDIATORS OF INFLAMMATION-INDUCED CANCINOGENESIS

It has generally been accepted that inflammation plays a critical role in carcinogenesis, and an inflammatory microenvironment is considered an essential component of almost all solid tumors (2,34). Many types of inflammation contribute to cancers, including inflammation caused by chronic infection by virus, bacteria, or parasites,

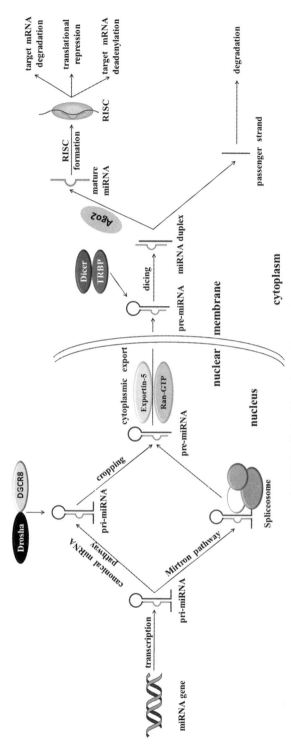

FIGURE 10.1 The miRNA biogenesis pathway.

chronic stimulation (such as chemical, physical, or metabolic), and auto-inflammation (such as inflammatory bowel diseases, ulcerative colitis, or chronic pancreatitis) (15,35). Activated inflammatory cells generate reactive oxygen species (ROSs) and reactive nitrogen species (RNSs) that are capable of inducing DNA damage and genomic instability (36). Free radicals coming from the inflammatory process may result in inactivation or repression of DNA repair genes, increase the expression of DNA methyltransferases, and lead to global genome hypermethylation (15). In addition, the proinflammatory stimuli, such as tumor necrosis factor (TNF), interleukin-1 (IL-1), IL-6, and chemokines lead to increased levels of cyclooxygenase 2 (COX-2) and 5-lipoxygenase (5-LOX), which contribute to tumorigenesis by augmenting ROS production (37,38).

Inflammatory signals lead to altered miRNA expression. Aberrant miRNA expression patterns have been found in some inflammatory conditions or diseases that are associated with increased risk of cancer. For example, differential expression of 11 miRNAs was detected in active ulcerative colitis associated with colon carcinoma. The expression of three miRNAs (miR-192, miR-375, and miR-422b) was decreased significantly, and expression of eight miRNAs (miR-16, miR-21, miR-23a, miR-24, miR-29a, miR-126, miR-195, and Let-7f) was increased significantly (39). Primary biliary cirrhosis is a chronic inflammatory autoimmune condition of the bile duct that carries an increased risk of liver cancer. A study confirmed that miR-122a and miR-26a were reduced and miR-328 and miR-299-5p were increased in primary biliary cirrhosis (40). However, these miRNAs are thought to regulate cell proliferation, apoptosis, inflammation, and oxidative stress, indicating that the alterations of these miRNAs may contribute to an inflammatory condition (15).

In pre-malignant cells, nuclear factor-κB (NF-κB), signal transducer and activator of transcription 3 (STAT3), or other transcription factors are activated by cytokines produced from tumor-infiltrating immune cells. These transcription factors then regulate a battery of genes, including miRNAs, that are critical to innate and adaptive immunity, and control numerous pro-tumorigenic processes, such as survival, proliferation, growth, and angiogenesis, which is crucial to tumor growth (41).

NF-κB is a transcriptional regulator consisting of reticuloendotheliosis (Rel) protein dimers that bind a DNA sequence motif known as the κB site. During the past decade, much accumulated evidence has supported the role of NF-κB in linking inflammation and tumorigenesis. Several miRNAs, such as miR-21 (42), miR-34a (43), miR-143 (44), miR-146 (45), and miR-301a (18), were induced by NF-κB (8). Furthermore, NF-κB-responsive miR-155 and miR-125b have a function in innate immune response. These miRNAs also target scores of genes encoding NF-κB, regulators, and effectors in the NF-κB signaling network, with the vast majority of them participating in positive or negative feedback loops. For example, miR-146a and miR-146b down-regulate IL-1 receptor–associated kinase 1 (IRAK1) and TNF receptor associated factor 6 (TRAF6) protein levels, which demonstrate miRNA's regulatory roles in the NF-κB pathway. Most recently, miR-199a is shown to regulate IKKβ, a known modulator of the tumor inflammatory microenvironment (Figure 10.2).

miR-21, which is located in a fragile region of chromosome 17q23.2, FRA17B, has been validated as a oncogene. Increased expression of miR-21 is found in several chronic inflammatory diseases, including ulcerative colitis, *H. pylori*

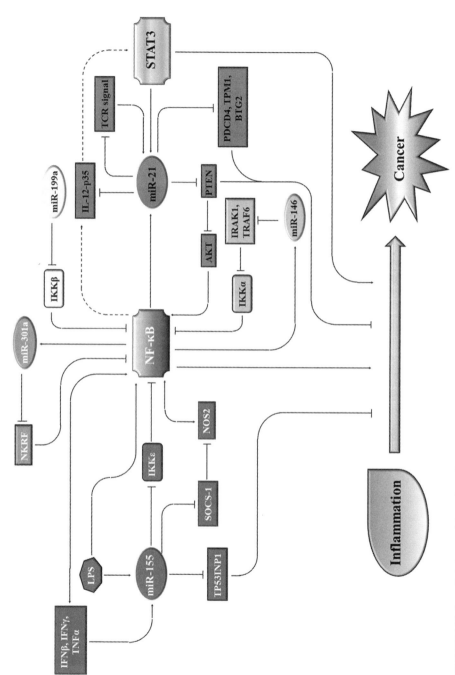

FIGURE 10.2 miRNAs participate in inflammation-induced cancinogenesis. (*See insert for color representation of the figure.*)

infection–related gastritis, and others (39,46). It has been shown that miR-21 can be induced by inflammatory stimuli: A pro-inflammatory cytokine, IL-6, induces the expression of miR-21 in an NF-κB- and STAT3-dependent manner (47). Upon T-cell receptor (TCR) stimulation, miR-21 induction is believed to be involved in a negative feedback loop regulating TCR signaling (48). The elevated levels of miR-21 in these inflammatory diseases may be partially responsible for the carcinogenesis. Several tumor suppressor genes have been identified as the targets of miR-21, such as programmed cell death 4 (PDCD4) (49), tropomyosin 1 (TPM1) (50), phosphatase and tensin homolog (PTEN) (51), and BTG family member 2 (BTG2) (52). Overexpression of miR-21 increases cell proliferation and inhibits apoptosis, whereas the inhibition of miR-21 causes tumor regression in xenograft models (53). Recently, IL-12–p35 was found to be regulated negatively by miR-21 in mouse models (54). In addition, the expression of miR-21 correlated negatively with IL-12–p35 and positively with IL-6 in human colon cancer tissues (55). This suggests that the connection between IL-6, IL-12–p35 and miR-21 plays a key role in human colon cancer and that miR-21 may contribute to inflammation-induced carcinogenesis (Figure 10.2).

miR-155, encoding within a region known as B-cell integration cluster (BIC) and identified originally as a frequent integration site for avian leucosis virus, is also an oncogenic miRNA (31). The pro-inflammatory conditions, including *H. pylori* and EBV (56) infections, as well as LPS treatment, lead to increased miR-155 expression (31,56–58). Inflammatory mediators such as TNF-α and IFN-β also stimulate the expression of miR-155 (59,60). miR-155 targets a key inhibitor of the inflammatory process, the suppressor of cytokine signaling 1 (SOCS-1) protein, demonstrating that it is a mediator of inflammatory signaling (61). It has been shown that the pro-apoptotic gene tumor protein 53-induced nuclear protein 1 (TP53INP1), which is a downstream gene of p53 signaling, is inhibited by miR-155 (62). The suppression of TP53INP1 is a likely mechanism for pro-tumorigenic functions of miR-155 and a possible mediator of inflammation-induced carcinogenesis. Increased miR-155 expression is found in the bone marrow of leukemic patients, and overexpression of miR-155 in mouse models causes hyperproliferation of B cells, a common hallmark of leukemia and lymphoma (63) (Figure 10.2).

Recently, miR-663 has attached attention due to its role not only as an anti-inflammatory miRNA but also as a tumor suppressor miRNA. The expression of miR-663 is lost in certain cancers, such as gastric or pancreatic cancer, and it induces mitotic catastrophe growth arrest when its expression is restored in these cells (64,65). Interestingly, accumulating evidence reveals an important role of miRNAs in drug resistance, and miRNA expression profiling can be used as a marker for distinguishing cancers that are resistant to certain drugs (66,67).

miRNAs INVOLVEMENT IN THE METASTASIS OF INFLAMMATION-RELATED CANCERS

Metastasis is a critical aspect of carcinogenesis, since most cancer mortality is associated with metastatic tumors rather than the primary tumor (68). A multistep process is included in cancer metastasis, for example, primary tumor cells invade adjacent tissue, enter the systemic circulation (intravasate), translocate through the vasculature, are arrested in distant capillaries, extravasate into the surrounding tissues, and finally, proliferate from initial microscopic growths (micrometastases) into macroscopic secondary tumors (69).

An increasing number of pro- and anti-metastatic miRNAs have been reported to regulate the metastasis of many inflammation-related cancers (16). miR-10b, a pro-metastatic miRNA, has been identified to promote metastasis in breast cancer (70). Ma et al. (70) found that miR-10b has invasive and metastatic properties but does not effect cell proliferation. RHOC, a well-characterized pro-metaststic Rho-family GTPase, is a downstream target of Homeobox D10 (HOXD10) (71), a direct target of miR-10b. Therefore, miR-10b promotes cancer metastasis through direct repression of the HOXD10 transcription factor and subsequent activation of the pro-metastasis gene, *RHOC*. In addition to the important role in tumor formation, miR-21 is also considered as a pro-metastatic miRNA and has a marked effect on tumor metastasis. Studies have shown that miR-21 can stimulate cell invasion and metastasis in several inflammation–cancer models both *in vitro* and *in vivo*, as in breast and colon cancer (72,73). Several targets of miR-21 are metastasis suppressors, including PDCD4, maspin, PTEN, and TPM1 (51,72,73). Among these targets, PDCD4 and maspin are known to suppress invasion of tumor cells, and maspin also correlates with improved prognosis (73–75). Furthermore, the expression of urokinase-type plasminogen activation receptor (uPAR), which breaks down the extracellular matrix (ECM) and hence promotes invasion and metastasis and correlates with poor prognosis, is regulated negatively by both PDCD4 and maspin (76). Tissue inhibitor of metalloprotease-3 (TIMP3) is an inhibitor of matrix metalloproteinases (MMPs). A recent study has shown that miR-21 inhibits the expression of TIMP3 and thereby de-represses MMPs activity and increases cancer cell invasion (77). Recently, the regulation pathway between NF-κB, miR-143, and the fibronectin type III domain containing 3B (FNDC3B) has been implicated in the promotion of the metastasis of HBV–related HCC (44). In this pathway mediated by NF-κB, miR-143 was regulated directly by NF-κB to repress FNDC3B gene expression. Down-regulation of FNDC3B enhanced the invasion and migration capability of hepatocarcinoma cells. let-7, a tumor-suppressing miRNA family, inhibits metastasis. The observation that let-7 was reduced in an inflammation-related cancer, lung cancer, was the first direct evidence for a role of let-7 in cancer (78). Several target genes of let-7, high mobility group A2 (HMGA2), RAS, and MYC, contribute to the epithelial–mesenchymal transition (EMT) through the RAS–MEK pathway (16,79). miR-146, which is involved in inflammation response via inflammatory pathways (such as NF-κB) and inflammatory cytokines (such as IL-6 and IL-8), is another antimetastatic miRNA. Studies revealed that miR-146 suppresses tumor metastasis through inhibition of both migration and invasion in an inflammation-related cancer, breast cancer (80,81) (Figure 10.3).

EMT, which describes the molecular reprogramming and phenotypic changes characterizing the conversion of polarized immotile epithelial cells to motile mesenchymal cells, is a prerequisite for cancer metastasis (82). Loss of tumor suppressor gene E-cadherin and activation of Vimentin promote EMT (16). Several transcription factors of the zinc finger family, such as Snail1, Slug, ZEB1, and ZEB2, are

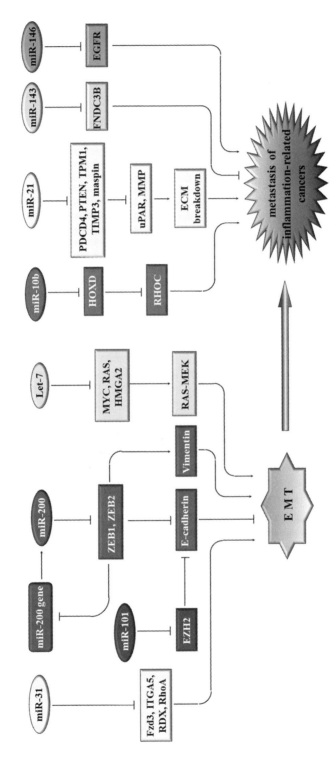

FIGURE 10.3 miRNAs involvement in the metastasis of inflammation-related cancers.

found to regulate this process (83). Recent progress demonstrates that miRNAs also participated in the regulation of EMT and that the miR-200 family (miR-200a, miR-200b, miR-200c, miR-141, and miR-429) has emerged as a key regulator of EMT (76,84,85). The ZEBs family and miR-200 family formed a double-negative feedback loop. On one hand, the ZEB1 and ZEB2 enter the nucleus to inhibit the transcription of miR-200 genes; on the other hand, the expression of ZEBs is directly repressed by the miR-200 family. In addition, the ZEBs activate Vimentin and inactivate E-cadherin. Therefore, the miR-200 family inhibits the ZEB1 and ZEB2 to prevent the EMT and metastasis by activating E-cadherin and inactivating Vimentin. Beyond miR-200 family, miR-101 also plays a major role in EMT through regulation of the enhancer of the zeste homolog (EZH2). EZH2 is a histone methyl transferase which mediates transcriptional silencing of the E-cadherin by trimethylation of H3 lysine 27 (86). It is up-regulated in several inflammation-related cancers, such as those of prostate, breast, and bladder. *miR-101* loss in metastatic prostate cancer leads to up-regulation of EZH2 (87). It has been verified that miR-31, which targets several pro-metastatic genes (such as Fzd3, ITGA5, M-RIP, MMP16, RDX, and RhoA), also regulates EMT and inhibits the metastasis of an inflammation-related cancer, breast cancer (16,88). Among the targets of miR-31, overexpression of Fzd3, ITGA5, RDX, or RhoA reverses, at least partially, miR-31-dependent metastasis-relevant phenotypes (88). Therefore, miR-31 impedes metastasis through post-transcriptionally silencing Fzd3, ITGA5, RDX, and RhoA (88) (Figure 10.3).

SINGLE-NUCLEOTIDE POLYMORPHISMS AMONG miRNAs AND INFLAMMATION-RELATED CANCER SUSCEPTIBILITY

Single-nucleotide polymorphisms (SNPs) are the most common variations in DNA sequences. They are responsible for the diversity among individuals, genome evolution, interindividual differences in drug response, and are associated with many diseases (89). Recent studies have suggested that SNPs in an miRNA sequence, miRNA binding sites of target genes, and miRNA biogenesis-related genes could potentially affect the structure or the expression level of mature miRNAs and disrupt miRNA–target interaction, which affects miRNA-mediated gene regulation (90–93). These miRNA-related SNPs play a substantial role in much inflammation-related cancer development, including HCC (94), lung cancer (95), and colorectal cancer (96).

SNPs in miRNA genes were reported to be involved in the alteration of miRNA processing and affect cancer susceptibility. miR-146a, elevated in several cancers, is oncogenic and induced by pro-inflammatory signals (97). A miR-146a SNP (rs2910164) has been studied thoroughly. This SNP reduced both the amount of pre-miR-146a and mature miR-146a, affected the Drosha/DGCR8 cropping step, and was associated with the risk of HCC, breast cancer, ovarian cancer, prostate cancer, and papillary thyroid carcinoma (90,94,98–100). miR-196a is another oncogenic miRNA regulating cell proliferation. Studies reported that a SNP rs11614913 in miR-196a-2 is associated with lung cancer prognosis and increased lung cancer risk (95,101).

SNPs in miRNA seed sequence or the binding sites of their targets may disrupt miRNA–target interaction, resulting in the deregulation of target gene expression, thereby contributing to cancer susceptibility (102). let-7, the first conserved miRNA identified, is a tumor-suppressing miRNA regulating the expression of a proto-oncogene KRAS. A SNP in one of the let-7-binding sites of KRAS, which inhibit the regulation of KRAS by let-7 and consequently lead to an overexpression of KRAS, has been identified to be associated with increased lung cancer risk (91). Another study showed that SNPs in miRNA-binding sites of nucleotide excision repair (NER) genes affect the DNA repair capacity and are important for modulating colorectal cancer risk (103). SNPs in miRNA-binding sites of other genes have also been reported to be associated with cancer risk (102,104).

Many genes have important roles in the processes of miRNA transcription and post-transcriptional maturation, such as Drosha, Dicer, DGCR8, Exportin-5, and Ran-GTP (21). SNPs in miRNA biogenesis-related genes may affect the processing of miRNAs, and are associated with cancer risk. GEMIN3 and GEMIN4 are core components of a large macromolecular complex that plays an essential role in pre-miRNA splicing and ribonucleoprotein assembly. Recent studies have identified that SNPs in GEMIN3 and GEMIN4 are associated with bladder and renal cell carcinoma risk (105).

VIRUS-ENCODED miRNAs AND INFLAMMATION-RELATED CANCERS

Since the late 1960s, six different human viruses—EBV, human papilloma virus (HPV), HBV, HCV, human T-cell lymphotrophic virus (HTLV) and Kaposi's sarcoma–associated virus (KSHV)—have been found to be associated with 10 to 15% of cancers (106). Among these cancer-related viruses, EBV and KSHV can encode miRNAs. Like the miRNAs of host cells, viral miRNAs are also involved in the development of inflammation-related cancers.

EBV, which is B lymphtropism, was the first virus identified to be associated with human cancer (107). Multiple human cancers are confirmed to be associated with EBV infection, including NPC, GC, Burkitt's lymphoma (BL), Hodgkin's B-cell lymphoma (HD), and post-transplantation lymphoproliferative disorders (PTLD) (108,109). The identification of five miRNAs in the B95-8 strain of EBV was the first proof supporting the existence of virus-encoded miRNAs (110). At present, a total of 25 EBV miRNA precursors with 44 mature miRNAs were known and mapped to the BHRF1 and BART regions of the EBV genome (111,112). In NPC, an EBV-associated carcinoma, the BART miRNAs have been implicated in regulating both viral and host targets for facilitating its latent infection and for furtively giving consistent stress to induce oncogenic signaling (112). miR-BART2 promotes the entry of the virus to latency through repressing the EBV DNA polymerase BALF5 expression (113). miR-BHRF1-3 shields EBV-infected B cells from cytotoxic T cells by targeting the interferon-induced CXCL11 (114). miR-BART2-5p helps virus escape from the recognition and consequent elimination by natural killer (NK) cells through reducing the expression of a stress-induced NK cell ligand, MICB (115). Another miRNA, miR-BART5, suppresses the apoptosis of virus-infected host cells through down-regulating the host cell gene *p53 up-regulated modulator of apoptosis* (PUMA) (116). EBV latent membrane proteins 1 and 2A (LMP1/2A) are critical to NPC development through their impact on cellular gene expression and cellular growth (117). Three EBV miRNAs—miR-BART1-5p, miR-BART16, and miRBART17-5p—down-regulate LMP1 expression (118), while miR-BART22 inhibits the expression of LMP2A (119). In addition, a recent report revealed a mechanism on host–virus interaction via miRNAs which supports the concept that miRNAs are not only involved in NPC pathogenesis but are also of great potential as biomarkers for clinical screening in a "liquid biopsy" because serological viral miRNAs are positively correlated with cellular EBV miRNAs (120).

KSHV, also known as human herpes virus 8 (HHV-8), was first isolated from Kaposi's sarcoma (KS) lesions in 1994 (121). Since then, KS and two other cancers, multicentric Castleman's disease (MCD) and primary effusion lymphoma (PEL), have been identified as being caused by KSHV (122). So far, 18 mature miRNAs, encoded by 12 pre-miRNAs, have been identified in the KSHV genome (123–125). Several of these miRNAs have been implicated in the pathogenesis of KSHV-induced malignancies. IL-6 and IL-10, secreted by KSHV-infected cells and other cells within the tumor microenvironment, collectively promote the growth of KSHV-infected tumor cell angiogenesis and suppress T-cell activation (126–129). A recent study has demonstrated that KSHV-miRNAs induce IL-6 and IL-10 secretion by murine macrophages and human myelomonocytic cells (130). This function is partially accomplished by miR-K12-3/7 through repression of a dominant-negative isoform of C/EBPβ (130), a transcriptional repressor of IL-6 and IL-10. miR-K12-10 reduces tumor necrosis factor-like weak inducer of apoptosis receptor (TWEAKR)-induced apoptosis through repressing TWEAKR expression (131). Cellular cyclin-dependent kinase inhibitor p21 is a key inducer of cell cycle arrest. miR-K12-1 targets p21 and attenuates cell cycle arrest induced by p53 activation (132). In addition, miR-K12-11 encoded by KSHV shows significant homology to cellular miR-155, which is an oncogenic miRNA. These two miRNAs share the same "seed" sequence (133). miR-K12-11 is demonstrated to function as an ortholog of cellular miR-155 (134). Like cellular miRNAs, SNPs within pri-miRNAs, pre-miRNAs, or mature miRNAs of KSHV have also been detected, some of which may affect mature miRNA processing and be associated with KS risk (135).

Overall, only a limited number of viral miRNAs has been characterized. The increasing knowledge of host–virus interaction via miRNAs will shed light on the underlying mechanisms of the formation and progression of virus infection–related tumors.

FUTURE PROSPECTS

It is gradually accepted that inflammation plays a role in many aspects of cancer, including tumorigenesis, invasion, and metastases. Studies about the molecular mechanisms linking inflammation and cancer poise new challenges to understand the fundamental biology of cancer and to develop strategies to treat and cure cancer. Burgeoning evidence is available to support the belief that miRNAs, which function as either oncogenes or tumor suppressor genes, have a substantial role in tumor initiation and progression of inflammation-related cancers. An increasing number of attempts have been made to establish miRNA expression signatures to classify tumors with higher accuracy. Unique expression profiles have been found in ovarian carcinoma (136), pancreatic cancer (137), breast cancer (138), HCC (139), chronic lymphocytic leukemia (140), and so on. In addition, miRNA profiles of human sera have been established for prostate cancer, colorectal cancer, lung cancer, and ovarian cancer (141). The number of miRNAs is much smaller than that of protein-coding genes in the human genome, yet they play specific functions in various biological processes via the regulation of many key genes. Therefore, the development of miRNA-based therapy with either miRNA inhibitors or miRNA mimics is practicable. Lentiviral vectors have been applied to deliver miRNAs for gene therapy of inflammation-related cancers. In the last decade, the knowledge "tree" of miRNAs has grown so tremendously that it will not be long until we see it bear "fruit": miRNA-based diagnostic tools, prognositic markers, and therapeutic agents.

REFERENCES

1. Balkwill F, Mantovani A. Inflammation and cancer: Back to Virchow? Lancet 2001;357:539–545.
2. Mantovani A, Allavena P, Sica A, Balkwill F. Cancer-related inflammation. Nature 2008;454: 436–444.
3. Colotta F, Allavena P, Sica A, Garlanda C, Mantovani A. Cancer-related inflammation, the seventh hallmark of cancer: links to genetic instability. Carcinogenesis 2009;30:1073–1081.
4. Hussain SP, Harris CC. Inflammation and cancer: an ancient link with novel potentials. Int J Cancer 2007;121:2373–2380.
5. Mantovani A. Cancer: Inflaming metastasis. Nature 2009;457:36–37.
6. Kumar M, Lu Z, Takwi AA, Chen W, Callander NS, Ramos KS, et al. Negative regulation of the tumor suppressor p53 gene by microRNAs. Oncogene 2011;30:843–853.
7. Bartel DP. MicroRNAs: genomics, biogenesis, mechanism, and function. Cell 2004;116:281–297.
8. Ma X, Becker Buscaglia LE, Barker JR, Li Y. MicroRNAs in NF-kappaB signaling. J Mol Cell Biol 2011;3:159–166.
9. Lu Z, Liu M, Stribinskis V, Klinge CM, Ramos KS, Colburn NH, et al. MicroRNA-21 promotes cell transformation by targeting the programmed cell death 4 gene. Oncogene 2008;27: 4373–4379.
10. Ma X, Kumar M, Choudhury SN, Becker Buscaglia LE, Barker JR, Kanakamedala K, et al. Loss of the miR-21 allele elevates the expression of its target genes and reduces tumorigenesis. Proc Natl Acad Sci USA 2011;108:10144–10149.
11. Takwi A, Li Y. The p53 pathway encounters the microRNA world. Curr Genomics 2009;10:194–197.
12. Buscaglia LE, Li Y. Apoptosis and the target genes of microRNA-21. Chin J Cancer 2011;30: 371–380.
13. Sonkoly E, Pivarcs A. microRNAs in inflammation. Int Rev Immunol 2009;28:535–561.
14. Gramantieri L, Fornari F, Callegari E, Sabbioni S, Lanza G, Croce CM, et al., MicroRNA involvement in hepatocellular carcinoma. J Cell Mol Med 2008;12:2189–2204.
15. Schetter AJ, Heegaard NH, Harris CC. Inflammation and cancer: interweaving microRNA, free radical, cytokine and p53 pathways. Carcinogenesis 2010;31:37–49.
16. Zhang H, Li Y, Lai M. The microRNA network and tumor metastasis. Oncogene 2010;29:937–948.

17. Jiang S, Zhang HW, Lu MH, He XH, Li Y, Gu H, et al. MicroRNA-155 functions as an OncomiR in breast cancer by targeting the suppressor of cytokine signaling 1 gene. Cancer Res 2010;70:3119–3127.

18. Lu Z, Li Y, Takwi A, Li B, Zhang J, Conklin DJ, et al. miR-301a as an NF-kappaB activator in pancreatic cancer cells. EMBO J 2011;30:57–67.

19. Zhang L, Deng T, Li X, Liu H, Zhou H, Ma J, et al. MicroRNA-141 is involved in a nasopharyngeal carcinoma-related genes network. Carcinogenesis 2010;31:559–566.

20. Lee RC, Feinbaum RL, Ambros V. The *C. elegans* heterochronic gene lin-4 encodes small RNAs with antisense complementarity to lin-14. Cell 1993;75:843–854.

21. Winter J, Jung S, Keller S, Diederichs S. Many roads to maturity: microRNA biogenesis pathways and their regulation. Nat Cell Biol 2009;11:228–234.

22. Ruby JG, Jan CH, Bartel DP. Intronic microRNA precursors that bypass Drosha processing. Nature 2007;448:83–86.

23. Okamura K, Hagen JW, Duan H, Tyler DM, Lai EC. The mirtron pathway generates microRNA-class regulatory RNAs in *Drosophila*. Cell 2007;130:89–100.

24. Berezikov E, Guryev V, van de Belt J, Wienholds E, Plasterk RH, Cuppen E. Phylogenetic shadowing and computational identification of human microRNA genes. Cell 2005;120:21–24.

25. Lim LP, Lau NC, Garrett-Engele P, Grimson A, Schelter JM, Castle J, et al. Microarray analysis shows that some microRNAs downregulate large numbers of target mRNAs. Nature 2005;433: 769–773.

26. Esquela-Kerscher A, Slack FJ. Oncomirs-microRNAs with a role in cancer. Nat Rev Cancer 2006;6:259–269.

27. Calin GA, Dumitru CD, Shimizu M, Bichi R, Zupo S, Noch E, et al. Frequent deletions and down-regulation of micro-RNA genes miR15 and miR16 at 13q14 in chronic lymphocytic leukemia. Proc Natl Acad Sci USA 2002;99:15524–15529.

28. Voorhoeve PM. MicroRNAs: Oncogenes, tumor suppressors or master regulators of cancer hetero-geneity? Biochim Biophys Acta 2010;1805:72–86.

29. Fabbri M, Bottoni A, Shimizu M, Spizzo R, Nicoloso MS, Rossi S, et al. Association of a microRNA/TP53 feedback circuitry with pathogenesis and outcome of B-cell chronic lymphocytic leukemia. JAMA 2011;305:59–67.

30. He L, He X, Lim LP, de Stanchina E, Xuan Z, Liang Y, et al. A microRNA component of the p53 tumour suppressor network. Nature 2007;447:1130–1134.

31. Faraoni I, Antonetti FR, Cardone J, Bonmassar E. miR-155 gene: a typical multifunctional microRNA. Biochim Biophys Acta 2009;1792:497–505.

32. Voorhoeve PM, le Sage C, Schrier M, Gillis AJ, Stoop H, Nagel R, et al. A genetic screen impli-cates miRNA-372 and miRNA-373 as oncogenes in testicular germ cell tumors. Cell 2006;124: 1169–1181.

33. He L, Thomson JM, Hemann MT, Hernando-Monge E, Mu D, Goodson S, et al. A microRNA polycistron as a potential human oncogene. Nature 2005;435:828–833.

34. Karin M. Nuclear factor-kappaB in cancer development and progression. Nature 2006;441: 431–436.

35. Medzhitov R. Origin and physiological roles of inflammation. Nature 2008;454:428–435.

36. Karin M, Greten FR. NF-kappaB: linking inflammation and immunity to cancer development and progression. Nat Rev Immunol 2005;5:749–759.

37. Aggarwal BB, Vijayalekshmi RV, Sung B. Targeting inflammatory pathways for prevention and therapy of cancer: short-term friend, long-term foe. Clin Cancer Res 2009;15:425–430.

38. Federico A, Morgillo F, Tuccillo C, Ciardiello F, Loguercio C. Chronic inflammation and oxidative stress in human carcinogenesis. Int J Cancer 2007;121:2381–2386.

39. Wu F, Zikusoka M, Trindade A, Dassopoulos T, Harris ML, Bayless TM, et al. MicroRNAs are dif-ferentially expressed in ulcerative colitis and alter expression of macrophage inflammatory peptide-2 alpha. Gastroenterology 2008;135:1624–1635 e24.

40. Padgett KA, Lan RY, Leung PC, Lleo A, Dawson K, Pfeiff J, et al. Primary biliary cirrhosis is associated with altered hepatic microRNA expression. J Autoimmun 2009;32:246–253.

41. Murdoch C, Muthana M, Coffelt SB, Lewis CE. The role of myeloid cells in the promotion of tumour angiogenesis. Nat Rev Cancer 2008;8:618–631.

42. Shin VY, Jin H, Ng EK, Cheng AS, Chong WW, Wong CY, et al. NF-kappaB targets miR-16 and miR-21 in gastric cancer: involvement of prostaglandin E receptors. Carcinogenesis 2011;32: 240–245.

43. Li J, Wang K, Chen X, Meng H, Song M, Wang Y, et al. Transcriptional activation of microRNA-34a by NF-kappa B in human esophageal cancer cells. BMC Mol Biol 2012;13:4.

44. Zhang X, Liu S, Hu T, He Y, Sun S. Up-regulated microRNA-143 transcribed by nuclear factor kappa B enhances hepatocarcinoma metastasis by repressing fibronectin expression. Hepatology 2009;50:490–499.

45. Perry MM, Williams AE, Tsitsiou E, Larner-Svensson HM, Lindsay MA. Divergent intracellular pathways regulate interleukin-1beta-induced miR-146a and miR-146b expression and chemokine release in human alveolar epithelial cells. FEBS Lett 2009;583:3349–3355.

46. Zhang Z, Li Z, Gao C, Chen P, Chen J, Liu W, et al. miR-21 plays a pivotal role in gastric cancer pathogenesis and progression. Lab Invest 2008;88:1358–1366.

47. Loffler D, Brocke-Heidrich K, Pfeifer G, Stocsits C, Hackermuller J, Kretzschmar AK, et al. Interleukin-6 dependent survival of multiple myeloma cells involves the Stat3-mediated induction of microRNA-21 through a highly conserved enhancer. Blood 2007;110:1330–1333.

48. Carissimi C, Fulci V, Macino G. MicroRNAs: novel regulators of immunity. Autoimmun Rev 2009;8:520–524.

49. Frankel LB, Christoffersen NR, Jacobsen A, Lindow M, Krogh A, Lund AH. Programmed cell death 4 (PDCD4) is an important functional target of the microRNA miR-21 in breast cancer cells. J Biol Chem 2008;283:1026–1033.

50. Zhu S, Si ML, Wu H, Mo YY. MicroRNA-21 targets the tumor suppressor gene tropomyosin 1 (TPM1). J Biol Chem 2007;282:14328–14336.

51. Meng F, Henson R, Wehbe-Janek H, Ghoshal K, Jacob ST, Patel T. MicroRNA-21 regulates expression of the PTEN tumor suppressor gene in human hepatocellular cancer. Gastroenterology 2007;133:647–658.

52. Liu M, Wu H, Liu T, Li Y, Wang F, Wan H, et al. Regulation of the cell cycle gene, BTG2, by miR-21 in human laryngeal carcinoma. Cell Res 2009;19:828–837.

53. Si ML, Zhu S, Wu H, Lu Z, Wu F, Mo YY. miR-21-mediated tumor growth. Oncogene 2007;26:2799–2803.

54. Lu TX, Munitz A, Rothenberg ME. MicroRNA-21 is up-regulated in allergic airway inflammation and regulates IL-12p35 expression. J Immunol 2009;182:4994–5002.

55. Schetter AJ, Nguyen GH, Bowman ED, Mathe EA, Yuen ST, Hawkes JE, et al. Association of inflammation-related and microRNA gene expression with cancer-specific mortality of colon adenocarcinoma. Clin Cancer Res 2009;15:5878–5887.

56. Motsch N, Pfuhl T, Mrazek J, Barth S, Grasser FA. Epstein–Barr virus-encoded latent membrane protein 1 (LMP1) induces the expression of the cellular microRNA miR-146a. RNA Biol 2007;4:131–137.

57. Xiao B, Liu Z, Li BS, Tang B, Li W, Guo G, et al. Induction of microRNA-155 during *Helicobacter pylori* infection and its negative regulatory role in the inflammatory response. J Infect Dis 2009;200:916–925.

58. Worm J, Stenvang J, Petri A, Frederiksen KS, Obad S, Elmen J, et al. Silencing of microRNA-155 in mice during acute inflammatory response leads to derepression of c/ebp beta and down-regulation of G-CSF. Nucleic Acids Res 2009;37:5784–5792.

59. O'Connell RM, Taganov KD, Boldin MP, Cheng G, Baltimore D. MicroRNA-155 is induced during the macrophage inflammatory response. Proc Natl Acad Sci USA 2007;104:1604–1609.

60. Tili E, Michaille JJ, Cimino A, Costinean S, Dumitru CD, Adair B, et al. Modulation of miR-155 and miR-125b levels following lipopolysaccharide/TNF-alpha stimulation and their possible roles in regulating the response to endotoxin shock. J Immunol 2007;179:5082–5089.

61. Cardoso AL, Guedes JR, Pereira de Almeida L, Pedroso de Lima MC. miR-155 modulates microglia-mediated immune response by down-regulating SOCS-1 and promoting cytokine and nitric oxide production. Immunology 2012;135:73–88.

62. Gironella M, Seux M, Xie MJ, Cano C, Tomasini R, Gommeaux J, et al. Tumor protein 53-induced nuclear protein 1 expression is repressed by miR-155, and its restoration inhibits pancreatic tumor development. Proc Natl Acad Sci USA 2007;104:16170–16175.

63. Costinean S, Zanesi N, Pekarsky Y, Tili E, Volinia S, Heerema N, et al. Pre-B cell proliferation and lymphoblastic leukemia/high-grade lymphoma in E(mu)-miR155 transgenic mice. Proc Natl Acad Sci USA 2006;103:7024–7029.

64. Pan J, Hu H, Zhou Z, Sun L, Peng L, Yu L, et al. Tumor-suppressive *mir-663* gene induces mitotic catastrophe growth arrest in human gastric cancer cells. Oncol Rep 2010;24:105–112.

65. Tili E, Michaille JJ. Resveratrol, microRNAs, inflammation, and cancer. J Nucleic Acids 2011;2011:102431.

66. Blower PE, Chung JH, Verducci JS, Lin S, Park JK, Dai Z, et al. MicroRNAs modulate the chemosensitivity of tumor cells. Mol Cancer Ther 2008;7:1–9.

67. Zheng T, Wang J, Chen X, Liu L. Role of microRNA in anticancer drug esistance. Int J Cancer 2009;126:2–10.

68. Hunter KW, Crawford NP, Alsarraj J. Mechanisms of metastasis. Breast Cancer Res 2008;10:S2.

69. Fidler IJ. The pathogenesis of cancer metastasis: the 'seed and soil' hypothesis revisited. Nat Rev Cancer 2003;3:453–458.

70. Ma L, Teruya-Feldstein J, Weinberg RA. Tumour invasion and metastasis initiated by microRNA-10b in breast cancer. Nature 2007;449:682–688.

71. Carrio M, Arderiu G, Myers C, Boudreau NJ. Homeobox D10 induces phenotypic reversion of breast tumor cells in a three-dimensional culture model. Cancer Res 2005;65:7177–7185.

72. Zhu S, Wu H, Wu F, Nie D, Sheng S, Mo YY. MicroRNA-21 targets tumor suppressor genes in invasion and metastasis. Cell Res 2008;18:350–359.

73. Asangani IA, Rasheed SA, Nikolova DA, Leupold JH, Colburn NH, Post S, et al. MicroRNA-21 (miR-21) post-transcriptionally downregulates tumor suppressor Pdcd4 and stimulates invasion, intravasation and metastasis in colorectal cancer. Oncogene 2008;27:2128–2136.

74. Nieves-Alicea R, Colburn NH, Simeone AM, Tari AM. Programmed cell death 4 inhibits breast cancer cell invasion by increasing tissue inhibitor of metalloproteinases-2 expression. Breast Cancer Res Treat 2009;114:203–209.

75. Lockett J, Yin S, Li X, Meng Y, Sheng S. Tumor suppressive maspin and epithelial homeostasis. J Cell Biochem 2006;97:651–660.

76. Bracken CP, Gregory PA, Khew-Goodall Y, Goodall GJ. The role of microRNAs in metastasis and epithelial–mesenchymal transition. Cell Mol Life Sci 2009;66:1682–1699.

77. Gabriely G, Wurdinger T, Kesari S, Esau CC, Burchard J, Linsley PS, et al. MicroRNA 21 promotes glioma invasion by targeting matrix metalloproteinase regulators. Mol Cell Biol 2008;28:5369–5380.

78. Takamizawa J, Konishi H, Yanagisawa K, Tomida S, Osada H, Endoh H, et al. Reduced expression of the let-7 microRNAs in human lung cancers in association with shortened postoperative survival. Cancer Res 2004;64:3753–3756.

79. Watanabe S, Ueda Y, Akaboshi S, Hino Y, Sekita Y, Nakao M. HMGA2 maintains oncogenic RAS-induced epithelial–mesenchymal transition in human pancreatic cancer cells. Am J Pathol 2009;174:854–868.

80. Bhaumik D, Scott GK, Schokrpur S, Patil CK, Campisi J, Benz CC. Expression of microRNA-146 suppresses NF-kappaB activity with reduction of metastatic potential in breast cancer cells. Oncogene 2008;27:5643–5647.

81. Hurst DR, Edmonds MD, Scott GK, Benz CC, Vaidya KS, Welch DR. Breast cancer metastasis suppressor 1 up-regulates miR-146, which suppresses breast cancer metastasis. Cancer Res 2009;69:1279–1283.

82. Acloque H, Thiery JP, Nieto MA. The physiology and pathology of the EMT. Meeting on the epithelial–mesenchymal transition. EMBO Rep 2008;9:322–326.

83. Baranwal S, Alahari SK. miRNA control of tumor cell invasion and metastasis. Int J Cancer 2010;126:1283–1290.

84. Gregory PA, Bert AG, Paterson EL, Barry SC, Tsykin A, Farshid G, et al. The miR-200 family and miR-205 regulate epithelial to mesenchymal transition by targeting ZEB1 and SIP1. Nat Cell Biol 2008;10:593–601.

85. Park SM, Gaur AB, Lengyel E, Peter ME. The miR-200 family determines the epithelial phenotype of cancer cells by targeting the E-cadherin repressors ZEB1 and ZEB2. Genes Dev 2008;22:894–907.

86. Cao Q, Yu J, Dhanasekaran SM, Kim JH, Mani RS, Tomlins SA, et al. Repression of E-cadherin by the polycomb group protein EZH2 in cancer. Oncogene 2008;27:7274–7284.

87. Varambally S, Cao Q, Mani RS, Shankar S, Wang X, Ateeq B, ct al. Genomic loss of microRNA-101 leads to overexpression of histone methyltransferase EZH2 in cancer. Science 2008;322: 1695–1699.

88. Valastyan S, Reinhardt F, Calogrias D, Szasz AM, Wang ZC, Brock JE, et al. A pleiotropically acting microRNA, miR-31, inhibits breast cancer metastasis. Cell 2009;137:1032–1046.

89. Shastry BS. SNPs: impact on gene function and phenotype. Methods Mol Biol 2009;578:3–22.

90. Jazdzewski K, Murray EL, Franssila K, Jarzab B, Schoenberg DR, de la Chapelle A. Common SNP in pre-miR-146a decreases mature miR expression and predisposes to papillary thyroid carcinoma. Proc Natl Acad Sci USA 2008;105:7269–7274.

91. Chin LJ, Ratner E, Leng S, Zhai R, Nallur S, Babar I, et al. A SNP in a let-7 microRNA complementary site in the KRAS 3′ untranslated region increases non-small cell lung cancer risk. Cancer Res 2008;68:8535–8540.

92. Mishra PJ, Banerjee D, Bertino JR. MiRSNPs or MiR-polymorphisms, new players in microRNA mediated regulation of the cell: introducing microRNA pharmacogenomics. Cell Cycle 2008;7:853–858.

93. Song FJ, Chen KX. Single-nucleotide polymorphisms among microRNA: big effects on cancer. Chin J Cancer 2011;30:381–391.

94. Xu T, Zhu Y, Wei QK, Yuan Y, Zhou F, Ge YY, et al. A functional polymorphism in the miR-146a gene is associated with the risk for hepatocellular carcinoma. Carcinogenesis 2008;29: 2126–2131.

95. Tian T, Shu Y, Chen J, Hu Z, Xu L, Jin G, et al. A functional genetic variant in microRNA-196a2 is associated with increased susceptibility of lung cancer in Chinese. Cancer Epidemiol Biomark Prev 2009;18:1183–1187.

96. Zhu L, Chu H, Gu D, Ma L, Shi D, Zhong D, et al. A functional polymorphism in miRNA-196a2 is associated with colorectal cancer risk in a Chinese population. DNA Cell Biol 2012;31:350–354.

97. Taganov KD, Boldin MP, Chang KJ, Baltimore D. NF-kappaB-dependent induction of microRNA miR-146, an inhibitor targeted to signaling proteins of innate immune responses. Proc Natl Acad Sci USA 2006;103:12481–12486.

98. Hu Z, Liang J, Wang Z, Tian T, Zhou X, Chen J, et al. Common genetic variants in pre-microRNAs were associated with increased risk of breast cancer in Chinese women. Hum Mutation 2009;30:79–84.

99. Shen J, Ambrosone CB, DiCioccio RA, Odunsi K, Lele SB, Zhao H. A functional polymorphism in the *miR-146a* gene and age of familial breast/ovarian cancer diagnosis. Carcinogenesis 2008;29:1963–1966.

100. Xu. B, Feng NH, Li PC, Tao J, Wu D, Zhang ZD, et al. A functional polymorphism in pre-miR-146a gene is associated with prostate cancer risk and mature miR-146a expression *in vivo*. Prostate 2010;70:467–472.

101. Hu Z, Chen J, Tian T, Zhou X, Gu H, Xu L, et al. Genetic variants of miRNA sequences and non-small cell lung cancer survival. J Clin Invest 2008;118:2600–2608.

102. Nicoloso MS, Sun H, Spizzo R, Kim H, Wickramasinghe P, Shimizu M, et al. Single-nucleotide polymorphisms inside microRNA target sites influence tumor susceptibility. Cancer Res 2010;70:2789–2798.

103. Naccarati A, Pardini B, Stefano L, Landi D, Slyskova J, Novotny J, et al. Polymorphisms in miRNA-binding sites of nucleotide excision repair genes and colorectal cancer risk. Carcinogenesis 2012;33:1346–1351.

104. Kontorovich T, Levy A, Korostishevsky M, Nir U, Friedman E. Single nucleotide polymorphisms in miRNA binding sites and miRNA genes as breast/ovarian cancer risk modifiers in Jewish high-risk women. Int J Cancer 2010;127:589–597.

105. Horikawa Y, Wood CG, Yang H, Zhao H, Ye Y, Gu J, et al. Single nucleotide polymorphisms of microRNA machinery genes modify the risk of renal cell carcinoma. Clin Cancer Res 2008;14:7956–7962.

106. Martin D, Gutkind JS. Human tumor-associated viruses and new insights into the molecular mechanisms of cancer. Oncogene 2008;27:S31–S42.

107. Epstein MA, Achong BG, Barr YM. Virus particles in cultured lymphoblasts from Burkitt's lymphoma. Lancet 1964;1:702–703.

108. Thorley-Lawson DA. EBV the prototypical human tumor virus: Just how bad is it? J Allergy Clin Immunol 2005;116:251–261.

109. Thorley-Lawson DA. Epstein–Barr virus: exploiting the immune system. Nat Rev Immunol 2001;1:75–82.

110. Pfeffer S, Zavolan M, Grasser FA, Chien M, Russo JJ, Ju J, et al. Identification of virus-encoded microRNAs. Science 2004;304:734–736.

111. Chen SJ, Chen GH, Chen YH, Liu CY, Chang KP, Chang YS, et al. Characterization of Epstein–Barr virus miRNAome in nasopharyngeal carcinoma by deep sequencing. PLoS One 2010;5: e12745.

112. He ML, Luo MX, Lin MC, Kung HF. MicroRNAs: potential diagnostic markers and therapeutic targets for EBV-associated nasopharyngeal carcinoma. Biochim Biophys Acta 2012;1825:1–10.

113. Barth S, Pfuhl T, Mamiani A, Ehses C, Roemer K, Kremmer E, et al. Epstein–Barr virus-encoded microRNA miR-BART2 down-regulates the viral DNA polymerase BALF5. Nucleic Acids Res 2008;36:666–675.

114. Xia T, O'Hara A, Araujo I, Barreto J, Carvalho E, Sapucaia JB, et al. EBV microRNAs in primary lymphomas and targeting of CXCL-11 by ebv-mir-BHRF1-3. Cancer Res 2008;68:1436–1442.

115. Nachmani D, Stern-Ginossar N, Sarid R, Mandelboim O. Diverse herpesvirus microRNAs target the stress-induced immune ligand MICB to escape recognition by natural killer cells. Cell Host Microbe 2009;5:376–385.

116. Choy EY, Siu KL, Kok KH, Lung RW, Tsang CM, To KF, et al. An Epstein–Barr virus-encoded microRNA targets PUMA to promote host cell survival. J Exp Med 2008;205:2551–2560.

117. Raab-Traub N. Epstein–Barr virus in the pathogenesis of NPC. Semin Cancer Biol 2002;12:431–441.

118. Cullen BR. Viral and cellular messenger RNA targets of viral microRNAs. Nature 2009;457:421–425.

119. Lung RW, Tong JH, Sung YM, Leung PS, Ng DC, Chau SL, et al. Modulation of LMP2A expression by a newly identified Epstein–Barr virus-encoded microRNA miR-BART22. Neoplasia 2009;11:1174–1184.

120. Wong AM, Kong KL, Tsang JW, Kwong DL, Guan XY. Profiling of Epstein–Barr virus-encoded microRNAs in nasopharyngeal carcinoma reveals potential biomarkers and oncomirs. Cancer 2011;118:698–710.

121. Chang Y, Cesarman E, Pessin MS, Lee F, Culpepper J, Knowles DM, et al. Identification of herpesvirus-like DNA sequences in AIDS-associated Kaposi's sarcoma. Science 1994;266: 1865–1869.

122. Wong EL, Damania B. Linking KSHV to human cancer. Curr Oncol Rep 2005;7:349–356.

123. Pfeffer S, Sewer A, Lagos-Quintana M, Sheridan R, Sander C, Grasser FA, et al. Identification of microRNAs of the herpesvirus family. Nat Methods 2005;2:269–276.

124. Cai X, Lu S, Zhang Z, Gonzalez CM, Damania B, Cullen BR. Kaposi's sarcoma–associated herpesvirus expresses an array of viral microRNAs in latently infected cells. Proc Natl Acad Sci USA 2005;102:5570–5575.

125. Samols MA, Hu J, Skalsky RL, Renne R. Cloning and identification of a microRNA cluster within the latency-associated region of Kaposi's sarcoma–associated herpesvirus. J Virol 2005;79: 9301–9305.

126. Jones KD, Aoki Y, Chang Y, Moore PS, Yarchoan R, Tosato G. Involvement of interleukin-10 (IL-10) and viral IL-6 in the spontaneous growth of Kaposi's sarcoma herpesvirus-associated infected primary effusion lymphoma cells. Blood 1999;94:2871–2879.

127. Aoki Y, Yarchoan R, Braun J, Iwamoto A, Tosato G. Viral and cellular cytokines in AIDS-related malignant lymphomatous effusions. Blood 2000;96:1599–1601.

128. Oksenhendler E, Carcelain G, Aoki Y, Boulanger E, Maillard A, Clauvel JP, et al. High levels of human herpesvirus 8 viral load, human interleukin-6, interleukin-10, and C reactive protein correlate with exacerbation of multicentric castleman disease in HIV-infected patients. Blood 2000;96: 2069–2073.

129. Cirone M, Lucania G, Aleandri S, Borgia G, Trivedi P, Cuomo L, et al. Suppression of dendritic cell differentiation through cytokines released by primary effusion lymphoma cells. Immunol Lett 2008;120:37–41.

130. Qin Z, Kearney P, Plaisance K, Parsons CH. Pivotal advance: Kaposi's sarcoma–associated herpesvirus (KSHV)-encoded microRNA specifically induce IL-6 and IL-10 secretion by macrophages and monocytes. J Leukoc Biol 2010;87:25–34.

131. Abend JR, Uldrick T, Ziegelbauer JM. Regulation of tumor necrosis factor-like weak inducer of apoptosis receptor protein (TWEAKR) expression by Kaposi's sarcoma–associated herpesvirus microRNA prevents TWEAK-induced apoptosis and inflammatory cytokine expression. J Virol 2010;84:12139–12151.

132. Gottwein E, Cullen BR. A human herpesvirus microRNA inhibits p21 expression and attenuates p21-mediated cell cycle arrest. J Virol 2010;84:5229–5237.

133. Lewis BP, Burge CB, Bartel DP. Conserved seed pairing, often flanked by adenosines, indicates that thousands of human genes are microRNA targets. Cell 2005;120:15–20.

134. Gottwein E, Mukherjee N, Sachse C, Frenzel C, Majoros WH, Chi JT, et al. A viral microRNA functions as an orthologue of cellular miR-155. Nature 2007;450:1096–1099.

135. Marshall V, Martro E, Labo N, Ray A, Wang D, Mbisa G, et al. Kaposi sarcoma (KS)-associated herpesvirus microRNA sequence analysis and KS risk in a European AIDS-KS case control study. J Infect Dis 2010;202:1126–1135.

136. Zhang L, Volinia S, Bonome T, Calin GA, Greshock J, Yang N, et al. Genomic and epigenetic alterations deregulate microRNA expression in human epithelial ovarian cancer. Proc Natl Acad Sci USA 2008;105:7004–7009.

137. Bloomston M, Frankel WL, Petrocca F, Volinia S, Alder H, Hagan JP, et al. MicroRNA expression patterns to differentiate pancreatic adenocarcinoma from normal pancreas and chronic pancreatitis. JAMA 2007;297:1901–1908.

138. Shimono Y, Zabala M, Cho RW, Lobo N, Dalerba P, Qian D, et al. Downregulation of miRNA-200c links breast cancer stem cells with normal stem cells. Cell 2009;138:592–603.

139. Murakami Y, Yasuda T, Saigo K, Urashima T, Toyoda H, Okanoue T, et al. Comprehensive analysis of microRNA expression patterns in hepatocellular carcinoma and non-tumorous tissues. Oncogene 2006;25:2537–2545.

140. Ward BP, Tsongalis GJ, Kaur P. MicroRNAs in chronic lymphocytic leukemia. Exp Mol Pathol 2011;90:173–178.

141. Lodes MJ, Caraballo M, Suciu D, Munro S, Kumar A, Anderson B. Detection of cancer with serum miRNAs on an oligonucleotide microarray. PLoS One 2009;4:e6229.

CHAPTER 11

INFLAMMATION AS A NICHE FOR TUMOR PROGRESSION

Futoshi Okada

Progression of tumor cells, that is, acquisition of malignant and stable phenotypes, is the primary cause of death in patients with cancer, not the growth of the primary tumor itself (1). It is important to understand the mechanisms of when, why, and how tumor cells gain such aggressive and irreversible phenotypes for reducing or preventing cancer death. Among the internal and external factors influencing tumor progression, inflammation is the most powerful intrinsic factor. The link between inflammation and tumor progression is evidenced by several observations: (1) inflammatory diseases increase tumor progression risk; (2) inflammatory cells and their derived mediators surround and are within the primary tumors; (3) induction of inflammatory cytokines and chemokines converts tumor cells into fully malignant cells; (4) inhibition of inflammatory mediators deters tumor progression; and (5) nonsteroidal anti-inflammatory drugs delay tumor progression in cancer patients (2,3).

A metaphor has been used to explain the inflammation-related carcinogenesis that inflammation may provide the "fuel" to support the flames if genetic damage is the match to light the "fire" of cancer (4). In this chapter I focus on the contents of the fuel for flames, following the recent evidence that the fuel also has a role, in a direct or indirect manner, as a "match," for inflammation is now recognized to be a fuel-equipped igniter and regions in which it has been located to be niches for tumor progression.

INFLAMMATION CONVERTS A PRIMARY TUMOR ENVIRONMENT TO A FAVORABLE SPOT FOR TUMOR PROGRESSION

Inflammation acts as a key regulator of tumor progression by aberrant mechanisms such as increased mutagenicity, accelerated DNA replication and cell proliferation, disruption of cell death machinery, enhanced angiogenesis, suppression of antitumor immunity, acquired resistance to chemo- or radiation therapy, and invasive and

Cancer and Inflammation Mechanisms: Chemical, Biological, and Clinical Aspects, First Edition.
Edited by Yusuke Hiraku, Shosuke Kawanishi, and Hiroshi Ohshima.
© 2014 John Wiley & Sons, Inc. Published 2014 by John Wiley & Sons, Inc.

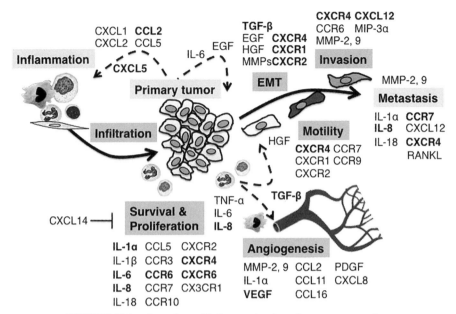

FIGURE 11.1 Overview of inflammation-based tumor progression.

metastatic abilities (2,5). An inflammatory environment also generates and selects malignant tumor variants from growing tumors primarily through induction of genomic instability, heterogeneity, and epigenetic alterations, which are induced basically by the mechanisms mentioned above. Traditionally, inflammation has been thought to contribute to tumor initiation by generating genotoxic substance, to tumor promotion by inducing cellular proliferation, and to tumor progression by enhancing angiogenesis and tissue invasion. However, accumulated data demonstrate that inflammation, in fact, continuously stimulates all of the processes of carcinogenesis, including tumor progression. How inflammatory cells and their derived molecules accelerate tumor progression is summarized in Figure 11.1.

Angiogenesis is crucial for tumor progression since it depends on whether tumor cells are adequately supplied with oxygen, nutrition, and growth and survival factors (6). The recruitment of endothelial cells and other blood vessel components is promoted by a number of prototypical pro-angiogenic mediators present ironically in the inflammatory environment. It is noteworthy that inflammatory cells, such as recruited macrophages and myeloid-derived suppressive cells, change themselves functionally into endothelial phenotypes and are incorporated into the newly formed vascular structure that supports tumor angiogenesis (6).

Antitumor immunity seems to consist of indeterminable host responses against tumor in its progression stage; namely, the balance between antitumor immunity and tumor progression-enhancing immunity is delicate in the primary tumor environment. Then immune responses to tumor cells, such as immune surveillance (immunoediting), arise for preventing and inhibiting tumor development; normally at this point, tumor antigen(s) on the tumor cell surfaces are null or negligible since the majority of antigenic tumor cells or precancerous cells have been recognized and eliminated by the host immunity.

However, antitumor responses, especially at the progression phase, do not function as expected; or if they have acted excessively, they are suppressed by potent immunosuppressive factors produced by tumor cells and/or inflammatory cells. Clinical observations show that unresolved immune responses, such as chronic inflammation, promote tumor progression. Concerning this, a problem of immunity-based acceleration of carcinogenesis was raised in the 1970s and termed *immunostimulation theory of tumor development* by Prehn and Lappe (7). Experimental studies also confirmed that uncontrollable and sustained inflammation accelerates tumor progression.

Epithelial–mesenchymal transition (EMT) has been suggested to be an essential step in tumor cell motility, invasion, and metastasis (1,3,6). EMT involves physiological processes that occur during normal embryogenesis or wound healing. When EMT occurs in tumor cells, it converts differentiated epithelial tumor cells into de-differentiated cells, which then show mesenchymal features. The molecular switch for EMT is basically driven by modulation of both E-cadherin (epithelial type) and Twist, Vimentine, and N-cadherin (mesenchymal type) (3). EMT in tumor cells is stimulated by inflammatory cells and its derived factors. EMT-inducible inflammatory cells are neutrophils, macrophages, and myeloid-derived suppressor cells, and their derivatives—transforming growth factor β (TGF-β), epidermal growth factor (EGF), hepatocyte growth factor (HGF), and chemokines (CCL5, CXCR4, CXCL1, CXCL8, and CXCR1)—are involved in tumor progression.

INFLAMMATORY CELL COMPONENTS NECESSARY TO ACCELERATE TUMOR PROGRESSION

Growing tumors are composed of tumor cells and stroma cells. Fibroblasts, myofibroblasts, endothelial cells, pericytes, and inflammatory cells (e.g., macrophages, monocytes, neutrophils, mast cells, dendritic cells, T and B lymphocytes, and platelets) come under the stroma cells. They create an environment in which inflammatory cells accumulate, soluble factors are secreted, and tissues called sustentacula are provided; all of them support tumor progression.

Clinically, tumor stroma accounts for up to 40% of infiltrated inflammatory cells, which then facilitate distant metastasis. Inflammatory components in tumor progression indicate an evident correlation between frequent infiltration of leukocytes into tumor tissues and poor outcome of patients with malignancies of various origins. However, the prognostic significance of infiltrated inflammatory cells into tumor mass remains controversial because the standard by which we can evaluate such infiltration of inflammatory cells in a variety of organ carcinogenesis and tumor progression has not yet been established. Presently, we merely grasp that inflammation contributes both positively and negatively to tumor progression in some situations.

Paradoxical roles of tumor-associated inflammation in tumor progression may depend on the following points: (1) pluripotent functions (i.e., genotoxicity, immunosuppression, anti-apoptotic activity, etc.) act simultaneously on one particular cell type, which may alter the following tumor growth and/or inflammatory reaction; (2) the activation and/or differentiation statuses of infiltrated cells could be changed, and in some cases the changes are reversible; and (3) infiltrated cellular components are altered sequentially, depending on the inflammatory or immunological conditions, rather than remaining stable in their differentiated phenotypes. For example, macrophages infiltrated into a tumor mass differentiate into two distinct phenotypes, M1 and M2. Although they derive from the same macrophage population, the two phenotypes have functionally apparent differences against tumor cells or differences in their secretion factors. The M1/M2 ratio in the tumor mass possibly determines the tumor fate (i.e., whether it will induce regression/dormant tumor cells or progress to be malignant).

The relation between cellular components and their roles in the induction of malignant tumor cells are given in Table 11.1. The prefix "tumor-associated" is added to those inflammatory cells that are particularly important for tumor progression (e.g., tumor-associated macrophage). Typical infiltrates in malignant tumor progression are detailed below.

Myeloid-derived suppressive cells (MDSCs) represent a heterogeneous population of immature myeloid cells composed of macrocytic/monocytic, granulocytic, or dendritic precursor cells or myeloid cells at earlier stages of differentiation. Significant increases in MDSCs in peripheral blood are a common feature for patients with advanced cancers. These are the two reasons why we assert that MDSCs are responsible for tumor metastasis. MDSCs are known to produce genotoxic reactive oxygen species (ROS) and nitric oxide (NO), inhibit antitumor immune responses, and promote angiogenesis in the primary tumor site. More important, MDSCs are recruited by tumor- and/or inflammatory cell–derived factors (IL-6, IL-1, and GM-CSF) from the bone marrow to a metastatic site to create a favorable pre-metastatic environment (1,5).

Neutrophils, which play an essential role in host defense by killing invading microorganisms, account for up to 70% of predominantly circulating leukocytes. Although neutrophils have been considered to have bacteriocidal functions, it is now clear that they also play a major role in tumor progression. An intense accumulation of intratumoral neutrophils can be associated with shorter survival in cancer patients. Neutrophilia is also linked to poorer prognosis in patients. An experimental study has shown that increased neutrophils in tumor-bearing mice aggravated the tumor cells since neutrophilia developed in the mice, and coimplantation of neutrophils with tumor cells induced experimental metastasis (8). From these, a dual function of neutrophils has been suggested. In tumor tissues, two different populations of neutrophils are present: a population that suppresses tumor development [i.e., tumor-associated neutrophils 1 (TAN1)], and a population that promotes tumor progression, TAN2. One of the factors that differentiate them into the two types is probably TGF-β, because reduction of TGF-β converts pro-tumorigenic TAN2 into the anti-tumorigenic TAN1 phenotype (5).

Monocytic leukocytes display remarkable plasticity and are highly heterogeneous; they function in distinct differentiation forms (monocytes, myeloid-derived suppressive cells, macrophages, and dendritic cells) and adopt different activation

TABLE 11.1 Modulation of Tumor Progression Features by Inflammatory Cells[a]

Inflammatory cells	Tumor growth	Apoptosis	DNA damage	DNA repair	Angiogenesis	MMPs	Antitumor immunity	ROS/iNOS	AM	EMT	Prognosis	Invasion/ metastasis
Neutrophils	○	●▲	●	▲	●	●	●▲	●	●	○	Poor	●
Monocytes	○		●		○	●	△	●	●	○		○
TAMs	●▲	▲			●	●	●▲	●	●	○	Poor	●
DCs					○		△					
MDSCs	●		○		●	●	▲	●	○	○	Poor	○
Mast cells	●											○
Adipocytes						○					Poor	○
T cells							△					
B cells							△					
CAFs	○		○		●	○	△	○				○
Ends					●							○
Platelets					●							

[a]Open circle; increased; closed circle; highly increased; open triangle; decreased; closed triangle; highly decreased; and circle with triangle; both increased and decreased. TAMs, tumor-associated macrophages; DCs, dendritic cells; MDSCs, myeloid-derived suppressive cells; CAFs, cancer-associated fibroblasts; Ends, endothelial cells; AM, arachidonic acid and its metabolites.

statuses in response to the tumor environment. On the one hand, monocytic lineage cells play antitumor immunosurveillance roles that limit tumor development; on the other hand, they orchestrate the inflammatory events during *de novo* carcinogenesis, participate in tumor immunostimulation, and contribute to the progression of tumors, which is actualized by tumor cell proliferation, promotion of cell motility, invasion and intravasation, angiogenesis, immunosuppression, and degradation of extracellular matrix (ECM).

Tumor-associated macrophages (TAMs) are also associated with tumor progression, which is known from the poor clinical outcome in the majority of TAM-positive cancer patients. Once recruited into tumor tissues, the macrophages differentiate into two types (M1 and M2), depending on the condition of environmental stimuli, including inflammatory cells and their derivative molecules. Macrophages activated by bacterial infection (M1), or by response to interferon γ (IFN-γ), granulocyte macrophage colony-stimulating factor (GM-CSF), or lipopolysaccharide enhance a type I inflammatory response, which leads to antitumor functioning. Conversely, macrophages activated by IL-4, IL-10, and IL-13 (M2) secrete immunosuppressive mediators (TGF-β, IL-10, and PGE2), growth and survival factors (M-CSF, EGF, IL-6, and CXCL8), angiogenic factors (VEGF, TGF-α, and PGE2), matrix metalloproteases (MMPs), and chemokines (CCL17, CCL18, and CCL22), which leads to the promotion of tumor progression. The two types are generally distinguished from their localization: Perivascular and migratory TAMs are the M1 phenotype, and sessile TAMs found at tumor–stroma borders and/or hypoxic regions are mostly the M2 phenotype. In a tumor mass, M2-polarized TAMs have been shown to express the fundamental aspects that accelerate tumor progression.

Fibroblasts present in tumor tissues are designated as cancer-associated fibroblasts (CAFs). CAFs are phenotypically and genetically very different from their normal fibroblast counterparts. Whereas normal organ-derived fibroblasts suppress tumorigenic conversion, tumor tissue–derived fibroblasts enhance the malignancy of tumor cells. Fibroblasts from normal tissue or weakly metastatic tumor tissues communicate well with adjacent cells through gap junctions, but those from highly metastatic tumor tissues do not (9). It is generally thought that fibroblasts switch their functions from tumor-suppressing to tumor-supporting during carcinogenesis. Thus, CAFs not only function as a component of tumor tissue architecture but also localize themselves at the invasion front of tumor tissues, being responsible for tumor cell growth and survival, including angiogenesis and invasion. Interestingly, compared to normal fibroblasts, CAFs undergo genetic alterations such as loss of heterozygosity/microsatellite instability, DNA hyper-/hypomethylation of several genes, and mutations of *Ras* or *Pten* genes.

The origin and outcome of CAFs are variable; currently, their sources are explained as follows (6): (1) local resting fibroblasts can be activated and recruited into a tumor mass by growth factors such as TGF-β, PDGF, and bFGF; (2) through epithelial- or endothelial-to-mesenchymal transition, pericytes, smooth muscle cells, or even tumor cells turn into CAFs; and (3) bone marrow–derived hematopoietic and mesenchymal stem cells differentiate into fibroblastic cells. In fact, it is reported that CAFs represent as much as 25% of the population of mesenchymal stem cells.

CAFs act as crucial mediators in initiating angiogenesis and lymphangiogenesis by modulating antitumor immunity and inflammation, and produce ECM and growth factors that lead to a conversion of the environment from normal to tumor-supporting, a process known as *tumor stromatogenesis*. The crosstalk between tumor cells and their modified stroma induces inflammation and, subsequently, tumor cell invasion, angiogenesis, and ultimately, metastasis (3).

A clear link is suspected between deregulated energy processing, as in obesity and metabolic disorders, and increased tumor aggressiveness accompanying poor patient outcome. Adipocytes are suspected to participate in tumor progression since adipose tissues induce an inflammatory environment while it works for physically composing tissue architecture, energy supply, and endocrine processes, although the precise mechanisms are not known yet. For further study, the following findings are accumulated: Infiltrated macrophages into the adipose tissue of a tumor mass may accelerate the initial events for tumor progression. Transplantation of tumor cells previously co-cultured with adipocytes causes more metastases than does that of tumor cells grown without adipocytes. One of the reasons for the difference is probably that the lipids transfer between adipocytes and tumor cells; the transfer might represent a key energy source (triglycerides and free fatty acids) for tumor cells. A report also shows that the tumor-associated adipocytes act as local regulators of tumor cell growth, invasion, and/or metastases (10).

INFLAMMATORY CELL–DERIVED FACTORS NECESSARY TO ACCELERATE TUMOR PROGRESSION

In an inflammatory environment, inflammatory cytokines, growth factors, chemokines, ROS, NO, reactive aldehydes, arachidonic acid metabolites, proteases, adhesion molecules, and so on, are induced; they accelerate the entire carcinogenic process. The inflammatory mediators induce direct and indirect genetic alterations, including activation of oncogenes and silencing tumor suppressor genes, and also support survival and proliferation of malignant cells. In other words, inflammation promotes both carcinogenesis and tumor progression (11). Major molecules secreted by inflammatory cells that facilitate tumor progression are given in Table 11.2.

Chemokines lead inflammatory cells to inflamed regions. This function of chemokines is of great importance in both physiological and pathological conditions. Chemokines and their receptors play a wide range of roles, recruiting leukocytes into tumors, promoting tumor growth in an autocrine and/or paracrine manner, regulating angiogenesis, promoting tumor metastasis via cell-to-cell adhesion, and so on (12). The homing theory was named after the phenomenon that metastasizing tumor cells are destined to a particular organ. The chemokine–ligand interaction will be one of the solutions for the phenomenon. For example, CXCR4 is the most frequently expressed chemokine receptor on tumor cells, and its ligand SDF-1/CXCL12 is expressed specifically in target organs. These interactions determine distant tumor metastasis. It is recently reported that chemokine receptors (CCR7 and CCR10) on tumor cells determine lymph node metastasis (12).

TABLE 11.2 Possible Tumor Progression-Inducible Features Secreted by Tumor Cells or Inflammatory Cells[a]

	Tumor cells	Neutrophils	Monocytes	TAMs	DCs	MDSCs	Mast cells	Adipocytes	T cells	B cells	CAFs	Ends	Platelets
AM	○			●			○		○	○	○	○	○
GM-CSF	●	○	○	○	○				○		○	○	○
M-CSF				● ●									
EGF			○	●		○	○				○		
FGF-2			○	○							● ●		
HGF			○	●		○							
IFN-α				○			○						
IFN-γ		○				●	○						
IGF-1								○	●				
IL-1α		○		○			○						
IL-1β		●	●	●		●	○	○					
IL-2				●					●				
IL-4				●			○	●	●				
IL-5				●				○	●				
IL-6		○	○	●	○	●	○	●	○	○	○		
IL-8	○	○	○	○	○	○	○		○	○	○ ○		
IL-10				● ◄		○	○	●	○				
IL-12				● ◄					○				
IL-13				●			○		○				
IL-17									○				
IL-18		○	●	○	○				○				
IL-23			●	●	●	○	○						
NGF				○	●	●	○						
PDGF	○		○			●							
RANKL	○							●	○				
TNF-α	○	●	●	●	●	●			○				
TGF-α		●		○		●	○		○				
TGF-β	●		○	●		●	○		○		●		
VEGF-A			○	●		○	●		○		●		
VEGF-C/D			○	○			●				●		

CCL-2
CCL-3
CCL-5
CCL-7
CCL-9
CCL-17
CCL-18
CCL-22
CCL-24
CX3 CL1
CXC L1
CXC L2
CXC L5
CXC L8
CXC L9
CXC L10
CXC L12
CXC L14
CXC L19
CXC R1
CXC R2
CXC R3
CXC R4
MMPs
ROS
NO

[a]Open circle, increased; closed circle, highly increased; open triangle, decreased; closed triangle, highly decreased; and circle with triangle, both increased and decreased. TAMs, tumor-associated macrophages; DCs, dendritic cells; MDSCs, myeloid-derived suppressive cells; CAFs, cancer-associated fibroblasts; Ends, endothelial cells; AM, arachidonic acid and its metabolites.

Inflammatory cell–derived cytokines not only stimulate cell growth but also act as transcriptional regulator on the tumor suppressor genes (e.g., *p53*). The mechanisms as to how pro- and anti-inflammatory cytokines deal with tumor progression vary: (1) accelerate tumor progression directly (TNF-α, IL-1α/β, IL-6, IL-18, and TGF-β); (2) suppress antitumor immunity (TGF-β and IL-10); and (3) augment immunostimulation (TNF-α and TGF-β).

Metabolism of arachidonic acid and its bioactive lipid metabolites is carried out by three distinct enzymes: cyclooxygenase, lipoxygenases, and cytochrome P450. Inflammation-related mediators such as prostaglandins and thromboxanes/leukotrienes are formed when arachidonic acid is metabolized by cyclooxygenases and lipoxygenases, respectively. Epoxy-eicosatrienoic acids are produced by cytochrome P450 from arachidonic acid. All these enzymes related to producing bioactive lipid metabolites are found in the inflammatory cells (neutrophils, platelets, and endothelial cells).

The inflammatory environment is characterized by sustained generation of ROS (superoxide, hydrogen peroxide, hypochlorous acid, singlet oxygen, and the hydroxyl radical) and reactive nitrogen intermediates (NO and peroxynitrite) (2). Both of these highly reactive substances damage nucleic acids, proteins, lipids, and carbohydrate of the cells. These molecules are produced mainly by infiltrated phagocytes and leukocytes and cause oxidative damage to DNA and nitration/nitrosation of DNA bases, which increases the risk for DNA mutations that may be nonrepairable and persist in subsequent generations (11,13). Accumulated studies in this field revealed that ROS and NO produced by inflammatory cells not only cause DNA damage directly but also indirectly by acting for deregulation of cell cycle checkpoints, cell proliferation and apoptosis, stimulation of angiogenesis, drive autophagy/mitophagy, or modification of gene/protein expressions; all these are critical steps to tumor progression (14).

ROS and NO are thus potent genotoxic agents that may increase the mutation rates and induce genomic instability by altering the genetic and biological functions: for example, inactivation of mismatch repair enzymes, inducing DNA damage in addition to enhancing the proliferation of mutated cells, and allowing tumor progression (15). Table 11.3 shows the possible relations between inflammatory cell–derived factors and their roles in phenotypic changes according to tumor progression.

It is evident that ROS released by activated phagocytes damage both nuclear and mitochondrial DNA. Mutations of the mitochondrial DNA (NADH dehydrogenase subunit 6 gene) induce deficiency in respiratory complex I activity and release the amounts of ROS that allow tumor cells to metastasize (16).

Alterations in DNA methylation patterns, especially hypermethylation of DNA, can be an effect of inflammation and are common in a variety of human cancers. Hypermethylation of DNA leads to transcriptional silencing of the genes related to cell cycle control, DNA repair, angiogenesis; hypermethylation of tumor suppressor genes enhances tumor malignancy (11). Recent studies reveal further that NO has a role in post-transcriptional modification through S-guanylation (17); that is, ROS and NO are the major factors for inflammation-based tumor progression.

Tumor-infiltrating phagocytes have ROS-mediated mutagenic activity, which will be one of the answers to the question of why malignant tumor cells arise from

TABLE 11.3 Modulation of Tumor Progression Features by Inflammatory Cell-Derived Factors[a]

Factor	Tumor cell growth	DNA damage	DNA repair	Angiogenesis	MMPs	Antitumor immunity	Prognosis	EMT	Invasion/metastasis	
									Clinical	Experimental
AM	○			●		▲	Poor	○		○
GM-CSF	○						Poor			○
M-CFS								●	●	○
EGF				●				●		
FGF	○			●				●	○	○
HGF	○			●				●	○	○
IFN-γ	△									
IL-1α	○			○					○	○
IL-1β	○			●		△		○	○	○
IL-2	△					●				
IL-6	○			○		△	Poor	○	○	○
IL-8	○			●	○	●	Poor	○	○	○
IL-10				○		○	Poor			○
IL-12				○					△	
IL-17	○						Poor			
IL-18	○						Poor			
IL-21	○									
IL-22	○									
IL-23	○									
MMPs	○			○			Poor		○	
PDGF	○			●						
RANKL				●				○	○	○
TGF-β	○			○		●	Poor	●	●	●
TNF-α	○	○	△	●	○	△	Poor	●	○	○
VEGF-A				●	○		Poor		●	●
VEGF-C/D							Poor		●	●
ROS	○	●	●	○			Poor		○	●
NO	○	●	●	○			Poor		○	●
CCL2	○			○			Poor			○
CCL5	○					○		●	○	○

(*continued*)

159

TABLE 11.3 (Continued)

Factor	Tumor cell growth	DNA damage	DNA repair	Angiogenesis	MMPs	Antitumor immunity	Prognosis	EMT	Invasion/metastasis Clinical	Invasion/metastasis Experimental
CCL11				○						
CCL16	○			○						
CCR1	○			○						
CCR2	○			○						
CCR3	○			○						
CCR5	○								●	
CCR6	●								●	●
CCR7	○						Poor		●	●
CCR9	○								●	
CCR10	○								○	○
CX3CR1	○							●	○	○
CXCL1	○			○						
CXCL2	○			○				○		
CXCL3	○			○						
CXCL4				△						
CXCL5				●			Poor			
CXCL6				○						
CXCL7				○						
CXCL8	○			●			Poor	●		○
CXCL9				△	○					○
CXCL10				△						○
CXCL11				△						○
CXCL12	●			○	○	○	Poor			○
CXCL16							Good			
CXCR1				●			Poor	●	●	○
CXCR2	○			△			Poor	●	○	●
CXCR3	●			●			Poor		●	○
CXCR4	●			●	○		Poor	●	●	●
CXCR5									○	○
CXCR6	●									
CXCR7				○						

160

primarily growing tumors. Phagocyte-derived ROS induce in tumor cells the motile and invasive capacity necessary for metastasis. Niitsu's group demonstrated that superoxide anion stimulates tumor metastasis through increased invasive ability. Moreover, they succeeded to determine protein kinase C (PKC) ζ as a molecular target for an ROS-induced motile phenotype of tumor cells (18).

EXPERIMENTAL MODEL OF INFLAMMATION-BASED TUMOR PROGRESSION

Although several lines of evidence show a close association of inflammation with tumor progression, there is hardly a solid animal model in which tumor progression accelerated by inflammation can be reproduced with certainty. We have developed a unique animal model in which progression of benign mouse regressive tumor cells can be observed consistently under concomitant inflammation. That model is described below.

Regressive clonal QR cells were obtained from clonal tumorigenic fibrosarcoma cells, BMT-11 cl-9, by exposing them *in vitro* to a mutagen (quercetin); they were nontumorigenic and nonmetastatic in normal syngeneic hosts (19). As shown in Figure 11.2A, QR cells did not develop tumor or form metastasis after subcutaneous $(2 \times 10^5$ cells) or intravenous $(1 \times 10^6$ cells) injection into mice (19). However, they were converted to grow lethally and acquired a metastatic phenotype after co-implanting with a piece of gelatin sponge that induced inflammation at implantation sites (20). The acquired tumorigenic and metastatic phenotypes, and other malignant phenotypes remained stable. Phenotypic and genetic alterations following inflammation-based QR cell progression are provided in Figure 11.2.

Migrating into the inserted gelatin sponge were predominantly neutrophils (Figure 11.2B). It was clear that inflammation definitely contributed to the progression of QR cells, since separately isolated inflammatory cells, mixed with QR cells and injected subcutaneously, converted the QR cells into malignant cells. We therefore use the model as inflammation-based tumor progression. To test the role of infiltrated neutrophils in the progression of QR cells, we eliminated neutrophils by administering antineutrophil antibody. The antibody-administered mice did not acquire malignant phenotypes (Figure 11.2C) (21). In contrast, nearly all of the arising tumors in the mice, non-treated or treated with control immunoglobulin, acquired malignant phenotypes. We confirmed the results by using integrin-β2 knockout mice (C57BL/6J^{Itgb2tm1Bay}, equivalent to CD18-deficient), in which the key adhesion molecule for the migration of neutrophils into an inflammatory region is absent (21). To determine the direct contribution of neutrophil-derived ROS to tumor progression, we used *gp91phox* gene (one of the major components of NADPH oxidase complex) knockout mice, which had reduced metastatic ability (22). To confirm whether phagocyte-derived ROS were actually involved in tumor progression, we isolated phagocytes from wild-type mice and transferred them into gp91$^{phox-/-}$ mice. As a result, wild-type-derived phagocytes enhanced the ability of metastasis in the mice. In contrast, the phagocytes obtained from knockout mice did not have such activity (22). Moreover, administration of aminoguanidine, a broad inhibitor

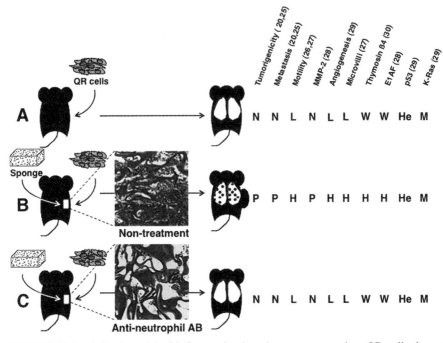

FIGURE 11.2 Animal model of inflammation-based tumor progression. QR cells do not form tumors in mice (A); however, they converted themselves to tumorigenic form and acquired metastatic ability after injection into pre-transplanted gelatin sponge, which induced inflammation at the site of transplantation (B). The inflammation-stimulated malignant progression of QR cells was inhibited by elimination of neutrophil infiltration by specific antibody administration (C). Phenotypic and genetic alterations in QR cells acquired after inflammation-based tumor progression are given in the right panel. N, negative; P, positive; W, weak; L, low; H, high; He, hemi-mutated; M, mutated.

of inducible nitric oxide synthase, partially but significantly suppressed malignant conversion in this model (23). These results show that ROS and NO, derived from infiltrated phagocytes, are an intrinsic factor for tumor malignant progression. The mouse model may recapitulate the typical inflammation-based carcinogenesis and tumor progression; thus, it is suitable for analyzing biological and genetic causes and for exploring preventive compounds for these processes.

CONCLUSIONS AND FUTURE PROSPECTS

Most of the molecules so far recognized as accelerators of tumor progression are among the inflammation-related factors. Based on that, we may conclude that an inflammatory environment is a niche for tumor progression.

It is important to note that inflammation, especially chronic inflammation, is an absolute cause of tumor development and progression, and it is accurately referred to as inflammation-related carcinogenesis. On the other hand, there is the indisputable fact that all chronic inflammations do not lead to carcinogenesis; for example, rheumatoid arthritis is not linked to cancer risk. It is reported that the inflammatory environment in rheumatoid arthritis patients shows p53 mutations at similar frequencies to those in digestive tract tumors arising from chronic inflammatory reaction (24). We may be able to determine the core of the inflammation that accelerates carcinogenesis and tumor progression by comparing the two typical types of inflammation.

REFERENCES

1. Toh B, Wang X, Keeble J, Sim WJ, Khoo K, Wong W-C, et al. Mesenchymal transition and dissemination of cancer cells is driven by myeloid-derived suppressor cells infiltrating the primary tumor. PLoS Biol 2011;9:e1001162.
2. Sethi G, Shanmugam MK, Ramachandran L, Kumar AP, Tergaonkar V. Multifaceted link between cancer and inflammation. Biosci Rep 2012;32:1–15.
3. Heinrich EL, Walser TC, Krysan K, Liclican EL, Grant JL, Rodriguez NL, et al. The inflammatory tumor microenvironment, epithelial mesenchymal transition and lung carcinogenesis. Cancer Microenviron 2012;5:5–18.
4. Balkwill F, Mantovani A. Inflammation and cancer: Back to Virchow? Lancet 2001;357:539–545.
5. Sansone P, Bromberg J. Environment, inflammation, and cancer. Curr Opin Genet Dev 2011;21: 80–85.
6. Mukaida N, Baba T. Chemokines in tumor development and progression. Exp Cell Res 2012;318: 95–102.
7. Prehn RT, Lappe MA. An immunostimulation theory of tumor development. Transplant Rev 1971;7:26–54.
8. Ishikawa M, Koga Y, Hosokawa M, Kobayashi H. Augmentation of B16 melanoma lung colony formation in C57BL/6 mice having marked granulocytosis. Int J Cancer 1986;37:919–924.
9. Hamada JI, Takeichi N, Kobayashi H. Metastatic capacity and intercellular communication between normal cells and metastatic cell clones derived from a rat mammary carcinoma. Cancer Res 1988;48:5129–5132.
10. Tan J, Buache E, Chenard M-P, Dali-Youcef N, Rio M-C. Adipocyte is a non-trivial, dynamic partner of breast cancer cells. Int J Dev Biol 2011;55:851–859.
11. Morrison WB. Inflammation and cancer: a comparative view. J Vet Intern Med 2012;26:18–31.
12. Allavena P, Germano G, Marchesi F, Mantovani A. Chemokines in cancer related inflammation. Exp Cell Res 2011;317:664–673.
13. Ohshima H, Sawa T, Akaike T. 8-Nitroguanine, a product of nitrative DNA damage caused by reactive nitrogen species: formation, occurrence, and implications in inflammation and carcinogenesis. Antioxid Redox Signal 2006;8:1033–1045.
14. Lisanti MP, Martinez-Outschoorn UE, Lin Z, Pavlides S, Whitaker-Menezes D, Pestell RG, et al. Hydrogen peroxide fuels aging, inflammation, cancer metabolism and metastasis. The seed and soil also needs "fertilizer." Cell Cycle 2011;10:2440–2449.
15. Guerra L, Guidi R, Frisan T. Do bacterial genotoxins contribute to chronic inflammation, genomic instability and tumor progression? FEBS J 2011;278:4577–4588.
16. Ishikawa K, Takenaga K, Akimoto M, Koshikawa N, Yamaguchi A, Imanishi H, et al. ROS generating mitochondrial DNA mutations can regulate tumor cell metastasis. Science 2008;320:661–664.
17. Sawa T, Zaki MH, Okamoto T, Akuta T, Tokutomi Y, Kim-Mitsuyama S, et al. Protein S-guanylation by the biological signal 8-nitroguanosine $3'$, $5'$-cyclic monophosphate. Nat Chem Biol 2007;3: 727–735.
18. Kuribayashi K, Nakamura K, Tanaka M, Sato T, Kato J, Sasaki K, et al. Essential role of protein kinase C zeta in transducing a motility signal induced by superoxide and a chemotactic peptide, fMLP. J Cell Biol 2007;176:1049–1060.

19. Ishikawa M, Okada F, Hamada J-I, Hosokawa M, Kobayashi H. Changes in the tumorigenic and metastatic properties of tumor cells treated with quercetin or 5-azacytidine. Int J Cancer 1987;39:338–342.

20. Okada F, Hosokawa M, Hamada J-I, Hasegawa J, Kato M, Mizutani M, et al. Malignant progression of a mouse fibrosarcoma by host cells reactive to a foreign body (gelatin sponge). Br J Cancer 1992;66:635–639.

21. Tazawa H, Okada F, Kobayashi T, Tada M, Mori Y, Une Y, et al. Infiltration of neutrophils is required for acquisition of metastatic phenotype of benign murine fibrosarcoma cells: implication of inflammation-associated carcinogenesis and tumor progression. Am J Pathol 2003;163:2221–2232.

22. Okada F, Kobayashi M, Tanaka H, Kobayashi T, Tazawa H, Iuchi Y, et al. The role of nicotinamide adenine dinucleotide phosphate oxidase-derived reactive oxygen species in the acquisition of metastatic ability of tumor cells. Am J Pathol 2006;169:294–302.

23. Okada F, Tazawa H, Kobayashi T, Kobayashi M, Hosokawa M. Involvement of reactive nitrogen oxides for acquisition of metastatic properties of benign tumors in a model of inflammation-based tumor progression. Nitric Oxide 2006;14:122–129.

24. Hussain SP, Harris CC. Inflammation and cancer: an ancient link with novel potentials. Int J Cancer 2007;121:2373–2380.

25. Okada F, Nakai K, Kobayashi T, Shibata T, Tagami S, Kawakami Y, et al. Inflammatory-cell-mediated tumour progression and minisatellite mutation correlate with the decrease of antioxidative enzymes in murine fibrosarcoma cells. Br J Cancer 1999;79:377–385.

26. Young MR, Okada F, Tada M, Hosokawa M, Kobayashi H Association of increased tumor cell responsiveness to prostaglandin E2 with more aggressive tumor behavior. Invasion Metastasis 1991;11:48–57.

27. Ren J, Hamada J, Okada F, Takeichi N, Morikawa K, Hosokawa M, et al. Correlation between the presence of microvilli and the growth or metastatic potential of tumor cells. Jpn J Cancer Res 1990;81:920–926.

28. Habelhah H, Okada F, Kobayashi M, Nakai K, Choi S, Hamada J, et al. Increased E1AF expression in mouse fibrosarcoma promotes metastasis through induction of MT1-MMP expression. Oncogene 1999;18:1771–1776.

29. Okada F. Inflammation and free radicals in tumor development and progression. Redox Rep 2002;7:357–368.

30. Kobayashi T, Okada F, Fujii N, Tomita N, Ito S, Tazawa H, et al. Thymosin β4 regulates motility and metastasis of malignant mouse fibrosarcoma cells. Am J Pathol 2002;160:869–882.

CHAPTER 12

HUMAN PAPILLOMAVIRUS AND CERVICAL CANCER

Kurt J. Sales

Cervical cancer is the second most common cancer in women worldwide (1–3). Statistics published by the International Agency for Research on Cancer (GLOBOCAN 2008) estimate that more than 493,243 women are newly diagnosed with cervical cancer each year, with more than 273,000 deaths occurring worldwide (1,3,4). The documented global cervical cancer prevalence is around 12% and is most widespread in sub-Saharan Africa (24%), Eastern Europe (21.4%), and Latin America (16.1%) (4,5). A staggering 85% of all cases of cervical cancer occur in developing countries. Since registrations of cervical cancer incidences on national databases are poor or inadequate in many of these countries, it is likely that these figures are underestimated and that the global incidence of morbidity and mortality as a result of cervical cancer is much higher. Significant epidemiological and laboratory evidence collected over the past four decades has established that infection of the cervical epithelium with oncogenic human papillomavirus (HPV) is the main causative agent for development of cervical cancer (6). Although HPV infection initiates the disease, cervical cancer is a multistep process and other factors are thought to contribute toward the etiology of the disease (7), since a large proportion of women infected with high-risk oncogenic HPV never develop neoplasms. In this chapter I provide an overview of cervical cancer and the role of HPV in regulating cervical carcinogenesis.

SYMPTOMS AND CLASSIFICATION OF CERVICAL CANCER

The diagnosis of cervical cancer is made after detection of irregular cells by cytology screening (Papanicolaou test or liquid cytology screening) and is confirmed by colposcopy and histology. Although cervical cancer can be asymptomatic, symptoms of advanced-stage disease may include abnormal vaginal bleeding, foul-smelling discharge, disruption of normal bladder or bowel function, and leg edema. Once

Cancer and Inflammation Mechanisms: Chemical, Biological, and Clinical Aspects, First Edition.
Edited by Yusuke Hiraku, Shosuke Kawanishi, and Hiroshi Ohshima.
© 2014 John Wiley & Sons, Inc. Published 2014 by John Wiley & Sons, Inc.

the basement membrane is invaded, the neoplastic process is termed *invasive*. The World Health Organization (WHO) classification system uses the terminology *dysplasia* and *carcinoma in situ* to refer to precursor lesions (8). In the 1960s, Richart introduced the term *cervical intraepithelial neoplasia* (CIN), which was later modified (9). The terms CIN grades 1 to 3 are used to describe preinvasive epithelial lesions or various categories of dysplasia and carcinoma *in situ*. *CIN grade 1* is equivalent to mild dysplasia in which undifferentiated cells (cells that are atypical with hyperchromatic nuclei, increased nuclear/cytoplasmic ratio, and mitotic index) occupy approximately the lower one-third of the epithelium. *CIN grade 2* is equivalent to moderate dysplasia, where undifferentiated cells replace two-thirds of the thickness of the normal epithelium. *CIN grade 3* denotes severe dysplasia and carcinoma *in situ*. Severe dysplasia describes a condition in which undifferentiated cells replace all but one or two of the most superficial cell layers of the cervical epithelium. When undifferentiated cells replace the entire surface of the epithelium, the diagnosis of carcinoma *in situ* is made. All degrees of dysplasia are preinvasive, meaning that the basement membrane (stromal–epithelial junction) remains intact. The Bethesda System of cytological diagnosis uses low-grade intraepithelial lesions (Box 1) to describe lesions referred to as dysplasia or CIN1 and high-grade intraepithelial lesions (Box 1) to denote severe dysplasia (CIN2 and CIN3) and carcinoma *in situ* (10). Carcinoma is defined according to the International Federation of Obstetricians and Gynecologists [FIGO; (11)] staging upon physical examination [Box 2 (12,13)].

The WHO recognizes three general categories of cervical cancer: squamous cell carcinoma, adenocarcinoma, and other epithelial tumors, including less common types, such as adenosquamous carcinoma, glassy cell carcinoma, and adenoid basal cell carcinoma as well as carcinoid-like and small-cell carcinoma (8). Approximately

---

BOX *1*

LOW-GRADE LESIONS

Low-grade lesions are referred to as low-grade squamous intraepithelial lesions (LSILs) or low-grade intraepithelial neoplasia (LGIN) and are characterized by disorganization of the basal cell and parabasal layers of the epithelium, crowding and overlapping of cells, and loss of polarity and koilocytosis (atypical nuclei and cytoplasmic cavitation), indicative of HPV infection. Low-grade squamous intraepithelial lesions are associated with infection by both low- and high-risk HPV types (Table 12.1), although low-risk types appear to predominate (10,14).

High-Grade Lesions

High-grade lesions are referred to as high-grade squamous intraepithelial lesions (HSILs) or high-grade intraepithelial neoplasia (HGIN) and are characterized by coarse granular chromatin, mitotic bodies, and loss of normal cell polarity and a high nuclear/cytoplasm ratio. The HPV types associated with it are classified as being almost exclusively of high oncogenic risk (Table 12.1) (10,14).

BOX 2

INTERNATIONAL FEDERATION OF OBSTETRICIANS AND GYNECOLOGISTS (FIGO) STAGING OF CERVICAL CARCINOMAS

I. Carcinoma confined to cervix.

IA. Preclinical carcinomas of the cervix (diagnosed by microscopy only).

IAi. Minimal microscopically evident stromal invasion.

IAii. Lesions detected microscopically that can be measured to a depth of not more than 5mm of invasion from the base of epithelium. The horizontal spread should not be more than 7 mm.

IB. Lesions of greater dimension than stage IAii.

IIA. Extends to upper two-thirds of the vagina.

IIB. Extends to paracervical tissue.

IIIA. Extends to lower one-third of the vagina.

IIIB. Pelvic sidewall extension or uretal obstruction on intravenous pyelogram.

IVA. Bladder or rectal mucosal involvement.

IVB. Distant metastasis.

80% of cervical carcinomas are squamous cell carcinoma. Adenocarcinomas and to lesser extent adenosquamous carcinomas and rarely sarcomas, lymphoma, and melanoma account for approximately 20% of the invasive cervical carcinoma and display a variety of types and subtypes (8,13).

Adenocarcinoma of the cervix arises from glandular epithelium lining the endocervical canal and the endocervical glands, and is considered to be multifocal in origin. The WHO has classified adenocarcinoma into five subtypes: (1) endocervical adenocarcinoma, (2) endometrial adenocarcinoma, (3) clear cell carcinoma, (4) adenoid cystic carcinoma, and (5) adenosquamous carcinoma. Squamous cell carcinomas arise in the metaplastic squamous cells lining the exocervix (12,13). Small stage I lesions are generally visible on speculum examination. Frequently, these give rise to barrel carcinoma, especially when from endocervical canal origin. Tumors are classified according to FIGO staging (Box 2).

HUMAN PAPILLOMAVIRUS AND CERVICAL CARCINOGENESIS

HPV Infection and Life Cycle

According to the International Committee on Taxonomy of Viruses (ICTV), papillomaviruses, belong to the Papillomaviridae family, comprising 29 genera and 189 papillomaviruses (PVs), including 120 human PVs, 69 nonmammalian PVs, 3 avian

TABLE 12.1   Classification of HPV Types Based on Their Oncogenic Potential

Low-risk HPV types	6, 11, 40, 42, 43, 44, 54, 61, 70, 72, 81, 83
High-risk HPV types	16, 18, 31, 33, 35, 39, 45, 51, 52, 56, 58, 59, 68, 73, 82
Probable high-risk HPV types	26, 53, 66

PVs, and 2 reptilian PVs (15). Human papillomaviruses preferentially infect the epithelia of the mucosa and skin, forming benign epithelial proliferations at the site of infection. As recently as 1970, the general view in the scientific community was that there was only one HPV species. Groundbreaking work conducted by zur Hausen and colleagues over the last four decades altered this perspective and led to the identification of more than 180 different papillomaviruses. These have been numbered sequentially in the order in which they were cloned from clinical lesions and are classified as genotypes on the basis of their DNA sequence (6). Several epidemiological studies over the years have associated HPV infection with cervical cancer. One of the most compelling laboratory studies correlating the association of HPV with cervical cancer was carried out in the 1990s. This study, conducted over 22 different countries on more than 1000 specimens of cervical squamous cell carcinoma, determined a 93% prevalence rate of HPV in cervical cancer, with HPV 16 being the predominant type (16). A follow-up reanalysis of the negative specimens from the same study using different polymerase chain reaction (PCR) primers raised this prevalence to 99.7% (17). The strong association of HPV and cervical cancer from these studies definitively established HPV as the main causative agent for the development of invasive cervical cancer and its precursor lesions.

HPV types are generally classified according to their oncogenic potential as described in Table 12.1. HPV types found to be most frequently associated with genital tract carcinoma and classified as being of highest oncogenic risk are HPV 16 (57%), 18 (16%), and 58, 33, 45, 31, 52, 35, 59 and 56 (18). HPV 16 and 18 are the most frequently detected types of HPVs, in adenocarcinoma and squamous cell carcinoma (19) and together are responsible for more than 70% of cervical cancers (1). Other groups of HPVs including types 5, 8, 9, 15, and 17, are genital tract viruses found to cause squamous intraepithelial lesions (SILs) of the cervix (16). Various other types of HPV are associated with SILs and invasive carcinomas of the cervix; however, these occur less frequently or are classified into low oncogenic risk groups.

Papillomaviruses are double-stranded DNA viruses approximately 8000 base pairs in length. The viral genome is divided into three sections: a noncoding upstream regulatory region (URR) of about 400 base pairs for regulation of transcription of viral proteins and virions called the *long control region* (LCR) (20); an early region downstream of the URR containing six open reading frames (ORF_1) encoding the E1, E2, E4, E5, E6, and E7 oncogenes (expressed early-on during infection); and a late region of two open reading frames encoding for the L1 major and L2 minor capsid proteins (expressed in the late phase of infection).

The primary route for HPV infection is via exposure of the anal, oral, or vaginal mucosa to virus present in saliva, seminal fluid, or in the infected partner's skin during

BOX 3

COFACTORS INVOLVED IN CERVICAL CANCER PROGRESSION

The precise contribution of cervical cancer–associated cofactors to disease onset and progression is still unclear. However, natural history studies have shown that viral persistence is key, and any set of factors that can facilitate persistence will enable disease onset and progression (25,26). Prolonged exposure to sex steroids has repeatedly been associated with HPV-mediated cancer. Oral contraceptives have been correlated with increased risk of cervical cancers (27), although the epidemiological data are slight. Nonetheless, animal models have shown clearly that estrogen can be a potential risk factor for HPV-induced cancer. In mice expressing HPV 16 under the control of the keratin 14 promoter, estradiol administration induces multistage progression of squamous cell carcinoma of the cervix and vagina, from squamous metaplasia through to LSILs and HSILs (28). Pregnancy, too, has been shown to create an environment permissive to HPV infection (29). Cigarette smoking and the accumulation of immune-suppressive tobacco carcinogens in the cervical mucus of smokers is thought to contribute to an accumulation of genetic mutation and to enhance the likelihood of cancer progression, although this association is seen for HSILs but not LSILs (27,30). In addition, the presence of other sexually transmitted infections, such as human immunodeficiency virus (HIV), herpes virus II (31), *Chlamydia trachomatis* (32), *Neisseria gonorrhea* (33), and bacterial vaginosis (34), are associated risk factors for cervical cancer development.

coitus. This results in urogenital or anorectal infection. HPV is highly contagious, with more than 65% of new sexual contacts at risk of infection. Unsurprisingly, the group with the highest risk of HPV infections is young adolescents and young sexually active adults. Although age and sexual behavior are regarded as the two major influences on genital HPV infection, several other risk factors thought to contribute toward HPV acquisition have been described (discussed in further detail in Box 3). In addition to their epithelia-tropic and mucosa-tropic preference, HPV DNA has also been found in nonepithelial and nonmucosal sites, including spermatozoa (21), placenta (22), and blood (23).

Acquisition of HPV infection occurs during or soon after sexual contact when the mucosal surface is permissive to infection. Here, HPV binds to the cell surface of basal keratinocytes, exposed through mild abrasion or microtrauma to the cervicovaginal epithelium. HPV attachment occurs via binding of L1 protein to heparin sulfate proteoglycan, present on the basal membrane, facilitating virus binding to a receptor expressed on the keratinocyte, followed by endocytosis (24). After transmission, the virus remains in the epithelial mucosa, where it is hidden from contact with the bloodstream and the innate immune system and thus manages to escape detection by evading immune defenses (discussed further in Box 4).

After viral entry and uncoating, the HPV genomic DNA is transported to the nucleus, where it is maintained at low copy number (50 to 100 copies per cell in the basal layer). The HPV genome is stabilized as a nonintegrated episome, which

BOX 4

MECHANISMS OF IMMUNE EVASION BY HPV

It is estimated that about 70% of all cases of HPV infection resolve spontaneously within one year (35) and about 90% resolve within two years (36). The resolution of these infections appears to be regulated by the host immune defenses (37), although the length of time taken to resolve infection suggests that the host response to infection is weak. This is due partially to the nonlytic mode of infection of HPV and the active mechanisms induced by the virus to induce immune evasion. Since no viremia or cytolysis is associated with initial viral infection of the cervix, there is no activation of the innate immune system, no migration of antigen-presenting cells, and no inflammation. However, the virus actively induces mechanisms to evade immune detection and ensure its success by deregulating the interferon pathway (38–40), reducing the number of Langerhans cells, inhibiting expression of MHC-1 molecules (41) and the down-regulation of pattern recognition receptors such as Toll-like receptor 9 (42), thereby allowing infection to proceed undetected for a considerable time. In immunosuppressed persons, the incidence and the progression of disease is accelerated. These observations illustrate that the virus goes to great lengths to camouflage itself, and despite a weak host-immune response, host defenses and cell-mediated immunity play a role in control of the infection (43).

can remain dormant or replicate to reinfect cells. After infection the cell divides and infects the suprabasal layers. Productive replication of HPV occurs coincident with the maturation cycle of the keratinocyte, which is described in detail in several reviews (38,44). Since no DNA polymerase is encoded by the HPV genome, it has to use the host's cellular machinery for replication. The critical role of the HPV E6 and E7 oncogenes is discussed in greater detail below; however, their contribution to cancer has been underscored by observations that most of the viral genome can be deleted in the process of chromosomal integration. However, E6 and E7 oncogenes are always retained, and transcription of their respective oncoprotein is always detectable in carcinomas and many cases of HSIL (45–48). E6 and E7 function is essential for reactivating cell division in differentiated cells in order for replication of the viral genome to take place. In the upper layers of the stratum spinosum and granulosum of the stratified epithelium, viral genome amplification is accelerated to thousands of copies per cell. As the cell differentiates, and following successful genome amplification, the viral capsid forms and the resulting virion can be released at the cell surface to infect other sites. Some of the low-risk HPV variants, such as HPV 6 and 11, manifest as benign epithelial proliferations—generally, condyloma acuminatum or genital warts (49). The oncogenic variants have the ability to integrate within the host genome (47,48), resulting in CIN. Although these have the capacity to resolve, under the influence of the E6 and E7 oncogenes, they can progress to precancerous lesions (CIN2 and 3) and invasive carcinoma. The general consensus has been that HPV infections are rapidly resolved and rarely lead to developments of SILs and that it takes persistent infection and years for HSILs to develop. Although persistent infection with high-risk types is the most crucial factor for CIN and cancer,

TABLE 12.2   HPV Oncogenes and Their Function

Gene	Function	Target factor
E1	Viral genome replication	Topoisomerase
E2	Viral genome replication	Brd4, C/EBP, p300/CBP,
	Gene transcription	SF2/ASF, Cdk2
	Viral genome maintenance	
E4	Viral assembly	Cytokeratin 8/18
E5	Viral assembly	EGFR, MHC1, PDGFR,
	Immune evasion	TRAIL receptor, FAS
	Regulation of proliferation and apoptosis	receptor
E6	Reactivation of cellular replication	p53, p300/CBP, E6AP, SP1,
	Proliferation, immortalization, inhibition of apoptosis	cMyc, FAK, caspase 8, BAX, BAK, PDZ domain
	Maintenance of viral genome	proteins
E7	Reactivation of cellular replication	RB, p107, p130, p21, p27,
	Proliferation, immortalisation, inhibition of apoptosis	CDK/cyclin
	Maintenance of viral genome	
L1	Major capsid protein	
L2	Minor capsid protein	

a longitudinal study of 603 female university students demonstrated that half of the biopsy-confirmed cases of CIN2 and CIN3 occurred within 14 months of initial HPV infection, indicating that HSILs can actually manifest early in young sexually active women (50). These findings are consistent with other studies on young women that have shown that SILs can develop early, within two years of initial infection.

HPV Oncogenes

HPV oncogene expression is regulated coincident with differentiation and maturation of the cervical epithelium. Some of the factors known to be regulated by HPV oncogenes are summarized in Table 12.2. The E1 and E2 oncogenes are involved in viral DNA replication, although some studies show that replication of HPV viral DNA can occur without E1 or E2 (51). In addition to the regulation of HPV viral oncogene transcription from the LCR, HPV16 E2 has been shown to transactivate cellular promoters to regulate splicing-related proteins within the cell (52,53). The E2 oncogene functions as a dimer, which binds to four conserved palindromic sequences in the LCR. Most reports describe E2 as a transcriptional repressor of HPV oncogenes; however, one of its more notable functions is to mediate viral genome segregation during cell division via interaction with mitotic bodies (54,55). For E2 to fulfill its role in replication, the host cell needs to be in S phase. Here, the E2 protein is stabilized and recruits the E1 helicase via protein–protein interaction to initiate viral genome replication (56). Loss of E1 and E2 ORFs, allowing integration of the E6 and E7 oncogenes from oncogenic high-risk HPVs, is regarded as one of the first steps in transformation. After integration, the E6 and E7 early genes are actively transcribed and the E6 and E7 oncoproteins, both of which are involved in neoplastic transformation, are actively expressed (57). *In vitro* studies have shown that both oncoproteins target different molecular pathways in the cell [reviewed in (58)]. The E6 oncoprotein from high-risk oncogenic variants activates telomerase (59) and binds to p53 protein and targets it for ubiquitin-dependent degradation (60), whereas E7 oncoprotein from high-risk HPV binds to family members of the retinoblastoma protein (Rb) and disrupts the complex between Rb and the E2F transcription factor family, which controls the expression of genes involved in cell-cycle progression (61). Deregulation of p53 and Rb by deletion or mutation is a hallmark of cancer (62).

The ability of oncogenic high-risk HPV variants to interfere with cell-cycle control appears to be crucial for immortalization of cells and triggers the early steps in malignant conversion (39,61). Animal studies using transgenic mouse models of cervical cancer have highlighted more specifically the contribution of individual oncogenes to tumorigenesis and have shown that the E7 protein induces high-grade cervical dysplasia and invasive tumors, whereas the E6 protein induces only low-grade cervical dysplasia. Furthermore, they have highlighted that both oncogenes work in synergy to promote cervical cancer progression since coexpression of the E6 and E7 proteins produces larger and more extensive tumors compared with those produced by E7 alone (63). Furthermore, E6 and E7 from high-risk oncogenic variants are thought to be responsible for the chromosomal aneuploidy observed in HSILs, by cooperating to generate abnormal centrosome numbers, which in turn derails normal mitosis, thus introducing genome instability (64). Although E6 and E7 oncogenes appear to be the main HPV genes involved in transformation, recent studies have highlighted a role for the E5 oncogene in immune cell modulation and tumorigenesis and the regulation of late viral functions (65). E5 has been associated with loss of surface MHC-1 expression in epithelial cells, leading to evasion of immune surveillance early-on during infection (66). E5 oncoprotein is thought to work in synergy with E4 oncoprotein, which has been shown to interact with cytoskeletal proteins and allow viral assembly. As the epithelial cell differentiates from the basal to the upper layers, the L1 and L2 major and minor capsid proteins become expressed (67). Although these are detected in less differentiated epithelia, fully processed mRNAs are only reported to occur in fully differentiated cells (67). Once expressed and fully processed, the viral capsid forms, packaging the DNA into a new virion, which can be released at the cell surface to mediate infection of naive cells.

Regulation of Cervical Cancer by HPV Oncogenes

Cancer is a multistep process characterized by several hallmarks acquired during tumor development: sustained proliferative signaling, evasion of growth suppressors, attainment of replicative immortality, angiogenesis, resistance to cell death, activation of invasion, and metastasis [reviewed in (68)]. Recent enabling characteristics and emerging hallmarks include genomic instability and mutation, alterations in energy metabolism, evasion of immune destruction, and inflammation (68). As highlighted earlier, cervical cancer progresses sequentially from normal cervix to CIN, to carcinoma *in situ*, to local invasion of the lymphatic system, and spread to the distant lymph nodes around the vessels on the pelvic wall. Metastasis remains the primary form of treatment failure in women with cervical cancer and the most frequent cause of death. For cells to invade and metastasize, the epithelial cells need to undergo a series of changes. In normal tissue the epithelium is present as a coordinated layer of cells that under normal conditions exhibit an apical-to-basal polarity. Adjacent cells are juxtaposed and held together by tight and adherens junctions. For cells to become mobile and metastasize, they need to break away from the primary site, and often do this by altering their phenotype and adopting a metastable, more mesenchyme phenotype (69). Appropriately termed *epithelial-to-mesenchymal transition* (EMT), metastable cells are characterized by mixed epithelial and mesenchymal markers, such as the presence of vimentin, loss of tight and adherens junctions, and loss of polarity, and are a characteristic of many solid tumors. When EMT transition is complete, the cells will exhibit a front–back polarity with no junctions and will be characterized by the presence of smooth muscle actin in the cytoskeleton, displaying increased scattering, migration, invasion, and resistance to anoikis (69).

In cervical cancer cells, HPV infection is regarded as a key factor in cellular transformation and EMT. HPV 16 E7 oncogene has been shown to cause molecular changes in cells, including the elimination of epithelial features, indicative of mesenchymal transition (70,71). In several model systems, including mouse models and *in vitro* studies, chemokine receptors such as CXCR4 and CCR7 have been shown to play a role in cell migration (72,73) and lymph node metastasis in cervical cancer (74). There is evidence that these receptors and their ability to promote metastasis are mediated by HPV E6 and E7 oncogenes, since treatment of cancer cells with antiviral agents, which inhibit E6 and E7 function, concomitantly reduces CXCR4 expression and function and metastases (75). For cells to move and metastasize, gross tissue remodeling events need to occur. Extracellular matrix needs to be degraded to facilitate cell movement, a process regulated by matrix metalloproteinases (MMPs). Up-regulation in MMP expression and activity is associated with progression of cervical cancer (76). Several studies have correlated transcription of HPV E6 and E7 with transcription of MMPs (77,78), suggesting that HPV oncogenes can drive tissue remodeling. Evidence from microarray analysis supports this, as HPV E6 and E7 oncoproteins have been shown to regulate several genes involved in tissue differentiation and remodeling, including genes involved in angiogenesis (79), inflammation, apoptosis, and tumor progression (80,81).

HPV Oncogenes as Regulators of Tumor-Associated Inflammation

Tumor-associated inflammation is an emerging hallmark of cancer. Around 20% of cancers occur as a result of inflammation, and cervical cancer in particular is regarded as a chronic inflammatory disease (38). The classical definition of inflammation refers to the mobilization of immune cells to the site of injury or infection, to promote repair or removal of the insult or invading pathogen. The hallmarks of inflammation—rubor (redness), calor (increased heat), tumor (swelling), dolor (pain), and function laesa or loss of function—were originally described by the Roman encyclopedist Aulus Cornelius Celsus (ca. 25 B.C–ca. A.D. 50) and later, Rodulph Virchow (82). These descriptions of an inflammatory tissue clearly define inflammation as a series of physiological events that involve extensive tissue remodeling across multiple cellular compartments and are orchestrated by complex networks of growth factors, cytokines, chemokines, and bioactive lipids.

Several reports have correlated the inflammatory cell infiltrate with HPV-induced high-grade lesions. Although HPV infection is not associated with inflammation at the onset, it is emerging that HPV oncogenes can regulate genes involved in tissue remodeling events in neoplastic cervical epithelial cells, which are critical regulators of the inflammatory processes in cancerous tissue. Resident HPV infections have been shown to promote the release of inflammatory mediators and cytokines from keratinocytes to alter the immune response and promote the infiltration of macrophages, lymphocytes, and natural killer cells (38). HPV-induced lesions have been shown to release the chemokine CCL2, which enhances macrophage recruitment into tumors (83). Infiltrating immune cells are considered to be potent regulators of disease progression, especially in the case of polarized tumor-associated macrophages and neutrophils (84). Once resident in the tissue, inflammatory cells are known to produce vast amounts of reactive oxygen species and nitric oxide. These have been shown to induce DNA damage and to contribute to the progression of the disease in high-grade cervical lesions. Furthermore, nitric oxide–induced DNA damage associated with HPV infection has been reported in cultured cervical epithelial cells (85) and cervical cancer biopsy specimens (86). In cultured cervical epithelial cells, exposure of HPV-infected cells to nitric oxide not only induces transcription of E6 and E7 oncogenes, but also decreases p53 and Rb protein levels and the mitotic index and increases DNA damage, double-stranded breaks, and mutagenic frequencies (85).

Immunohistochemistry conducted on cervical cancer biopsy specimens has shown that expression of nitric oxide synthase, the enzyme responsible for production of nitric oxide, is present in both the epithelial and immune compartments in HPV-positive cancer lesions (86). These findings indicate that nitric oxide, produced by the immune infiltrate and epithelium, is a potent molecular cofactor in regulating cervical cancer (85). There is some evidence that nitric oxide can promote HPV persistence, although it has not been shown to enhance viral replication in cervical cancer cells *in vitro* (85). Nonetheless, once established, persistent infections can promote further alterations in the release of inflammatory cytokines, which in turn can alter immune cell infiltration and inflammation. Alterations in immune responsiveness and elevated systemic levels of inflammatory cytokines have been observed in older women with persistent HPV infection (87,88). It is likely that this elevated systemic inflammation can drive tumor progression.

Emerging evidence shows that HPV 16 E5, E6, and E7 oncogenes can drive tumor-associated inflammation in immortalized cells by inducing the inflammatory cyclooxygenase (COX)–prostaglandin (PG) axis, by elevating expression of the immediate early oncogene COX-2 and expression of the E-series prostaglandin receptors (PTGERs), such as PTGER2 and PTGER4 (65,89,90). Cyclooxygenases, of which there are two isoforms in humans, COX-1 and COX-2, catalyze the rate-limiting conversion of arachidonic acid, derived by deesterification of plasma membrane phospholipids or dietary polyunsaturated fatty acids, to eicosanoids (prostaglandins, thromboxanes, and prostacyclins). Many chronic inflammatory diseases, including allergy, asthma, atherosclerosis, autoimmunity, transplant rejection, metabolic and degenerative diseases, and cancer (91), are associated with up-regulation in COX enzyme expression and aberrant biosynthesis of pro-inflammatory eicosanoids. Elevated COX enzyme expression and biosynthesis and signaling of prostaglandins have been observed in cervical cancers and are considered a key modulator of tumor progression (92–94). In the past two decades, numerous studies have highlighted the fact that prostaglandins, produced as a consequence of elevated COX enzyme expression, can promote extensive tissue remodeling within tumors by evoking all the classical hallmarks of cancer: cellular proliferation, angiogenesis, inhibition of apoptosis, and alteration in vascular permeability, to allow immune cell extravasation from the vasculature and inflammation (95,96). These hallmarks of cancer in cervical cancer cells have now all been shown to be driven by HPV oncogenes via the induction of potent pro-inflammatory pathways (65,89,90).

THERAPEUTIC INTERVENTIONS

The global burden of HPV-associated disease is thought to be on the order of 60 million cases per year (97). Recently, prophylactic vaccines for HPV 16 and 18—Cervarix (GlaxoSmithKline)—and for HPV 6, 11, 16, and 18—Gardasil (Merck & Co.)—have been developed. Vaccine-induced protection against HPV infection occurs via IgG-neutralizing antibodies that prevent HPV entry into the basal cell. This is achieved by preventing the conformational change of the virus and subsequent inhibition of binding to its receptor on the basal cell, effectively preventing infection. Although the antibodies produced by the host immune response to vaccination is type-specific, there is a degree of cross-protection against HPV types genetically and antigenically closely related to the vaccine type (98). However, prophylactic vaccines cannot be used therapeutically and are no good for women with HPV infection or neoplastic lesions, highlighting the need for adequate cytology screening for all women. Over the years, several alternative intervention strategies have been proposed for infected women; these include molecular intervention strategies to inhibit viral replication by targeting E1 and E2 oncogenes [discussed in [99]] or E6/E7 oncogene, to reverse the effects of these high-risk oncogenes on the disruption to tumor suppressor pathways. These gene-silencing approaches using RNA interference to promote growth inhibition of cervical cancer cells have demonstrated some success *in vitro* (100–102). Strategies aimed at inhibiting oncogenes which are up-regulated by HPV E6 and E7 oncogenes have also been suggested (103), as well as inhibitors of viral replication and repackaging at the end of the life cycle.

Current therapies for HPV infections are mainly surgical, by scissor excision, ablative by cryotherapy, laser therapy, or electrosurgery (104). Some nonsurgical interventions are available in the form of a topical gel or cream and are used extensively for genital warts [reviewed in (105)]. These kill cells on contact, eliminating the virus. One of the major issues for developing countries, where the majority of infections occur, is affordability of treatment. Current prophylactic vaccination is out of reach of the majority of women in Africa, for example, as they are not available at a national health service level but are only available at substantial cost via private medical practices. To this end, other means of prevention of virus entry are needed in poorer countries. Condoms are ineffective at preventing transmission completely, so it would appear that topical treatments such as microbicides or virocides may be the only other alternative. Recent evidence from rodent models indicates that this might be a feasible intervention. For example, the sulfate polysaccharide carrageenan, extracted from seaweed, has been shown to inhibit virus binding to heparin sulfate proteoglycans (106). Since carrageenan is already listed among the various ingredients in lubricant gels used for vaginal examination in clinics, it could be a naturally available compound that is worth exploring further for women in developing countries.

REFERENCES

1. Arbyn M, Castellsague X, de Sanjose S, Bruni L, Saraiya M, Bray F, et al. Worldwide burden of cervical cancer in 2008. Ann Oncol 2011;22:2675–2686.
2. Anorlu RI. Cervical cancer: the sub-Saharan African perspective. Reprod Health Matters 2008;16:41–49.
3. Ferlay J, Shin HR, Bray F, Forman D, Mathers C, Parkin DM. Estimates of worldwide burden of cancer in 2008: GLOBOCAN 2008. Int J Cancer 2010;127:2893–2917.
4. Jemal A, Bray F, Center MM, Ferlay J, Ward E, Forman D. Global cancer statistics. CA Cancer J Clin 2011;61:69–90.
5. Bruni L, Diaz M, Castellsague X, Ferrer E, Bosch FX, de Sanjose S. Cervical human papillomavirus prevalence in 5 continents: meta-analysis of 1 million women with normal cytological findings. J Infect Dis 2010;202:1789–1799.
6. zur Hausen H. Papillomaviruses in the causation of human cancers: a brief historical account. Virology 2009;384:260–265.
7. Al-Daraji WI, Smith JH. Infection and cervical neoplasia: facts and fiction. Int J Clin Exp Pathol 2009;2:48–64.
8. Scully RE, Bonfiglio TA, Kurman RJ, Silverberg SG, Wilkinson EJ. Histological Typing of Female Genital Tract Tumors. Berlin: Springer-Verlag; 1994.
9. Richart RM. A modified terminology for cervical intraepithelial neoplasia. Obstet Gynecol 1990;75:131–133.
10. Luff RD. The Bethesda System for reporting cervical/vaginal cytologic diagnoses: report of the 1991 Bethesda workshop. The Bethesda System Editorial Committee. Hum Pathol 1992;23:719–721.
11. FIGO. T.N.M. *Atlas*, 3rd ed. (2nd rev. ed), Heidelberg, Germany: Springer-Verlag; 1992.
12. Hempling RE. Preinvasive Lesions of the Cervix: Diagnosis and Management., 2nd ed. Boston: Little, Brown; 1995.
13. Blake PR. *Tumors of the* Uterine Cervix and Corpus Uteri. London: Chapman & Hall: 1995.
14. Lungu O, Sun XW, Felix J, Richart RM, Silverstein S, Wright TC, Jr. Relationship of human papillomavirus type to grade of cervical intraepithelial neoplasia. JAMA 1992;267:2493–2496.
15. Bernard HU, Burk RD, Chen Z, van Doorslaer K, Hausen H, de Villiers EM. Classification of papillomaviruses (PVs) based on 189 PV types and proposal of taxonomic amendments. Virology 2010;401:70–79.
16. Bosch FX, Manos MM, Munoz N, Sherman M, Jansen AM, Peto J, et al. Prevalence of human papillomavirus in cervical cancer: a worldwide perspective. International biological study on cervical cancer (IBSCC) Study Group. J Natl Cancer Inst 1995;87:796–802.
17. Walboomers JM, Jacobs MV, Manos MM, Bosch FX, Kummer JA, Shah KV, et al. Human papillomavirus is a necessary cause of invasive cervical cancer worldwide. J Pathol 1999;189:12–19.
18. Li N, Franceschi S, Howell-Jones R, Snijders PJ, Clifford GM. Human papillomavirus type distribution in 30,848 invasive cervical cancers worldwide: variation by geographical region, histological type and year of publication. Int J Cancer 2011;128:927–935.

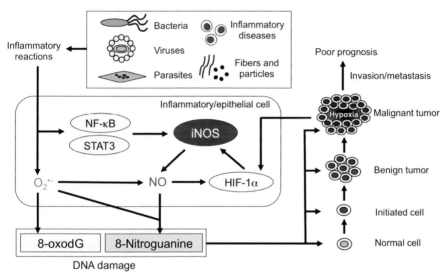

FIGURE 4.4 Proposed mechanisms of inflammation-related carcinogenesis via nitrative DNA damage.

Cancer and Inflammation Mechanisms: Chemical, Biological, and Clinical Aspects, First Edition.
Edited by Yusuke Hiraku, Shosuke Kawanishi, and Hiroshi Ohshima.
© 2014 John Wiley & Sons, Inc. Published 2014 by John Wiley & Sons, Inc.

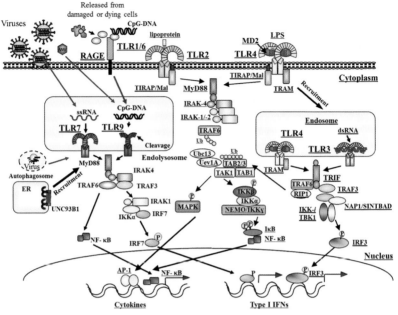

FIGURE 7.1 TLR signaling pathways. TLR4–MD2 complex and a heterodimer of TLR1/6 and TLR2 recognize the LPS and the lipoproteins on the cell surface, respectively, thereby recruiting MyD88 and TIR domains containing adaptor protein (TIRAP) to the TLR, and a complex of IRAKs and TRAF6 is subsequently formed. TRAF6 acts as an E3 ubiquitin ligase and catalyzes formation of a K63-linked polyubiquitin chain on TRAF6 itself and generation of an unconjugated polyubiquitin chain with an E2 ubiquitin ligase complex of Ubc13 and Uev1A. Ubiquitination activates a complex of TAK1, TAB1, and TAB2/3, resulting in the phosphorylation of NF-κB essential modulator (NEMO) and activation of an IKK complex. Degradation of the phosphorylated IκB by the ubiquitin proteasome system allows the translocation of NF-κB to the nucleus, where it crusades the expression of cytokine genes. Simultaneously, TAK1 activates MAP kinase cascades, leading to the activation of activator protein 1 (AP-1), which is also decisive for the induction of cytokine genes. LPS induces translocation of TLR4 to the endosome together with TRIF-related adaptor molecule (TRAM). TLR3 is present in the endosome and recognizes dsRNA. TLR3 and TLR4 activate TRIF-dependent signaling, which activates NF-κB and IRF3, resulting in the induction of pro-inflammatory cytokine genes and type I IFNs. TRAF6 and receptor-interacting protein 1 (RIP1) activate NF-κB, whereas TRAF3 is responsible for the phosphorylation of IRF3 by TBK1/IKK-i. Nucleosome assembly protein 1 (NAP1) and (similar to NAP1) TBK1 adaptor (SINTBAD) are required for the activation of TBK1/IKK-i. Phosphorylated IRF3 translocates into the nucleus to induce expression of type I IFN genes. Viral ssRNA and CpG DNA are recognized by TLR7 and TLR9, respectively. Viruses that have entered the cytoplasm are engulfed by autophagosomes and deliver viral nucleic acids to the endolysosome. An HMGB1-DNA complex released from damaged cells is captured by a receptor of advanced glycation end products (RAGE). Autoantibodies recognizing self-DNA or -RNA bind to FcγRIIa. LL37, an antimicrobial peptide, associates with endogenous DNA. These proteins are responsible for the delivery of endogenous nucleic acids to endolysosomes. Stimulation with ligands or infection by viruses induces trafficking of TLR7 and TLR9 from the ER to the endolysosome via UNC93B1. TLR9 undergoes cleavage by proteases present in the endolysosome. A complex of MyD88, IRAK-4, TRAF6, TRAF3, IRAK-1, IKK-α, and IRF7 is recruited to the TLR. Phosphorylated IRF7 translocates into the nucleus and up-regulates the expression of type I IFN genes.

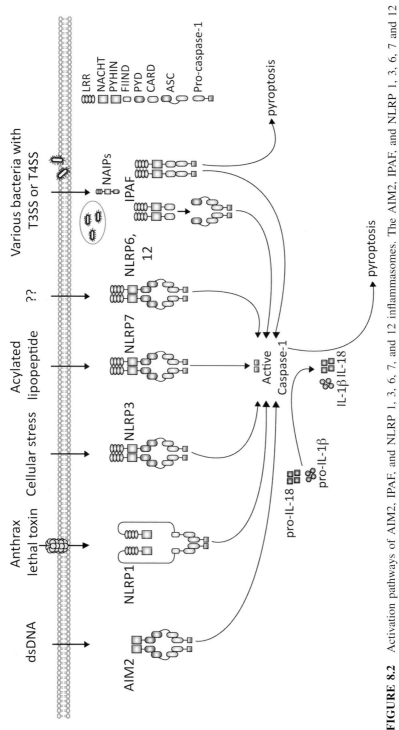

FIGURE 8.2 Activation pathways of AIM2, IPAF, and NLRP 1, 3, 6, 7, and 12 inflammasomes. The AIM2, IPAF, and NLRP 1, 3, 6, 7 and 12 inflammasomes are activated by distinct microbial triggers or cellular pathways. Activation of the IPAF inflammasome is mediated by NAIP proteins of the NLR family. Inflammasome assembly culminates in the activation of caspase-1, the caspase-1-directed maturation and secretion of IL-1β and IL-18, and pyroptotic cell death.

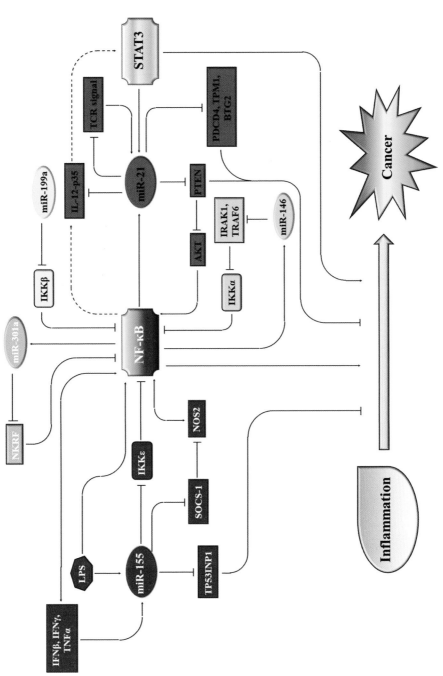

FIGURE 10.2 miRNAs participate in inflammation-induced cancinogenesis.

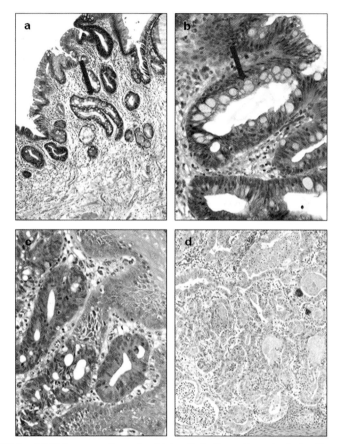

FIGURE 15.1 (a) Columnar cells and goblet cells in Barrett's metaplasia. The goblet cells are characterized by mucus in their cytoplasm (arrow). HE × 100. (b) Low-grade intraepithelial neoplasia. Slight architectural irregularities. Nuclear atypia. Some goblet cells are preserved (arrow). HE × 250. (c) High-grade intraepithelial neoplasia. More pronounced architectural irregularities and nuclear atypia. HE × 250. (d) Intramucsosal Barrett's carcinoma. HE × 100.

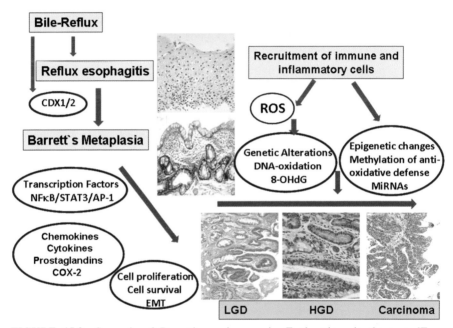

FIGURE 15.2 Synopsis of Barrett's carcinogenesis. Explanations in the text. [From Poehlmann et al. (7)].

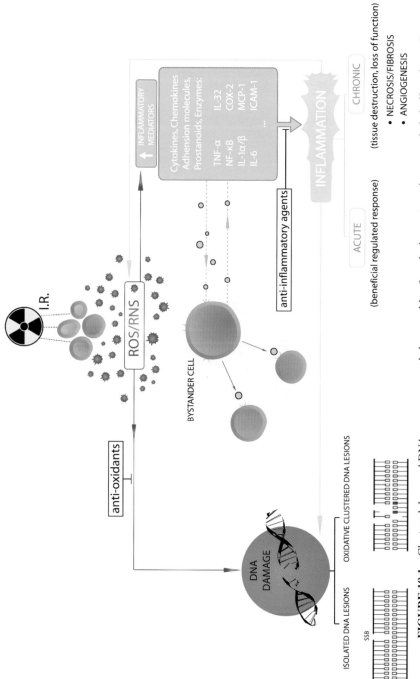

FIGURE 18.1 Clustered damaged DNA areas are regarded as capable of producing mutagenic or even lethal effects in a cell.

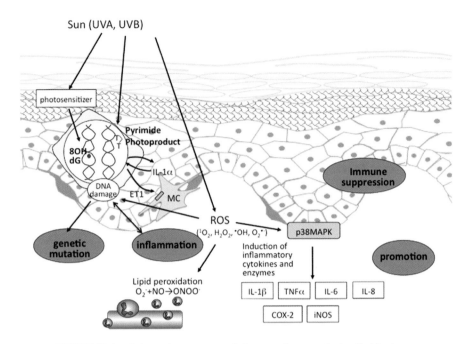

FIGURE 19.1   Schematic summary of photocarcinogenesis detailed in the text.

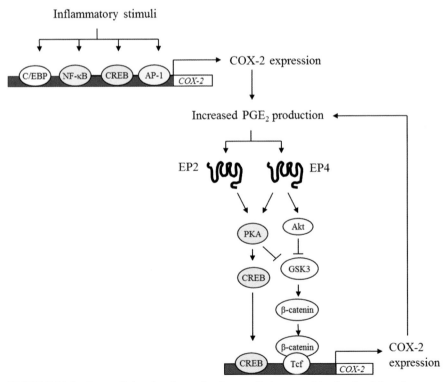

FIGURE 22.2   Intracellular signal transduction mediating a positive feedback loop between COX-2 and an EP receptor.

19. Apple RJ, Erlich HA, Klitz W, Manos MM, Becker TM, Wheeler CM. HLA DR-DQ associations with cervical carcinoma show papillomavirus-type specificity. Nat Genet 1994;6:157–162.

20. Turek LP. The structure, function, and regulation of papillomaviral genes in infection and cervical cancer. Adv Virus Res 1994;44:305–356.

21. Kaspersen MD, Larsen PB, Ingerslev HJ, Fedder J, Petersen GB, Bonde J, et al. Identification of multiple HPV types on spermatozoa from human sperm donors. PLoS One 2011;6:e18095.

22. Weyn C, Thomas D, Jani J, Guizani M, Donner C, Van Rysselberge M, et al. Evidence of human papillomavirus in the placenta. J Infect Dis 2011;203:341–343.

23. Sathish N, Abraham P, Peedicayil A, Sridharan G, John S, Shaji RV, et al. HPV DNA in plasma of patients with cervical carcinoma. J Clin Virol 2004;31:204–209.

24. Kines RC, Thompson CD, Lowy DR, Schiller JT, Day PM. The initial steps leading to papillomavirus infection occur on the basement membrane prior to cell surface binding. Proc Natl Acad Sci USA 2009;106:20458–20463.

25. Remmink AJ, Walboomers JM, Helmerhorst TJ, Voorhorst FJ, Rozendaal L, Risse EK, Meijer CJ, Kenemans P. The presence of persistent high-risk HPV genotypes in dysplastic cervical lesions is associated with progressive disease: natural history up to 36 months. Int J Cancer 1995;61:306–311.

26. Londesborough P, Ho L, Terry G, Cuzick J, Wheeler C, Singer A. Human papillomavirus genotype as a predictor of persistence and development of high-grade lesions in women with minor cervical abnormalities. Int J Cancer 1996;69:364–368.

27. Brisson J, Morin C, Fortier M, Roy M, Bouchard C, Leclerc J, et al. Risk factors for cervical intraepithelial neoplasia: differences between low- and high-grade lesions. Am J Epidemiol 1994;140:700–710.

28. Arbeit JM, Howley PM, Hanahan D. Chronic estrogen-induced cervical and vaginal squamous carcinogenesis in human papillomavirus type 16 transgenic mice. Proc Natl Acad Sci USA 1996;93:2930–2935.

29. Schneider A, Hotz M, Gissmann L. Increased prevalence of human papillomaviruses in the lower genital tract of pregnant women. Int J Cancer 1987;40:198–201.

30. Sasson IM, Haley NJ, Hoffmann D, Wynder EL, Hellberg D, Nilsson S. Cigarette smoking and neoplasia of the uterine cervix: smoke constituents in cervical mucus. N Engl J Med 1985;312:315–316.

31. Smith JS, Herrero R, Bosetti C, Munoz N, Bosch FX, Eluf-Neto J, et al. Herpes simplex virus-2 as a human papillomavirus cofactor in the etiology of invasive cervical cancer. J Natl Cancer Inst 2002;94:1604–1613.

32. Anttila T, Saikku P, Koskela P, Bloigu A, Dillner J, Ikaheimo I, et al. Serotypes of *Chlamydia trachomatis* and risk for development of cervical squamous cell carcinoma. JAMA 2001;285:47–51.

33. Castle PE, Giuliano AR. Chapter 4: Genital tract infections, cervical inflammation, and antioxidant nutrients: assessing their roles as human papillomavirus cofactors. J Natl Cancer Inst Monogr 2003;29–34.

34. Castle PE, Hillier SL, Rabe LK, Hildesheim A, Herrero R, Bratti MC, et al. An association of cervical inflammation with high-grade cervical neoplasia in women infected with oncogenic human papillomavirus (HPV). Cancer Epidemiol Biomark Prev 2001;10:1021–1027.

35. zur Hausen H. Papillomaviruses in human cancers. Proc Assoc Am Physicians 1999;111:581–587.

36. Ho GY, Bierman R, Beardsley L, Chang CJ, Burk RD. Natural history of cervicovaginal papillomavirus infection in young women. N Engl J Med 1998;338:423–428.

37. Frazer I. Correlating immunity with protection for HPV infection. Int J Infect Dis 2007;11 (Suppl 2):S10–S16.

38. Boccardo E, Lepique AP, Villa LL. The role of inflammation in HPV carcinogenesis. Carcinogenesis 2010;31:1905–1912.

39. Stanley MA, Pett MR, Coleman N. HPV: from infection to cancer. Biochem Soc Trans 2007;35:1456–1460.

40. Stanley MA. Epithelial cell responses to infection with human papillomavirus. Clin Microbiol Rev 2012;25:215–222.

41. Bottley G, Watherston OG, Hiew YL, Norrild B, Cook GP, Blair GE. High-risk human papillomavirus E7 expression reduces cell-surface MHC class I molecules and increases susceptibility to natural killer cells. Oncogene 2008;27:1794–1799.

42. Hasan UA, Bates E, Takeshita F, Biliato A, Accardi R, Bouvard V, et al. TLR9 expression and function is abolished by the cervical cancer-associated human papillomavirus type 16. J Immunol 2007;178:3186–3197.

43. Benton EC, Arends MJ. Human Papillomavirus in the Immunosuppressed. Leeds, UK: Leeds University Press; 1996.

44. Kajitani N, Satsuka A, Kawate A, Sakai H. Productive lifecycle of human papillomaviruses that depends upon squamous epithelial differentiation. Front Microbiol 2012;3:152.

45. Schwarz E, Freese UK, Gissmann L, Mayer W, Roggenbuck B, Stremlau A, et al. Structure and transcription of human papillomavirus sequences in cervical carcinoma cells. Nature 1985;314:111–114.

46. Cone RW, Minson AC, Smith MR, McDougall JK. Conservation of HPV-16 E6/E7 ORF sequences in a cervical carcinoma. J Med Virol 1992;37:99–107.

47. Daniel B, Mukherjee G, Seshadri L, Vallikad E, Krishna S. Changes in the physical state and expression of human papillomavirus type 16 in the progression of cervical intraepithelial neoplasia lesions analysed by PCR. J Gen Virol 1995;76 (Pt 10):2589–2593.

48. Mincheva A, Gissmann L, zur Hausen H. Chromosomal integration sites of human papillomavirus DNA in three cervical cancer cell lines mapped by *in situ* hybridization. Med Microbiol Immunol 1987;176:245–256.

49. Gissmann L, deVilliers EM, zur Hausen H. Analysis of human genital warts (condylomata acuminata) and other genital tumors for human papillomavirus type 6 DNA. Int J Cancer 1982;29:143–146.

50. Winer RL, Kiviat NB, Hughes JP, Adam DE, Lee SK, Kuypers JM, et al. Development and duration of human papillomavirus lesions, after initial infection. J Infect Dis 2005;191:731–738.

51. Pittayakhajonwut D, Angeletti PC. Viral trans-factor independent replication of human papillomavirus genomes. Virol J 2010;7:123.

52. Mole S, McFarlane M, Chuen-Im T, Milligan SG, Millan D, Graham SV. RNA splicing factors regulated by HPV16 during cervical tumour progression. J Pathol 2009;219:383–391.

53. Mole S, Milligan SG, Graham SV. Human papillomavirus type 16 E2 protein transcriptionally activates the promoter of a key cellular splicing factor, SF2/ASF. J Virol 2009;83:357–367.

54. Parish JL, Bean AM, Park RB, Androphy EJ. ChlR1 is required for loading papillomavirus E2 onto mitotic chromosomes and viral genome maintenance. Mol Cell 2006;24:867–876.

55. Parish JL, Kowalczyk A, Chen HT, Roeder GE, Sessions R, Buckle M, et al. E2 proteins from high- and low-risk human papillomavirus types differ in their ability to bind p53 and induce apoptotic cell death. J Virol 2006;80:4580–4590.

56. Johansson C, Graham SV, Dornan ES, Morgan IM. The human papillomavirus 16 E2 protein is stabilised in S phase. Virology 2009;394:194–199.

57. Middleton K, Peh W, Southern S, Griffin H, Sotlar K, Nakahara T, et al. Organization of human papillomavirus productive cycle during neoplastic progression provides a basis for selection of diagnostic markers. J Virol 2003;77:10186–10201.

58. McLaughlin-Drubin ME, Meyers J, Munger K. Cancer associated human papillomaviruses. Curr Opin Virol 2012;2:459–466.

59. Klingelhutz AJ, Foster SA, McDougall JK. Telomerase activation by the E6 gene product of human papillomavirus type 16. Nature 1996;380:79–82.

60. Rapp L, Chen JJ. The papillomavirus E6 proteins. Biochim Biophys Acta 1998;1378:F1–F19.

61. Dyson N, Guida P, Munger K, Harlow E. Homologous sequences in adenovirus E1A and human papillomavirus E7 proteins mediate interaction with the same set of cellular proteins. J Virol 1992;66:6893–6902.

62. Sherr CJ. Cell cycle control and cancer. Harvey Lect 2000;96:73–92.

63. Riley RR, Duensing S, Brake T, Munger K, Lambert PF, Arbeit JM. Dissection of human papillomavirus E6 and E7 function in transgenic mouse models of cervical carcinogenesis. Cancer Res 2003;63:4862–4871.

64. Duensing S, Lee LY, Duensing A, Basile J, Piboonniyom S, Gonzalez S, et al. The human papillomavirus type 16 E6 and E7 oncoproteins cooperate to induce mitotic defects and genomic instability by uncoupling centrosome duplication from the cell division cycle. Proc Natl Acad Sci USA 2000;97:10002–10007.

65. Oh JM, Kim SH, Lee YI, Seo M, Kim SY, Song YS, et al. Human papillomavirus E5 protein induces expression of the EP4 subtype of prostaglandin E2 receptor in cyclic AMP response element-dependent pathways in cervical cancer cells. Carcinogenesis 2009;30:141–149.

66. Venuti A, Paolini F, Nasir L, Corteggio A, Roperto S, Campo MS, et al. Papillomavirus E5: the smallest oncoprotein with many functions. Mol Cancer 2011;10:140.

67. Milligan SG, Veerapraditsin T, Ahamet B, Mole S, Graham SV. Analysis of novel human papillomavirus type 16 late mRNAs in differentiated W12 cervical epithelial cells. Virology 2007;360:172–181.

68. Hanahan D, Weinberg RA. Hallmarks of cancer: the next generation. Cell 2011;144:646–674.

69. Lee JM, Dedhar S, Kalluri R, Thompson EW. The epithelial–mesenchymal transition: new insights in signaling, development, and disease. J Cell Biol 2006;172:973–981.

70. Hellner K, Mar J, Fang F, Quackenbush J, Munger K. HPV16 E7 oncogene expression in normal human epithelial cells causes molecular changes indicative of an epithelial to mesenchymal transition. Virology 2009;391:57–63.

71. Geiger T, Sabanay H, Kravchenko-Balasha N, Geiger B, Levitzki A. Anomalous features of EMT during keratinocyte transformation. PLoS One 2008;3:e1574.

72. Pan MR, Hou MF, Chang HC, Hung WC. Cyclooxygenase-2 up-regulates CCR7 via EP2/EP4 receptor signaling pathways to enhance lymphatic invasion of breast cancer cells. J Biol Chem 2008;283:11155–11163.

73. Li JY, Ou ZL, Yu SJ, Gu XL, Yang C, Chen AX, et al. The chemokine receptor CCR4 promotes tumor growth and lung metastasis in breast cancer. Breast Cancer Res Treat 2012;131:837–848.

74. Kodama J, Hasengaowa, Kusumoto T, Seki N, Matsuo T, et al. Association of CXCR4 and CCR7 chemokine receptor expression and lymph node metastasis in human cervical cancer. Ann Oncol 2007;18:70–76.

75. Amine A, Rivera S, Opolon P, Dekkal M, Biard DS, Bouamar H, et al. Novel anti-metastatic action of cidofovir mediated by inhibition of E6/E7, CXCR4 and Rho/ROCK signaling in HPV tumor cells. PLoS One 2009;4:e5018.

76. Libra M, Scalisi A, Vella N, Clementi S, Sorio R, Stivala F, et al. Uterine cervical carcinoma: role of matrix metalloproteinases (review). Int J Oncol 2009;34:897–903.

77. Nuovo GJ, MacConnell PB, Simsir A, Valea F, French DL. Correlation of the *in situ* detection of polymerase chain reaction-amplified metalloproteinase complementary DNAs and their inhibitors with prognosis in cervical carcinoma. Cancer Res 1995;55:267–275.

78. da Silva Cardeal LB, Brohem CA, Correa TC, Winnischofer SM, Nakano F, Boccardo E, et al. Higher expression and activity of metalloproteinases in human cervical carcinoma cell lines is associated with HPV presence. Biochem Cell Biol 2006;84:713–719.

79. Xi L, Wang S, Wang C, Xu Q, Li P, Tian X, et al. The pro-angiogenic factors stimulated by human papillomavirus type 16 E6 and E7 protein in C33A and human fibroblasts. Oncol Rep 2009;21:25–31.

80. Duffy CL, Phillips SL, Klingelhutz AJ. Microarray analysis identifies differentiation-associated genes regulated by human papillomavirus type 16 E6. Virology 2003;314:196–205.

81. Kuner R, Vogt M, Sultmann H, Buness A, Dymalla S, Bulkescher J, et al. Identification of cellular targets for the human papillomavirus E6 and E7 oncogenes by RNA interference and transcriptome analyses. J Mol Med 2007;85:1253–1262.

82. Larhammar D. Evolution of neuropeptide Y, peptide YY and pancreatic polypeptide. Regul pept 1996;62:1–11.

83. Pahler JC, Tazzyman S, Erez N, Chen YY, Murdoch C, Nozawa H, et al. Plasticity in tumor-promoting inflammation: impairment of macrophage recruitment evokes a compensatory neutrophil response. Neoplasia 2008;10:329–340.

84. Balkwill F, Mantovani A. Inflammation and cancer: back to Virchow? Lancet 2001;357:539–545.

85. Wei L, Gravitt PE, Song H, Maldonado AM, Ozbun MA. Nitric oxide induces early viral transcription coincident with increased DNA damage and mutation rates in human papillomavirus-infected cells. Cancer Res 2009;69:4878–4884.

86. Hiraku Y, Tabata T, Ma N, Murata M, Ding X, Kawanishi S. Nitrative and oxidative DNA damage in cervical intraepithelial neoplasia associated with human papilloma virus infection. Cancer Sci 2007;98:964–972.

87. Garcia-Pineres AJ, Hildesheim A, Herrero R, Trivett M, Williams M, Atmetlla I, et al. Persistent human papillomavirus infection is associated with a generalized decrease in immune responsiveness in older women. Cancer Res 2006;66:11070–11076.

88. Kemp TJ, Hildesheim A, Garcia-Pineres A, Williams MC, Shearer GM, Rodriguez AC, et al. Elevated systemic levels of inflammatory cytokines in older women with persistent cervical human papillomavirus infection. Cancer Epidemiol Biomark Prev 2010;19:1954–1959.

89. Subbaramaiah K, Dannenberg AJ. Cyclooxygenase-2 transcription is regulated by human papillomavirus 16 E6 and E7 oncoproteins: evidence of a corepressor/coactivator exchange. Cancer Res 2007;67:3976–3985.

90. Oh JM, Kim SH, Cho EA, Song YS, Kim WH, Juhnn YS. Human papillomavirus type 16 E5 protein inhibits hydrogen-peroxide-induced apoptosis by stimulating ubiquitin-proteasome-mediated degradation of Bax in human cervical cancer cells. Carcinogenesis 2010;31:402–410.

91. Wymann MP, Schneiter R. Lipid signalling in disease. Nat Rev Mol Cell Biol 2008;9:162–176.

92. Ryu HS, Chang KH, Yang HW, Kim MS, Kwon HC, Oh KS. High cyclooxygenase-2 expression in stage IB cervical cancer with lymph node metastasis or parametrial invasion. Gynecol Oncol 2000;76:320–325.

93. Sales KJ, Katz AA, Davis M, Hinz S, Soeters RP, Hofmeyr MD, et al. Cyclooxygenase-2 expression and prostaglandin E(2) synthesis are up-regulated in carcinomas of the cervix: a possible autocrine/paracrine regulation of neoplastic cell function via EP2/EP4 receptors. J Clin Endocrinol Metab 2001;86:2243–2249.

94. Sales KJ, Katz AA, Howard B, Soeters RP, Millar RP, Jabbour HN. Cyclooxygenase-1 is up-regulated in cervical carcinomas: autocrine/paracrine regulation of cyclooxygenase-2, prostaglandin e receptors, and angiogenic factors by cyclooxygenase-1. Cancer Res 2002;62:424–432.

95. Jabbour HN, Sales KJ, Catalano RD, Norman JE. Inflammatory pathways in female reproductive health and disease. Reproduction 2009;138:903–919.

96. Sales KJ, Jabbour HN. Cyclooxygenase enzymes and prostaglandins in pathology of the endometrium. Reproduction 2003;126:559–567.

97. Koutsky L. Epidemiology of genital human papillomavirus infection. Am J Med 1997;102:3–8.

98. Bonanni P, Boccalini S, Bechini A. Efficacy, duration of immunity and cross protection after HPV vaccination: a review of the evidence. Vaccine 2009;27 Suppl 1:A46–A53.

99. Stanley MA. Genital human papillomavirus infections: current and prospective therapies. J Gen Virol 2012;93:681–691.

100. Lea JS, Sunaga N, Sato M, Kalahasti G, Miller DS, Minna JD, et al. Silencing of HPV 18 oncoproteins With RNA interference causes growth inhibition of cervical cancer cells. Reprod Sci 2007;14:20–28.

101. Jiang M, Milner J. Selective silencing of viral gene expression in HPV-positive human cervical carcinoma cells treated with siRNA, a primer of RNA interference. Oncogene 2002;21:6041–6048.

102. Butz K, Ristriani T, Hengstermann A, Denk C, Scheffner M, Hoppe-Seyler F. siRNA targeting of the viral E6 oncogene efficiently kills human papillomavirus-positive cancer cells. Oncogene 2003;22:5938–5945.

103. Kuroda M, Kiyono T, Oikawa K, Yoshida K, Mukai K. The human papillomavirus E6 and E7 inducible oncogene, hWAPL, exhibits potential as a therapeutic target. Br J Cancer 2005;92:290–293.

104. Sonnex C, Lacey CJ. The treatment of human papillomavirus lesions of the lower genital tract. Best Pract Res Clin Obstet Gynaecol 2001;15:801–816.

105. Viera MH, Amini S, Huo R, Konda S, Block S, Berman B. Herpes simplex virus and human papillomavirus genital infections: new and investigational therapeutic options. Int J Dermatol 2010;49:733–749.

106. Roberts JN, Buck CB, Thompson CD, Kines R, Bernardo M, Choyke PL, et al. Genital transmission of HPV in a mouse model is potentiated by nonoxynol-9 and inhibited by carrageenan. Nature 2007;13:857–861.

CHAPTER 13

HEPATITIS VIRUSES AND HEPATOCELLULAR CARCINOMA

Wai-Kay Seto, Ching-Lung Lai, and Man-Fung Yuen

Liver cancer is currently the seventh most common cancer worldwide (10.8% of all cancers). Its estimated age-standardized mortality rate ranks third for both genders (9.9%) and ranks second for men (14.5%) among all cancers (1). Hepatocellular carcinoma (HCC) is the most common form of liver cancer, with approximately 78% associated with chronic hepatitis B virus (HBV) and hepatitis C virus (HCV) infections (2), roughly in the ratio of 7 to 3, respectively. Among cirrhotic patients, the five-year cumulative incidence of HCC for HBV- and HCV-infected persons is 15% and 30%, respectively (3), much higher than that seen in other chronic liver diseases. Although surveillance programs based on ultrasonography and α-fetoprotein measurement are moderately effective in improving HCC detection at an earlier stage, (4), surveillance coverage still needs improvement, even in developed countries (5), since less than 50% of HCCs detected by screening are resectable (4). In addition, the prevalence of HCC in developed countries has increased significantly over the past decade, which could be explained partially by an aging HCV-infected cohort (6).

In this chapter we discuss the pathogenesis of HBV- and HCV-related HCC, concentrating particularly on three factors: the virus, the host genome, and the environment. All three factors are not mutually exclusive, with host–viral interactions playing critical roles in hepatocarcinogenesis.

HBV

The Virus

Chronic hepatitis B affects 400 million people worldwide (7), with the risk of HCC among HBV-infected subjects compared to the normal population increased 223-fold (8). Although the majority of subjects with HBV-related HCCs have underlying cirrhosis, HCC could also develop in noncirrhotic chronic hepatitis B subjects and can even develop after hepatitis B surface antigen (HBsAg) seroclearance (9).

Cancer and Inflammation Mechanisms: Chemical, Biological, and Clinical Aspects, First Edition.
Edited by Yusuke Hiraku, Shosuke Kawanishi, and Hiroshi Ohshima.

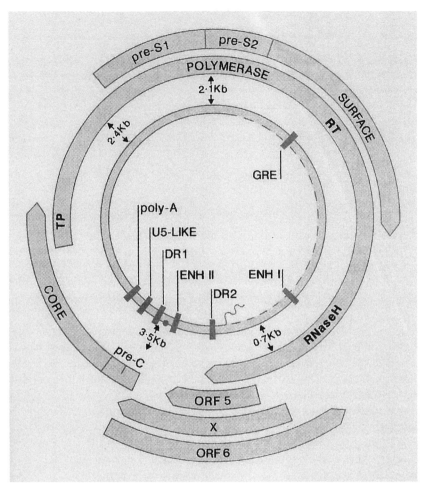

FIGURE 13.1   Hepatitis B virus genome. Regions associated with HCC development include the X gene, pre-S domain, the core promoter region, and ENH II. ORF, opening reading frame; RT, reverse transcriptase; ENH, enhancer; GRE, glucocorticoid responsive element. (Reproduced from Lau JY, Wright TL, Molecular virology and pathogenesis of hepatitis B, The Lancet, vol. 342, pp. 1311–1340, copyright © 1993, with permission from Elsevier Limited.)

Viral factors play a predominant role in the pathogenesis of HBV-related HCC. Replication of HBV is regulated primarily by the reverse transcriptase domain in the polymerase gene, one of the six open reading frames found in the HBV genome (10) (Figure 13.1). HBV has oncogenic properties (11), and the partial double-stranded nature of the virus allows it to integrate into the host genome, resulting in multiple chromosomal instability and promoting hepatocarcinogenesis (12,13). Tumor intrahepatic HBV DNA is mainly in the form of covalently closed circular DNA (cccDNA) (14,15), which is also the main viral reservoir maintaining chronic infectivity.

Other components of HBV genome also contribute to hepatocarcinogenesis. The open reading frame X (Figure 13.1) has aroused the most interest. The X gene is expressed in 70% of HBV-related HCC and in 95% of chronic hepatitis B patients with cirrhosis or dysplasia (16). The X gene could be expressed in HBV-related HCC tumorous cells even in the absence of HBV replication (17), with antibodies to the X gene found more frequently in the sera of HBV-related HCC patients than in chronic hepatitis B patients without HCC (18). The X gene is a multifunctional regulator of different tumorigenesis mechanisms and is involved in the signal transduction of multiple oncogenic pathways (19), in associated epigenetic changes, in influencing apoptosis (18), and in the modulation of angiogenesis (20). Natural mutants of the X gene have been described abundantly, and it is likely that such X-gene mutations are also important in hepatocarcinogenesis (21). Other genes predisposing to HCC development include deletions in the pre-S domain of the HBV surface gene (22), core promoter mutants (T1762/A1764) (23,24), and the T1653 mutation at the enhancer II region (25) (Figure 13.1).

In a clinical setting, natural history studies have shown high (>2000 IU/mL) and moderate (60 to 2000 IU/mL) serum HBV DNA levels to be associated with increased incidence of HBV-related HCC (25–27). Moderate levels of viremia already predispose to HCC even when serum alanine aminotransferase (ALT) is normal (28). Serum HBV DNA levels at the time of tumor resection also influence the rate of HCC recurrence (29). A high viral load contributes to increased inflammation, repeated and sustained cycles of inflammation, necrosis, and regeneration enable telomere erosion and promote oncogenesis (30).

There are other viral factors associated with HCC development. Among the two HBV genotypes commonly seen in Asian patients, genotype C is associated with a higher risk of HCC than is genotype B (24,31). Hepatitis B e antigen (HBeAg)-positivity has been identified as a possible risk factor, although this is likely to be related to the higher viral load and genotype (32). A recent study demonstrated high quantitative serum HBsAg levels to be predictive of HCC development, especially among patients with low serum HBV DNA levels (<2000 IU/mL) (33). HBsAg production is closely linked to the integration of HBV sequences into the host genome and may play a role in promoting genomic instability, resulting in carcinogenesis (34).

Several clinical scores have since been developed for the prediction of HBV-related HCC. These are shown in Table 13.1 (35–37).

Host–Viral Interactions

Signaling pathways activated via host–viral interactions play an important role in HCC carcinogenesis (30). Pathways known to be involved in HBV-related HCC are shown in Table 13.2. An important signal transduction pathway is the mitogen-activated protein kinase (MAPK) cascade (38), which involves the activation of multiple transducers, including Ras and Junas kinase/signal transducer and activator of transcription (Jak/Stat) (39). The MAPK pathway is also closely linked to HBV integration with the host genome (11). The HBV X gene is involved in the regulation of several pathways, including β-catenin (40), p53 (41), NF-κB (42), and Pin1 (43). Other signaling pathways involved include retinoblastoma and cyclin D1 (44).

TABLE 13.1   Clinical Parameters Used in Scoring Systems for Prediction of HBV-Related HCC

Yuen et al. (35)
Male gender
Increased age
Increased HBV DNA levels
Presence of core promoter mutations
Presence of cirrhosis

Wong et al. (36)
Increased age
Decreased serum albumin
Increased serum bilirubin
Increased HBV DNA levels
Presence of cirrhosis

Yang et al. (37)
Male gender
Increased age
Increased ALT levels
HBeAg positivity
Increased HBV DNA levels

The Host Genome

HBV-related HCC is well known to have a familial predeposition (45). However, the identification of genes increasing susceptibility to HBV-related HCC has not been well investigated. Recent developments in high-throughput genomic technology have enabled the use of genome-wide association studies to identify host-genomic factors associated with susceptibility to various diseases. The utilization of genome-wide association studies has identified a novel single-nucleotide polymorphism (SNP), rs17401966, in the KIF1B gene associated with HBV-related HCC in northern Chinese persons (46). However, this result has not be reproduced in subsequent validation studies involving other East Asian populations (47,48). Other genetic variants

TABLE 13.2   Major Signaling Pathways Involved in the Pathogenesis of HCC

Signaling pathway	HBV	HCV
Mitogen-activated protein kinase	✓	✓
Retinoblastoma	✓	✓
p53	✓	✓
Cyclin D1	✓	✓
Wnt//Frizzled/β-catenin	✓	✓
NF-κB	✓	✓
Pin1	✓	
Transforming growth factor beta (TGF-β)		✓
Insulin-like growth factor-2 (IGF-2)		✓

associated with HBV-related HCC include estrogen receptor α polymorphisms (49), various human leucocyte antigen locus (50,51), and mutations in the ARID1A (52) and IRF2 genes (53). The carcinogenesis of HCC is a multistep process requiring multiple genetic aberrations, with the relative significance of each gene in different populations and the interactions between different genes requiring further study.

HCV

The Virus

Chronic HCV infection currently affects 180 million people worldwide and is an increasingly important cause of liver-related morbidity and mortality in developed countries (54). Unlike HBV, the majority of cases are infected during adulthood, with a more rapid rate of disease progression associated independently with an older age of infection (55). HCV has a considerable degree of genetic heterogenicity (56), partially contributing to the lack of an effective vaccine for the virus, hindering efforts to control the disease, especially among high-risk groups.

The risk of HCC is increased 17-fold in HCV-infected patients compared to HCV-negative controls (57). Unlike HBV, HCV (Figure 13.2) is a single-stranded RNA virus with no DNA intermediate and no reverse-transcriptase capabilities, and hence is unable to integrate into the host genome (58). Liver carcinogenesis in HCV is thus achieved primarily through indirect mechanisms, although recent evidence has emerged to support a possible direct virologic influence. These mechanisms are not mutually exclusive.

Indirect Carcinogenic Mechanisms   Although occasionally seen in patients with advanced fibrosis (59), the large majority of HCV-related HCC occurs in cases with established cirrhosis. Since liver cirrhosis is the end result of repeated inflammatory damage and fibrogenesis, HCV-related carcinogenesis has often been linked with the persistent inflammatory damage associated with HCV. Continuous cycles of immune-mediated hepatocytes death and its subsequent regeneration would result in increased mutations promoting hepatocarcinogenesis (60).

Chronic HCV infection is also characterized by an imbalance of CD4 T-helper cells, with T-helper (Th)2 cytokines involved in the down-regulation of antitumor activity. In HCV-related HCC, there is Th2 cell predominance and a loss of Th1 cells (61), with Th2-like cytokines, including interleukin (IL)-4, IL-8, and IL-10, found in abundance in the HCV-related HCC microenvironment (62).

Oxidative DNA stress also promotes HCV-related hepatocarcinogenesis, as illustrated by increased intrahepatic oxidative stress found in HCV-related HCC (63). The metabolism of nitric oxide has also been shown to contribute to tumor progression (64,65). Oxidative stress is closely associated with immune-mediated apoptosis, which holds a key role in the pathogenesis of HCV-related HCC (66). Pro-apoptotic signals induced by HCV replication result in hepatocyte destruction, with the compensatory proliferation and reinfection creating an unstable inflammatory environment and increased oxidative stress (64).

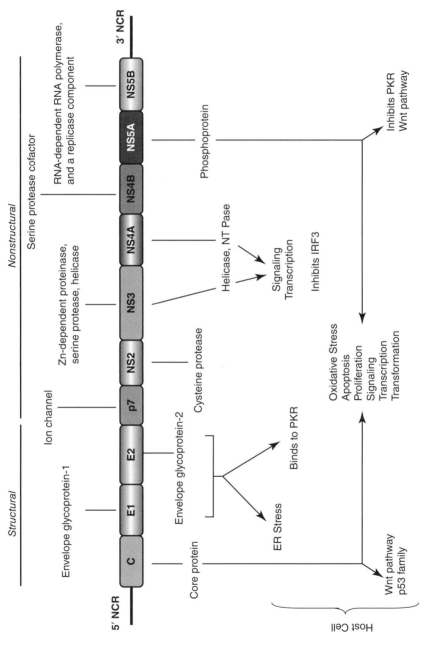

FIGURE 13.2 Hepatitis C virus genome. Regions directly or indirectly associated with HCC development include C (core), E2, NS3, NS5A, and NS5B. (Reproduced from Castello G et al., HCV-related hepatocellular carcinoma: from chronic inflammation to cancer, Clinical Immunology, vol. 134, pp. 237–250, copyright © 2010, with permission from Elsevier Limited.)

Direct Carcinogenic Mechanisms   The lack of viral integration seen in HBV infection would suggest that HCV is not directly carcinogenic. However, there is now evidence that HCV viral proteins (Figure 13.2) have a direct carcinogenic effect through their interaction with associated signaling pathways of various tumor-suppression genes and oncogenes (60). The viral protein NS5B could be involved in down-regulation of the tumor suppressor protein retinoblastoma (Rb) (67). Viral protein NS3 could also interact with oncogene p53 (68). There is evidence showing that the core (69), E2, and NS5A may also be involved in HCC pathogenesis (70). These interactions could lead to increased oxidative DNA stress and genomic instability, eventually promoting carcinogenesis.

Host–Viral Interactions

Involvement of the host genome through epigenetic and signal transduction alterations has been reported extensively in HCV-related HCC (30) and is illustrated in Table 13.2. Similar to HBV, the MAPK cascade has a well-established role in HCV-related carcinogenesis and is responsible for multiple related genetic alternations (39,71). Other pathways, including retinoblastoma, p53, cyclin D1, β-catenin, NF-κB, and transcription growth factor-β (TGF-β) (44,72–74), also contribute to malignant transformation in chronic HCV infection. Unlike HBV, the lipid metabolism-related pathway insulin growth factor-2 (IGF-2) (75) is also involved in HCV-related hepatocarcinogenesis. The host cell factor microRNA (miR)-122, actively involved in HCV replication, is also involved in the pathogenesis of HCC by the regulation of p53 and cyclin G1 (76).

The Host Genome

A Japanese GWAS located an SNP, rs2596542, found along the MICA region of chromosome 6, to be strongly associated with HCV-induced HCC (77). Unfortunately, the control group of this study did not consist of subjects with HCV-related cirrhosis, and hence it is not be possible to distinguish if this risk allele is associated independently with HCC or is associated with cirrhosis that subsequently leads to hepatocarcinogenesis. Other genetic variants found to be associated with HCV-related HCC include epidermal growth factor gene polymorphisms (78,79), interleukin 28 (IL-28) polymorphisms (80,81), and the glutathione S-transferase gene (82). Further studies are needed to delineate whether any interaction exists between these genetic variants, as well as their relationship with variants related to the progression of liver fibrosis and cirrhosis (83,84).

ENVIRONMENTAL AND METABOLIC FACTORS AND INTERACTION WITH HBV AND HCV

Multiple environmental factors are associated with HCC development. The best known are aflatoxins and alcohol, the latter through the development of cirrhosis (85). More recent data have suggested dietary factors, including increased intake of coffee

(86), unsaturated fatty acids, and fish (87) to be protective against HCC. Nevertheless, it is still undetermined whether these dietary factors have synergistic effects with HBV and HCV. Studies concerning coffee and fish intake are so far concentrated in the Japanese population, where the majority of HCCs are HCV-related, and hence validation epidemiological studies in other ethnic groups, including HBV-infected populations, are needed. Although smoking can lead to HCC development (88), any synergistic interaction with HBV or HCV remains controversial (89).

Although diabetes is associated with the development of HCV-related HCC (90), it may be due to the higher prevalence of diabetes among HCV-infected persons than healthy subjects (2). Data on diabetes and HBV-related HCC are less consistent (91,92).

CONCLUDING REMARKS

Chronic HBV and HCV infections together affect more than 550 million persons worldwide; hence, HCC will remain an important global health issue for at least the next few decades. The pathogenesis of both HBV- and HCV-related HCC is multifactorial, and despite recent emerging evidence highlighting key mechanisms of hepatocarcinogenesis, the degree of interaction between factors remains unclear. Although genomic sequencing has identified different genetic variants associated with HBV- and HCV-related HCC, their relative significance at the functional expression level remains undetermined. Further research is thus paramount in identifying potential molecular targets for therapy and in reducing mortalities and morbidities related to HCC.

REFERENCES

1. International Agency for Research on Cancer. GLOBOCAN 2008. Estimated age-standardised incidence and mortality rates. 2008 (cited Aug. 27, 2012); Available from: http://globocan.iarc.fr/factsheet.asp#BOTH.
2. El-Serag HB. Epidemiology of viral hepatitis and hepatocellular carcinoma. Gastroenterology 2012;142:1264–1273 e1261.
3. Fattovich G, Stroffolini T, Zagni I, Donato F. Hepatocellular carcinoma in cirrhosis: incidence and risk factors. Gastroenterology 2004;127:S35–S50.
4. Zhang BH, Yang BH, Tang ZY. Randomized controlled trial of screening for hepatocellular carcinoma. J Cancer Res Clin Oncol 2004;130:417–422.
5. Davila JA, Henderson L, Kramer JR, Kanwal F, Richardson PA, Duan Z, et al. Utilization of surveillance for hepatocellular carcinoma among hepatitis C virus–infected veterans in the United States. Ann Intern Med 2011;154:85–93.
6. Kanwal F, Hoang T, Kramer JR, Asch SM, Goetz MB, Zeringue A, et al. Increasing prevalence of HCC and cirrhosis in patients with chronic hepatitis C virus infection. Gastroenterology 2011;140:1182–1188 e1181.
7. Lai CL, Yuen MF. The natural history and treatment of chronic hepatitis B: a critical evaluation of standard treatment criteria and end points. Ann Intern Med 2007;147:58–61.
8. Beasley RP, Hwang LY, Lin CC, Chien CS. Hepatocellular carcinoma and hepatitis B virus: a prospective study of 22 707 men in Taiwan. Lancet 1981;2:1129–1133.

9. Yuen MF, Wong DK, Fung J, Ip P, But D, Hung I, et al. HBsAg seroclearance in chronic hepatitis B in Asian patients: replicative level and risk of hepatocellular carcinoma. Gastroenterology 2008;135:1192–1199.

10. Lau JY, Wright TL. Molecular virology and pathogenesis of hepatitis B. Lancet 1993;342:1335–1340.

11. Fung J, Lai CL, Yuen MF. Hepatitis B and C virus–related carcinogenesis. Clin Microbiol Infect 2009;15:964–970.

12. Brechot C, Pourcel C, Louise A, Rain B, Tiollais P. Presence of integrated hepatitis B virus DNA sequences in cellular DNA of human hepatocellular carcinoma. Nature 1980;286:533–535.

13. Bonilla Guerrero R, Roberts LR. The role of hepatitis B virus integrations in the pathogenesis of human hepatocellular carcinoma. J Hepatol 2005;42:760–777.

14. Wong DK, Yuen MF, Poon RT, Yuen JC, Fung J, Lai CL. Quantification of hepatitis B virus covalently closed circular DNA in patients with hepatocellular carcinoma. J Hepatol 2006;45:553–559.

15. Wong DK, Huang FY, Lai CL, Poon RT, Seto WK, Fung J, Hung IF, et al. Occult hepatitis B infection and HBV replicative activity in patients with cryptogenic cause of hepatocellular carcinoma. Hepatology 2011;54:829–836.

16. Rossner MT. Review: hepatitis B virus X-gene product: a promiscuous transcriptional activator. J Med Virol 1992;36:101–117.

17. Peng Z, Zhang Y, Gu W, Wang Z, Li D, Zhang F, et al. Integration of the hepatitis B virus X fragment in hepatocellular carcinoma and its effects on the expression of multiple molecules: a key to the cell cycle and apoptosis. Int J Oncol 2005;26:467–473.

18. Ng SA, Lee C. Hepatitis B virus X gene and hepatocarcinogenesis. J Gastroenterol 2011;46:974–990.

19. Cross JC, Wen P, Rutter WJ. Transactivation by hepatitis B virus X protein is promiscuous and dependent on mitogen-activated cellular serine/threonine kinases. Proc Natl Acad Sci USA 1993;90:8078–8082.

20. Liu LP, Liang HF, Chen XP, Zhang WG, Yang SL, Xu T, et al. The role of NF-kappaB in hepatitis B virus X protein-mediated upregulation of VEGF and MMPs. Cancer Invest 2010;28:443–451.

21. Sirma H, Giannini C, Poussin K, Paterlini P, Kremsdorf D, Brechot C. Hepatitis B virus X mutants, present in hepatocellular carcinoma tissue abrogate both the antiproliferative and transactivation effects of HBx. Oncogene 1999;18:4848–4859.

22. Yeung P, Wong DK, Lai CL, Fung J, Seto WK, Yuen MF. Association of hepatitis B virus pre-S deletions with the development of hepatocellular carcinoma in chronic hepatitis B. J Infect Diss 2011;203:646–654.

23. Kao JH, Chen PJ, Lai MY, Chen DS. Basal core promoter mutations of hepatitis B virus increase the risk of hepatocellular carcinoma in hepatitis B carriers. Gastroenterology 2003;124:327–334.

24. Yuen MF, Tanaka Y, Mizokami M, Yuen JC, Wong DK, Yuan HJ, et al. Role of hepatitis B virus genotypes Ba and C, core promoter and precore mutations on hepatocellular carcinoma: a case control study. Carcinogenesis 2004;25:1593–1598.

25. Yuen MF, Tanaka Y, Shinkai N, Poon RT, But DY, Fong DY, et al. Risk for hepatocellular carcinoma with respect to hepatitis B virus genotypes B/C, specific mutations of enhancer II/core promoter/precore regions and HBV DNA levels. Gut 2008;57:98–102.

26. Chen CJ, Yang HI, Su J, Jen CL, You SL, Lu SN, et al. Risk of hepatocellular carcinoma across a biological gradient of serum hepatitis B virus DNA level. JAMA 2006;295:65–73.

27. Chen JD, Yang HI, Iloeje UH, You SL, Lu SN, Wang LY, et al. Carriers of inactive hepatitis B virus are still at risk for hepatocellular carcinoma and liver-related death. Gastroenterology 2010;138:1747–1754.

28. Yuen MF, Yuan HJ, Wong DK, Yuen JC, Wong WM, Chan AO, et al. Prognostic determinants for chronic hepatitis B in Asians: therapeutic implications. Gut 2005;54:1610–1614.

29. Hung IF, Poon RT, Lai CL, Fung J, Fan ST, Yuen MF. Recurrence of hepatitis B–related hepatocellular carcinoma is associated with high viral load at the time of resection. Am J Gastroenterol 2008;103:1663–1673.

30. Farazi PA, DePinho RA. Hepatocellular carcinoma pathogenesis: from genes to environment. Nat Rev Cancer 2006;6:674–687.

31. Chan HL, Hui AY, Wong ML, Tse AM, Hung LC, Wong VW, et al. Genotype C hepatitis B virus infection is associated with an increased risk of hepatocellular carcinoma. Gut 2004;53:1494–1498.

32. Yang HI, Lu SN, Liaw YF, You SL, Sun CA, Wang LY, et al. Hepatitis B e antigen and the risk of hepatocellular carcinoma. N Engl J Med 2002;347:168–174.

33. Tseng TC, Liu CJ, Yang HC, Su TH, Wang CC, Chen CL, et al. High levels of hepatitis B surface antigen increase risk of hepatocellular carcinoma in patients with low HBV load. Gastroenterology 2012;142:1140–1149 e1143; quiz e1113–1144.

34. Kao JH, Chen PJ, Chen DS. Recent advances in the research of hepatitis B virus–related hepatocellular carcinoma: epidemiologic and molecular biological aspects. Adv Cancer Res 2010;108: 21–72.

35. Yuen MF, Tanaka Y, Fong DY, Fung J, Wong DK, Yuen JC, et al. Independent risk factors and predictive score for the development of hepatocellular carcinoma in chronic hepatitis B. J Hepatol 2009;50:80–88.

36. Wong VW, Chan SL, Mo F, Chan TC, Loong HH, Wong GL, et al. Clinical scoring system to predict hepatocellular carcinoma in chronic hepatitis B carriers. J Clin Oncol 2010;28:1660–1665.

37. Yang HI, Yuen MF, Chan HL, Han KH, Chen PJ, Kim DY, et al. Risk estimation for hepatocellular carcinoma in chronic hepatitis B (REACH-B): development and validation of a predictive score. Lancet Oncol 2011;12:568–574.

38. Chin R, Earnest-Silveira L, Koeberlein B, Franz S, Zentgraf H, Dong X, et al. Modulation of MAPK pathways and cell cycle by replicating hepatitis B virus: factors contributing to hepatocarcinogenesis. J Hepatol 2007;47:325–337.

39. Calvisi DF, Ladu S, Gorden A, Farina M, Conner EA, Lee JS, et al. Ubiquitous activation of Ras and Jak/Stat pathways in human HCC. Gastroenterology 2006;130:1117–1128.

40. Cha MY, Kim CM, Park YM, Ryu WS. Hepatitis B virus X protein is essential for the activation of Wnt/beta-catenin signaling in hepatoma cells. Hepatology 2004;39:1683–1693.

41. Ueda H, Ullrich SJ, Gangemi JD, Kappel CA, Ngo L, Feitelson MA, et al. Functional inactivation but not structural mutation of p53 causes liver cancer. Nat Genet 1995;9:41–47.

42. Chiao PJ, Na R, Niu J, Sclabas GM, Dong Q, Curley SA. Role of Rel/NF-kappaB transcription factors in apoptosis of human hepatocellular carcinoma cells. Cancer 2002;95:1696–1705.

43. Pang R, Lee TK, Poon RT, Fan ST, Wong KB, Kwong YL, et al. Pin1 interacts with a specific serine-proline motif of hepatitis B virus X-protein to enhance hepatocarcinogenesis. Gastroenterology 2007;132:1088–1103.

44. Edamoto Y, Hara A, Biernat W, Terracciano L, Cathomas G, Riehle HM, et al. Alterations of RB1, p53 and Wnt pathways in hepatocellular carcinomas associated with hepatitis C, hepatitis B and alcoholic liver cirrhosis. Int J Cancer 2003;106:334–341.

45. Yu MW, Chang HC, Liaw YF, Lin SM, Lee SD, Liu CJ, et al. Familial risk of hepatocellular carcinoma among chronic hepatitis B carriers and their relatives. J Natl Cancer Inst 2000;92:1159–1164.

46. Zhang H, Zhai Y, Hu Z, Wu C, Qian J, Jia W, et al. Genome-wide association study identifies 1p36.22 as a new susceptibility locus for hepatocellular carcinoma in chronic hepatitis B virus carriers. Nat Genet 2010;42:755–758.

47. Chan KY, Wong CM, Kwan JS, Lee JM, Cheung KW, Yuen MF, et al. Genome-wide association study of hepatocellular carcinoma in southern Chinese patients with chronic hepatitis B virus infection. PLoS One 2011;6:e28798.

48. Sawai H, Nishida N, Mbarek H, Matsuda K, Mawatari Y, Yamaoka M, et al. No association for Chinese HBV-related hepatocellular carcinoma susceptibility SNP in other East Asian populations. BMC Med Genet 2012;13:47.

49. Zhai Y, Zhou G, Deng G, Xie W, Dong X, Zhang X, et al. Estrogen receptor alpha polymorphisms associated with susceptibility to hepatocellular carcinoma in hepatitis B virus carriers. Gastroenterology 2006;130:2001–2009.

50. Clifford RJ, Zhang J, Meerzaman DM, Lyu MS, Hu Y, Cultraro CM, et al. Genetic variations at loci involved in the immune response are risk factors for hepatocellular carcinoma. Hepatology 2010;52:2034–2043.

51. Hu L, Zhai X, Liu J, Chu M, Pan S, Jiang J, et al. Genetic variants in human leukocyte antigen/DP–DQ influence both hepatitis B virus clearance and hepatocellular carcinoma development. Hepatology 2012;55:1426–1431.

52. Huang J, Deng Q, Wang Q, Li KY, Dai JH, Li N, et al. Exome sequencing of hepatitis B virus–associated hepatocellular carcinoma. Nat Genet 2012;44:1117–1121.

53. Guichard C, Amaddeo G, Imbeaud S, Ladeiro Y, Pelletier L, Maad IB, et al. Integrated analysis of somatic mutations and focal copy-number changes identifies key genes and pathways in hepatocellular carcinoma. Nat Genet 2012;44:694–698.

54. Rosen HR. Clinical practice. Chronic hepatitis C infection. N Engl J Med 2011;364:2429–2438.

55. Minola E, Prati D, Suter F, Maggiolo F, Caprioli F, Sonzogni A, et al. Age at infection affects the long-term outcome of transfusion-associated chronic hepatitis C. Blood 2002;99:4588–4591.

56. Simmonds P. Viral heterogeneity of the hepatitis C virus. J Hepatol 1999;31 (Suppl 1):54–60.

57. Donato F, Tagger A, Gelatti U, Parrinello G, Boffetta P, Albertini A, et al. Alcohol and hepatocellular carcinoma: the effect of lifetime intake and hepatitis virus infections in men and women. Am J Epidemiol 2002;155:323–331.

58. Castello G, Scala S, Palmieri G, Curley SA, Izzo F. HCV-related hepatocellular carcinoma: From chronic inflammation to cancer. Clin Immunol 2010;134:237–250.

59. Lok AS, Everhart JE, Wright EC, Di Bisceglie AM, Kim HY, Sterling RK, et al. Maintenance peginterferon therapy and other factors associated with hepatocellular carcinoma in patients with advanced hepatitis C. Gastroenterology 2011;140:840–849.

60. Lemon SM, McGivern DR. Is hepatitis C virus carcinogenic? Gastroenterology 2012;142:1274–1278.

61. Matsui T, Nagai H, Sumino Y, Miki K. Relationship of peripheral blood CD4-positive T cells to carcinogenesis in patients with HCV-related chronic hepatitis and liver cirrhosis. Cancer Chemother Pharmacol 2008;62:401–406.

62. Budhu A, Wang XW. The role of cytokines in hepatocellular carcinoma. J Leukoc Biol 2006; 80:1197–1213.

63. Maki A, Kono H, Gupta M, Asakawa M, Suzuki T, Matsuda M, et al. Predictive power of biomarkers of oxidative stress and inflammation in patients with hepatitis C virus–associated hepatocellular carcinoma. Ann Surg Oncol 2007;14:1182–1190.

64. Joyce MA, Walters KA, Lamb SE, Yeh MM, Zhu LF, Kneteman N, et al. HCV induces oxidative and ER stress, and sensitizes infected cells to apoptosis in SCID/Alb-uPA mice. PLoS Pathog 2009;5:e1000291.

65. Machida K, Tsukamoto H, Liu JC, Han YP, Govindarajan S, Lai MM, et al. c-Jun mediates hepatitis C virus hepatocarcinogenesis through signal transducer and activator of transcription 3 and nitric oxide-dependent impairment of oxidative DNA repair. Hepatology 2010;52:480–492.

66. Qiu W, Wang X, Leibowitz B, Yang W, Zhang L, Yu J. PUMA-mediated apoptosis drives chemical hepatocarcinogenesis in mice. Hepatology 2011;54:1249–1258.

67. Munakata T, Nakamura M, Liang Y, Li K, Lemon SM. Down-regulation of the retinoblastoma tumor suppressor by the hepatitis C virus NS5B RNA-dependent RNA polymerase. Proc Natl Acad Sci USA 2005;102:18159–18164.

68. McGivern DR, Lemon SM. Virus-specific mechanisms of carcinogenesis in hepatitis C virus associated liver cancer. Oncogene 2011;30:1969–1983.

69. Moriya K, Fujie H, Shintani Y, Yotsuyanagi H, Tsutsumi T, Ishibashi K, et al. The core protein of hepatitis C virus induces hepatocellular carcinoma in transgenic mice. Nat Med 1998;4:1065–1067.

70. Hung CH, Chen CH, Lee CM, Wu CM, Hu TH, Wang JH, et al. Association of amino acid variations in the NS5A and E2-PePHD region of hepatitis C virus 1b with hepatocellular carcinoma. J Viral Hepat 2008;15:58–65.

71. Branda M, Wands JR. Signal transduction cascades and hepatitis B and C related hepatocellular carcinoma. Hepatology 2006;43:891–902.

72. Pavio N, Battaglia S, Boucreux D, Arnulf B, Sobesky R, Hermine O, et al. Hepatitis C virus core variants isolated from liver tumor but not from adjacent non-tumor tissue interact with Smad3 and inhibit the TGF-beta pathway. Oncogene 2005;24:6119–6132.

73. Munakata T, Liang Y, Kim S, McGivern DR, Huibregtse J, Nomoto A, et al. Hepatitis C virus induces E6AP-dependent degradation of the retinoblastoma protein. PLoS Pathog 2007;3:1335–1347.

74. Jiang YF, He B, Li NP, Ma J, Gong GZ, Zhang M. The oncogenic role of NS5A of hepatitis C virus is mediated by up-regulation of survivin gene expression in the hepatocellular cell through p53 and NF-kappaB pathways. Cell Biol Int 2011;35:1225–1232.

75. Nardone G, Romano M, Calabro A, Pedone PV, de Sio I, Persico M, et al. Activation of fetal promoters of insulinlike growth factors II gene in hepatitis C virus-related chronic hepatitis, cirrhosis, and hepatocellular carcinoma. Hepatology 1996;23:1304–1312.

76. Fornari F, Gramantieri L, Giovannini C, Veronese A, Ferracin M, Sabbioni S, et al. MiR 122/cyclin G1 interaction modulates p53 activity and affects doxorubicin sensitivity of human hepatocarcinoma cells. Cancer Res 2009;69:5761–5767.

77. Kumar V, Kato N, Urabe Y, Takahashi A, Muroyama R, Hosono N, et al. Genome-wide association study identifies a susceptibility locus for HCV-induced hepatocellular carcinoma. Nat Genet 2011;43:455–458.

78. Abu Dayyeh BK, Yang M, Fuchs BC, Karl DL, Yamada S, Sninsky JJ, et al. A functional polymorphism in the epidermal growth factor gene is associated with risk for hepatocellular carcinoma. Gastroenterology 2011;141:141–149.

79. Tanabe KK, Lemoine A, Finkelstein DM, Kawasaki H, Fujii T, Chung RT, et al. Epidermal growth factor gene functional polymorphism and the risk of hepatocellular carcinoma in patients with cirrhosis. JAMA 2008;299:53–60.

80. Fabris C, Falleti E, Cussigh A, Bitetto D, Fontanini E, Bignulin S, et al. IL-28B rs12979860 C/T allele distribution in patients with liver cirrhosis: role in the course of chronic viral hepatitis and the development of HCC. J Hepatol 2011;54:716–722.

81. Joshita S, Umemura T, Katsuyama Y, Ichikawa Y, Kimura T, Morita S, et al. Association of IL28B gene polymorphism with development of hepatocellular carcinoma in Japanese patients with chronic hepatitis C virus infection. Hum Immunol 2012;73:298–300.

82. White DL, Li D, Nurgalieva Z, El-Serag HB. Genetic variants of glutathione S-transferase as possible risk factors for hepatocellular carcinoma: a HuGE systematic review and meta-analysis. Am J Epidemiol 2008;167:377–389.

83. Hartmann D, Srivastava U, Thaler M, Kleinhans KN, N'Kontchou G, Scheffold A, et al. Telomerase gene mutations are associated with cirrhosis formation. Hepatology 2011;53:1608–1617.

84. Patin E, Kutalik Z, Guergnon J, Bibert S, Nalpas B, Jouanguy E, et al. Genome-wide association study identifies variants associated with progression of liver fibrosis from HCV infection. Gastroenterology 2012;143:1244–1252.

85. Chuang SC, La Vecchia C, Boffetta P. Liver cancer: descriptive epidemiology and risk factors other than HBV and HCV infection. Cancer Lett 2009;286:9–14.

86. Inoue M, Yoshimi I, Sobue T, Tsugane S. Influence of coffee drinking on subsequent risk of hepatocellular carcinoma: a prospective study in Japan. J Natl Cancer Inst 2005;97:293–300.

87. Sawada N, Inoue M, Iwasaki M, Sasazuki S, Shimazu T, Yamaji T, et al. Consumption of n-3 fatty acids and fish reduces risk of hepatocellular carcinoma. Gastroenterology 2012;142:1468–1475.

88. Gandini S, Botteri E, Iodice S, Boniol M, Lowenfels AB, Maisonneuve P, et al. Tobacco smoking and cancer: a meta-analysis. Int J Cancer 2008;122:155–164.

89. Mori M, Hara M, Wada I, Hara T, Yamamoto K, Honda M, et al. Prospective study of hepatitis B and C viral infections, cigarette smoking, alcohol consumption, and other factors associated with hepatocellular carcinoma risk in Japan. Am J Epidemiol 2000;151:131–139.

90. Davila JA, Morgan RO, Shaib Y, McGlynn KA, El-Serag HB. Diabetes increases the risk of hepatocellular carcinoma in the United States: a population based case control study. Gut 2005;54:533–539.

91. Chao LT, Wu CF, Sung FY, Lin CL, Liu CJ, Huang CJ, et al. Insulin, glucose and hepatocellular carcinoma risk in male hepatitis B carriers: results from 17-year follow-up of a population-based cohort. Carcinogenesis 2011;32:876–881.

92. Li Q, Li WW, Yang X, Fan WB, Yu JH, Xie SS, et al. Type 2 diabetes and hepatocellular carcinoma: a case–control study in patients with chronic hepatitis B. Int J Cancer 2012;131:1197–1202.

CHAPTER 14

EPSTEIN–BARR VIRUS AND NASOPHARYNGEAL CARCINOMA

Xiaoying Zhou, Xue Xiao, Fu Chen, Tingting Huang, and Zhe Zhang

Epstein–Barr virus (EBV), or human herpes virus 4, is an important, unique, and intriguing human pathogen. Humans are the only natural host for EBV. As a member of the herpes virus family, EBV infection is seen worldwide, and it sustains asymptomatic lifelong infection.

EBV is transmitted via salivary contact. During acute infection, EBV infects and replicates primarily in the stratified squamous epithelium of the oropharynx (1). EBV infection of B lymphocytes is thought to occur in the oropharyngeal lymphoid organs. In normal healthy carriers, EBV persists in circulating memory B cells (2,3); any disruption in this condition results in virus-associated B-cell tumors, such as with Burkitt's lymphoma and Hodgkin's and non-Hodgkin's lymphomas. The carcinogenesis of epithelial-cell tumors with EBV infection, such as nasopharyngeal carcinoma (NPC) and EBV-positive gastric carcinoma, is less well understood.

Although NPC has been reported worldwide, most cases are found in Southeast Asia and southern China. For reasons that remain unclear, Asians are more susceptible to NPC than the other races. Dietary habits and environmental factors, including the ingestion of salted fish and preserved food and exposure to smoke or chemical pollutants, could be major contributors to the pathogenic process.

EBV contributes to tumorigenesis in NPC. EBV has been classified as a group I carcinogen by the International Agency for Research on Cancer (4). The presence of EBV in NPC has been well documented since 1970 (5); almost every undifferentiated NPC case is positive for EBV (6). However, evidence for a role of EBV in the pathogenesis of NPC is not well understood and is controversial. The multistep process of nasopharyngeal carcinogesis is considered a consequence of the aberrant establishment of latent EBV infection in epithelial cells with premalignant genetic changes. Early genetic changes may predispose epithelial cells to EBV infection or persistent latent infection. The expression of latent genes in EBV-infected cells may enhance its transformation capacity, and subsequent clonal expansion may result in

Cancer and Inflammation Mechanisms: Chemical, Biological, and Clinical Aspects, First Edition.
Edited by Yusuke Hiraku, Shosuke Kawanishi, and Hiroshi Ohshima.

rapid progression to invasive carcinoma. In this chapter we focus on the relationship between EBV and the carcinogenicity of NPC.

MOLECULAR BIOLOGY OF EBV

EBV Genome

EBV has a toroid-shaped protein core wrapped with double-stranded DNA, a nucleocapsid with 162 capsomeres, a protein tegument between the nucleocapsid and envelope, and an outer envelope with external glycoprotein spikes (7). In the infectious virion, the viral genome is present as a linear, double-stranded DNA of 172 kilo base pairs (kbp). In infected cells, the viral genome persists primarily as an extrachromosomal episome. Episome formation is mediated by a set of 0.5-kbp terminal repetitive (TR) sequences located at each end of the linear molecule and results in a TR region with a variable number of repeats. Individual infection events lead to episomes that differ in number of TRs. Analysis of the TR region by Southern blot hybridization provides evidence of the clonality of the viral genomes and, by implication, of the cell population harboring the virus (7). Since the EBV genome was sequenced from an EBV DNA library digested with the BamHI restriction enzyme, the open reading frames, genes, and sites for transcription or RNA processing frequently reference specific BamHI fragments, from A to Z, in descending order of fragment size (8).

With polymorphisms in the EBV nuclear antigen 2 (EBNA2) gene, EBV can be genotyped into EBV1 and EBV2 genotypes. EBV1 infection is more prevalent in Europe and the United States, while EBV2 infection is prevalent in central Africa, in New Guinea, and among Alaskan native people (9). These two genotypes also differ in their *in vitro*-transforming efficiency: EBV2 transforms B cells less efficiently than does EBV1 *in vitro*, and EBV2 lymphoblastoid cell lines are less viable than EBV1 lines (10). The differences in transforming efficiency of the two virus subtypes may also relate to divergence in the EBNA2 sequences (11).

Natural History of EBV Infection

As a ubiquitous pathogen, EBV causes infection in almost all populations in the world. In many developing countries, primary infection with EBV normally occurs early in life and is asymptomatic or is associated with minor, nonspecific illnesses such as low-grade fever or sore throat. In more industrialized countries of the world, primary infection is often delayed until adolescence or early adulthood and is associated with a self-limiting clinical syndrome, infectious mononucleosis. The clinical features are highly variable, from mild to serious symptoms. In either circumstance, primary EBV infection is followed by lifelong persistence of the infection that is usually asymptomatic (12,13).

Early in the course of primary infection, EBV infects B lymphocytes. The main reservoir of EBV in humans is memory B lymphocytes. EBV does not usually replicate in B lymphocytes. Instead, following infection, EBV establishes latent infection in memory B lymphocytes, characterized by the limited expression of a subset of latent virus genes without production of virions. These latently infected B lymphocytes, termed *lymphoblastoid cells*, are transformed or immortalized and can replicate indefinitely.

The receptor for EBV in B lymphocytes is CD21, also known as the C3d complement receptor or CR2 (14). EBV infection of B lymphocytes is initiated by the major virus-encoded glycoprotein gp350 binding CD21 (15,16). CD21 is the only B-cell membrane protein that binds gp350/220. Other cells, such as epithelial cells, T cells, and natural killer cells, can become latently infected with EBV (17,18). The viral genome can be found in epithelial tumor cells in NPC, epithelial lesions in oral hairy leukoplakia and Hodgkin's disease, and peripheral T-cell tumors (18–20). Epithelial cells lack CD21; the EBV receptor on epithelial cells that initiates EBV infection has not yet been identified.

Patterns of Latent Infection

During latent EBV infection, only a few EBV-encoded genes are expressed: the six EBNAs, two latent membrane proteins (LMPs), two EBV-encoded RNAs (EBERs), and transcripts from the BamHI A region of the genome. EBV latent gene expression in various EBV-associated malignancies and EBV-derived cell lines has led to the identification of four distinct latency programmes that result from differential promoter activity and are influenced by host cell factors:

> *Latency type 0:* a controversial latency designation, with a putative role in EBV persistence in B cells, where infected cells express undetectable latent mRNA or proteins (21).
>
> *Latency type 1:* typically seen in tissues from patients with Burkitt's lymphoma. Exhibits expression of Qp-promoter-induced EBNA1, EBERs, and BamHI-A rightward transcripts (BARTs) but not all other EBNAs or the LMPs (LMP1, LMP2A, and LMP2B).
>
> *Latency type 2:* seen in patients with EBV-positive Hodgkin's lymphoma, peripheral T-cell lymphomas, EBV-positive gastric cancer, and NPC. Exhibits expression of EBERs, BARTs, and Qp-promoter-driven EBNA1, LMP1, and LMP2.
>
> *Latency type 3:* seen in post transplant lymphoproliferative disease, infectious mononucleosis, and lymphoblastoid cell lines. Exhibits expression of the full spectrum of EBV-associated latency products, including EBNAs 1, 2, 3A, 3B, 3C, and -LP and all LMPs, as well as EBERs and BARTs.

EBV Latent Gene Function

Among the nearly 100 genes encoded by the genome, a limited set of genes, called *latent genes*, are expressed constitutively during EBV infection. These include six nuclear antigens (EBNA1, 2, 3A, 3B, 3C, and -LP), three LMPs (LMP 1, 2A, and 2B), two EBERs, and BARTs. Only two of these genes, EBNA1 and LMP1, are expressed

TABLE 14.1   EBV Latent Proteins

	Required for B-cell transformation	Gene expression in nasopharyngeal carcinoma	Functions
EBNA1	Yes	+	Episomal maintenance
			Up-regulates viral genes
EBNA2	Yes	–	Up-regulates viral and cellular genes
EBNA3	-A, -C: yes;	–	Inhibits EBNA-2 activity
	-B: no		Up-regulates cellular genes
EBNA-LP	Probably	–	Augments EBNA-2 activity
LMP1	Yes	+	CD40 signaling
			Activates NF-κB,c-Jun terminal kinase
			Up-regulates multiple cellular genes
			Oncogene
LMP2	No	+	Prevents EBV reactivation from latency
EBERs	No	+	Oncogene
BARTs	No	+	Maintains and regulates viral latent proteins
			Contributes to pathogenesis

in lytic infection. Five of these genes are required for *in vitro* EBV transformation of B cells (Table 14.1). Despite various studies, the functions of the latent genes need further study.

EBNA1   Among all the latent genes, EBNA1 is expressed in both latent and lytic infection and can be detected in all EBV-associated tumors. As a ubiquitous DNA-binding nuclear phosphoprotein, ENBA1 binds DNA via two domains: a core domain that does not exactly bind DNA but resembles the DNA-binding domain of the papillomavirus E2 protein, and a flanking domain that binds DNA (22,23). At the replication origin oriP, EBNA1 binding can activate viral DNA replication and tether the viral episome to the cellular chromosomes (24,25). At the BamHI C or W promoter (Cp or Wp), EBNA1 binding can transactivate Cp/Wp and up-regulate its own expression (26,27). At the BamHI Q promoter (Qp), EBNA1 binds to two sequences downstream of the transcriptional start site to repress Qp activity and inhibit transcription initiation (28). EBNA1 is a regulator of the expression of other latent viral genes, such as LMP1 and EBNA2, and of cellular gene expression via changes in binding promoters (29,30). Increasing clues indicate that EBNA1 plays an important role in tumorigenesis by various mechanisms (31).

A glycine–glycine–alanine (gly-gly-ala) repeat sequence separates the EBNA1 protein into amino- and carboxy-terminal domains. It functions as a *cis*-acting inhibitor of major histocompability complex class I–restricted presentation and protein degradation via the ubiquitin/proteosome pathway. EBNA1 is therefore allowed to avoid immune destruction by class I–restricted cytotoxic T cells and degradation of protein (32,33).

EBNA2 EBNA2 protein is required for B-cell transformation (34). It acts as a main transactivator, switching transcription from the Wp promoter, used initially after infection, to the Cp promoter. EBNA2 can tether to the host RBP-Jκ protein and is considered similar to a Notch receptor to activate genes, thus achieving B-cell immortalization (35,36). EBNA2 can up-regulate both cellular and viral genes, including CD21, CD23, c-*fgr*, LMP1, and LMP2A, and function in the process of virus–host interactions, including B-cell activation, cell-cycle entry, mitotic checkpoint disruption, and chromosomal instability (37,38).

EBNA3 Family The EBNA3 family contains three members, EBNA3A, -3B, and -3C (also termed EBNA3, EBNA4, and EBNA6). However, the family shows weak homology among members. EBNA3A and -3C are essential and -3B is dispensable in B-cell transformation (39,40).

Efficient association with RBP-Jκ is greater with the EBNA3 family than with EBNA2, so EBNA3 may inhibit EBNA2 from binding RBP-Jκ, thus restraining EBNA2-mediated transactivation. The family regulates cellular and viral gene expression. EBNA3C may up-regulate the expression of CD21 and LMP1 to target p53 and modulate its transcriptional and apoptotic activities (41–43). In addition to regulating transcription, EBNA3C may have a role in chromatin remodeling. Although EBNA3B is not essential for transformation, it is involved in inducing the expression of vimentin and CD40 (44).

EBNA-LP EBNA5, also called EBNA-leader protein (EBNA-LP), has multiple isoforms, depending on the number of BamHI W repeats of an EBV isolate (45). The gene plays a key role in EBV-induced B-cell transformation and growth, transcriptional activation, or repression of cellular genes with the association of other genes; for examples, it cooperates with EBNA2 to stimulate the expression of latent membrane proteins for the G0-to-G1 transition during EBV immortalization of resting human B lymphocytes (46–50).

LMP1 The membrane protein LMP1 has three domains: six transmembrane loops, an amino-terminal cytoplastic tail, and a long carboxy-terminal cytoplasmic region (51). The C-terminal region has two domains, C-terminal activation regions 1 and 2 (CTAR1 and CTAR2). CTAR1 contains a PXQXT motif and binds to tumor necrosis factor receptor (TNFR)-associated factors (52). Thus, LMP1 can function as a mimic of the TNFR family member CD40 to induce the proliferation and differentiation of B cells (53). CTAR2 possesses a YYD motif and interacts with the death domain–containing protein TRADD. Specific protein combination and interaction lead to various signaling, such as NF-κB, mitogen-activated protein kinase (MAPK), c-Jun N-terminal kinase (JNK), and phosphatidylinositol 3-kinase pathways (54,55). The CTAR3 domain, located between CTAR1 and 2, may recruit JAK3, to result in STAT3 activation (54).

LMP1 is essential for EBV-mediated B-cell transformation. It can cause rodent fibroblast transformation, resulting in tumors in nude mice (55–57). As an oncogene, it changes the cell phenotype for altered cell morphologic features, enhanced cell invasion, resistance to growth inhibition caused by culture medium with low serum, induction of apoptosis, loss of anchorage-dependence growth in soft agar, and contact inhibition. In addition, LMP1 induces an epithelial–mesenchymal transition (EMT) and reprogram EBV-infected epithelial cells to acquire stem cell characteristics (58–61).

In NPC, LMP1 is detected at the protein level in 50 to 65% of tumor cases but at the mRNA level in almost all NPC cases (62,63). A 30-bp deletion of LMP1 changes the cell phenotype from nononcogenic to oncogenic. Restoring the 30-bp sequence can reverse the transformation ability (64). NPC may also feature a point mutation of LMP1, inducing loss of an *XhoI* restriction site (65).

LMP1 has two forms: normal B-cell-associated LMP1 (B-LMP1) and NPC-associated LMP1 (NPC-LMP1). B-LMP1 can promote B-lymphocyte transformation and modulate the immunogenicity of EBV-infected cells by up-regulating the NF-κB subunit RelB, thus activating the antigen-presenting cell function. In contrast, NPC-LMP1 induces immune escape by reducing the antigen presentation to CD4⁺ T cells by increased production of interleukin 10 (IL-10) (66). Thus, LMP1-positive NPC cases show fast progression and lymph-node metastasis but have good prognosis, whereas LMP1-negative NPC cases show an increased tendency to metastasis (62).

LMP2   LMP2 encodes two proteins, LMP2A and LMP2B. Neither is required for B-cell transformation *in vitro* (67). The proteins have similar structures, except that LMP2A has an additional 119 amino-acid cytoplasmic amino-terminal domain (68).

LMP2A is an EBV oncoprotein expressed in NPC, gastric carcinoma, and Hodgkin's lymphoma, for example. It is detected at the protein level in about 50% of NPC cases and at the mRNA level in more than 95% of cases (69). Its amino-terminal domain contains important motifs: an immunoreceptor tyrosine-based activation motif with two tyrosines located at position Y74/Y85, a YEEA motif with a tyrosine located at Y112, and two PPPPY motifs (67,70–73). These motifs can recruit Lyn, Syk, and the ubiquitin ligases Nedd4/Itchy, thus stimulating calcium mobilization and protein phosphorylation or degraduation. These processes lead to the blocking of B-cell-receptor signaling and activation of the PI3K/Akt, Wnt/β-catenin, and ERK-MAPK signaling pathways, thus inducing anchorage-independent growth and cell adhesion and motility and inhibiting cell differentiation (55,71). Also, like LMP1, LMP2A can stimulate EMT and reprogram EBV-infected cells to acquire stem-cell-like characteristics (74).

The detailed function of LMP2B is not fully understood. Previous studies indicated that LMP2B regulates the function of LMP2A negatively, then prevents the switch from latent to lytic EBV replication (75,76).

EBERs   The nontranslated EBERs, including EBER1 and EBER2, are highly transcribed in latent infection but are not required for EBV-transformed B cells. EBERs can form complexes with several cellular proteins, such as La, EAP/L22, and protein kinase R (PKR) (77–80). The best-studied combination is the EBER–PKR complex. In *in vitro* assays, the EBER–PKR complex blocked protein synthesis by inhibiting PKR activation and blocking the phosphorylation of the protein synthesis initiation factor eIF2α. In addition, EBERs can induce the expression of IL-10 in Burkitt's lymphoma cells, IL-6 in B-cells, IL-9 in T cells, and insulin-like growth factor 1 in NPC and gastric carcinoma.

The oncogenic role of EBERs has been well studied in the Burkitt's-lymphoma cell line Akata (81). The gene expression of EBERs in EBV-negative Akata cells could restore their growth in soft agar, tumorigenicity in SCID mice, and resistance to apoptotic inducers and increase the expression of oncoprotein bcl-2, retained in parental EBV-positive Akata cells originally and lost in EBV-negative subclones.

BARTs BARTs, also referred to as complementary strand transcripts, are abundant EBV-encoded microRNAs (miRNAs). BARTs were first identified in NPC and later in EBV-associated tumors, and the expression is higher in epithelial tissues than in B lymphocytes (82), which suggests that BARTs might contribute to EBV-driven epithelial malignancy. BARTs produce two clusters of miRNAs (clusters 1 and 2) and an individual miRNA, miR-BART2 (83). BARTs have different splicing forms; four major forms (BARF0, RPMS1, RPMS1A, and A73) and two minor forms (RK-BARF0 and RB3) are well identified. BARTs play important roles in maintaining and regulating latent viral proteins and in regulating apoptosis and immune evasion, thus contributing greatly to NPC pathogenesis (83). According to miRNA functions, negative regulation of the expression and activity of specific miRNAs may be a therapeutic target in NPC.

EBV AND NPC

NPC is a unique head-and-neck cancer with distinct geographic distribution: high frequency in southern China [annual incidence about 20 to 30 in 100,000 (84)] but rare in the rest of the world. It has a complex etiology, including genetic susceptibility, EBV infection, and environmental factors (85).

According to the World Health Organization classification, NPC is histologically classified into three types by degree of differentiation: type I, keratinizing squamous cell carcinoma (25% of cases in North America; ca. 2% in southern China); type II, nonkeratinizing squamous cell carcinoma (12% of cases in North America; ca. 3% in southern China); and type III, undifferentiated carcinoma (63% of cases in North America; 95% in southern China) (86,87). Type III NPC accounts for more than 97% of NPC cases in southern China (88), and with type II NPC, is particularly associated with EBV infection.

The first link of EBV infection and NPC was from serology data in 1966 showing the association of EBV antibody titers with the malignancy (89). EBV DNA was detected in tumor biopsies, which made the correlation speculated more likely (5). In 1973, key evidence demonstrated the existence of the EBV genome in epithelial tumor cells but not in infiltrating lymphoid cells (6). The association of the virus with type I NPC is controversial. Some studies have detected EBV in all squamous-cell NPC (90,91), and others did not (92–94). After comparing the EBV-positive ratio in squamous-cell NPC in high- and low-incidence areas, Niedobitek et al. concluded that squamous-cell NPC may show geographical variability in its association with EBV infection. These observations also suggest that unlike undifferentiated NPCs, squamous-cell NPCs represent a pathogenetically heterogeneous group of tumors (92). Despite the well-established presence of EBV infection in NPC since 1973, the timing and route of EBV infection and its role in NPC remain unclear.

EBV Infection in NP Cells

EBV Entry into NP Cells   EBV virions are transmitted through saliva into oral-pharynx and nasopharynx epithelial cells, assumed to be the site of viral replication contributing to virus persistence (92). However, how the virus enters epithelial cells is less well defined. The EBV envelope glycoproteins gp350 and gp42 binding to the CD21 receptor on the surface of B cells and human leukocyte antigen class II molecules, respectively, are the main mechanism of infection of B cells (16,95). However, in general, most EBV-positive epithelial cells show low or negative expression of CD21, which suggests a CD21-independent mechanism (96). Two pathways have been proposed: direct cell-to-cell contact of the apical surface of the epithelium with EBV-infected salivary B cells and the basolateral membrane entry of cell-free virions (97), which are distinct from infection of B cells. The polymeric immunoglobulin-A receptor of epithelia can mediate the internalization of EBV (98). The other mechanism shows that the integrins αV-$\beta 6$ and -$\beta 8$ can be specific receptors binding gHgL (glycoproteins from EBV), which triggers the virus fusing with the epithelial-cell membrane (99–101). In addition, integrins interact with BMRF2 (a multispanning EBV membrane protein), which induces significant EBV infection (102).

Whether EBV infection in nasopharygeal epithelial cells is an early or relatively late event has been discussed for years because of the difficulty in identifying and obtaining the premalignant NPC material for detailed study. In 1995, EBV was detected in preinvasive NPC lesions, including carcinoma *in situ* and dysplasia, so EBV infection was considered an early event in the development of NPC (103). However, later studies suggested that genetic alterations are early events, before EBV infection in NPC development. Loss of chromosomes 3p and 9p are early events in NPC development, and EBV infection was not detected in low-grade NP lesions harboring the loss of heterozygosity of 3p and 9p. The high-grade dysplastic nasopharyngeal epithelium is infected with clonal EBV later. By expressing viral oncogenes such as LMP1 and other gene products, the high-grade dysplastic nasopharyngeal epithelium is then transformed into invasive cancer cells (103,104). Increasing evidence suggests that the sensitivity of EBV entering epithelial cells and persistent latent infection depend on genetically premodified epithelial cells (105,106).

EBV and NP Carcinogenesis

The carcinogenesis of NPC is a multistep procedure with multiple factors, including environmental carcinogenesis, genetic and epigenetic alterations, and EBV latent infection. EBV DNA is detected with a monoclonal pattern in all NPC tumor cells, so NPC may be a clonal expansion of single EBV-infected progenitor cells and indicate a causal role of EBV in tumor (7). In NPC, the virus is in the form of an episome and is not integrated into the host genome. EBV adopts a specific form of latent infection: Only EBNA1, LMP1, LMP2, and EBER, BARF1, and several BARTs can be detected (107), which are responsible for malignant transformation, thus resulting in NPC (104).

LMP1 and LMP2 activate several transcriptional pathways and modulate the expression of various oncogenes and tumor suppressor genes (108–115) involved in cellular proliferation and metastasis leading to tumor development. Example pathways are resistance to apoptosis and promoting tumor cell motility and angiogenesis (116–118). A key aspect of these two viral proteins is inducing EMT via reprogramming EBV-infected nasopharyngeal epithelial cells to acquire stem-cell or progenitor-like cell characteristics (58,74).

The EBV nuclear antigen EBNA1 is required for persistence of EBV genomes in latency and is the only EBV protein expressed in all EBV-associated tumors, which suggests its contribution to initiating transformation and/or maintenance of the transformed state. The antigen can bind the EBV genome to host chromosomes and thus mediate equal partitioning of viral DNA into daughter cells during cell division for a role in immune evasion (119,120). In addition, EBNA1 can alter cellular pathways in multiple ways that probably contribute to cell immortalization and malignant transformation.

The most abundant viral transcripts found in NPC cells are EBERs (121), with as many as 10^5 to 10^7 copies per cell (122). EBERs induce the expression of insulin-like growth factor 1, which acts as an autocrine growth factor, in EBV-negative NPC-derived cell lines. Moreover, the growth factor is necessary for growth of the EBV-positive NPC-derived cell line C666-1 and is consistently expressed in NPC. EBERs may play an important role in the development of NPC *in vivo* (123). BARF1 can immortalize primate epithelial cells and enhance their growth (124). Cooperation with Ras transformed the human nasopharyngeal epithelial cell line NP69 and induced tumorigenicity in nude mice (125). In addition, the major functions of BARTs in NPC include modulating apoptosis, regulating the expression of LMP1, and contributing to immune evasion (126–129).

EBV as a Biomarker for NPC

NPC patients have high levels of broad-spectrum anti-EBV antibodies, especially immunoglobulin A (IgA), compared with healthy carriers and patients with other head-and-neck diseases. Studies in mainland China and Taiwan have shown the feasibility of using IgA serology for population screening (130,131). Such screening has identified increased EBV-specific antibody titers in high-incidence areas; in particular, IgA antibodies to the EBV capsid antigen and early antigens have been useful in diagnosis and monitoring the effectiveness of therapy (132), even though these markers lack specificity (133). The predictive value of EBV serology for NPC was strongly supported by a report in Zhongshan City in southern China. The long-term study (16 years) revealed a window of about 37 months immediately preceding clinical onset when the level of the IgA viral caspid antigen was elevated and maintained (134).

In contrast, the sensitivity and specificity of EBV DNA in plasma or serum of NPC patients for detecting and monitoring are greater with real-time PCR assay. The quantification of circulating cell-free EBV DNA is well associated with tumor burden

(135,136). Plasma EBV DNA load before therapy is an independent prognostic factor of overall survival in NPC and is strongly associated with the tumor–node–metastasis stage (137). Post-treatment EBV DNA levels are also powerful predictors of recurrence and survival and reflect residual tumor load after therapy (105,138). Moreover, plasma EBV DNA clearance rate was recently found to be a novel prognostic marker for guiding therapy (139).

In addition to detecting EBV DNA load in circulation, detecting EBV DNA in saliva can be a feasible and noninvasive method for early diagnosis of NPC (140).

EBV as a Therapeutic Target for NPC

Radiation therapy remains the standard treatment for NPC because of its high degree of sensitivity (141). With intensity-modulated radiation therapy, the outcome of NPC patients has improved and conferred a high rate of locoregional control and favorable toxicity profile (142,143). Regrettably, most newly diagnosed NPC patients are in advanced stages, with distal metastasis, commonly with lymph-node involvement in the neck, which is the main reason for treatment failure (144). Concurrent chemotherapy with radiation therapy provides a significant benefit in locoregional and distant control. Novel therapies targeting EBV are being developed, including the prevention of viral oncogene expression, induction of lytic form of EBV, and enhancement of the host immune response to virally encoded antigens.

EBV-Based Gene Therapy   The EBV genome can be detected in NPC cells, which provides a tumor specific and potential therapeutic target. The EBV oriP, constructed in a replication-deficient adenovirus vector to drive p53, enhances cytotoxicity specifically in EBV-positive tumor cells (145). Overexpression of the epidermal growth factor receptor (EGFR), common in primary NPC tumors, can be induced by LMP1 and is associated with poor prognosis (146). Cetuximab, the monoclonal antibody against EGRF, inhibits the growth of NPC cells and demonstrates a promising clinical outcome when combined with chemotherapy (147). Meanwhile, targeting LMP1 mRNA DNAzymes and regulation of EBV miRNA levels have been explored to interfere in specific EBV gene function and signaling pathways (126,148–151). In EBV latent-infected NPC, only a portion of viral genes are constitutively expressed; the other genes may remain inactivated because of epigenetic regulation (152). Reactivating silenced EBV genes in tumors by the demethylated agent azacitidine might facilitate immune-mediated destruction of tumor cells (153,154). Moreover, *in vivo* study showed that activation of pharmacologic agents or exogenous expression of BZLF1 and BRLF1 (two viral immediate-early proteins) in EBV-positive tumor cells induced the lytic form of EBV infection and inhibited tumor growth (155,156). The lytic induction allows for ganciclovir-mediated "bystander killing" (157), thus helping to target adjacent latent-infected cells. The evidence supports further clinical trials.

EBV-Based Immunotherapy   EBV-infected tumor cells express viral antigens, thus inspiring treatment strategies based on enhancing the host immune response to viral proteins. Only a few EBV proteins are expressed in NPC: EBNA1, LMP1 and

LMP2. The immunogenes are weak but can still induce a T-lymphocyte response (158–160). The plasma EBV burden increases with a low level of cytotoxic T-lymphocytes (CTLs) of human leukocyte antigen class I in NPC patients, which may explain the host immune evasion of EBV replication. Adoptive transfer of autologous EBV-specific CTLs reduces the plasma EBV load, which may be a feasible immune intervention (149). CTLs stimulated with EBV lymphoblastoid cell lines were used to treat NPC at an advanced stage, thus controlling tumor progression and increasing LMP2-specific immunity. Immunotherapeutic intervention may be a realistic treatment option for NPC (161,162). Allogeneic EBV-specific CTLs were reported to boost long-term LMP2-specific immunity in an NPC patient and induce an antitumor effect (163). Inducing T-cell response by immunization with LMP2A-peptide-pulsed dendritic cells also led to tumor reduction (164). To boost CD4$^+$ (targeting EBNA1) and CD8$^+$ (targeting LMP2A) T-cell immunity, a therapeutic vaccine for NPC targeting EBV is being developed (165,166). EBV-specific immunotherapy together with the other standard therapies for NPC may be beneficial.

REFERENCES

1. Sixbey JW, Nedrud JG, Raab-Traub N, Hanes RA, Pagano JS. Epstein–Barr virus replication in oropharyngeal epithelial cells. N Engl J Med 1984;310:1225–1230.
2. Babcock GJ, Decker LL, Volk M, Thorley-Lawson DA. EBV persistence in memory B cells *in vivo*. Immunity 1998;9:395–404.
3. Gerber P, Lucas S, Nonoyama M, Perlin E, Goldstein LI. Oral excretion of Epstein–Barr virus by healthy subjects and patients with infectious mononucleosis. Lancet 1972;2:988–989.
4. Parkin DM. The global health burden of infection-associated cancers in the year 2002. Int J Cancer 2006;118:3030–3044.
5. zur Hausen H, Schulte-Holthausen H, Klein G, Henle W, Henle G, Clifford P, et al. EBV DNA in biopsies of Burkitt tumours and anaplastic carcinomas of the nasopharynx. Nature 1970;228:1056–1058.
6. Wolf H, zur Hausen H, Becker V. EB viral genomes in epithelial nasopharyngeal carcinoma cells. Nat New Biol 1973;244:245–247.
7. Raab-Traub N, Flynn K. The structure of the termini of the Epstein–Barr virus as a marker of clonal cellular proliferation. Cell 1986;47:883–889.
8. Young LS, Murray PG. Epstein–Barr virus and oncogenesis: from latent genes to tumours. Oncogene 2003;22:5108–5121.
9. Zimber U, Adldinger HK, Lenoir GM, Vuillaume M, Knebel-Doeberitz MV, Laux G, et al. Geographical prevalence of two types of Epstein–Barr virus. Virology 1986;154:56–66.
10. Buisson M, Morand P, Genoulaz O, Bourgeat MJ, Micoud M, Seigneurin JM. Changes in the dominant Epstein–Barr virus type during human immunodeficiency virus infection. J Gen Virol 1994;75 (Pt 2):431–437.
11. Rickinson AB, Young LS, Rowe M. Influence of the Epstein–Barr virus nuclear antigen EBNA 2 on the growth phenotype of virus-transformed B cells. J Virol 1987;61:1310–1317.
12. Lopes V, Young LS, Murray PG. Epstein–Barr virus–associated cancers: aetiology and treatment. Herpes 2003;10:78–82.
13. Shah KM, Young LS. Epstein–Barr virus and carcinogenesis: beyond Burkitt's lymphoma. Clin Microbiol Infect 2009;15:982–988.
14. Fingeroth JD, Weis JJ, Tedder TF, Strominger JL, Biro PA, Fearon DT. Epstein–Barr virus receptor of human B lymphocytes is the C3d receptor CR2. Proc Natl Acad Sci USA 1984;81:4510–4514.
15. Tanner J, Weis J, Fearon D, Whang Y, Kieff E. Epstein–Barr virus gp350/220 binding to the B lymphocyte C3d receptor mediates adsorption, capping, and endocytosis. Cell 1987;50:203–213.

16. Nemerow GR, Mold C, Schwend VK, Tollefson V, Cooper NR. Identification of gp350 as the viral glycoprotein mediating attachment of Epstein–Barr virus (EBV) to the EBV/C3d receptor of B cells: sequence homology of gp350 and C3 complement fragment C3d. J Virol 1987;61:1416–1420.

17. Yoshiyama H, Shimizu N, Takada K. Persistent Epstein–Barr virus infection in a human T-cell line: unique program of latent virus expression. EMBO J 1995;14:3706–3711.

18. Nonoyama M, Huang CH, Pagano JS, Klein G, Singh S. DNA of Epstein–Barr virus detected in tissue of Burkitt's lymphoma and nasopharyngeal carcinoma. Proc Natl Acad Sci USA 1973;70:3265–3268.

19. Niedobitek G, Young LS, Lau R, Brooks L, Greenspan D, Greenspan JS, et al. Epstein–Barr virus infection in oral hairy leukoplakia: virus replication in the absence of a detectable latent phase. J Gen Virol 1991;72 (Pt 12):3035–3046.

20. Ambinder RF, Mann RB. Detection and characterization of Epstein–Barr virus in clinical specimens. Am J Pathol 1994;145:239–252.

21. Miyashita EM, Yang B, Babcock GJ, Thorley-Lawson DA. Identification of the site of Epstein–Barr virus persistence *in vivo* as a resting B cell. J Virol 1997;71:4882–4891.

22. Bochkarev A, Barwell JA, Pfuetzner RA, Bochkareva E, Frappier L, Edwards AM. Crystal structure of the DNA-binding domain of the Epstein–Barr virus origin-binding protein, EBNA1, bound to DNA. Cell 1996;84:791–800.

23. Bochkarev A, Barwell JA, Pfuetzner RA, Furey W, Jr., Edwards AM, Frappier L. Crystal structure of the DNA-binding domain of the Epstein–Barr virus origin-binding protein EBNA 1. Cell 1995;83: 39–46.

24. Yates JL, Camiolo SM, Ali S, Ying A. Comparison of the EBNA1 proteins of Epstein–Barr virus and herpesvirus papio in sequence and function. Virology 1996;222:1–13.

25. Marechal V, Dehee A, Chikhi-Brachet R, Piolot T, Coppey-Moisan M, Nicolas JC. Mapping EBNA-1 domains involved in binding to metaphase chromosomes. J Virol 1999;73:4385–4392.

26. Altmann M, Pich D, Ruiss R, Wang J, Sugden B, Hammerschmidt W. Transcriptional activation by EBV nuclear antigen 1 is essential for the expression of EBV's transforming genes. Proc Natl Acad Sci USA 2006;103:14188–14193.

27. Yates J, Warren N, Reisman D, Sugden B. A *cis*-acting element from the Epstein–Barr viral genome that permits stable replication of recombinant plasmids in latently infected cells. Proc Natl Acad Sci USA 1984;81:3806–3810.

28. Sung NS, Wilson J, Davenport M, Sista ND, Pagano JS. Reciprocal regulation of the Epstein–Barr virus BamHI-F promoter by EBNA-1 and an E2F transcription factor. Mol Cell Biol 1994;14:7144–7152.

29. Gahn TA, Sugden B. An EBNA-1-dependent enhancer acts from a distance of 10 kilobase pairs to increase expression of the Epstein–Barr virus LMP gene. J Virol 1995;69:2633–2636.

30. Canaan A, Haviv I, Urban AE, Schulz VP, Hartman S, Zhang Z, et al. EBNA1 regulates cellular gene expression by binding cellular promoters. Proc Natl Acad Sci USA 2009;106:22421–22426.

31. Wilson JB, Bell JL, Levine AJ. Expression of Epstein–Barr virus nuclear antigen-1 induces B cell neoplasia in transgenic mice. EMBO J 1996;15:3117–3126.

32. Levitskaya J, Coram M, Levitsky V, Imreh S, Steigerwald-Mullen PM, Klein G, et al. Inhibition of antigen processing by the internal repeat region of the Epstein–Barr virus nuclear antigen-1. Nature 1995;375:685–688.

33. Apcher S, Daskalogianni C, Manoury B, Fahraeus R. Epstein–Barr virus-encoded EBNA1 interference with MHC class I antigen presentation reveals a close correlation between mRNA translation initiation and antigen presentation. PLoS Pathog 2010;6:e1001151.

34. Hammerschmidt W, Sugden B. Genetic analysis of immortalizing functions of Epstein–Barr virus in human B lymphocytes. Nature 1989;340:393–397.

35. Grossman SR, Johannsen E, Tong X, Yalamanchili R, Kieff E. The Epstein–Barr virus nuclear antigen 2 transactivator is directed to response elements by the J kappa recombination signal binding protein. Proc Natl Acad Sci USA 1994;91:7568–7572.

36. Sakai T, Taniguchi Y, Tamura K, Minoguchi S, Fukuhara T, Strobl LJ, et al. Functional replacement of the intracellular region of the Notch1 receptor by Epstein–Barr virus nuclear antigen 2. J Virol 1998;72:6034–6039.

37. Maier S, Staffler G, Hartmann A, Hock J, Henning K, Grabusic K, et al. Cellular target genes of Epstein–Barr virus nuclear antigen 2. J Virol 2006;80:9761–9771.

38. Pan SH, Tai CC, Lin CS, Hsu WB, Chou SF, Lai CC, et al. Epstein–Barr virus nuclear antigen 2 disrupts mitotic checkpoint and causes chromosomal instability. Carcinogenesis 2009;30: 366–375.

39. Tomkinson B, Kieff E. Use of second-site homologous recombination to demonstrate that Epstein–Barr virus nuclear protein 3B is not important for lymphocyte infection or growth transformation *in vitro*. J Virol 1992;66:2893–2903.

40. Tomkinson B, Robertson E, Kieff E. Epstein–Barr virus nuclear proteins EBNA-3A and EBNA-3C are essential for B-lymphocyte growth transformation. J Virol 1993;67:2014–2025.

41. Allday MJ, Crawford DH, Thomas JA. Epstein–Barr virus (EBV) nuclear antigen 6 induces expression of the EBV latent membrane protein and an activated phenotype in Raji cells. J Gen Virol 1993;74 (Pt 3):361–369.

42. Yi F, Saha A, Murakami M, Kumar P, Knight JS, Cai Q, et al. Epstein–Barr virus nuclear antigen 3C targets p53 and modulates its transcriptional and apoptotic activities. Virology 2009;388: 236–247.

43. Wang F, Gregory C, Sample C, Rowe M, Liebowitz D, Murray R, et al. Epstein–Barr virus latent membrane protein (LMP1) and nuclear proteins 2 and 3C are effectors of phenotypic changes in B lymphocytes: EBNA-2 and LMP1 cooperatively induce CD23. J Virol 1990;64:2309–2318.

44. Silins SL, Sculley TB. Modulation of vimentin, the CD40 activation antigen and Burkitt's lymphoma antigen (CD77) by the Epstein–Barr virus nuclear antigen EBNA-4. Virology 1994;202: 16–24.

45. Dillner J, Kallin B, Alexander H, Ernberg I, Uno M, Ono Y, et al. An Epstein–Barr virus (EBV)-determined nuclear antigen (EBNA5) partly encoded by the transformation-associated Bam WYH region of EBV DNA: preferential expression in lymphoblastoid cell lines. Proc Natl Acad Sci USA 1986;83:6641–6645.

46. Sinclair AJ, Palmero I, Peters G, Farrell PJ. EBNA-2 and EBNA-LP cooperate to cause G0 to G1 transition during immortalization of resting human B lymphocytes by Epstein–Barr virus. EMBO J 1994;13:3321–3328.

47. Allan GJ, Inman GJ, Parker BD, Rowe DT, Farrell PJ. Cell growth effects of Epstein–Barr virus leader protein. J Gen Virol 1992;73 (Pt 6):1547–1551.

48. Mannick JB, Cohen JI, Birkenbach M, Marchini A, Kieff E. The Epstein–Barr virus nuclear protein encoded by the leader of the EBNA RNAs is important in B-lymphocyte transformation. J Virol 1991;65:6826–6837.

49. Peng R, Moses SC, Tan J, Kremmer E, Ling PD. The Epstein–Barr virus EBNA-LP protein preferentially coactivates EBNA2-mediated stimulation of latent membrane proteins expressed from the viral divergent promoter. J Virol 2005;79:4492–4505.

50. Portal D, Rosendorff A, Kieff E. Epstein–Barr nuclear antigen leader protein coactivates transcription through interaction with histone deacetylase 4. Proc Natl Acad Sci USA 2006;103: 19278–19283.

51. Young LS, Rickinson AB. Epstein–Barr virus: 40 years on. Nat Rev Cancer 2004;4:757–768.

52. Higuchi M, Izumi KM, Kieff E. Epstein–Barr virus latent-infection membrane proteins are palmitoylated and raft-associated: protein 1 binds to the cytoskeleton through TNF receptor cytoplasmic factors. Proc Natl Acad Sci USA 2001;98:4675–4680.

53. Uchida J, Yasui T, Takaoka-Shichijo Y, Muraoka M, Kulwichit W, Raab-Traub N, et al. Mimicry of CD40 signals by Epstein–Barr virus LMP1 in B lymphocyte responses. Science 1999;286: 300–303.

54. Devergne O, Hatzivassiliou E, Izumi KM, Kaye KM, Kleijnen MF, Kieff E, et al. Association of TRAF1, TRAF2, and TRAF3 with an Epstein–Barr virus LMP1 domain important for B-lymphocyte transformation: role in NF-kappaB activation. Mol Cell Biol 1996;16:7098–7108.

55. Dawson CW, Port RJ, Young LS. The role of the EBV-encoded latent membrane proteins LMP1 and LMP2 in the pathogenesis of nasopharyngeal carcinoma (NPC). Semin Cancer Biol 2012;22:144–153.

56. Dawson CW, Rickinson AB, Young LS. Epstein–Barr virus latent membrane protein inhibits human epithelial cell differentiation. Nature 1990;344:777–780.

57. Wang D, Liebowitz D, Kieff E. An EBV membrane protein expressed in immortalized lymphocytes transforms established rodent cells. Cell 1985;43:831–840.

58. Horikawa T, Yang J, Kondo S, Yoshizaki T, Joab I, Furukawa M, et al. Twist and epithelial–mesenchymal transition are induced by the EBV oncoprotein latent membrane protein 1 and are associated with metastatic nasopharyngeal carcinoma. Cancer Res 2007;67:1970–1978.

59. Sides MD, Klingsberg RC, Shan B, Gordon KA, Nguyen HT, Lin Z, et al. The Epstein–Barr virus latent membrane protein 1 and transforming growth factor–beta1 synergistically induce epithelial–mesenchymal transition in lung epithelial cells. Am J Respir Cell Mol Biol 2011;44: 852–862.

60. Horikawa T, Yoshizaki T, Kondo S, Furukawa M, Kaizaki Y, Pagano JS. Epstein–Barr virus latent membrane protein 1 induces Snail and epithelial–mesenchymal transition in metastatic nasopharyngeal carcinoma. Br J Cancer 2011;104:1160–1167.

61. Kondo S, Wakisaka N, Muramatsu M, Zen Y, Endo K, Murono S, et al. Epstein–Barr virus latent membrane protein 1 induces cancer stem/progenitor-like cells in nasopharyngeal epithelial cell lines. J Virol 2011;85:11255–11264.

62. Hu LF, Chen F, Zhen QF, Zhang YW, Luo Y, Zheng X, et al. Differences in the growth pattern and clinical course of EBV-LMP1 expressing and non-expressing nasopharyngeal carcinomas. Eur J Cancer 1995;31A:658–660.

63. Ozyar E, Ayhan A, Korcum AF, Atahan IL. Prognostic role of Epstein–Barr virus latent membrane protein-1 and interleukin-10 expression in patients with nasopharyngeal carcinoma. Cancer Invest 2004;22:483–491.

64. Li SN, Chang YS, Liu ST. Effect of a 10-amino acid deletion on the oncogenic activity of latent membrane protein 1 of Epstein–Barr virus. Oncogene 1996;12:2129–2135.

65. Chen ML, Tsai CN, Liang CL, Shu CH, Huang CR, Sulitzeanu D, et al. Cloning and characterization of the latent membrane protein (LMP) of a specific Epstein–Barr virus variant derived from the nasopharyngeal carcinoma in the Taiwanese population. Oncogene 1992;7: 2131–2140.

66. Pai S, O'Sullivan B, Abdul-Jabbar I, Peng J, Connoly G, Khanna R, et al. Nasopharyngeal carcinoma-associated Epstein–Barr virus-encoded oncogene latent membrane protein 1 potentiates regulatory T-cell function. Immunol Cell Biol 2007;85:370–377.

67. Longnecker R. Epstein–Barr virus latency: LMP2, a regulator or means for Epstein–Barr virus persistence? Adv Cancer Res 2000;79:175–200.

68. Longnecker R, Kieff E. A second Epstein–Barr virus membrane protein (LMP2) is expressed in latent infection and colocalizes with LMP1. J Virol 1990;64:2319–2326.

69. Busson P, McCoy R, Sadler R, Gilligan K, Tursz T, Raab-Traub N. Consistent transcription of the Epstein–Barr virus LMP2 gene in nasopharyngeal carcinoma. J Virol 1992;66:3257–3262.

70. Fruehling S, Swart R, Dolwick KM, Kremmer E, Longnecker R. Tyrosine 112 of latent membrane protein 2A is essential for protein tyrosine kinase loading and regulation of Epstein–Barr virus latency. J Virol 1998;72:7796–7806.

71. Pang MF, Lin KW, Peh SC. The signaling pathways of Epstein–Barr virus-encoded latent membrane protein 2A (LMP2A) in latency and cancer. Cell Mol Biol Lett 2009;14:222–247.

72. Cambier JC. New nomenclature for the Reth motif (or ARH1/TAM/ARAM/YXXL). Immunol Today 1995;16:110.

73. Ikeda A, Caldwell RG, Longnecker R, Ikeda M. Itchy, a Nedd4 ubiquitin ligase, downregulates latent membrane protein 2A activity in B-cell signaling. J Virol 2003;77:5529–5534.

74. Kong QL, Hu LJ, Cao JY, Huang YJ, Xu LH, Liang Y, et al. Epstein–Barr virus-encoded LMP2A induces an epithelial–mesenchymal transition and increases the number of side population stem-like cancer cells in nasopharyngeal carcinoma. PLoS Pathog 2010;6:e1000940.

75. Rovedo M, Longnecker R. Epstein–Barr virus latent membrane protein 2B (LMP2B) modulates LMP2A activity. J Virol 2007;81:84–94.

76. Rechsteiner MP, Berger C, Zauner L, Sigrist JA, Weber M, Longnecker R, et al. Latent membrane protein 2B regulates susceptibility to induction of lytic Epstein–Barr virus infection. J Virol 2008;82:1739–1747.

77. Clarke PA, Schwemmle M, Schickinger J, Hilse K, Clemens MJ. Binding of Epstein–Barr virus small RNA EBER-1 to the double-stranded RNA-activated protein kinase DAI. Nucleic Acids Res 1991;19:243–248.

78. Toczyski DP, Matera AG, Ward DC, Steitz JA. The Epstein–Barr virus (EBV) small RNA EBER1 binds and relocalizes ribosomal protein L22 in EBV-infected human B lymphocytes. Proc Natl Acad Sci USA 1994;91:3463–3467.

79. Toczyski DP, Steitz JA. EAP, a highly conserved cellular protein associated with Epstein–Barr virus small RNAs (EBERs). EMBO J 1991;10:459–466.

80. Lerner MR, Andrews NC, Miller G, Steitz JA. Two small RNAs encoded by Epstein–Barr virus and complexed with protein are precipitated by antibodies from patients with systemic lupus erythematosus. Proc Natl Acad Sci USA 1981;78:805–809.

81. Komano J, Maruo S, Kurozumi K, Oda T, Takada K. Oncogenic role of Epstein–Barr virus-encoded RNAs in Burkitt's lymphoma cell line Akata. J Virol 1999;73:9827–9831.

82. Gilligan KJ, Rajadurai P, Lin JC, Busson P, Abdel-Hamid M, Prasad U, et al. Expression of the Epstein–Barr virus BamHI A fragment in nasopharyngeal carcinoma: evidence for a viral protein expressed *in vivo*. J Virol 1991;65:6252–6259.

83. Lo AK, Dawson CW, Jin DY, Lo KW. The pathological roles of BART miRNAs in nasopharyngeal carcinoma. J Pathol 2012;227:392–403.

84. Yu MC, Yuan JM. Epidemiology of nasopharyngeal carcinoma. Semin Cancer Biol 2002;12: 421–429.

85. Tao Q, Chan AT. Nasopharyngeal carcinoma: molecular pathogenesis and therapeutic developments. Expert Rev Mol Med 2007;9:1–24.

86. Shanmugaratnam K, Sobin LH. The World Health Organization histological classification of tumours of the upper respiratory tract and ear: a commentary on the second edition. Cancer 1993;71:2689–2697.

87. Marks JE, Phillips JL, Menck HR. The National Cancer Data Base report on the relationship of race and national origin to the histology of nasopharyngeal carcinoma. Cancer 1998;83:582–588.

88. Lo KW, To KF, Huang DP. Focus on nasopharyngeal carcinoma. Cancer Cell 2004;5:423–428.

89. Old LJ, Boyse EA, Oettgen HF, Harven ED, Geering G, Williamson B, et al. Precipitating antibody in human serum to an antigen present in cultured Burkitt's lymphoma cells. Proc Natl Acad Sci USA 1966;56:1699–1704.

90. Pathmanathan R, Prasad U, Chandrika G, Sadler R, Flynn K, Raab-Traub N. Undifferentiated, nonkeratinizing, and squamous cell carcinoma of the nasopharynx. Variants of Epstein–Barr virus-infected neoplasia. Am J Pathol 1995;146:1355–1367.

91. Raab-Traub N, Flynn K, Pearson G, Huang A, Levine P, Lanier A, et al. The differentiated form of nasopharyngeal carcinoma contains Epstein–Barr virus DNA. Int J Cancer 1987;39:25–29.

92. Niedobitek G, Agathanggelou A, Barber P, Smallman LA, Jones EL, Young LS. P53 overexpression and Epstein–Barr virus infection in undifferentiated and squamous cell nasopharyngeal carcinomas. J Pathol 1993;170:457–461.

93. Niedobitek G, Hansmann ML, Herbst H, Young LS, Dienemann D, Hartmann CA, et al. Epstein–Barr virus and carcinomas: undifferentiated carcinomas but not squamous cell carcinomas of the nasopharynx are regularly associated with the virus. J Pathol 1991;165:17–24.

94. Klein G, Giovanella BC, Lindahl T, Fialkow PJ, Singh S, Stehlin JS. Direct evidence for the presence of Epstein–Barr virus DNA and nuclear antigen in malignant epithelial cells from patients with poorly differentiated carcinoma of the nasopharynx. Proc Natl Acad Sci USA 1974;71: 4737–4741.

95. Borza CM, Hutt-Fletcher LM. Alternate replication in B cells and epithelial cells switches tropism of Epstein–Barr virus. Nat Med 2002;8:594–599.

96. Fingeroth JD, Diamond ME, Sage DR, Hayman J, Yates JL. CD21-dependent infection of an epithelial cell line, 293, by Epstein–Barr virus. J Virol 1999;73:2115–2125.

97. Tugizov SM, Berline JW, Palefsky JM. Epstein–Barr virus infection of polarized tongue and nasopharyngeal epithelial cells. Nat Med 2003;9:307–314.

98. Gan YJ, Chodosh J, Morgan A, Sixbey JW. Epithelial cell polarization is a determinant in the infectious outcome of immunoglobulin A-mediated entry by Epstein–Barr virus. J Virol 1997;71: 519–526.

99. Borza CM, Morgan AJ, Turk SM, Hutt-Fletcher LM. Use of gHgL for attachment of Epstein–Barr virus to epithelial cells compromises infection. J Virol 2004;78:5007–5014.

100. Burman A, Clark S, Abrescia NG, Fry EE, Stuart DI, Jackson T. Specificity of the VP1 GH loop of foot-and-mouth disease virus for alphav integrins. J Virol 2006;80:9798–9810.

101. Mu D, Cambier S, Fjellbirkeland L, Baron JL, Munger JS, Kawakatsu H, et al. The integrin alpha(v)beta8 mediates epithelial homeostasis through MT1-MMP-dependent activation of TGF-beta1. J Cell Biol 2002;157:493–507.

102. Xiao J, Palefsky JM, Herrera R, Tugizov SM. Characterization of the Epstein–Barr virus glycoprotein BMRF-2. Virology 2007;359:382–396.

103. Pathmanathan R, Prasad U, Sadler R, Flynn K, Raab-Traub N. Clonal proliferations of cells infected with Epstein–Barr virus in preinvasive lesions related to nasopharyngeal carcinoma. N Engl J Med 1995;333:693–698.

104. Raab-Traub N. Epstein–Barr virus in the pathogenesis of NPC. Semin Cancer Biol 2002;12:431–441.

105. Chan AT, Lo YM, Zee B, Chan LY, Ma BB, Leung SF, et al. Plasma Epstein–Barr virus DNA and residual disease after radiotherapy for undifferentiated nasopharyngeal carcinoma. J Natl Cancer Inst 2002;94:1614–1619.

106. Chan AS, To KF, Lo KW, Mak KF, Pak W, Chiu B, et al. High frequency of chromosome 3p deletion in histologically normal nasopharyngeal epithelia from southern Chinese. Cancer Res 2000;60:5365–5370.

107. Herrmann K, Niedobitek G. Epstein–Barr virus-associated carcinomas: facts and fiction. J Pathol 2003;199:140–145.

108. Kieser A, Kilger E, Gires O, Ueffing M, Kolch W, Hammerschmidt W. Epstein–Barr virus latent membrane protein-1 triggers AP-1 activity via the c-Jun N-terminal kinase cascade. EMBO J 1997;16:6478–6485.

109. Izumi KM, Kieff ED. The Epstein–Barr virus oncogene product latent membrane protein 1 engages the tumor necrosis factor receptor–associated death domain protein to mediate B lymphocyte growth transformation and activate NF-kappaB. Proc Natl Acad Sci USA 1997;94:12592–12597.

110. Eliopoulos AG, Young LS. Activation of the cJun N-terminal kinase (JNK) pathway by the Epstein–Barr virus-encoded latent membrane protein 1 (LMP1). Oncogene 1998;16:1731–1742.

111. Chen H, Lee JM, Zong Y, Borowitz M, Ng MH, Ambinder RF, et al. Linkage between STAT regulation and Epstein–Barr virus gene expression in tumors. J Virol 2001;75:2929–2937.

112. Wu HC, Lu TY, Lee JJ, Hwang JK, Lin YJ, Wang CK, et al. MDM2 expression in EBV-infected nasopharyngeal carcinoma cells. Lab Invest 2004;84:1547–1556.

113. Sheu LF, Chen A, Lee HS, Hsu HY, Yu DS. Cooperative interactions among p53, bcl-2 and Epstein–Barr virus latent membrane protein 1 in nasopharyngeal carcinoma cells. Pathol Int 2004;54:475–485.

114. Man C, Rosa J, Lee LT, Lee VH, Chow BK, Lo KW, et al. Latent membrane protein 1 suppresses RASSF1A expression, disrupts microtubule structures and induces chromosomal aberrations in human epithelial cells. Oncogene 2007;26:3069–3080.

115. Fukuda M, Longnecker R. Epstein–Barr virus latent membrane protein 2A mediates transformation through constitutive activation of the Ras/PI3-K/Akt Pathway. J Virol 2007;81:9299–9306.

116. Allen MD, Young LS, Dawson CW. The Epstein–Barr virus-encoded LMP2A and LMP2B proteins promote epithelial cell spreading and motility. J Virol 2005;79:1789–1802.

117. Yoshizaki T, Horikawa T, Qing-Chun R, Wakisaka N, Takeshita H, Sheen TS, et al. Induction of interleukin-8 by Epstein–Barr virus latent membrane protein-1 and its correlation to angiogenesis in nasopharyngeal carcinoma. Clin Cancer Res 2001;7:1946–1951.

118. Morris MA, Dawson CW, Young LS. Role of the Epstein–Barr virus-encoded latent membrane protein-1, LMP1, in the pathogenesis of nasopharyngeal carcinoma. Future Oncol 2009;5:811–825.

119. Lo AK, Lo KW, Tsao SW, Wong HL, Hui JW, To KF, et al. Epstein–Barr virus infection alters cellular signal cascades in human nasopharyngeal epithelial cells. Neoplasia 2006;8:173–180.

120. Yates JL, Warren N, Sugden B. Stable replication of plasmids derived from Epstein–Barr virus in various mammalian cells. Nature 1985;313:812–815.

121. Wu TC, Mann RB, Epstein JI, MacMahon E, Lee WA, Charache P, et al. Abundant expression of EBER1 small nuclear RNA in nasopharyngeal carcinoma: a morphologically distinctive target for detection of Epstein–Barr virus in formalin-fixed paraffin-embedded carcinoma specimens. Am J Pathol 1991;138:1461–1469.

122. Howe JG, Shu MD. Epstein–Barr virus small RNA (EBER) genes: unique transcription units that combine RNA polymerase II and III promoter elements. Cell 1989;57:825–834.

123. Iwakiri D, Sheen TS, Chen JY, Huang DP, Takada K. Epstein–Barr virus-encoded small RNA induces insulin-like growth factor 1 and supports growth of nasopharyngeal carcinoma-derived cell lines. Oncogene 2005;24:1767–1773.

124. Wei MX, de Turenne-Tessier M, Decaussin G, Benet G, Ooka T. Establishment of a monkey kidney epithelial cell line with the BARF1 open reading frame from Epstein–Barr virus. Oncogene 1997;14:3073–3081.

125. Jiang R, Cabras G, Sheng W, Zeng Y, Ooka T. Synergism of BARF1 with Ras induces malignant transformation in primary primate epithelial cells and human nasopharyngeal epithelial cells. Neoplasia 2009;11:964–973.

126. Lo AK, To KF, Lo KW, Lung RW, Hui JW, Liao G, et al. Modulation of LMP1 protein expression by EBV-encoded microRNAs. Proc Natl Acad Sci USA 2007;104:16164–16169.

127. Marquitz AR, Mathur A, Nam CS, Raab-Traub N. The Epstein–Barr virus BART microRNAs target the pro-apoptotic protein Bim. Virology 2011;412:392–400.

128. Lisnic VJ, Krmpotic A, Jonjic S. Modulation of natural killer cell activity by viruses. Curr Opin Microbiol 2010;13:530–539.

129. Sofos E, Pescosolido MF, Quintos JB, Abuelo D, Gunn S, Hovanes K, et al. A novel familial 11p15.4 microduplication associated with intellectual disability, dysmorphic features, and obesity with involvement of the ZNF214 gene. Am J Med Genet A 2012;158A:50–58.

130. Zong YS, Sham JS, Ng MH, Ou XT, Guo YQ, Zheng SA, et al. Immunoglobulin A against viral capsid antigen of Epstein–Barr virus and indirect mirror examination of the nasopharynx in the detection of asymptomatic nasopharyngeal carcinoma. Cancer 1992;69:3–7.

131. Chien YC, Chen JY, Liu MY, et al. Serologic markers of Epstein–Barr virus infection and nasopharyngeal carcinoma in Taiwanese men. N Engl J Med 2001;345:1877–1882.

132. Zeng Y. Seroepidemiological studies on nasopharyngeal carcinoma in China. Adv Cancer Res 1985;44:121–138.

133. Altun M, Fandi A, Dupuis O, Cvitkovic E, Krajina Z, Eschwege F. Undifferentiated nasopharyngeal cancer (UCNT): current diagnostic and therapeutic aspects. Int J Radiat Oncol Biol Phys 1995;32:859–877.

134. Ji MF, Wang DK, Yu YL, Guo YQ, Liang JS, Cheng WM, et al. Sustained elevation of Epstein–Barr virus antibody levels preceding clinical onset of nasopharyngeal carcinoma. Br J Cancer 2007;96:623–630.

135. Lo YM, Chan LY, Lo KW, Leung SF, Zhang J, Chan AT, et al. Quantitative analysis of cell-free Epstein–Barr virus DNA in plasma of patients with nasopharyngeal carcinoma. Cancer Res 1999;59:1188–1191.

136. Lo YM. Quantitative analysis of Epstein–Barr virus DNA in plasma and serum: applications to tumor detection and monitoring. Ann N Y Acad Sci 2001;945:68–72.

137. Leung SF, Zee B, Ma BB, Hui EP, Mo F, Lai M, et al. Plasma Epstein–Barr viral deoxyribonucleic acid quantitation complements tumor-node-metastasis staging prognostication in nasopharyngeal carcinoma. J Clin Oncol 2006;24:5414–5418.

138. Lin JC, Wang WY, Chen KY, Wei YH, Liang WM, Jan JS, et al. Quantification of plasma Epstein–Barr virus DNA in patients with advanced nasopharyngeal carcinoma. N Engl J Med 2004;350:2461–2470.

139. Wang WY, Twu CW, Chen HH, Jan JS, Jiang RS, Chao JY, et al. Plasma EBV DNA clearance rate as a novel prognostic marker for metastatic/recurrent nasopharyngeal carcinoma. Clin Cancer Res 2010;16:1016–1024.

140. Pow EH, Law MY, Tsang PC, Perera RA, Kwong DL. Salivary Epstein–Barr virus DNA level in patients with nasopharyngeal carcinoma following radiotherapy. Oral Oncol 2011;47:879–882.

141. Lee AW, Poon YF, Foo W, Law SC, Cheung FK, Chan DK, et al. Retrospective analysis of 5037 patients with nasopharyngeal carcinoma treated during 1976-1985: overall survival and patterns of failure. Int J Radiat Oncol Biol Phys 1992;23:261–270.

142. Kam MK, Teo PM, Chau RM, Cheung KY, Choi PH, Kwan WH, et al. Treatment of nasopharyngeal carcinoma with intensity-modulated radiotherapy: the Hong Kong experience. Int J Radiat Oncol Biol Phys 2004;60:1440–1450.

143. Wolden SL, Chen WC, Pfister DG, Kraus DH, Berry SL, Zelefsky MJ. Intensity-modulated radiation therapy (IMRT) for nasopharynx cancer: update of the Memorial Sloan–Kettering experience. Int J Radiat Oncol Biol Phys 2006;64:57–62.

144. Vokes EE, Liebowitz DN, Weichselbaum RR. Nasopharyngeal carcinoma. Lancet 1997;350: 1087–1091.

145. Li JH, Chia M, Shi W, Ngo D, Strathdee CA, Huang D, et al. Tumor-targeted gene therapy for nasopharyngeal carcinoma. Cancer Res 2002;62:171–178.

146. Rumack BH. Hydrocarbon ingestions: an opinion. Bull Natl Clgh Poison Control Cent 1976:2–5.

147. Chan AT, Hsu MM, Goh BC, Hui EP, Liu TW, Millward MJ, et al. Multicenter, phase II study of cetuximab in combination with carboplatin in patients with recurrent or metastatic nasopharyngeal carcinoma. J Clin Oncol 2005;23:3568–3576.

148. Micheau C. What's new in histological classification and recognition of nasopharyngeal carcinoma (N.P.C.). Pathol Res Pract 1986;181:249–253.

149. Chua D, Huang J, Zheng B, Lau SY, Luk W, Kwong DL, et al. Adoptive transfer of autologous Epstein–Barr virus–specific cytotoxic T cells for nasopharyngeal carcinoma. Int J Cancer 2001;94:73–80.

150. Cosmopoulos K, Pegtel M, Hawkins J, Moffett H, Novina C, Middeldorp J, et al. Comprehensive profiling of Epstein–Barr virus microRNAs in nasopharyngeal carcinoma. J Virol 2009;83:2357–2367.

151. Yang L, Lu Z, Ma X, Cao Y, Sun LQ. A therapeutic approach to nasopharyngeal carcinomas by DNAzymes targeting EBV LMP-1 gene. Molecules 2010;15:6127–6139.

152. Ambinder RF, Robertson KD, Tao Q. DNA methylation and the Epstein–Barr virus. Semin Cancer Biol 1999;9:369–375.

153. Ambinder RF, Robertson KD, Moore SM, Yang J. Epstein–Barr virus as a therapeutic target in Hodgkin's disease and nasopharyngeal carcinoma. Semin Cancer Biol 1996;7:217–226.

154. Chan AT, Tao Q, Robertson KD, Flinn IW, Mann RB, Klencke B, et al. Azacitidine induces demethylation of the Epstein–Barr virus genome in tumors. J Clin Oncol 2004;22:1373–1381.

155. Feng WH, Westphal E, Mauser A, Raab-Traub N, Gulley ML, Busson P, et al. Use of adenovirus vectors expressing Epstein–Barr virus (EBV) immediate-early protein BZLF1 or BRLF1 to treat EBV-positive tumors. J Virol 2002;76:10951–10959.

156. Feng WH, Israel B, Raab-Traub N, Busson P, Kenney SC. Chemotherapy induces lytic EBV replication and confers ganciclovir susceptibility to EBV-positive epithelial cell tumors. Cancer Res 2002;62:1920–1926.

157. Freeman SM, Abboud CN, Whartenby KA, Packman CH, Koeplin DS, Moolten FL, et al. The "bystander effect": tumor regression when a fraction of the tumor mass is genetically modified. Cancer Res 1993;53:5274–5283.

158. Khanna R, Busson P, Burrows SR, Raffoux C, Moss DJ, Nicholls JM, et al. Molecular characterization of antigen-processing function in nasopharyngeal carcinoma (NPC): evidence for efficient presentation of Epstein–Barr virus cytotoxic T-cell epitopes by NPC cells. Cancer Res 1998;58:310–314.

159. Rickinson AB, Moss DJ. Human cytotoxic T lymphocyte responses to Epstein–Barr virus infection. Annu Rev Immunol 1997;15:405–431.

160. Meij P, Leen A, Rickinson AB, Verkoeijen S, Vervoort MB, Bloemena E, et al. Identification and prevalence of CD8(+) T-cell responses directed against Epstein–Barr virus–encoded latent membrane protein 1 and latent membrane protein 2. Int J Cancer 2002;99:93–99.

161. Straathof KC, Bollard CM, Popat U, Huls MH, Lopez T, Morriss MC, et al. Treatment of nasopharyngeal carcinoma with Epstein–Barr virus–specific T lymphocytes. Blood 2005;105: 1898–1904.

162. Comoli P, Pedrazzoli P, Maccario R, Basso S, Carminati O, Labirio M, et al. Cell therapy of stage IV nasopharyngeal carcinoma with autologous Epstein–Barr virus–targeted cytotoxic T lymphocytes. J Clin Oncol 2005;23:8942–8949.

163. Comoli P, De Palma R, Siena S, Nocera A, Basso S, Del Galdo F, et al. Adoptive transfer of allogeneic Epstein–Barr virus (EBV)-specific cytotoxic T cells with *in vitro* antitumor activity boosts LMP2-specific immune response in a patient with EBV-related nasopharyngeal carcinoma. Ann Oncol 2004;15:113–117.

164. Lin CL, Lo WF, Lee TH, Ren Y, Hwang SL, et al. Immunization with Epstein–Barr Virus (EBV) peptide-pulsed dendritic cells induces functional CD8+ T-cell immunity and may lead

to tumor regression in patients with EBV-positive nasopharyngeal carcinoma. Cancer Res 2002;62:6952–6958.

165. Smith C, Tsang J, Beagley L, Chua D, Lee V, Li V, et al. Effective treatment of metastatic forms of Epstein–Barr virus–associated nasopharyngeal carcinoma with a novel adenovirus-based adoptive immunotherapy. Cancer Res 2012;72:1116–1125.

166. Taylor GS, Haigh TA, Gudgeon NH, Phelps RJ, Lee SP, Steven NM, et al. Dual stimulation of Epstein–Barr virus (EBV)–specific CD4+- and CD8+-T-cell responses by a chimeric antigen construct: potential therapeutic vaccine for EBV-positive nasopharyngeal carcinoma. J Virol 2004;78:768–778.

CHAPTER 15

BARRETT'S ESOPHAGUS AND ESOPHAGEAL CANCER

Albert Roessner and Angela Poehlmann

In typical cases, esophageal adenocarcinoma (EA) develops via the well-known sequence from reflux esophagitis, Barrett's metaplasia, preneoplastic dysplasia, and adenocarcinoma. According to the most reliable analyses, the prevalence of Barrett's esophagus (BE) is about 1.6%, as shown by one particular analysis from the Netherlands (1). The annual risk of progression to EA as estimated in the United States amounts to 0.4% (2). Besides histologically proved preneoplastic dysplasia, older age and male gender are considered predictors of malignant transformation. In the last few years, many data on molecular alterations in Barrett's epithelium, dysplasia, and cancer have been published. These data have been evaluated in current reviews 3–8. This means that distinct molecular events of the carcinogenic process from metaplasia to adenocarcinoma have begun unraveling. However, the complexity of the cellular pathogenesis from normal squamous epithelium of the esophagus via columnar cell metaplasia, intestinal metaplasia, and different degrees of dysplasia to adenocarcinoma still raises many questions. In clinical practice, the assessment of biological behavior still depends on histological evaluation.

Therefore, in this chapter we cover the following topics:

1. The continuing paramount importance of the histological diagnosis of Barrett's cancer and its prestages.

2. The possible role of molecular predictors for estimation of the malignant potential of the prestages. This includes a brief discussion of the current experience with "biomarkers" for the diagnosis of Barrett's carcinogenesis, addressing the question of why their practical values have proven limited so far.

3. The relationship between inflammation and Barrett's carcinogenesis, particularly for reflux esophagitis, including the role of genetic and epigenetic alterations in Barrett's carcinogenesis and the role of inflammation-associated reactive oxygen and nitrogen species (ROSs/RNSs).

Cancer and Inflammation Mechanisms: Chemical, Biological, and Clinical Aspects, First Edition.
Edited by Yusuke Hiraku, Shosuke Kawanishi, and Hiroshi Ohshima.

HISTOLOGICAL DIAGNOSIS OF BARRETT'S-RELATED DYSPLASIA AND BARRETT'S-ASSOCIATED CANCER

In contrast to some types of stomach carcinoma which reveal mutations in the E-cadherin gene and inherited colon carcinoma with alterations in mismatch repair genes or alterations in the APC suppressor gene, no defined specific genetic alterations have been observed in Barrett's adenocarcinoma. Therefore, estimation of the risk of the prestages to develop esophageal cancer in practice depends exclusively on the histopathological diagnosis (9). However, the different degrees of intraepithelial neoplasia may be similar to each other and are sometimes not easy to diagnose (Figure 15.1).

Dysplasia in the gastrointestinal tract has been defined in the context of classifying the prestages of adenocarcinoma in chronic inflammatory bowel disease as an unequivocal neoplastic epithelium strictly confined within the basement membrane of the gland from which it arises (10). This definition has also been applied to other regions in the gastrointestinal tract. Dysplasia in the gastrointestinal tract is usually classified as negative, indefinite, or positive for dysplasia. If the epithelium in the biopsy under assessment is positive for dysplasia, it should be divided further into low-grade dysplasia (LGD) and high-grade dysplasia (HGD). In this two-tier system, LGD includes the category of moderate dysplasia from a three-tier system.

The term *indefinite* for dysplasia includes alterations suspected of being dysplasia but which cannot be clearly differentiated from regenerative changes or changes in association with severe inflammation. The histopathological picture of dysplasia in Barrett's esophagus has often been described (11–13).

The principal histological definition of dysplasia is relatively clear and familiar to every diagnostic histopathologist. However, the problem lies in the inter- and even intraobserver variations (14–17). Montgomery et al. reported on 12 specialized gastrointestinal pathologists who diagnosed 125 biopsies (16). When a four-tier system was employed (no dysplasia, indefinite, LGD, HGD), the κ index was 0.43. The κ index is a statistical measure of interobserver agreement. If the observers are in complete agreement, κ is 1; if there is no agreement, κ is 0. A study involving 20 general pathologists showed a large variation even when diagnosing nondysplastic mucosa to be separated from LGD and HGD (15). Ormsby et al. showed that even experienced gastrointestinal pathologists frequently disagree on a diagnosis of HGD versus intramucosal adenocarcinoma (17). The differences in the diagnostic assessment of dysplasia in the gastrointestinal tract are even more pronounced when pathologists from Japan and Western countries are involved. Japanese pathologists classified all lesions as carcinoma, whereas Western pathologists recognized carcinomas in only 10 to 67% of cases (18). A recent investigation of Coco et al. on the interobserver variability in the diagnosis of crypt dysplasia in Barrett's esophagus showed moderate agreement among six pathologists (κ = 0.44) (19). The highest levels of agreement were observed for lesions at the low ands high ends of the spectrum (i.e., Barrett's epithelium without dysplasia and HGD). The authors concluded from this study that crypt dysplasia in BE can be histologically diagnosed reliably with a moderate level of interobserver agreement.

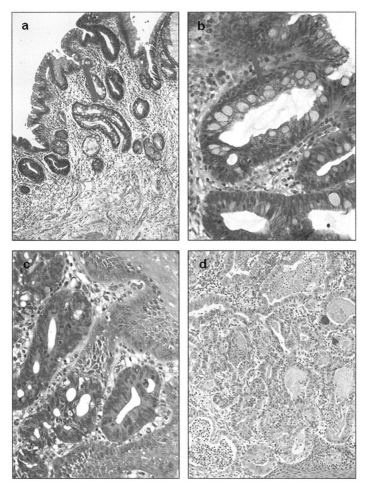

FIGURE 15.1   (a) Columnar cells and goblet cells in Barrett's metaplasia. The goblet cells are characterized by mucus in their cytoplasm (arrow). HE × 100. (b) Low-grade intraepithelial neoplasia. Slight architectural irregularities. Nuclear atypia. Some goblet cells are preserved (arrow). HE × 250. (c) High-grade intraepithelial neoplasia. More pronounced architectural irregularities and nuclear atypia. HE × 250. (d) Intramucsosal Barrett's carcinoma. HE × 100. (*See insert for color representation of the figure.*)

To minimize the various assessments of dysplasia in the gastrointestinal tract, the Vienna classification of epithelial neoplasia of the digestive tract has introduced a five-tier system. It avoids the term *dysplasia* and replaces it with *intraepithelial neoplasia*. The Vienna classification has some advantages, but it obviously cannot solve the principal limitations regarding the histological diagnosis of dysplasia, irrespective of the use of a four- or five-tier system or any other histological system. There remains a certain degree of inter- and intraobserver variability. Good results achieved with the Vienna classification for diagnosing Barrett's dysplasia have been published by Kaye et al. (20).

ROLE OF MOLECULAR PREDICTORS FOR BARRETT'S CARCINOGENESIS

The manifold impressive data from the current literature show that many molecular steps in Barrett's carcinogenesis have been unraveled, which leads to an increasingly detailed molecular understanding of the carcinogenesis process. Many of these studies point to a biomarker potential of the molecular factors investigated, as discussed for overexpression of Erb-B2 in EA (21), hypermethylation of the tumor suppressor p16 in tissue samples with dysplasia (22), Ki67 proliferation rate increasing with degree of dysplasia (23), alterations in p53 tumor suppressor gene (24) and hypermethylation of the pro-apoptotic factor DAPK (25). Furthermore, DNA aneuploidy has been established as a biomarker for progression in Barrett's carcinogenesis (26). cDNA microarray technology and the combination of different hypermethylation biomarkers also seem to be useful. Of these candidates serving as biomarkers, most notably aneuploidy, proliferation rate estimated by Ki67, and p53 immunohistology seem to be of some practical value (13, 27). On the other hand, regarding assessment of the prestages of Barrett's adenocarcinoma, useful biomarkers are still not widely used in clinical practice, as pointed out by Jankowski and Odze (9). This means that in relation to the many impressive molecular biological data revealing the complex pattern of carcinogenesis, the clinically exploitable results are still limited. Therefore, despite its shortcomings due to intra- and interobserver variation and possible sampling error, histopathology will continue to be the decisive biomarker for assessing Barrett's carcinogenesis (9).

There seem to be some definable reasons for this not yet completely satisfying situation:

1. The reliability of the molecular biological results is hampered by the generally small number of cases available for study. A multicenter cooperative study, including subjects undergoing intensive surveillance, would be desirable (27).

2. The majority of clinical evaluation studies are usually confined to a single or a few molecular biological factors. Considering the complexity of the molecular circuit of the cell and its alterations in carcinogenesis, a systematic synchronous evaluation of groups of molecular–biological factors in a logical order [e.g., according to the hallmarks of cancer (28)], would be advisable for establishing the practical applicability of the molecular investigations. The six hallmarks of cancer are (1) sustaining proliferative signaling, (2) evading growth suppressors, (3) activating invasion and metastasis, (4) enabling replicative immortality, (5) inducing angiogenesis, and (6) resisting cell death.

3. It can be assumed that tumor sample acquisition is also of importance. So far, the vast majority of the investigations have been performed on tissue samples that were homogenized for molecular analysis without controlling the composition of the tissue in detail microscopically. Practical histodiagnostic experience shows that only tiny groups of metaplastic cells often reveal dysplastic features. The majority of the samples may be composed of squamous epithelium,

columnar cells, and sometimes specialized intestinal epithelium with goblet cells but no dysplasia. For the experienced histopathologist, it is possible to evaluate these diverse and complex cytological structures separately. However, for molecular analysis, these tissue complexes, which play an absolutely different biological role, may all be mixed up in one homogenate. Realistically, not many reasonable data can be expected from an analysis of such a "tissue integral." In a painstaking study on clonality at the crypt level, Leedham et al. (29) reported considerable clonal heterogeneity at the crypt level. This means that a precise microdissection of the various preneoplastic tissue components might be a prerequisite for reliable molecular results. For practical clinical use, this method would be slow and expensive. Therefore, the question arises if such a biologically reliable approach would be realistic for use in clinical routine investigations. In a recent review, Prasad et al. also stressed that well-designed multicenter studies at least would be required to translate improved knowledge of Barrett's carcinogenesis into clinically significant progress on predictive testing (27).

THE LINK BETWEEN BARRETT'S CARCINOGENESIS AND INFLAMMATION

Inflammation is the host response to tissue injury, with a complex network of cellular reactions and chemical signals. Normally, it is self-limiting, but under certain conditions it can take a prolonged course, and this may be a factor in paving the way to carcinogenesis (30). The cellular and molecular pathways linking inflammation and cancer have been unraveled increasingly in the past few years. An accumulating number of molecular factors have emerged from these investigations as attractive targets for cancer therapy (31).

In the first instance, inflammation can act as a classical tumor promoter, increasing cancer risk and promoting tumor progression (32). In addition to promoting effects, increasing evidence suggests that chronic inflammation can also generate tumor-initiating DNA alterations (33,34). However, whether or not inflammation is causally linked to tumor initiation remains to be clarified (31). It is of particular importance that inflammation can trigger the production of reactive oxygen species released preferentially by neutrophils and macrophages (35). ROSs are toxic to the DNA molecule, resulting in oxidized DNA bases, the most important of which is 8-hydroxy-2′-deoxyguanosine. Furthermore, in inflammation, mutagenic enzymes such as activation-induced cytidine deaminase (AID) can be induced (36,37). Recently, it has also been shown that inflammation can induce epigenetic alterations. Hahn et al. have shown that methylation of polycomb target genes in intestinal cancer is mediated by inflammation (38). Inflammation and hypoxia can result in epigenetic repression of DNA mismatch repair genes in inflammatory bowel disease (IBD)-associated colorectal cancer (39). Inflammation can down-regulate mismatch repair proteins (40). Decreased levels of the hMLH1 protein have been detected by immunohistochemistry in gastric epithelium in patients with *Helicobacter pylori*–induced gastritis (41).

In conclusion, the well known fact that inflammatory diseases in different organs can promote cancer development has a molecular basis which is being increasingly unraveled. The most important factors are ROSs, cytokines, and chemokines. Inflammatory mediators can also down-regulate important factors in the DNA damage repair system (42). In view of the numerous molecular links between cancer and inflammation, it has been suggested that inflammation be considered as the seventh hallmark of cancer (40) in addition to the well known six hallmarks of cancer described by Hanahan and Weinberg (28).

Gastroesophageal reflux disease (GERD) and BE are the major risk factors responsible for the development of EA. Both are associated with inflammation of the esophageal squamous epithelium, a condition called *reflux esophagitis*. Therefore, EA and Barrett's metaplasia are common in the context of inflammation as a result of acid and bile reflux. Chronic inflammation leads to DNA damage and altered expression of genes involved in cellular proliferation and inhibition of apoptosis. Inflammatory key players in Barrett's carcinogenesis also include ROSs, activation of kinase pathways and transcription factors (CDX2, NF-κB), and production of cytokines and inflammatory enzymes. Current research activities highlight the link between reflux-induced inflammation culminating in genetic and epigenetic alterations and Barrett's carcinogenesis. Genetic and epigenetic events affect the cell cycle and therefore the DNA damage checkpoints, and this leads to growth sufficiency and the ignoring of antigrowth signals, which are important hallmarks of cancers. In Barrett's carcinogenesis, the principal genetic and epigenetic alterations are comparable to those known from other epithelial malignancies: loss of p16 gene expression by deletion or hypermethylation (epigenetic silencing), loss of p53 expression by mutation and deletion, increase in cyclin expression, and losses of *Rb*, *APC*, and various chromosomal loci.

INFLAMMATION AND GENETIC ALTERATIONS IN BARRETT'S CARCINOGENESIS

It is not yet completely clear whether chronic inflammation alone is sufficient to initiate carcinogenesis (31). Data obtained in the last few years have shown that cancer-related inflammation can induce genetic alterations and genetic instability by inflammatory mediators with development of genetic alterations in cancer cells (40). As mentioned earlier, ROS-mediated DNA damage is an important factor in carcinogenesis (43). The damage to the DNA molecule by ROSs can result in altered transcription, genomic instability, and replication errors (44,45). However, in addition to these directly damaging effects, a variety of mechanisms affecting DNA damage checkpoints, including DNA repair systems and cell-cycle control, may also play a role. Genetic instability can occur at the chromosomal and nucleotide levels. Instability at the nucleotide level is due to faulty DNA repair pathways and includes the instability of microsatellite repeat sequences (microsatellite instability) caused by defects in the mismatch repair pathway. More details of genetic alterations in Barrett's carcinogenesis have been summarized recently (7).

INFLAMMATION AND EPIGENETIC ALTERATIONS IN BARRETT'S CARCINOGENESIS

The fundamental role of epigenetics in cancer has long been established (46,47). The most important epigenetic alterations in carcinogenesis are aberrant DNA methylation and histone modifications, usually occurring at the N-terminal tails of histones protruding from nucleosomes (48,49). In addition to the initiation of DNA damage and genetic alterations, chronic inflammation is considered an important factor triggering epigenetic alterations (35). Accelerated age-related CpG island hypermethylation has been observed in ulcerative colitis (50). Recently, it has been shown that *Helicobacter pylori* infection promotes methylation and silencing of trefoil factor 2, leading to gastric tumor development in mice and humans (51). Hypermethylation of the p16 promoter has long been observed during neoplastic progression in BE (22). Recently, it was observed that epigenetic alteration of the Wnt inhibitory factor-1 promoter occurs early in the carcinogenesis of BE (52). The aberrant DNA methylation pattern in BE and EA has been found to show similarities (53). It was assumed that methylation biomarkers can also be used for the prediction of progression of BE (54). Wang et al. (55) have found that promoter hypermethylation of p16 and APC predicts neoplastic progression in BE. O^6-Methylguanine-DNA methyltransferase (MGMT) is a cellular DNA repair protein that can protect cells from the mutagenic and carcinogenic effects of alkylating agents. It is inactivated in the pathogenesis of various gastrointestinal cancers. Kuester et al. (56) could show that aberrant promoter methylation of MGMT is a frequent and early event during Barrett's carcinogenesis, and may represent a candidate marker for improved diagnosis in early Barrett's adenocarcinoma.

CONCLUDING ASSESSMENT OF INFLAMMATION AND BARRETT'S CARCINOGENESIS

Reflux esophagitis leading to Barrett's carcinogenesis is one of the most frequent, clinically important, and pathogenically well documented examples of a chronic inflammatory disease contributing to tumor promotion and, with high probability, also to tumor initiation. As the histological steps of Barrett's carcinogenesis have been defined very precisely, this disease can almost be considered a model for describing the molecular pathogenetic steps of inflammation-associated cancer. The various groups of molecular players of inflammation-associated Barrett's carcinogenesis are depicted in Figure 15.2: In the figure the morphological sequence of Barrett's carcinogenesis is depicted in gray rectangles, the molecular events in blue ellipses. Bile reflux triggers reflux esophagitis. Immune and inflammatory cells are recruited. The homeobox genes CDX1 and CDX2 trigger the development of Barrett's metaplasia. The expression of transcription factors is activated. They coordinate different inflammatory mediators (i.e., cytokines, etc.). In this manner, cell proliferation is increased and EMT (epithelial–mesenchymal transition) is activated. Inflammatory cells produce ROSs, which cause genetic alterations, mainly DNA oxidation. The inflammatory

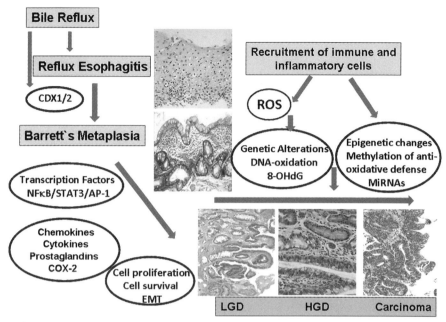

FIGURE 15.2 Synopsis of Barrett's carcinogenesis. Explanations in the text. [From Poehlmann et al. (7)]. (*See insert for color representation of the figure.*)

cells can trigger epigentic alterations such as methylation of antioxidative defense genes. MicroRNAs are also considered important mediators in inflammation-induced cancer.

REFERENCES

1. de Jonge PJ, van BM, Looman CW, Casparie MK, Meijer GA, Kuipers EJ. Risk of malignant progression in patients with Barrett's oesophagus: a Dutch nationwide cohort study. Gut 2010;59: 1030–1036.
2. Wood RK, Yang YX. Barrett's esophagus in 2008: an update. Keio J Med 2008; 57:132–138.
3. Buttar NS, Wang KK. Mechanisms of disease: carcinogenesis I Barrett's esophagus. Nat Clin Pract Gastroenterol Hepatol 2004;1:106–112.
4. McManus DT, Olaru A, Meltzer SJ. Biomarkers of esophageal adenocarcinoma and Barrett's esophagus. Cancer Res 2004;64:1561–1569.
5. Conteduca V, Sansonno D, Ingravallo G, Marangi S, Russi S, Lauletta G, et al. Barrett's esophagus and esophageal cancer: an overview. Int J Oncol 2012;41:414–424.
6. Spechler SJ. Barrett's esophagus: a molecular perspective. Curr Gastroenterol Rep 2005;7: 177–181.
7. Poehlmann A, Kuester D, Malfertheiner P, Guenther T, Roessner A. Inflammation and Barrett's carcinogenesis. Pathol Res Pract 2012;208:269–280.
8. Vieth M, Langer C, Neumann H, Tabuko K. Barrett's esophagus: practical issues for daily routine diagnosis. Pathol Res Pract 2012;208:261–268.
9. Jankowski JA, Odze RD. Biomarkers in gastroenterology: between hope and hype comes histopathology. Am J Gastroenterol 2009;104:1093–1096.

10. Riddell RH, Goldman H, Ransohoff DF, Appelman HD, Fenoglio CM, Haggitt RC, et al. Dysplasia in inflammatory bowel disease: standardized classification with provisional clinical applications. Hum Pathol 1983;14:931–968.
11. Goldblum JR, Barrett's esophagus and Barrett's-related dysplasia. Mod Pathol 2003;16:316–324.
12. Guindi M, Riddell RH. Histology of Barrett's esophagus and dysplasia. Gastro enterol Clin N Am 2003;13:349–368.
13. Flejou JF. Barrett's oesophagus: from metaplasia to dysplasia and cancer. Gut 2005;54:6–12.
14. Reid BJ, Haggitt RC, Rubin CE, Roth G, Surawicz CM, Van Belle G, et al. Observer variation in the diagnosis of dysplasia in Barrett's esophagus. Hum Pathol 1988;19:166–178.
15. Alikhan M, Rex D, Khan A, Rahmani E, Cummings O, Ulbright TM. Variable pathologic interpretation of columnar lined esophagus by general pathologists in community practice. Gastrointest Endosc 1999;50:23–26.
16. Montgomery E, Bronner MP, Goldblum JP, Greenson JK, Haber MM, Hart J, et al. Reproducibility of the diagnosis of dysplasia in Barrett's esophagus: a reaffirmation. Hum Pathol 2001;32:368–378.
17. Ormsby AH, Petras RE, Henricks WH, Rice TW, Rybicki LA, Richter JE, et al. Observer variation in the diagnosis of superficial oesophageal adenocarcinoma. Gut 2002;51:671–676.
18. Schlemper RJ, Riddell RH, Kato Y, Borchard F, Cooper HS, Dawsey SM, et al. The Vienna classification of gastrointestinal epithelial neoplasia. Gut 2000;47:251–255.
19. Coco DP, Goldblum JR, Hornick JL, Lauwers GY, Montgomery E, Srivastava A, et al. Interobserver variability in the diagnosis of crypt dysplasia in Barrett esophagus. Am J Surg Pathol 2011;35:45–54.
20. Kaye PV, Haider SA, Ilyas M, James PD, Soomro I, Faisal W, et al. Barrett's dysplasia and the Vienna classification: reproducibility, prediction, progression and impact of consensus reporting and p53 immunohistochemistry. Histopathology 2009;54:699–712.
21. Yentz S, Wang TD. Molecular imaging for guiding oncologic prognosis and therapy in esophageal adenocarcinoma. Hosp Pract 2011;39:97–106.
22. Klump B, Hsieh CJ, Holzmann K, Gregor M, Porschen R. Hypermethylation of the CDKN2/p16 promoter during neoplastic progression in Barrett's esophagus. Gastroenterology 1998;115:1381–1386.
23. Volkweis BS, Gurski RR, Meurer L, Pretto GG, Mazzini GS, Edelweiss MI. Ki-67 antigen overexpression is associated with the metaplasia–adenocarcinoma sequence in Barrett's esophagus. Gastroenterol Res Pract 2012;639–748.
24. Hollstein MC, Metcalf RA, Welsh JA, Montesano R, and Harris CC. Frequent mutation of the p53 gene in human esophageal cancer. Proc Natl Acad Sci USA 1990;87:9958–9961.
25. Kuester D, Dar AA, Moskaluk CC, Krueger S, Meyer F, Hartig R, et al. Early involvement of death-associated protein kinase promoter hypermethylation in the carcinogenesis of Barrett's esophageal adenocarcinoma and its association with clinical progression. Neoplasia 2007;9:236–245.
26. Souza RF. The molecular basis of carcinogenesis in Barrett's esophagus. J Gastrointest Surg 2010;14:937–940.
27. Prasad GA, Bansal A, Sharma P, Wang KK. Predictors of progression in Barrett's esophagus: current knowledge and future directions. Am J Gastroenterol 2010;105:1490–1502.
28. Hanahan D, Weinberg RA. The hallmarks of cancer. Cell 2000;100:57–70.
29. Leedham SJ, Preston SL, McDonald SA, Elia G, Bhandari P. Poller D, Harrison R, et al. Individual crypt genetic heterogeneitiy and the origin of metaplastic glandular epithelium in human Barrett's oesophagus. Gut 2008;57:1041–1048.
30. Coussens LM, Werb Z. Inflammation and cancer. Nature 2002; 420:860–667.
31. Grivennikov SI, Karin M. Inflammation and oncogenesis: a vicious connection. Curr Opin Genet Dev 2010;20:65–71.
32. Mantovani A, Allavena P, Sica A, Balkwill F. Cancer-related inflammation. Nature 2008;454: 436–444.
33. Meira LB, Bugni JM, Green SL, Lee CW, Pang B, Borenshtein D, et al. DNA damage induced by chronic inflammation contributes to colon carcinogenesis in mice. J Clin Invest 2008;118: 2516–2525.
34. Hofseth LJ, Khan MA, Ambrose M, Nikolayeva O, Xu-Welliver M, Kartalou M, et al. The adaptive imbalance in base excision-repair enzymes generates microsatellite instability in chronic inflammation. J Clin Invest 2003;112:1887–1894.
35. Kundu JK, Surh YJ. Inflammation: gearing the journey to cancer. Mutat Res 2008;659:15–30.

36. Takai A, Toyoshima T, Uemura M, Kitawaki Y, Marusawa H, Hiai H, et al. A novel mouse model of hepatocarcinogenesis triggered by AID causing deleterious p53 mutations. Oncogene 2009;28:469–478.

37. Matsumoto Y, Marusawa H, Kinoshita K, Endo Y, Kou T, Morisawa T, et al. *Helicobacter pylori* infection triggers aberrant expression of activation-induced cytidine deaminase in gastric epithelium. Nat Med 2007;13:470–476.

38. Hahn MA, Hahn T, Lee DH, Esworthy RS, Kim BW, Riggs AD, et al. Methylation of polycomb target genes in intestinal cancer is mediated by inflammation. Cancer Res 2008;68:10280–10289.

39. Edwards RA, Witherspoon M, Wang K, Afrasiabi K, Pham T, Birnbaumer L, et al. Epigenetic repression of DNA mismatch repair by inflammation and hypoxia in inflammatory bowel disease–associated colorectal cancer. Cancer Res 2009;69:6423–6429.

40. Colotta F, Allavena P, Sica A, Garlanda C, Mantovani A. Cancer-related inflammation, the seventh hallmark of cancer: links to genetic instability. Carcinogenesis 2009;30:1073–1081.

41. Mirzaee V, Molaei M, Shalmani HM, Zali MR. *Helicobacter pylori* infection and expression of DNA mismatch repair proteins. World J Gastroenterol 2008;14:6717–6721.

42. Porta C, Larghi P, Rimoldi M, Totaro MG, Allavena P, Mantovani A, et al. Cellular and molecular pathways linking inflammation and cancer. Immunobiology 2009;214:761–777.

43. Kawanishi S, Hiraku Y. Oxidative and nitrative DNA damage as biomarker for carcinogenesis with special reference to inflammation. Antioxid Redox Signal 2006;8:1047–1058.

44. Marnett LJ, Plastaras JP. Endogenous DNA damage and mutation. Trends Genet.2001;17:214–221.

45. De BR, Van LN. Endogenous DNA damage in humans: a review of quantitative data. Mutagenesis 2004;19:169–185.

46. Jones PA, Baylin SB. The fundamental role of epigenetic events in cancer. Nat Rev Genet 2002;3:415–428.

47. Feinberg AP, Tycko B. The history of cancer epigenetics. Nat Rev Cancer 2004;4:143–153.

48. Herceg Z, Ushijima T. Epigenetics is a fascinating field of modern biology. Adv Genet 2010;71:xi–xii.

49. Herceg Z, Ushijima T. Introduction: epigenetics and cancer. Adv Genet; 2010;70:1–23.

50. Issa JP, Ahuja N, Toyota M, Bronner MP, Brentnall TA. Accelerated age-related CpG island methylation in ulcerative colitis. Cancer Res 2001;61:3573–3577.

51. Peterson AJ, Menheniott TR, O'Connor L, Walduck AK, Fox JG, Kawakami K, et al. *Helicobacter pylori* infection promotes methylation and silencing of trefoil factor 2, leading to gastric tumor development in mice and humans. Gastroenterology 2010;1396:2005–2017.

52. Clement G, Guilleret I, He B, Yagui-Beltran A, Lin YC, You L, et al. Epigenetic alteration of the Wnt inhibitory factor-1 promoter occurs early in the carcinogenesis of Barrett's esophagus. Cancer Sci 2008;99:46–53.

53. Smith E, De Young NJ, Pavey SJ, Hayward NK, Nancarrow DJ, Whiteman DC, et al. Similarity of aberrant DNA methylation in Barrett's esophagus and esophageal adenocarcinoma. Mol Cancer 2008;7:75.

54. Jin Z, Cheng Y, Gu W, Zheng Y, Sato F, Mori Y, et al. A multicenter, double-blinded validation study of methylation biomarkers for progression prediction in Barrett's esophagus. Cancer Res 2009;69:4112–4115.

55. Wang JS, Guo M, Montgomery EA, Thompson RE, Cosby H, Hicks L, et al. DNA promoter hypermethylation of p16 and APC predicts neoplastic progression in Barrett's esophagus. Am J Gastroenterol 2009;104:2153–2160.

56. Kuester D, El-Rifai W, Peng D, Ruemmele P, Kroeckel I, Peters B, et al. Silencing of MGMT expression by promoter hypermethylation in the metaplasia–dysplasia–carcinoma sequence of Barrett's esophagus. Cancer Lett 2009;275:117–126.

CHAPTER 16

ASBESTOS-INDUCED CHRONIC INFLAMMATION AND CANCER

Andrea Napolitano, Sandro Jube, Giovanni Gaudino,
Harvey I. Pass, Michele Carbone, and Haining Yang

The links between chronic inflammation and cancer are numerous, complex, and bidirectional. On one hand, chronic inflammation is a common denominator of environmental and behavioral factors that are considered critical initiators of most human cancers, such as pathogens, obesity, tobacco, alcohol, benzene, arsenic, vinyl chloride, asbestos, and others (1). On the other hand, chronic inflammation is also present in the microenvironment of tumors that are not epidemiologically related to inflammation; cancer-related inflammation is indeed now regarded as a new hallmark of cancer (2) and a decisive player in all the stages of tumor development (3). In this chapter we focus specifically on asbestos-induced chronic inflammation and carcinogenesis, discussing possible mechanisms of action of asbestos exposure, and concluding with suggestions on potential therapeutic targets.

ASBESTOS AND ASBESTOS-LIKE FIBERS: OLD AND NEW HAZARDS

One of the most notorious carcinogens in the lung and pleura, *asbestos* is a nonspecific term commonly used to describe six types of naturally occurring fibrous silicate minerals used commercially (4). Asbestos fibers are divided into two major groups, serpentine and amphibole, and distinguished further based on their chemical composition (e.g., relative concentrations of iron, aluminum, calcium, and magnesium) and crystalline structure (5). *Serpentine asbestos* is chrysotile (white asbestos), and *amphibole asbestos* includes crocidolite (blue asbestos), amosite (brown asbestos), anthophyllite, actinolite, and tremolite.

The carcinogenic potential of asbestos fibers has been linked to their geometry, size, and chemical composition. For example, fibers with a length greater than 4 μm

Cancer and Inflammation Mechanisms: Chemical, Biological, and Clinical Aspects, First Edition.
Edited by Yusuke Hiraku, Shosuke Kawanishi, and Hiroshi Ohshima.
© 2014 John Wiley & Sons, Inc. Published 2014 by John Wiley & Sons, Inc.

have increased potential to cause pleural inflammation, probably due to the inability of alveolar macrophages to phagocytize the longer fibers properly (6). Carcinogenicity is also influenced by the biopersistence of the asbestos fibers: amphibole fibers have the tendency to persist at sites of tumor development, with the fiber concentration increasing with prolonged exposure (7), whereas serpentine fibers are usually cleared rapidly from the lung (7), a finding that appears related to the higher risk of malignancies in people exposed to amphiboles (8). In addition, the chemical composition of the fibers can play an important role in determining the inflammatory and carcinogenic potential, with the iron content playing a potential role due to increased production of reactive oxygen species (ROSs) and nitrogen species (RNSs) (9–11).

As expected, asbestos-related diseases have a significant higher incidence in people with occupational exposure, such as miners, insulators, plumbers, and workers in asbestos industries. Nevertheless, it is well documented that naturally occurring deposits of minerals containing asbestos or asbestos-like fibers are sources of a significantly higher exposure compared to the environmental background; examples are the serpentinite and tremolite deposits in New Caledonia (12,13) and the tremolite deposits in Turkey, Greece, Corsica, and Cyprus (14).

Moreover, asbestos used commercially represents only a very small portion of all natural potentially carcinogenic fibers. A total of about 390 fibrous minerals occur in nature, and only six of these fibers are included in the asbestos nomenclature. The other fibrous minerals, at times called "asbestos-like," are not subject to restrictive regulation, since they have not been used commercially. As a consequence, the number of cancers related to nonoccupational exposure to asbestos may well include cancers associated to the many types of fibers present in nature.

One of best-characterized asbestos-like fibers, the zeolite mineral *erionite*, displays physical characteristics similar to those of asbestos (15), is a carcinogen more potent than asbestos itself (16), and has been associated epidemiologically with an increased risk of asbestos-related malignancies in several villages in Cappadocia, Turkey (17). Worldwide, there are known erionite deposits in Germany, New Zealand, Russia, Japan, Kenya, and Sardinia (Italy). Deposits are also present in at least 12 U.S. states, including Oregon, California, Montana, and South and North Dakota (18). Similarly, three different types of fibrous fluoroedenite (an amphibole that is not regulated as asbestos) show chemical reactivity, such as production of ROSs and RNSs comparable with or even higher than that of regulated asbestos minerals (19).

Moreover, the recent commercial use of artificial fibers such as carbon nanotubes has raised concerns about their potential to cause human disease and cancer. Multiwalled carbon nanotubes have been shown to cause chronic mesothelial inflammation and tumor initiation both *in vitro* and *in vivo*, with features very reminiscent of asbestos carcinogenesis (20–22), thus constituting a potential novel class of carcinogenic compounds.

All this evidence points toward the need for new regulatory and preventive measures to safeguard the general population that might be exposed to asbestos or natural and artificial asbestos-like fibers.

GENES MAY MODULATE THE RISK OF ASBESTOS-RELATED DISEASES

Asbestos exposure has been associated with a group of diseases collectively referred to as *asbestos-related diseases* (ARDs), which can be divided into malignant and benign. The most common malignant ARDs are malignant mesothelioma (MM), lung and laryngeal cancer, among the benign ARDs, the most common are asbestosis, a form of diffuse interstitial lung fibrosis that can be the cause of death in some of affected persons; diffuse pleural thickening; pleural plaques; and benign pleural effusions (4). Benign asbestos-related pleural diseases are not premalignant conditions; they are, however, markers of significant asbestos exposure and therefore of an increased risk of cancer (23,24). Beside the classical ARDs, asbestos has also been shown to cause global alterations in the immune system, resulting in autoimmunity (25,26) and possibly reducing antitumor immunity, by influencing $CD4^+$ T cells, regulatory T cells, $CD8^+$ cytotoxic T cells, and natural killer cells [reviewed extensively in (27)].

Although the relationship between asbestos exposure and pleuropulmonary diseases is well established, many questions remain regarding the pathogenesis of these conditions. To date, the various outcomes of asbestos exposure are impossible to predict. For example, only about 5% of asbestos miners with prolonged asbestos exposure develop malignant mesothelioma, and the content of lung fibers does not show a dose–response relationship with cancer incidence (5).

The presence of additional environmental and genetic factors that influence asbestos pathogenesis could explain why some people exposed to asbestos develop diseases and cancer and others do not. During the last decade, the associations between single-nucleotide polymorphisms (SNPs) in genes involved in inflammation, DNA repair, xenobiotic and oxidative metabolism, and ARDs have been studied extensively, with various degrees of statistical significance (28–33).

Among the different ARDs, we proposed an important role of genetic predisposition in malignant mesothelioma, based on the relatively low incidence of this cancer in heavily exposed persons and on familial aggregation in Cappadocia and in some U.S. families (28,34,35). Our hypothesis was verified by our discovery that germline *BAP1* mutations cause a novel cancer syndrome, characterized by a high incidence of pleural and peritoneal malignant mesothelioma, uveal melanoma, and possibly other cancers (36), including cutaneous melanocytic tumors (37). We are currently investigating the possible interactions between this genetic predisposition and increased susceptibility to fiber-induced inflammation and carcinogenesis.

ASBESTOS, INFLAMMATION, AND CARCINOGENESIS

The best recognized asbestos-associated malignancies are malignant mesothelioma (MM), lung cancer, and laryngeal cancer. Recently, the International Agency for Research on Cancer concluded that there is also sufficient evidence for a causal association between exposure to asbestos and ovarian cancer (38,39).

Inhalation is the most common way of exposure to asbestos fibers, but it is clear that fibers can translocate from the lung interstitium via pulmonary lymph flow from where they can reach the bloodstream (40). Here we use MM—the archetypal asbestos-induced cancer—as our model to describe the molecular features of fiber-related inflammation and carcinogenesis, considering that differences might hold true for cancers in other anatomical localizations.

MM is a primary tumor arising from the mesothelial cell linings of the serous membranes, most commonly involving the pleural and peritoneal spaces and, less frequently, the pericardium and the tunica vaginalis of the testis. Approximately 70% of MM cases have been linked to occupational and environmental exposure to asbestos and/or erionite fibers, with a 5 to 10% incidence in exposed persons. The commercial use of asbestos fibers have been banned in the United States and most Western European countries; however, in rapidly industrializing countries, such as India, China, and Brazil, the use of asbestos is still unrestricted and the incidence of MM is expected to rise in the next decades (41). In the United States alone, it has been estimated that more than 25 million people have been exposed to asbestos (either occupational of environmental exposure), whereas the number of those exposed to erionite in the environment is still unknown (5). MM is highly aggressive, with a median survival of one year from diagnosis, with a latency from exposure to cancer development usually spanning several decades (5). Beside asbestos, other potential risk factors include radiation therapy (42) and infection with Simian Virus 40 (SV40), a DNA virus that causes MM in animals and acts as a cofactor for asbestos carcinogenesis (43).

In vitro experiments show that asbestos is cytotoxic, rather than carcinogenic, to mesothelial and lung epithelial cells (44–46), leading to an apparent paradox, in view of the epidemiological evidence of its causative role in most MMs. During the last decade, light has been shed on the molecular mechanisms linking asbestos exposure to MM pathogenesis, and the paradox was finally explained.

After inhalation, most of the asbestos fibers deposit along the airways. Only a small percentage reaches the alveoli, where they come in contact with alveolar macrophages (47). Asbestos deposition in tissues is constantly associated with inflammatory infiltrate (48,49), largely constituted by these phagocytic macrophages that internalize fibers—or try to—in a process called *frustrated phagocytosis* (50,51). Upon particle phagocytosis, NADPH oxidase generates ROSs that in turn activate the NALP3 inflammasome, a macromolecular complex critical for cytokine secretion (52). Notably, also, carbon nanotubes have been shown to activate the NALP3 inflammasome and cause the release of inflammatory mediators (53,54). ROS production is not the only mechanism behind NALP3 activation; asbestos and nanoparticles may also activate the inflammasome through the release of cathepsin B from lysosome to cytoplasm, stimulation of the $P2X_7$ receptor and activation of Src and Syk kinases (55). Among the various cytokines released by alveolar macrophages, interleukin (IL)-1β and tumor necrosis factor (TNF)-α have been proposed as key mediators of asbestos-induced carcinogenesis. Both were shown to inhibit asbestos-induced cytotoxicity of mesothelial cells, possibly via a nuclear factor κ-light-chain enhancer of an activated B-cell (NF-κB)-dependent pathway, enhancing asbestos carcinogenicity (49,56).

In addition to the activation of cytokines, sterile inflammation is always associated with active secretion by inflammatory cells as well as with passive release by necrotic cells, of molecules referred to collectively as damage-associated molecular patterns (DAMPs). The common feature of DAMP molecules is to function as endogenous activators of Toll-like receptors (TLRs), a class of receptors that play a key role in the innate immune system, first identified in pathogen-related inflammation. Well-recognized DAMP molecules include high-mobility group box protein 1 (HMGB1), several heat shock proteins, nucleic acids, and purine metabolites (57). We have shown that HMGB1 plays a crucial role both in MM onset (58) and progression (59). Asbestos-exposed human primary mesothelial cells activate poly(ADP-ribose) polymerase, secrete H_2O_2, deplete ATP, and translocate HMGB1 from the nucleus to the cytoplasm and into the extracellular space. The release of HMGB1 induces macrophages to secrete TNF-α, which protects mesothelial cells from asbestos-induced cell death (58). Macrophages can also actively secrete HMGB1, thus reinforcing the signals in an autocrine fashion (60).

The relative contributions of NALP3 and HMGB1 in asbestos-induced chronic inflammation and MM pathogenesis are being investigated. Recently, it has been shown that early inflammatory reactions triggered by asbestos are NALP3-dependent; however, NALP3 is not critical in the chronic development of asbestos-induced mesothelioma (61), suggesting that asbestos-induced chronic inflammation and MM onset might be the result of orchestrated events in which NALP3 is an important, but nonessential, trigger of HMGB1 and possibly other mediators necessary in the transformation of mesothelial cells.

Besides activating a chronic inflammatory response, asbestos deregulates cell physiology in a number of direct and indirect ways. Compared to other carcinogens (e.g., cigarette smoke), asbestos is known to be a weak mutagen (62). Although it has been shown that asbestos can cause DNA damage by direct mechanical interference with chromosome segregation during mitosis (63) and single-stranded breaks (64), most of its damage is thought to be indirectly mediated by external (i.e., macrophage-released) or internal ROS production, while the iron contained in asbestos fibers behaves like a catalyst for free-radical generation (65,66). Moreover, asbestos has been shown to be capable of adsorbing proteins (67) such as vitronectin, a component of extracellular matrix, and this property might be responsible for increased phagocytosis by mesothelial cells and of increased intracellular oxidation (68). Finally, mitochondria have also been shown to be a target of asbestos fibers and an important source of oxidative species (69). Notably, the hypothesis of oxidative stress as a clastogenic factor in mesothelial carcinogenesis is confirmed by the evidence that in rodents, iron saccharate can cause mesothelioma with the same CDKN2A/2B homozygous deletion (70) observed in human MM (71). Therefore, it is not surprising that cells exposed to asbestos respond to oxidative stress by up-regulating antioxidant enzymes such as superoxide dismutases (72).

Asbestos fibers induce several dramatic cell-signaling events, either directly or via production of ROSs and RNSs, leading eventually to increased cell survival. Oxidative stress and protein kinase C induce expression of activator protein-1 (AP-1) (73,74); fibers activate growth factor receptors (e.g., epidermal and vascular endothelial growth factor receptors, EGFR and VEGFR) directly, which in turn activates extracellular signal-regulated kinase (ERK) and phosphoinositide 3-kinase (PI3K)/AKT pathways (75–77). As already mentioned, the pleiotropic transcription factor NF-κB plays a crucial role in fiber-induced carcinogenesis (78–80). Moreover, glycoproteins involved in integrin and CD44 regulation, including osteopontin, become overexpressed (81), along with fibulin-3 (FBLN3) (82). FBLN3 is a chief component of basement membranes of epithelial and endothelial cells and interacts with tissue inhibitor of metalloproteinase-3 (TIMP-3), collagen XVIII/endostatin, hepatitis B virus-encoded X antigen, and elastin monomer tropoelastin to stabilize basement membranes. FBLN3 mediates cell-to-cell and cell-to-matrix communication, is inversely related to cell growth, and has variable angiogenic effects (83,84).

Finally, chronic inflammation-induced carcinogenesis has been associated with epigenetic modifications. In patients with MM, a significant association has been found between asbestos exposure and methylation of specific loci (85,86); moreover, overall methylation profiles were associated significantly with asbestos exposure burden (87).

In conclusion, asbestos fibers exert their carcinogenetic activity via two principal mechanisms (Figure 16.1):

1. By creating a chronic inflammatory milieu that provides cells with important pro-survival signals (e.g., TNF-α).
2. By causing both direct (mechanical) and indirect (via ROS/RNS production) DNA damage and profound alterations in signaling pathways.
3. The possible role of asbestos in influencing antitumor immunity remains speculative.

TARGETING ASBESTOS-INDUCED CHRONIC INFLAMMATION

Given the established role of chronic inflammation in asbestos-related cancers and the long latency time between fiber exposure and cancer formation, asbestos-induced inflammation is an appealing therapeutic target for cancer prevention (88). Notably, daily treatment with a common anti-inflammatory drug, aspirin, was shown to reduce the long-term risk of death due to cancer (89) and the short-term risk of cancer incidence and mortality (90).

In regard to asbestos-induced cancers, no preventive trials have been set up for 20 years, although there have been considerable advances in cancer chemoprevention (91). Preclinical evaluation of potential chemopreventative drugs has to be thorough, since the potential effects of a chemoprevention clinical trial in asbestos-exposed persons would probably be observed only after several years of treatment. Adequate *in vivo* models of asbestos-induced cancers are therefore particularly needed. A recently developed transgenic murine model of asbestos-induced MM, called Mex-TAg (mesothelin-driven transgenic SV40 T antigen), has been validate as a MM model (92,93), and tested as a potential *in vivo* model for chemopreventive drugs (92,94).

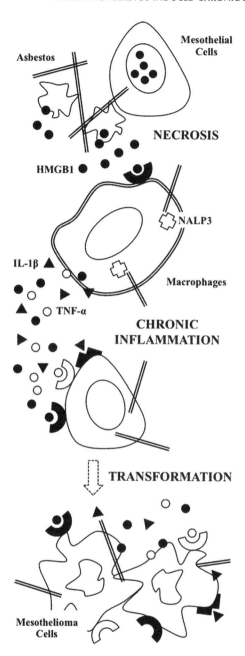

FIGURE 16.1   Asbestos causes necrotic death of mesothelial cells, leading to the release of HMGB1 into the extracellular space. In macrophages, asbestos activates NALP3 inflammasome and, together with HMGB1, induces secretion of numerous cytokines, including TNF-α and IL-1β. The chronic presence of these inflammatory pro-survival mediators promotes transformation of mesothelial cells with asbestos-induced DNA damage and perturbation of signaling pathways. Established tumors secrete HMGB1 and cytokines to further promote cell growth.

Given the insight in the molecular pathways involved in asbestos-induced malignancies, it is tempting to speculate that HMGB1, NALP3, TNF-α, and IL-1β may serve as potential targets for inhibition of asbestos-induced inflammation and carcinogenesis. Treatment with IL-1 receptor antagonist (95) or soluble TNF receptors (96) can protect mice from developing silica-induced pulmonary fibrosis, a condition very similar to asbestosis. Approved drugs targeting TNF-α and IL-1 are already available (e.g., infliximab and anakinra, respectively); glyburide, a common sulfonylurea drug for type 2 diabetes, inhibits the NALP3 inflammasome (97); and specific molecules that target HMGB1 include anti-HMGB1 and anti-receptor for advanced glycation end product (RAGE) monoclonal antibodies, recombinant HMGB1 box A, and ethyl pyruvate, among others (98).

By a greater understanding of asbestos-induced chronic inflammation, we expect increased efforts in the chemoprevention of asbestos-induced malignancies in the near future. In parallel, further studies should be aimed at identifying high-risk persons, by integrating information on fiber exposure, genetic susceptibility, and biomarkers of chronic inflammation.

REFERENCES

1. Aggarwal BB, Vijayalekshmi RV, Sung B. Targeting inflammatory pathways for prevention and therapy of cancer: short-term friend, long-term foe. Clin Cancer Res 2009;15:425–430.
2. Colotta F, Allavena P, Sica A, Garlanda C, Mantovani A. Cancer-related inflammation, the seventh hallmark of cancer: links to genetic instability. Carcinogenesis 2009;30:1073–1081.
3. Grivennikov SI, Greten FR, Karin M. Immunity, inflammation, and cancer. Cell 2010;140:883–899.
4. Carbone M, Kratzke RA, Testa JR. The pathogenesis of mesothelioma. Semin Oncol 2002;29:2–17.
5. Carbone M, Ly BH, Dodson RF, Pagano I, Morris PT, Dogan UA, et al. Malignant mesothelioma: facts, myths, and hypotheses. J Cell Physiol 2012;227:44–58.
6. Schinwald A, Murphy FA, Prina-Mello A, Poland CA, Byrne F, Movia D, et al. The threshold length for fiber-induced acute pleural inflammation: shedding light on the early events in asbestos-induced mesothelioma. Toxicol Sci 2012;128:461–470.
7. Britton M. The epidemiology of mesothelioma. Semin Oncol 2002;29:18–25.
8. McDonald JC. Epidemiology of malignant mesothelioma: an outline. Ann Occup Hyg 2010; 54:851–857.
9. Shannahan JH, Ghio AJ, Schladweiler MC, McGee JK, Richards JH, Gavett SH, et al. The role of iron in Libby amphibole-induced acute lung injury and inflammation. Inhal Toxicol 2011;23:313–323.
10. Jiang L, Akatsuka S, Nagai H, Chew SH, Ohara H, Okazaki Y, et al. Iron overload signature in chrysotile-induced malignant mesothelioma. J Pathol 2012;228:366–377.
11. Choe N, Tanaka S, Kagan E. Asbestos fibers and interleukin-1 upregulate the formation of reactive nitrogen species in rat pleural mesothelial cells. Am J Respir Cell Mol Biol 1998;19:226–236.
12. Baumann F, Maurizot P, Mangeas M, Ambrosi JP, Douwes J, Robineau B. Pleural mesothelioma in New Caledonia: associations with environmental risk factors. Environ Health Perspect 2011;119:695–700.
13. Luce D, Bugel I, Goldberg P, Goldberg M, Salomon C, Billon-Galland MA, et al. Environmental exposure to tremolite and respiratory cancer in New Caledonia: a case–control study. Am J Epidemiol 2000;151:259–265.
14. Constantopoulos SH. Environmental mesothelioma associated with tremolite asbestos: lessons from the experiences of Turkey, Greece, Corsica, New Caledonia and Cyprus. Regul Toxicol Pharmacol 2008;52:S110–S115.
15. Dogan AU, Dogan M, Hoskins JA. Erionite series minerals: mineralogical and carcinogenic properties. Environ Geochem Health 2008;30:367–381.

16. Bertino P, Marconi A, Palumbo L, Bruni BM, Barbone D, Germano S, et al. Erionite and asbestos differently cause transformation of human mesothelial cells. Int J Cancer 2007;121:12–20.

17. Carbone M, Emri S, Dogan AU, Steele I, Tuncer M, Pass HI, et al. A mesothelioma epidemic in Cappadocia: scientific developments and unexpected social outcomes. Nat Rev Cancer 2007;7:147–154.

18. Carbone M, Baris YI, Bertino P, Brass B, Comertpay S, Dogan AU, et al. Erionite exposure in North Dakota and Turkish villages with mesothelioma. Proc Natl Acad Sci USA 2011;108:13618–13623.

19. Fantauzzi M, Pacella A, Fournier J, Gianfagna A, Andreozzi GB, Rossi A. Surface chemistry and surface reactivity of fibrous amphiboles that are not regulated as asbestos. Anal Bioanal Chem 2012;404:821–833.

20. Xu J, Futakuchi M, Shimizu H, Alexander DB, Yanagihara K, Fukamachi K, et al. Multi-walled carbon nanotubes translocate into the pleural cavity and induce visceral mesothelial proliferation in rats. Cancer Sci 2012;103:2045–2050.

21. Murphy FA, Poland CA, Duffin R, Donaldson K. Length-dependent pleural inflammation and parietal pleural responses after deposition of carbon nanotubes in the pulmonary airspaces of mice. Nanotoxicology 2012.

22. Nagai H, Okazaki Y, Chew SH, Misawa N, Yamashita Y, Akatsuka S, et al. Diameter and rigidity of multiwalled carbon nanotubes are critical factors in mesothelial injury and carcinogenesis. Proc Natl Acad Sci USA 2011;108:E1330–E1338.

23. Greillier L, Astoul P. Mesothelioma and asbestos-related pleural diseases. Respiration 2008;76:1–15.

24. King C, Mayes D, Dorsey DA. Benign asbestos-related pleural disease. Dis Mon 2011;57:27–39.

25. Pfau JC, Sentissi JJ, Li S, Calderon-Garciduenas L, Brown JM, Blake DJ. Asbestos-induced autoimmunity in C57BL/6 mice. J Immunotoxicol 2008;5:129–137.

26. Pfau JC, Sentissi JJ, Weller G, Putnam EA. Assessment of autoimmune responses associated with asbestos exposure in Libby, Montana, USA. Environ Health Perspect 2005;113:25–30.

27. Kumagai-Takei N, Maeda M, Chen Y, Matsuzaki H, Lee S, Nishimura Y, et al. Asbestos induces reduction of tumor immunity. Clin Dev Immunol 2011;2011:481439.

28. Neri M, Ugolini D, Dianzani I, Gemignani F, Landi S, Cesario A, et al. Genetic susceptibility to malignant pleural mesothelioma and other asbestos-associated diseases. Mutat Res 2008;659:126–136.

29. Betti M, Ferrante D, Padoan M, Guarrera S, Giordano M, Aspesi A, et al. XRCC1 and ERCC1 variants modify malignant mesothelioma risk: a case–control study. Mutat Res 2011;708:11–20.

30. Wei S, Wang LE, McHugh MK, Han Y, Xiong M, Amos CI, et al. Genome-wide gene-environment interaction analysis for asbestos exposure in lung cancer susceptibility. Carcinogenesis 2012;33:1531–1537.

31. Helmig S, Aliahmadi N, Schneider J. Tumour necrosis factor-alpha gene polymorphisms in asbestos-induced diseases. Biomarkers 2010;15:400–409.

32. Helmig S, Belwe A, Schneider J. Association of transforming growth factor beta1 gene polymorphisms and asbestos-induced fibrosis and tumors. J Investig Med 2009;57:655–661.

33. Hirvonen A, Tuimala J, Ollikainen T, Linnainmaa K, Kinnula V. Manganese superoxide dismutase genotypes and asbestos-associated pulmonary disorders. Cancer Lett 2002;178:71–74.

34. Dogan AU, Baris YI, Dogan M, Emri S, Steele I, Elmishad AG, Carbone M. Genetic predisposition to fiber carcinogenesis causes a mesothelioma epidemic in Turkey. Cancer Res 2006;66:5063–5068.

35. Roushdy-Hammady I, Siegel J, Emri S, Testa JR, Carbone M. Genetic-susceptibility factor and malignant mesothelioma in the Cappadocian region of Turkey. Lancet 2001;357:444–445.

36. Testa JR, Cheung M, Pei J, Below JE, Tan Y, Sementino E, et al. Germline BAP1 mutations predispose to malignant mesothelioma. Nat Genet 2011;43:1022–1025.

37. Carbone M, Korb Ferris L, Baumann F, Napolitano A, Lum CA, Flores EG, et al. BAP1 cancer syndrome: malignant mesothelioma, uveal and cutaneous melanoma, and MBAITs. J Transl Med 2012;10:179.

38. Straif K, Benbrahim-Tallaa L, Baan R, Grosse Y, Secretan B, El Ghissassi F, et al. A review of human carcinogens—part C: metals, arsenic, dusts, and fibres. Lancet Oncol 2009;10:453–454.

39. Camargo MC, Stayner LT, Straif K, Reina M, Al-Alem U, Demers PA, et al. Occupational exposure to asbestos and ovarian cancer: a meta-analysis. Environ Health Perspect 2011;119:1211–1217.

40. Miserocchi G, Sancini G, Mantegazza F, Chiappino G. Translocation pathways for inhaled asbestos fibers. Environ Health 2008;7:4.

41. Burki T. Health experts concerned over India's asbestos industry. Lancet 2010;375:626–627.

42. Goodman JE, Nascarella MA, Valberg PA. Ionizing radiation: a risk factor for mesothelioma. Cancer Causes Control 2009;20:1237–1254.

43. Kroczynska B, Cutrone R, Bocchetta M, Yang H, Elmishad AG, Vacek P, et al. Crocidolite asbestos and SV40 are cocarcinogens in human mesothelial cells and in causing mesothelioma in hamsters. Proc Natl Acad Sci USA 2006;103:14128–14133.

44. Kinnula VL, Aalto K, Raivio KO, Walles S, Linnainmaa K. Cytotoxicity of oxidants and asbestos fibers in cultured human mesothelial cells. Free Radic Biol Med 1994;16:169–176.

45. Dong H, Buard A, Renier A, Levy F, Saint-Etienne L, Jaurand MC. Role of oxygen derivatives in the cytotoxicity and DNA damage produced by asbestos on rat pleural mesothelial cells *in vitro*. Carcinogenesis 1994;15:1251–1255.

46. Broaddus VC, Yang L, Scavo LM, Ernst JD, Boylan AM. Crocidolite asbestos induces apoptosis of pleural mesothelial cells: role of reactive oxygen species and poly(ADP-ribosyl) polymerase. Environ Health Perspect 1997;105 Suppl 5:1147–1152.

47. Brody AR, Hill LH, Adkins B, Jr., O'Connor RW. Chrysotile asbestos inhalation in rats: deposition pattern and reaction of alveolar epithelium and pulmonary macrophages. Am Rev Respir Dis 1981;123:670–679.

48. Hillegass JM, Shukla A, Lathrop SA, MacPherson MB, Beuschel SL, Butnor KJ, et al. Inflammation precedes the development of human malignant mesotheliomas in a SCID mouse xenograft model. Ann NY Acad Sci 2010;1203:7–14.

49. Yang H, Bocchetta M, Kroczynska B, Elmishad AG, Chen Y, Liu Z, et al. TNF-alpha inhibits asbestos-induced cytotoxicity via a NF-kappaB-dependent pathway, a possible mechanism for asbestos-induced oncogenesis. Proc Natl Acad Sci USA 2006;103:10397–10402.

50. McLemore T, Corson M, Mace M, Arnott M, Jenkins T, Snodgrass D, et al. Phagocytosis of asbestos fibers by human pulmonary alveolar macrophages. Cancer Lett 1979;6:183–192.

51. McLemore TL, Mace ML, Jr., Roggli V, Marshall MV, Lawrence EC, Wilson RK, et al. Asbestos body phagocytosis by human free alveolar macrophages. Cancer Lett 1980;9:85–93.

52. Dostert C, Petrilli V, Van Bruggen R, Steele C, Mossman BT, et al. Innate immune activation through Nalp3 inflammasome sensing of asbestos and silica. Science 2008;320:674–677.

53. Yazdi AS, Guarda G, Riteau N, Drexler SK, Tardivel A, Couillin I, et al. Nanoparticles activate the NLR pyrin domain containing 3 (Nlrp3) inflammasome and cause pulmonary inflammation through release of IL-1alpha and IL-1beta. Proc Natl Acad Sci USA 2010;107:19449–19454.

54. Murphy FA, Schinwald A, Poland CA, Donaldson K. The mechanism of pleural inflammation by long carbon nanotubes: interaction of long fibres with macrophages stimulates them to amplify pro-inflammatory responses in mesothelial cells. Part Fibre Toxicol 2012;9:8.

55. Palomaki J, Valimaki E, Sund J, Vippola M, Clausen PA, Jensen KA, et al. Long, needle-like carbon nanotubes and asbestos activate the NLRP3 inflammasome through a similar mechanism. ACS Nano 2011;5:6861–6870.

56. Wang Y, Faux SP, Hallden G, Kirn DH, Houghton CE, Lemoine NR, et al. Interleukin-1beta and tumour necrosis factor-alpha promote the transformation of human immortalised mesothelial cells by erionite. Int J Oncol 2004;25:173–178.

57. Piccinini AM, Midwood KS. DAMPening inflammation by modulating TLR signalling. Mediators Inflamm 2010;2010.

58. Yang H, Rivera Z, Jube S, Nasu M, Bertino P, Goparaju C, et al. Programmed necrosis induced by asbestos in human mesothelial cells causes high-mobility group box 1 protein release and resultant inflammation. Proc Natl Acad Sci USA 2010;107:12611–12616.

59. Jube S, Rivera ZS, Bianchi ME, Powers A, Wang E, Pagano I, et al. Cancer cell secretion of the DAMP protein HMGB1 supports progression in malignant mesothelioma. Cancer Res 2012;72:3290–3301.

60. Pisetsky DS, Jiang W. Role of Toll-like receptors in HMGB1 release from macrophages. Ann NY Acad Sci 2007;1109:58–65.

61. Chow MT, Tschopp J, Moller A, Smyth MJ. NLRP3 promotes inflammation-induced skin cancer but is dispensable for asbestos-induced mesothelioma. Immunol Cell Biol 2012;90:983–986.

62. Sugarbaker DJ, Richards WG, Gordon GJ, Dong L, De Rienzo A, Maulik G, et al. Transcriptome sequencing of malignant pleural mesothelioma tumors. Proc Natl Acad Sci USA 2008;105:3521–3526.

63. Wang NS, Jaurand MC, Magne L, Kheuang L, Pinchon MC, Bignon J. The interactions between asbestos fibers and metaphase chromosomes of rat pleural mesothelial cells in culture: a scanning and transmission electron microscopic study. Am J Pathol 1987;126:343–349.

64. Ollikainen T, Linnainmaa K, Kinnula VL. DNA single strand breaks induced by asbestos fibers in human pleural mesothelial cells *in vitro*. Environ Mol Mutagen 1999;33:153–160.

65. Simeonova PP, Luster MI. Iron and reactive oxygen species in the asbestos-induced tumor necrosis factor-alpha response from alveolar macrophages. Am J Respir Cell Mol Biol 1995;12:676–683.

66. Kamp DW, Israbian VA, Preusen SE, Zhang CX, Weitzman SA. Asbestos causes DNA strand breaks in cultured pulmonary epithelial cells: role of iron-catalyzed free radicals. Am J Physiol 1995;268:L471–L480.

67. Scheule RK, Holian A. Modification of asbestos bioactivity for the alveolar macrophage by selective protein adsorption. Am J Respir Cell Mol Biol 1990;2:441–448.

68. Wu J, Liu W, Koenig K, Idell S, Broaddus VC. Vitronectin adsorption to chrysotile asbestos increases fiber phagocytosis and toxicity for mesothelial cells. Am J Physiol Lung Cell Mol Physiol 2000;279:L916–L923.

69. Huang SX, Partridge MA, Ghandhi SA, Davidson MM, Amundson SA, Hei TK. Mitochondria-derived reactive intermediate species mediate asbestos-induced genotoxicity and oxidative stress-responsive signaling pathways. Environ Health Perspect 2012;120:840–847.

70. Hu Q, Akatsuka S, Yamashita Y, Ohara H, Nagai H, Okazaki Y, et al. Homozygous deletion of CDKN2A/2B is a hallmark of iron-induced high-grade rat mesothelioma. Lab Invest 2010;90:360–373.

71. Xio S, Li D, Vijg J, Sugarbaker DJ, Corson JM, Fletcher JA. Codeletion of p15 and p16 in primary malignant mesothelioma. Oncogene 1995;11:511–515.

72. Janssen YM, Marsh JP, Absher MP, Hemenway D, Vacek PM, Leslie KO, et al. Expression of antioxidant enzymes in rat lungs after inhalation of asbestos or silica. J Biol Chem 1992;267:10625–10630.

73. Heintz NH, Janssen YM, Mossman BT. Persistent induction of c-fos and c-jun expression by asbestos. Proc Natl Acad Sci USA 1993;90:3299–3303.

74. Fung H, Quinlan TR, Janssen YM, Timblin CR, Marsh JP, Heintz NH, et al. Inhibition of protein kinase C prevents asbestos-induced c-fos and c-jun proto-oncogene expression in mesothelial cells. Cancer Res 1997;57:3101–3105.

75. Wang H, Gillis A, Zhao C, Lee E, Wu J, Zhang F, et al. Crocidolite asbestos-induced signal pathway dysregulation in mesothelial cells. Mutat Res 2011;723:171–176.

76. Scapoli L, Ramos-Nino ME, Martinelli M, Mossman BT. Src-dependent ERK5 and Src/EGFR-dependent ERK1/2 activation is required for cell proliferation by asbestos. Oncogene 2004;23:805–813.

77. Tamminen JA, Myllarniemi M, Hyytiainen M, Keski-Oja J, Koli K. Asbestos exposure induces alveolar epithelial cell plasticity through MAPK/Erk signaling. J Cell Biochem 2012;113:2234–2247.

78. Janssen YM, Driscoll KE, Howard B, Quinlan TR, Treadwell M, Barchowsky A, et al. Asbestos causes translocation of p65 protein and increases NF-kappa B DNA binding activity in rat lung epithelial and pleural mesothelial cells. Am J Pathol 1997;151:389–401.

79. Janssen YM, Barchowsky A, Treadwell M, Driscoll KE, Mossman BT. Asbestos induces nuclear factor kappa B (NF-kappa B) DNA-binding activity and NF-kappa B-dependent gene expression in tracheal epithelial cells. Proc Natl Acad Sci USA 1995;92:8458–8462.

80. Oettinger R, Drumm K, Knorst M, Krinyak P, Smolarski R, Kienast K. Production of reactive oxygen intermediates by human macrophages exposed to soot particles and asbestos fibers and increase in NF-kappa B p50/p105 mRNA. Lung 1999;177:343–354.

81. Pass HI, Lott D, Lonardo F, Harbut M, Liu Z, Tang N, et al. Asbestos exposure, pleural mesothelioma, and serum osteopontin levels. N Engl J Med 2005;353:1564–1573.

82. Pass HI, Levin SM, Harbut MR, Melamed J, Chiriboga L, Donington J, et al. Fibulin-3 as a blood and effusion biomarker for pleural mesothelioma. N Engl J Med 2012;367:1417–1427.

83. Klenotic PA, Munier FL, Marmorstein LY, Anand-Apte B. Tissue inhibitor of metalloproteinases-3 (TIMP-3) is a binding partner of epithelial growth factor-containing fibulin-like extracellular matrix protein 1 (EFEMP1). Implications for macular degenerations. J Biol Chem 2004;279:30469–30473.

84. Segade F. Molecular evolution of the fibulins: implications on the functionality of the elastic fibulins. Gene 2010;464:17–31.

85. Tsou JA, Galler JS, Wali A, Ye W, Siegmund KD, Groshen S, et al. DNA methylation profile of 28 potential marker loci in malignant mesothelioma. Lung Cancer 2007;58:220–230.

86. Christensen BC, Godleski JJ, Marsit CJ, Houseman EA, Lopez-Fagundo CY, Longacker JL, et al. Asbestos exposure predicts cell cycle control gene promoter methylation in pleural mesothelioma. Carcinogenesis 2008;29:1555–1559.

87. Christensen BC, Houseman EA, Godleski JJ, Marsit CJ, Longacker JL, Roelofs CR, et al. Epigenetic profiles distinguish pleural mesothelioma from normal pleura and predict lung asbestos burden and clinical outcome. Cancer Res 2009;69:227–234.

88. Carbone M, Yang H. Molecular pathways: targeting mechanisms of asbestos and erionite carcinogenesis in mesothelioma. Clin Cancer Res 2012;18:598–604.

89. Rothwell PM, Fowkes FG, Belch JF, Ogawa H, Warlow CP, Meade TW. Effect of daily aspirin on long-term risk of death due to cancer: analysis of individual patient data from randomised trials. Lancet 2011;377:31–41.

90. Rothwell PM, Price JF, Fowkes FG, Zanchetti A, Roncaglioni MC, Tognoni G, et al. Short-term effects of daily aspirin on cancer incidence, mortality, and non-vascular death: analysis of the time course of risks and benefits in 51 randomised controlled trials. Lancet 2012;379:1602–1612.

91. Neri M, Ugolini D, Boccia S, Canessa PA, Cesario A, Leoncini G, et al. Chemoprevention of asbestos-linked cancers: a systematic review. Anticancer Res 2012;32:1005–1013.

92. Robinson C, Walsh A, Larma I, O'Halloran S, Nowak AK, Lake RA. MexTAg mice exposed to asbestos develop cancer that faithfully replicates key features of the pathogenesis of human mesothelioma. Eur J Cancer 2011;47:151–161.

93. Robinson C, van Bruggen I, Segal A, Dunham M, Sherwood A, Koentgen F, et al. A novel SV40 TAg transgenic model of asbestos-induced mesothelioma: malignant transformation is dose dependent. Cancer Res 2006;66:10786–10794.

94. Robinson C, Woo S, Walsh A, Nowak AK, Lake RA. The antioxidants vitamins A and E and selenium do not reduce the incidence of asbestos-induced disease in a mouse model of mesothelioma. Nutr Cancer 2012;64:315–322.

95. Piguet PF, Vesin C, Grau GE, Thompson RC. Interleukin 1 receptor antagonist (IL-1ra) prevents or cures pulmonary fibrosis elicited in mice by bleomycin or silica. Cytokine 1993;5:57–61.

96. Piguet PF, Vesin C. Treatment by human recombinant soluble TNF receptor of pulmonary fibrosis induced by bleomycin or silica in mice. Eur Respir J 1994;7:515–518.

97. Lamkanfi M, Mueller JL, Vitari AC, Misaghi S, Fedorova A, Deshayes K, et al. Glyburide inhibits the cryopyrin/Nalp3 inflammasome. J Cell Biol 2009;187:61–70.

98. Ellerman JE, Brown CK, de Vera M, Zeh HJ, Billiar T, Rubartelli A, et al. Masquerader: high mobility group box-1 and cancer. Clin Cancer Res 2007;13:2836–2848.

CHAPTER 17

NANOMATERIALS

Yiqun Mo, Rong Wan, David J. Tollerud, and Qunwei Zhang

Nanotechnologies include the design, characterization, production, and application of structures, devices, and systems by controlling shape and size at the nanometer scale (1,2). These technologies improve our lives directly in areas as diverse as engineering, information technology, and medicine, and are expected to become one of the key technologies in the twenty-first century. According to the U.S. National Nanotechnology Initiative (NNI), *nanomaterial* (NM) is defined as a small object that behaves as an entire unit in terms of its transport and properties, with at least one dimension in the approximate range 1 to 100 nm frequently with atomic/molecular precision. NMs are the building blocks of this new nanotechnology and include a range of different morphologies, including nanoparticles (NPs), nanotubes, nanowires, nanofiber, nanodots, and a range of spherical or aggregated dendritic forms. Generally speaking, the NMs can be divided into three categories: organic, inorganic, and hybrid particles (3). The most widely used are inorganic NMs, which includes carbon nanotubes, fullerene particles, and metal and metal oxide NMs.

With the development of nanotechnology, a large number of NMs will be developed and produced as new formulations with surface properties to meet novel demands. The importance of nanotechnology for sustaining economic growth is well recognized, but it is also clear that safeguarding the growth of this technology will require public understanding of the societal benefits offered by the technology as well as the health risks associated with its development and use (1–9). However, whenever a new technology arises, there are gaps in the knowledge base relating to its potential health and safety hazards and risks. In some cases, extrapolating from what is already known can easily fill this knowledge gap, but in fields such as nanotechnology, new data must be generated.

The increased development and use of NMs for various industries could lead to increased exposure, affecting human health, and the environment (4–9). Our current knowledge of health effects of NMs is limited but suggests that they may exert adverse effects at their portal of entry, such as lung, skin, and gastrointestinal tract (7–9). The existing studies show a general trend for increased toxicity as particulate size decrease. The enhanced toxicity is associated with many factors, including

Cancer and Inflammation Mechanisms: Chemical, Biological, and Clinical Aspects, First Edition.
Edited by Yusuke Hiraku, Shosuke Kawanishi, and Hiroshi Ohshima.
© 2014 John Wiley & Sons, Inc. Published 2014 by John Wiley & Sons, Inc.

increased surface area, ability to pass through cellular barriers, interaction with subcellular structures, and increased ability to activate macrophages or neutrophils and stimulate the release of inflammatory mediators. NMs have widely varying physical and chemical properties that can affect their *in vivo* properties. Scale, surface area/mass ratio, and surface charge can all have consequences on the interaction of NMs with a host organism. NM size has been implicated to have a number of important biological effects, including interaction with the reticuloendothelial system (RES), adjuvant properties, and ease of phagocytosis. For example, a previous study demonstrated that particles 50 nm or greater are more readily taken up by macrophages than are smaller equivalents (10). Similarly, surface charge has been shown to affect toxicity, clearance, immune cell stimulation, and the binding of plasma proteins to the NMs (11). In addition, some NMs may be carcinogenic because of the material, such as metals, used to make them. Certain metal particles can also generate reactive oxygen species (ROSs) that can cause oxidative stress and DNA damage (12–15). The scope of this chapter is to review the biological effects of NMs, such as oxidative stress, acute and chronic inflammation, genotoxic, carcinogenic and epigenetic effects, and their potential mechanisms.

POTENTIAL HEALTH EFFECTS OF NMs

Any technology needs careful evaluation regarding its sustainability and risk perception before being introduced to the marketplace and into the product chain (16,17). Inert materials can become reactive just by making them smaller. For example, TiO_2 is a so-called "inert" material that had been regarded as a nontoxic dust and indeed had been served as an innocuous control dust in many studies on the toxicology of particles. However, chemical studies have demonstrated that even TiO_2 in nanosize can become chemically active, due to the fact that small size does not allow a surface of only oxygen atoms (18). These results suggest that NMs may need to be reassessed since their physical state is substantially different from that of existing materials in larger particles (16–21).

In addition, present environmental and occupational exposure limits for various particles are based on mass only and do not take nanoparticle size into account. Very little information is available regarding the safety of manufactured NMs. Will NMs be like silica or asbestos, both of which are recognized as etiology for silicosis, asbestosis, and even lung cancer, respectively? The process of these diseases is characterized by inflammation, apoptosis, or progressive cell survival, and finally, fibrosis or cancer (22). Although the mechanism is not described fully, reducing the concentration of NMs should be effective to reduce toxicity. The pathology of NMs may share the same results as larger particles at the portal of entry, but NMs may diffuse into distal areas in alveoli that microparticles do not reach (23–25). Evidence of the risk and hazard of NMs has recently been investigated. For example, the uptake of NMs, their inflammatory effects, and the corresponding expression of cell adhesion molecules have been reported (1–8,23–25). Thus, it is becoming increasingly important to understand the potential health effects of various NMs on the human body.

OXIDATIVE STRESS AND NMs

Oxidative stress has been defined as an imbalance between oxidant and antioxidant in favor of the former, resulting in an overall increase in cellular levels of ROSs (26,27). ROSs are oxygen-containing free radicals such as superoxide, hydrogen peroxide (H_2O_2), and hydroxyl radical. There are four main potential sources of superoxide and H_2O_2: NADPH oxidase (28–31), xanthine oxidase (32), NO synthase (28–31), and mitochondria (33). Production of ROSs and, subsequently, oxidative stress and damage are shown to be involved in the pathogenesis of several chronic diseases including cancer.

It has been well documented that exposure to various particles *in vitro* or *in vivo*, such as ambient particulate matter (PM), diesel particles, asbestos, and silica, induces oxidative stress (22–25,34–39). Substantial results also show that exposure to some types of NMs at various doses causes an increase in ROS generation, which could lead to oxidative stress. For example, exposure to ^{60}C fullerenes, nanocarbon, and various metal nanoparticles (Fe, Ni, Co, Zn, Cu, and TiO_2) caused ROS generation and oxidative stress (40–48). NM-induced oxidative stress plays an important role in a number of active processes in cells, such as apoptosis, DNA adduct formation, and pro-inflammatory gene expression (40–42,45,48). The following three factors may be involved in the NM-induced ROS generation: (1) active redox cycling on the surface of NMs, particularly the metal-based NPs (37,49); (2) oxidative groups functionalized on NMs (46–48); and (3) particle–cell interactions, especially in the lungs, where there is a rich pool of ROS producers such as inflammatory phagocytes, neutrophils, and macrophages (40–48). In addition, ROSs, in particular H_2O_2, are now recognized as important signaling molecules. They may influence cell proliferation, cell death, and the expression of genes, or may be involved in the activation of several signaling pathways (50). Studies have shown that H_2O_2 activates the transcription factor nuclear factor-κB (NF-κB). NF-κB usually resides in the cytoplasm of the cell in association with an inhibitory protein, Iκ-B, but addition of H_2O_2 to cells results in the dissociation of NF-κB from Iκ-B and translocation of NF-κB to the nucleus (50). NF-κB may up-regulate the expression of inflammatory mediators, cytokines, and adhesion molecules. Therefore, oxidative stress is a central hypothetical mechanism for the adverse effects of NMs. Several studies have shown that the biological effects of NMs on intracellular signaling and induction of pro-inflammatory mediators from lung cells during exposure to NMs induce ROS generation (14,15,37).

NF-κB and AP-1 are two major redox-sensitive transcription factors that are activated through PKC and MAP kinase pathways and mediate an altered gene expression response to oxidative stress (23–25,37,41,43–48). Previous studies have shown that intracellular calcium acts as a key signaling mechanism which, through interaction with a number of proteins such as calmodulin and enzymes such as protein kinases, is able to regulate the activation of a number of key transcription factors (50–52), including the nuclear factors in activated T cells and NF-κB (51,52). In addition, ROS activates a series of cytokine cascades, which include an up-regulation of interleukins, kinases, and tumor necrosis factor (TNF-α) pro-inflammatory signaling processes as a counter reaction to oxidative stress (25,26,36). ROSs may also activate receptor tyrosine kinases and mitogen-activated protein kinases (48). Sustained activation of these pathways by elevated NM-induced ROS production may lead to carcinogenesis caused by NM exposure.

INFLAMMATION AND NMs

Inflammation is a normal biological response of the body to various assaults, including microorganisms, injuries, dusts, drugs, and other chemicals. Under normal circumstances, inflammation will subside and resolve itself in a person through a series of tightly regulated responses (53). However, when deregulation occurs, inflammation can lead to inflammatory disorders and diseases, including various pulmonary lung diseases, dermatitis, arthritis, and inflammatory bowel diseases (54–56). The elicitation of an inflammatory response has been shown to be connected to the oxidative stress in the cell via pro-inflammatory gene transcription (50). In addition, there may be oxidation of lipid species within the cell that lead to the production of pro-inflammatory eicosanoids (26). Exposure to some NMs, such as carbon nanotubes, carbon black, fullerenes, silica, and metal or metal oxide NPs, has been shown to cause pulmonary inflammation and/or fibrosis after intratracheal installation (19,21,40,45) or inhalation exposure (41,44). Studies also showed increased pro-inflammatory gene expression in cells with exposure to some NMs at various doses (44–48). Pro-inflammatory pathways such as the MAP kinases are oxidative stress-responsive, play a role in gene expression, and are activated by some NMs (46–48). The redox-responsive NF-κB and AP-1 transcription factors have also been reported to be activated in NM-exposed cells (17,23–26,36,49,50). IL-1β is one pro-inflammatory cytokine thought to be involved in the initiation of the inflammatory process and thus contributes to acute and chronic inflammation (57–60). There are three well-known constituents of the IL-1 gene family: IL-1α, IL-1β, and IL-1 receptor antagonist (IL-1RA), which all bind to the IL-1 receptor with similar affinity (59). Although IL-1α remains in the cytosol or is expressed at the cell membrane, IL-1β is released by a tightly controlled process in which caspase-1-mediated cleavage of pro-IL-β is a rate-limiting step (60). Inflammasome complexes control the regulated cleavage of pro-IL-1β and also other procytokines by assembling a multicomponent protein platform that leads to the activation of procaspase-1 (61–64).

Several proteins have been shown to have the ability to initiate the formation of an inflammasome complex, including the NOD-like receptor (NLR) proteins NLRP1, NLRP3, and NLRP4 and the pyrin and HIN 200 domain-containing protein AIM2 (61–64). The NLRP3 inflammasome, composed of the NLRP3, cardinal, the adaptor apoptosis-associated Speck-like protein containing a C-terminal caspase recruitment domain (ASC), and caspase-1, is implicated in the production of mature IL-1β in response to a variety of signals (61–65). There are several danger signals that activate the NLRP3 inflammasome; for example, pathogen-associated molecular patterns (PAMPs) and other pathogen-associated molecules (such as bacterial pore-forming toxin) (64,65), environmental irritants (e.g., silica, asbestos) (66–71), and damage-associated molecules (DAMPs) such as extracellular ATP (62,66,70). Because NLRP3 seems to sense a larger variety of stimuli of a diverse physiochemical nature and also plays a pivotal role in many inflammatory diseases, it has been a particularly attractive research target. In fact, several studies have shown that exposure to asbestos, silica, and aluminum salts caused IL-1β secretion through activation of NLRP3 (66–73). Further, exposure to some NPs, such as nanocarbon and nano-TiO_2, resulted in pulmonary inflammation through IL-1β release (71,72).

Chronic inflammation plays a pivotal role in the development of many diseases, including heart disease, Alzheimer's disease, arthritis, and some forms of cancer (74–77). Consequently, the ability to detect and treat such inflammation is critical in the treatment and prevention of these conditions. Chronic inflammation has been thought to play an important role in foreign body carcinogenesis induced by various particles (26,39,49,53,54,74,75). This type of inflammation is characterized by persistent release of cytokines and oxidants from macrophages. Pro-inflammatory cytokines change intracellular signaling pathways of epithelial–mesothelial cells and promote carcinogenesis by suppressing apoptosis (39,49,53,75). Previous studies showed that lung overload in the rat is a phenomenon associated with the overloading of lung clearance mechanisms, leading to a rapid accumulation of NMs in the lungs, resulting in chronic inflammation (7,9,25,50). Exposure to nanoparticles such as nano-Co, nano-Ni, nanocarbon, and nano-TiO_2, causes severe lung inflammation (36,40,41,44,45,52). This in turn is likely to generate fibro-proliferative changes, including alveolar cell proliferation (hyperplasia), the conversion of cells to cell types not normally associated with the specific lung location (metaplasia). The consequence of respiratory exposure to some NMs may cause local chronic inflammation and tumor formation (neoplasia) (25,49,50).

GENOTOXICITY OF NMs

It is well known that one initiating pathogenic effect of oxidative stress is inflammation, which can also cause DNA damage and genotoxicity (14,78,79). The association between genotoxicity and cancer is also well known. For example, the carcinogenic effects of ionizing radiation, ultraviolet radiation, and many chemical carcinogens are based on their ability to cause DNA damage and consequent gene mutation (78–80). There are several excellent reviews regarding metals, oxidative stress, and cancer (13,14,78–80). It is generally accepted that excessive generation of ROSs, which overwhelms the antioxidant defense system, can oxidize cellular biomolecules (14). Free radicals also cause oxidative modifications in DNA, including strand breaks and base oxidations. Among oxidative DNA damage products, 8-oxo-7,8-dihydro-2′-deoxyguanosine (8-oxodG) is probably the most studied, due to its relative ease of measurement and pre-mutagenic potential (14,78–80). Elevated 8-oxodG has been noted in numerous tumors, strongly implicating such damage in the etiology of cancer (13,80).

A number of studies have demonstrated that some metal NMs cause chromosome aberrations, DNA strand breaks, oxidative stress DNA damage, and DNA mutations (42,44,48,81–84). For example, recent studies have demonstrated that exposure to nontoxic doses of nano-Co caused oxidative stress, which further resulted in DNA damage. Nano-Co-induced DNA damage may further activate the ataxia telangiectasia mutant (ATM) pathway and increase the phophorylation of p53 and Rad 51 protein expression (48). For DNA damage, DNA double-strand breaks (DSBs) are considered to be among the most lethal forms of DNA damage, severely compromising genomic stability. Phosphorylation of the H2AX event is one of the best-established chromatin modifications linked to DNA damage and repair. The results also showed that rapid phosphorylation of H2AX is an early cellular response to nano-Co-induced DNA damage. ATM acts as a sensor of ROSs and oxidative damage of DNA. In turn, ATM induces signaling through multiple pathways, thereby coordinating acute-phase response with cell-cycle checkpoint control and repair of oxidative stress. p53 is a key effector molecule that is activated by nano-Co-induced DNA damage, while Rad 51 may be involved in assisting in the repair of nano-Co-induced DNA DSBs (Figure 17.1) (48).

Although currently available data on NMs genotoxicity *in vivo* indicate a potential for DNA damage by exposure to some NMs, primarily in rodents, these *in vivo* studies are rare and inconsistent because of a lack of standard test methods. Because of their small size, NMs can penetrate nuclei and hence interact directly with the structure and function of genomic DNA (81,85), or they can release metal ion directly into the cell. NMs can not only interact directly with the target cell but also can interact with other cells, causing the production of an environment conducive to the accumulation of genetic damage and proliferation. This is typically through the induction of chronic inflammation leading to persistent oxidative stress from inflammatory cells such as macrophages and polymorphonuclear leukocytes as well as the secretion of various pro-survival and proliferation signaling factors.

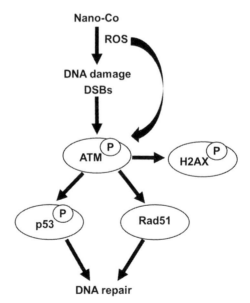

FIGURE 17.1 Schematic diagram of the signaling pathways that may be involved in nano-Co-induced DNA damage. Exposure of cells to nano-Co caused oxidative stress, which may further result in DNA DSBs and activate ataxia telangiectasia mutant (ATM). Increased phosphorylation of ATM may mediate the phosphorylation of p53 and H2AX and up-regulation of Rad51 expression. [Modified from Wan et al. (48).]

CARCINOGENICITY OF NMs

Chemical carcinogenesis is a multistep process involving both DNA damage and increase in cell proliferation (25,50). Some particles, such as asbestos, have been recognized as a carcinogen by the International Agency for Research on Cancer (IARC), and silica and diesel exhaust particulate are identified as potential carcinogens. However, when considering the potential carcinogenicity of NMs, it is much more complex because NMs may have a complex structure and display multiple functional groups at the surface charged or even chemically reactive. Many factors may be involved in NM-induced carcinogenic effects, such as the material from which NMs are originally made, size, shape, chemical composition, and oxidative potential. Carbon nanotubes (CNTs) represent an important class of engineered NMs. Depending on the number of walls, there are three forms of CNTs: single-walled carbon nanotubes (SWCNTs), double-walled carbon nanotubes (DWCNTs), and multiwalled carbon nanotubes (MWCNTs) (86,87). MWCNTs have been widely used in the industry and biomedicine, including supercapacitors, batteries, structural materials in the automotive and aerospace industries, electronics, pharmaceutics, bioengineering, medical devices, and biomedicine (86,87).

Concerns have been raised regarding potential carcinogenicity of MWCNTs because of their physical properties that are analogous to those of asbestos fibers, such as high aspect ratio (length/diameter), nanoscale diameter, micrometer length, fiberlike shape, and durability (86–90). Recent studies have demonstrated that exposure to CNTs induced malignant transformation and tumorigenesis of human lung epithelial cells (91,92). Other *in vivo* studies showed that exposure to MWCNTs caused significant oxidative stress, lung inflammation, and pleural inflammation. The initial *in vivo* studies showed that MWCNTs have the same toxicological and pathogenic behavior as asbestos, therefore that exposure to some types of MWCNTs could cause the retention of MWCNTs in the parietal pleura leading to the initiation of inflammation and pleural pathology such as mesothelioma (93). Recently, comprehensive studies showed that the toxic effects of nonfunctionalized MWCNTs on the human mesothelial cells were associated with their diameter-dependent piercing of the cell membrane (94). Intraperitoneal injection of thin MWCNTs (diameter less than 50 nm with high crystallinity) caused greater subsequent inflammation and mesotheliomagenicity than those of thick MWCNTs (diameter about 150 nm or tangled with a diameter of about 2 to 20 nm), although both thin and thick MWCNTs have similar effects on macrophages. Furthermore, they also found that the toxic effects of MWCNTs on the mesothelial cells are different from those of asbestos. MWCNTs can pierce mesothelial cell plasma and nuclear membranes directly, whereas asbestos fibers are internalized by mesothelial cells via encapsulation by vesicular membrane structure (94).

In addition, exposure of p53 heterozygous mice to MWCNTs caused a dose–response mesothelioma induction (95). These results suggest that some NMs, such as MWCNTs, have potential carcinogenic effects. In addition, due to the unique and diverse physicochemical properties of NMs, their carcinogenic properties may differ from those of the corresponding bulk materials. For example, it is well known that titanium by itself is chemically inert. When the particles become progressively smaller, their relative surface area, in turn, becomes progressively larger, promoting the interaction of this surface with the environment. Previous studies showed that TiO_2 NPs induced single- and double-strand DNA breaks, chromosomal damage, and inflammation, all of which increase the risk for cancer (81). NMs may also accumulate in different organs because the body has no effective way to eliminate them. Due to their small size, NMs may penetrate cells and accumulate within otherwise privileged DNA or cellular compartments such as mitochondria and endoplasmic reticulum where their presence may have deleterious effects.

Some metal NPs damage the DNA indirectly, not by passing through the barrier but by generating signaling molecules within the barrier cells that are then transmitted to cause damage in cells on the other side of the barrier (96). This may be a potentially new route of genotoxicity that may drive a carcinogenic effect. In general, different particulate carcinogens show a similar pattern of underlying processes (i.e., oxidative stress, inflammation, proliferation, and mutation), leading to cancer. The clear implication is that inflammation is the primary driver of the carcinogenesis. In additional, some people may be more sensitive than others to NM exposure. Previous results suggest that because of the potential risk of cancer or genetic disorders, especially for persons exposed occupationally to high concentrations of NMs, it may be prudent to limit their ingestion through nonessential drug additives, food colors, and so on. The slow clearance of NMs may lead to accumulation in tissues or organs, which may cause chronic inflammation and cell proliferation, resulting in benign and malignant tumors.

EPIGENETIC EFFECTS OF NMs

Epigenetics is defined as heritable changes in gene expression where the changes do not alter DNA sequence but are mitotically and transgenerationally heritable and provide an extra layer of transcriptional control that regulates the way that genes are expressed. Epigenetics refers to the processes of DNA methylation, histone modification, and expression of small, noncoding RNAs. Animal and epidemiological studies have demonstrated that exposure to air pollution causes modification and associated inheritable changes of DNA (97–100). Those findings support the hypothesis that mutagenic volatile chemicals adsorbed onto airborne particle pollutants may induce somatic and germ-line mutations. Therefore, the persistent effects of air pollution have a more pronounced effect on the phenotype rather than on the genotype (101–103). In addition, it is well documented that environmental factors, including heavy metal (As, Cr, Cd, Ni) exposure, can lead to human diseases, the most frequent and significant being cancer (104,105). Epigenetic changes, either DNA methylation or histone modification, have been linked to heavy metal exposure and in some cases to tumor formation (106,107).

Because epigenetic changes can occur at any time in a person's lifetime and can be passed on to future generations, the exact mechanisms by which metals induce these changes are of great significance. A few mechanisms have been proposed, such as oxidative stress induction, inflammation, and disruption of cell metabolism, but a more comprehensive approach is needed to address this issue (108–110). For example, studies have shown that exposure to quantum dots at very low levels caused global histone hypoacetylation in human breast carcinoma cells (111). Global hypoacetylation was suggested to induce a global reduction in gene transcription, including antiapoptotic or tumor suppressor genes. Exposure to nano-SiO$_2$ was also found to cause hypomethylation in the DNA of human HaCaT cells, and this effect was not seen with the same concentration of micro-size SiO$_2$ (112). However, the detailed mechanism is still unclear. The mechanisms that induce epigenetic changes as well as the changes themselves could be used as biomarkers for the prognosis and diagnosis of diseases. The understanding of epigenetic changes caused by NM exposure will also be of great significance in evaluating their potential toxic and carcinogenic effects. Undoubtedly, advances in the understanding of epigenetics and its involvement in the development of pathological conditions are critical for preventing the potential health effects of exposure to various NMs, especially metal NPs.

CONCLUDING REMARKS

Although a number of studies have demonstrated the potential genotoxicity and carcinogenicity of NMs and potential mechanisms (Figure 17.2), especially MWCNTs and metal NPs, detailed mechanistic studies on NM genotoxicity and carcinogenicity *in vivo*, especially on the mammalian system, are limited. In addition, long-term *in vitro* and *in vivo* exposure is needed to best understand the potential genototoxic

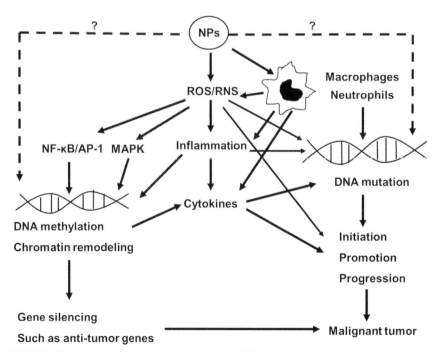

FIGURE 17.2 Potential mechanistic linking of NP exposure to genotoxic or epigenetic effects. Exposure to NPs may lead to oxidative stress due to increased production of ROSs and downstream signaling responses that promote persistent inflammation and produce carcinogenesis through gentoxic or epigenetic effects. NPs, nanoparticles; ROS/RNS, reactive oxygen species/reactive nitrogen species.

and carcinogenic effects of NMs. The growth of the field of nanotechnology and the increase in the population exposed to NMs warrant a thorough understanding of potential mechanisms of NM toxicity and genotoxicity for appropriate safety assessment and identification of exposure biomarkers.

Acknowledgments

This work was partially supported by CTSPGP 20018 and IRIG 50753 from the University of Louisville, and T32-ES011564 and ES01443 from the National Institutes of Health.

REFERENCES

1. Maynard AD, Aitken RJ, Butz T, Colvin V, Donaldson K, Oberdorster G, et al. Safe handing of nanotechnology. Nature 2006;444:267 269.
2. Kubik T, Bogunia-Kubik K, Sugusaka M. Nanotechnology on duty in medical application. Curr Pharm Biotechnol 2005;6:17–33.
3. Makarucha AJ, Todorova N, Yarovsky I. Nanomaterials in biological environment: a review of computer modeling studies. Eur Biophys J 2011;40:103–115.
4. Owen R, Depledge M. Nanotechnology and the environment: risk and rewards. Mar Pollut Bull 2005;50:609–612.
5. Chow JC, Watson JG, Savage N, Solomon CJ, Cheng YS, McMurry PH, et al. Nanoparticles and the environment. J Air Waste Manage Assoc 2005;55:1411–1417.
6. Biswas P, Wu CY. Critical review: nanoparticles and the environment. J Air Waste Manag Assoc 2005;55:708–746.
7. Oberdoster G, Maynard A, Donaldson K, Castranova V, Fitzpatrick, Ausman K, et al. A report from the ILSI Research Foundation/Risk Science Institute Nanomaterial Toxicity Screening Working. Principle for characterizing the potential human health effects from exposure to nanomaterials: elements of a screening strategy. Part Fibre Toxicol 2005;2:8.
8. Colvin VL. The potential environmental impact of engineered nanomaterials. Nat Biotechnol 2003;21:1166–1170.
9. Oberdorster G, Oberdoster E, Oberdorster L. Nanotoxicology: An emerging discipline evolving from studies of ultrafine particles. Environ Health Perspect 2005;113:823–839.
10. Metz S, Bonaterra G, Rudelius M, Settles M, Rummeny EJ, Daldrup-Link HE. Capacity of human monocytes to phagocytose approved iron oxide MR contrast agents *in vitro*. Eur Radiol 2004;14:1851–1818.
11. Serda RE, Godin B, Blanco E, Chiappini C, Ferrari M. Multi-stage delivery nanoparticle systems for therapeutic applications. Biochim Biophys Acta 2011;1810:317–329.
12. Valko M, Morris H, Cronin MTD. Metal, toxicity, and oxidative stress. Curr Med Chem 2005;12:1161–1208.
13. Costa M, Salnikow K, Sutherland JE, Broday L, Peng W, Zhang Q, Kluz T. The role of oxidative stress in nickel and chromate genotoxicity. Mol Cell Biochem 2004;134/135:265–275.
14. Valko M, Rhodes CJ, Moncol J, Izakovic M, Mazur M. Free radicals, metals and antioxidants in oxidative stress-induced cancer. Chem Biol Interact 2006;160:1–40.
15. Leonard SS, Harris GK, Shi X. Metal-induced oxidative stress and signal transduction. Free Radic Biol Med 2004;37:1921–1942.
16. Born PJA. Particle toxicology: from coal mining to nanotechnology. Inhal Toxicol 2002;14:311–314.
17. Donaldson K, Stone V, Tran CL, Kreyling W, Borm PJ. Nanotoxicology. Occup Environ Med 2004;61:727–728.
18. Jefferson DA, Tilley EEM. The structure and physical chemistry of nanoparticles. In: Maynard RL, Howard CV, editors. Particulate Matter: Properties and Effects upon Health. Oxford, UK: BIOS Scientific Publishers; 1999. pp. s63–s84.

19. Lam CW, James JT, McCluskey R, Hunter RL. Pulmonary toxicity of single-wall carbon nanotubes in mice 7 and 90 days after intracheal instillation. Toxicol Sci 2004;17:126–134.
20. Kreyling WG, Semmler M, Moller W. Dosimetry and toxicology of ultrafine particles. J Aerosol Med 2004;17:140–152.
21. Warheit DB, Laurence BR, Reed KL, Roach DH, Reynolds GAM, Webb TR. Comparative pulmonary toxicity assessment of single-wall carbon nanotubes in rats. Toxicol Sci 2004;77:117–125.
22. Borm PJ, Tran L, Donaldson K. The carcinogenic action of crystalline silica: a review of the evidence supporting secondary inflammation-driven genotoxicity as a principal mechanism. Crit Rev Toxicol 2011;41:756–770.
23. Nel A, Xia T, Madler L, Li N. Toxic potential of materials at the nanolevel. Science 2006; 311:622–627.
24. Li N, Xia T, Nel AE. The role of oxidative stress in ambient particulate matter–induced lung diseases and its implications in the toxicity of engineered nanoparticles. Free Radic Biol Med 2008;44:1689–1699.
25. Donaldson K, Poland CA. Inhaled nanoparticles and lung cancer: what we can learn from conventional particle toxicology. Swiss Med Wkly 2012;142:w13547. doi:10.4414/smw.2012.13547.
26. Klaunig JE, Kamendulis LM. The role of oxidative stress in carcinogenesis. Annu Rev Pharmacol Toxicol 2004;44:239–267.
27. Kliment CR, Oury TD. Oxidative stress, extracellular matrix targets, and idiopathic pulmonary fibrosis. Free Radic Biol Med 2010;49:707–717.
28. Bokoch GM, Knaus UG. NADPH oxidases: not just for leukocytes anymore! Trends Biochem Sci 2003;28:502–508.
29. Cave AC, Brewer AC, Narayanapanicker N, Ray R, Grieve DJ, Walker S, et al. NADPH oxidase in cardiovascular health and disease. Antioxid Redox Signal 2006;8:691–728.
30. Cross AR, Segal AW. The NADPH oxidase of professional phagocytes: prototype of the NOX electron transport chain systems. Biochim Biophys Acta 2004;1657:1–22.
31. Rueckschloss U, Duerrschmidt N, Morawietz H. NADPH oxidase in endothelial cells: impact on atherosclerosis. Antioxid Redox Signal 2003;5:171–180.
32. Kelley EE, Khoo NK, Hundley NJ, Malik UZ, Freeman BA, Tarpey MM. Hydrogen peroxide is the major oxidant product of xanthine oxidase. Free Radic Biol Med 2010;48:493–498.
33. Murphy MP. Modulating mitochondrial intracellular location as a redox signal. Sci Signal 2012;5:pe39.
34. Ghio AJ, Carraway MS, Madden MC. Composition of air pollution particles and oxidative stress in cells, tissues, and living systems. J Toxicol Environ Health B Crit Rev 2012;15:1–21.
35. Aust AE, Cook PM, Dodson RF. Morphological and chemical mechanisms of elongated mineral particle toxicities. Toxicol Environ Health B Crit Rev 2011;14:40–75.
36. Møller P, Jacobsen NR, Folkmann JK, Danielsen PH, Mikkelsen L, Hemmingsen JG, et al. Role of oxidative damage in toxicity of particulates. Free Radic Res 2010;44:1–46.
37. Xia T, Li N, Nel AE. Potential health impact of nanoparticles. Annu Rev Public Health 2009; 30:137–150.
38. Yang W, Omaye ST. Air pollutants, oxidative stress and human health. Mutat Res 2009;674:45–54.
39. Valavanidis A, Fiotakis K, Vlachogianni T. Airborne particulate matter and human health: toxicological assessment and importance of size and composition of particles for oxidative damage and carcinogenic mechanisms. J Environ Sci Health C Environ Carcinog Ecotoxicol Rev 2008;26: 339–362.
40. Folkmann JK, Risom L, Jacobsen NR, Wallin H, Loft S, Møller P. Oxidatively damaged DNA in rats exposed by oral gavage to C60 fullerenes and single-walled carbon nanotubes. Environ Health Perspect 2009;117:703–708.
41. Wessels A, Van Berlo D, Boots AW, Gerloff K, Scherbart AM, Cassee FR, et al. Oxidative stress and DNA damage responses in rat and mouse lung to inhaled carbon nanoparticles. Nanotoxicology 2011;5:66–78.
42. Ramkumar KM, Manjula C, Gnanakumar G, Kanjwal MA, Sekar TV, Paulmurugan R, et al. Oxidative stress–mediated cytotoxicity and apoptosis induction by TiO_2 nanofibers in HeLa cells. Eur J Pharm Biopharm 2012;81:324–333.
43. Shvedova AA, Pietroiusti A, Fadeel B, Kagan VE. Mechanisms of carbon nanotube–induced toxicity: focus on oxidative stress. Toxicol Appl Pharmacol 2012;261:121–133.

44. Srinivas A, Rao PJ, Selvam G, Goparaju Λ, Murthy BP, Reddy NP. Oxidative stress and inflammatory responses of rat following acute inhalation exposure to iron oxide nanoparticles. Hum Exp Toxicol 2012;31:1113–1131.

45. Zhang Q, Kusaka Y, Sato K, Nakakuki K, Kohyama N, Donaldson K. Differences in the extent of inflammation caused by intratracheal exposure to three ultrafine metals: role of free radicals. J Toxicol Environ Health A 1998;53:423–438.

46. Wan R, Mo Y, Zhang X, Chien S, Tollerud DJ, Zhang Q. Matrix metalloproteinase-2 and -9 are induced differently by metal nanoparticles in human monocytes: the role of oxidative stress and protein tyrosine kinase activation. Toxicol Appl Pharmacol 2008;233:276–285.

47. Yu M, Mo Y, Wan R, Chien S, Zhang X, Zhang Q. Regulation of plasminogen activator inhibitor-1 expression in endothelial cells with exposure to metal nanoparticles. Toxicol Lett 2010;195: 82–89.

48. Wan R, Mo Y, Feng L, Chien S, Tollerud DJ, Zhang Q. DNA damage caused by metal nanoparticles: involvement of oxidative stress and activation of ATM. Chem Res Toxicol 2012;25:1402–1411.

49. Knaapen AM, Borm PJA, Albrecht C, Schins RPF Knaapen AM, Borm PJA, et al. Inhaled particles and lung cancer. Part A: Mechanisms. Int J Cancer 2004;109:799–809.

50. Mena S, Ortega A, Estrela JM. Oxidative stress in environmental-induced carcinogenesis. Mutat Res 2009;674:36–44.

51. Brown DM, Donaldson K, Borm PJ, Schins RP, Dehnhardt M, Gilmour P, et al. Calcium and ROS-mediated activation of transcription factors and TNF-alpha cytokine gene expression in macrophages exposed to ultrafine particles. Am J Physiol Lung Cell Mol Physiol 2004;286:L344–353.

52. Barlow PG, Clouter-Baker A, Donaldson K, Maccallum J, Stone V. Carbon black nanoparticles induce type II epithelial cells to release chemotaxins for alveolar macrophages. Part Fibre Toxicol 2005;2:11.

53. Stevenson R, Hueber AJ, Hutton A, McInnes IB, Graham D. Nanoparticles and inflammation. Scientific World J 2011;11:1300–1312.

54. Thannickal VJ, Toews GB, White ES, Lynch JP 3rd, Martinez FJ. Mechanisms of pulmonary fibrosis. Annu Rev Med 2004;55:395–417.

55. Engel MA, Khalil M, Neurath MF. Highlights in inflammatory bowel disease—from bench to bedside. Clin Chem Lab Med 2012;50:1229–1235.

56. Oyoshi MK, He R, Li Y, Mondal S, Yoon J, Afshar R, et al. Leukotriene b4-driven neutrophil recruitment to the skin is essential for allergic skin inflammation. Immunity 2012;37:747–758.

57. van de Veezrdonk FL, Netea MG, Dinarello CA, Joosten LA. Inflammasome activation and IL-1β and IL-18 processing during infection. Trends Immunol 2011;32:110–116.

58. Kapoor M, Martel-Pelletier J, Lajeunesse D, Pelletier JP, Fahmi H. Role of proinflammatory cytokines in the pathophysiology of osteoarthritis. Nat Rev Rheumatol 2011;7:33–42.

59. Dinarello CA. Interleukin-1. Cytokine Growth Factor Rev 1997;8:253–265.

60. Dinarello CA. Interleukin-1beta and the autoinflammatory diseases. N Engl J Med 2009;360: 2467–2470.

61. Martinon F, Mayor A, Tschopp J. The inflammasomes: guardians of the body. Annu Rev Immunol 2009;27:229–265.

62. Mariathasan S, Weiss DS, Newton K, McBride J, O'Rourke K, Roose-Girma M, et al. Cryopyrin activates the inflammasome in response to toxins and ATP. Nature 2006;440:228–232.

63. Tschopp J, Schroder K. NLRP3 inflammasome activation: The convergence of multiple signalling pathways on ROS production? Nat Rev Immunol 2010;10:210–215.

64. Bauernfeind F, Bartok E, Rieger A, Franchi L, Núñez G, Hornung V. Cutting edge: reactive oxygen species inhibitors block priming, but not activation, of the NLRP3 inflammasome. J Immunol 2011;187:613–617.

65. Saïd-Sadier N, Padilla E, Langsley G, Ojcius DM. *Aspergillus fumigatus* stimulates the NLRP3 inflammasome through a pathway requiring ROS production and the Syk tyrosine kinase. PLoS One 2010;5:e10008.

66. Hornung V, Bauernfeind F, Halle A, Samstad EO, Kono H, Rock KL, et al. Silica crystals and aluminum salts activate the NALP3 inflammasome through phagosomal destabilization. Nat Immunol 2008;9:847–856.

67. Riteau N, Gasse P, Fauconnier L, Gombault A, Couegnat M, Fick L, et al. Extracellular ATP is a danger signal activating P2×7 receptor in lung inflammation and fibrosis. Am J Respir Crit Care Med 2010;182:774–783.

68. Dostert C, Pétrilli V, Van Bruggen R, Steele C, Mossman BT, Tschopp J. Innate immune activation through Nalp3 inflammasome sensing of asbestos and silica. Science 2008;320:674–677.

69. Cassel SL, Eisenbarth SC, Iyer SS, Sadler JJ, Colegio OR, Tephly LA, et al. The Nalp3 inflammasome is essential for the development of silicosis. Proc Natl Acad Sci USA 2008;105:9035–9040.

70. Franchi L, Eigenbrod T, Núñez G. Cutting edge: TNF-alpha mediates sensitization to ATP and silica via the NLRP3 inflammasome in the absence of microbial stimulation. J Immunol 2009;183: 792–796.

71. Yazdi AS, Guarda G, Riteau N, Drexler SK, Tardivel A, Couillin I, et al. Nanoparticles activate the NLR pyrin domain containing 3 (Nlrp3) inflammasome and cause pulmonary inflammation through release of IL-1α and IL-1β. Proc Natl Acad Sci USA 2010;107:19449–19454.

72. Reisetter AC, Stebounova LV, Baltrusaitis J, Powers L, Gupta A, Grassian VH, et al. Induction of inflammasome-dependent pyroptosis by carbon black nanoparticles. J Biol Chem 2011; 286:21844–21852.

73. Franchi L, Núñez G. The NLRP3 inflammasome is critical for alum-mediated IL-1β secretion but dispensable for adjuvant activity. Eur J Immunol 2008;38:2085–2089.

74. Rosanna DP, Salvatore C. Reactive oxygen species, inflammation, and lung diseases. Curr Pharm Des 2012;18:3889–900.

75. Vendramini-Costa DB, Carvalho JE. Molecular link mechanisms between inflammation and cancer. Curr Pharm Des 2012;18:3831–3852.

76. Al Ghouleh I, Khoo NK, Knaus UG, Griendling KK, Touyz RM, Thannickal VJ, et al. Oxidases and peroxidases in cardiovascular and lung disease: new concepts in reactive oxygen species signaling. Free Radic Biol Med 2011;51:1271–88.

77. Kapoor M, Martel-Pelletier J, Lajeunesse D, Pelletier JP, Fahmi H. Role of proinflammatory cytokines in the pathophysiology of osteoarthritis. Nat Rev Rheumatol 2011;7:33–42.

78. Pulido MD, Parrish AR. Metal-induced apoptosis: mechanism. Mutat Res 2003;533:227–241.

79. Galanis A, Karapetsas A, Sandaltzopoulos R. Metal-induced carcinogenesis, oxidative stress and hypoxia signaling. Mutat Res 2009;674:31–35.

80. Risom L, Moller P, Loft S. Oxidative stress-induced DNA damage by particulate air pollution. Mutat Res 2005;592:119–137.

81. Trouiller B, Reliene R, Westbrook A, Solaimani P, Schiestl RH. Titanium dioxide nanoparticles induce DNA damage and genetic instability *in vivo* in mice. Cancer Res 2009;69:8784–8789.

82. Tsaousi A, Jones E, Case CP. The *in vitro* genotoxicity of orthopaedic ceramic (Al_2O_3) and metal (CoCr alloy) particles. Mutat Res 2010;697:1–9.

83. Petković J, Zegura B, Stevanović M, Drnovšck N, Uskoković D, Novak S, et al. DNA damage and alterations in expression of DNA damage responsive genes induced by TiO_2 nanoparticles in human hepatoma HepG2 cells. Nanotoxicology 2011;5:341–353.

84. Anas A, Akita H, Harashima H, Itoh T, Ishikawa M, Biju V. Photosensitized breakage and damage of DNA by CdSe–ZnS quantum dots. J Phys Chem B 2008;112:10005–10011.

85. Xie H, Mason MM, Wise JP. Genotoxicity of metal nanoparticles. Rev Environ Health 2011;26:251–268.

86. Aschberger K, Johnston HJ, Stone V, Aitken RJ, Hankin SM, Peters SA, et al. Review of carbon nanotubes toxicity and exposure: appraisal of human health risk assessment based on open literature. Crit Rev Toxicol 2010;40:759–790.

87. Guo NL, Wan YW, Denvir J, Porter DW, Pacurari M, Wolfarth MG, et al. Multiwalled carbon nanotube-induced gene signatures in the mouse lung: potential predictive value for human lung cancer risk and prognosis. J Toxicol Environ Health A 2012;75:1129–1153.

88. Kobayashi N, Naya M, Ema M, Endoh S, Maru J, Mizuno K, et al. Biological response and morphological assessment of individually dispersed multiwall carbon nanotubes in the lung after intratracheal instillation in rats. Toxicology 2010;276:143–153.

89. Becker H, Herzberg F, Schulte A, Kolossa-Gehring M. The carcinogenic potential of nanomaterials, their release from products and options for regulating them. Int J Hyg Environ Health 2011;214:231–238.

90. Kisin ER, Murray AR, Sargent L, Lowry D, Chirila M, Siegrist KJ, et al. Genotoxicity of carbon nanofibers: Are they potentially more or less dangerous than carbon nanotubes or asbestos? Toxicol Appl Pharmacol 2011;252:1–10.

91. Nagai H, Toyokuni S. Biopersistent fiber-induced inflammation and carcinogenesis: lessons learned from asbestos toward safety of fibrous nanomaterials. Arch Biochem Biophys 2010;502:1–7.

92. Wang L, Luanpitpong S, Castranova V, Tse W, Lu Y, Pongrakhananon V, et al. Carbon nanotubes induce malignant transformation and tumorigenesis of human lung epithelial cells. Nano Lett 2011;11:2796–2803.

93. Poland CA, Duffin R, Kinloch I, Maynard A, Wallace WA, Seaton A, et al. Carbon nanotubes introduced into the abdominal cavity of mice show asbestos-like pathogenicity in a pilot study. Nat Nanotechnol 2008;3:423–428.

94. Nagai H, Okazaki Y, Chew SH, Misawa N, Yamashita Y, Akatsuka S, et al. Diameter and rigidity of multiwalled carbon nanotubes are critical factors in mesothelial injury and carcinogenesis. Proc Natl Acad Sci USA. 2011;108:E1330–E1338.

95. Takagi A, Hirose A, Futakuchi M, Tsuda H, Kanno J. Dose-dependent mesothelioma induction by intraperitoneal administration of multi-wall carbon nanotubes in p53 heterozygous mice. Cancer Sci 2012;103:1440–1444.

96. Sood A, Salih S, Roh D, Lacharme-Lora L, Parry M, Hardiman B, et al. Signalling of DNA damage and cytokines across cell barriers exposed to nanoparticles depends on barrier thickness. Nat Nanotechnol 2011;6:824–833.

97. Somers CM, McCarry BE, Malek F, Quinn JS. Reduction of particulate air pollution lowers the risk of heritable mutations in mice. Science 2004;304:1008–1010.

98. Samet JM, DeMarini DM, Malling HV. Biomedicine: Do airborne particles induce heritable mutations? Science 2004;304:971–972.

99. Baccarelli A, Wright RO, Bollati V, Tarantini L, Litonjua AA, Suh HH, et al. Rapid DNA methylation changes after exposure to traffic particles. Am J Respir Crit Care Med 2009;179:572–578.

100. Nawrot TS, Adcock I. The detrimental health effects of traffic-related air pollution: a role for DNA methylation? Am J Respir Crit Care Med 2009;179:523–524.

101. Vineis P, Husgafvel-Pursiainen K. Air pollution and cancer: biomarker studies in human populations. Carcinogenesis 2005;26:1846–1855.

102. Reliene R, Hlavacova A, Mahadevan B, Baird WM, Schiestl RH. Diesel exhaust particles cause increased levels of DNA deletions after transplacental exposure in mice. Mutat Res 2005;570: 245–252.

103. Yauk CL, Berndt ML, Williams A, Rowan-Carroll A, Douglas GR, Stämpfli MR. Mainstream tobacco smoke causes paternal germ-line DNA mutation. Cancer Res 2007;67:5103–5106.

104. International Agency for Research on Cancer. IARC Monographs on the Evaluation of Carcinogenic Risks to Humans, vol. 86, Cobalt in Hard Metal and Cobalt Sulfate, Gallium Arsenide, Indium Phosphide and Vanadium Pentoxide. Lyon, France: IARC Press; 2006.

105. International Agency for Research on Cancer. IARC Monographs on the Evaluation of Carcinogenic Risks to Humans, vol. 100C, Arsenic, Metals, Fibres, and Dusts. Lyon, France: IARC Press; 2012.

106. Arita A, Costa M. Epigenetics in metal carcinogenesis: nickel, arsenic, chromium and cadmium. Metallomics 2009;1:222–228.

107. Chervona Y, Costa M. The control of histone methylation and gene expression by oxidative stress, hypoxia, and metals. Free Radic Biol Med 2012;53:1041–1047.

108. Valinluck V, Sowers LC. Inflammation-mediated cytosine damage: a mechanistic link between inflammation and the epigenetic alterations in human cancers. Cancer Res 2007;67:5583–5586.

109. Baccarelli A, Ghosh S. Environmental exposures, epigenetics and cardiovascular disease. Curr Opin Clin Nutr Metab Care 2012;15:323–329.

110. Dik S, Scheepers PT, Godderis L. Effects of environmental stressors on histone modifications and their relevance to carcinogenesis: a systematic review. Crit Rev Toxicol 2012;42:491–500.

111. Choi AO, Brown SE, Szyf M, Maysinger D. Quantum dot–induced epigenetic and genotoxic changes in human breast cancer cells. J Mol Med (Berl) 2008;86:291–302.

112. Gong C, Tao G, Yang L, Liu J, Liu Q, Zhuang Z. SiO$_2$ nanoparticles induce global genomic hypomethylation in HaCaT cells. Biochem Biophys Res Commun 2010;397:397–400.

CHAPTER 18

INFLAMMATORY PATHWAYS OF RADIATION-INDUCED TISSUE INJURY

Danae A. Laskaratou, Ifigeneia V. Mavragani, and Alexandros G. Georgakilas

Ionizing radiation (IR), whether it is of low or high linear energy transfer (LET), has been the subject of many studies, especially when it comes to interaction with biological tissue and human beings. The impact and effects of radiation on cells and tissues is a complicated function of various parameters, such as energy, dose delivered, and dose rate, all contributing separately and synergistically to the *relative biological effectiveness* (RBE), which is characteristic for each radiation. Whereas the effects of high-level radiation exposure are delineated quite clearly, low doses (<1 Gy = 100 rad) cause stochastic effects that do not necessarily follow a linear model, and as such, they are still not easily determined (1).

Inflammation is a process of the innate immune system of the body, triggered primarily by harmful stimuli, such as pathogen infection, tissue injury and/or IR. It is a finely regulated, complex response, during which a cascade of fully orchestrated events takes place, engaging a plethora of various inflammatory cells and proteins in the process (2). The aim of this rapid response is normally the elimination of microbes and host cells, the repair of affected tissue, the initiation of wound healing, and therefore the restoration of homeostasis. Nevertheless, however beneficial for the body such a response may be, in some cases of disorder it has been observed that self-inflicted damage can be of even greater significance than the microbial target itself. In other words, a controlled and properly synchronized inflammatory reaction is of vital importance to the organism contributing to, but in some cases also deregulating the homeostasis and well-being of a system (3,4). Inflammation is closely linked to radiation treatment, since, together with the area of interest, healthy tissue is inevitably irradiated and injured. Inflammatory response is initiated by IR, with the principal function of controlling damage and repairing lesions. This applies to both directly damaged cells and unirradiated ones that either happen to be in close

Cancer and Inflammation Mechanisms: Chemical, Biological, and Clinical Aspects, First Edition.
Edited by Yusuke Hiraku, Shosuke Kawanishi, and Hiroshi Ohshima.
© 2014 John Wiley & Sons, Inc. Published 2014 by John Wiley & Sons, Inc.

proximity (short- and long-range bystander effects) or are descendants of irradiated cells (genomic instability) (5). A problematic situation starts to develop when inflammation persists long after radiation treatment is completed and subsequently turns from acute response to chronic late effect(s). The inflammation is no longer self-limited and leads to an overproduction of reactive oxygen and nitrogen species (ROSs and RNSs), which in turn results into chronic stress, DNA and RNA damage, mutation, and carcinogenesis. In the following pages we discuss the current status of knowledge on the role(s) of inflammatory responses in the induction of radiation effects, especially cell and tissue injury.

INTERACTION OF IR WITH CELLS AND TISSUES

IR is energy transmitted via cosmic rays, α-particles (the nucleus of the helium atom), β-particles (high-speed electrons), x-rays, γ-rays, neutrons, protons, and other heavy ions, and generally any charged particle moving at relativistic speeds. These radiation types have the ability to penetrate living cells or tissue and result in the transfer of "radiation energy" to the biological material. The ionization of atoms and molecules caused by the absorbed energy of the IR can break chemical bonds of molecules, such as water and different biologically essential macromolecules, such as DNA (6) and membrane lipids or proteins (7,8).

When IR is emitted or absorbed by an atom, a subatomic particle such as a proton, electron, or neutron can be released. As an example, when an electron passes through a tissue or a cell, it releases energy by interacting with the electrons of nearby molecules. This energy is absorbed by the adjacent atoms, and as a result there can be either excitation (elevation of an electron in energy level above an arbitrary baseline energy state) or ionization (release of an electron from the atom). When radiation releases its energy to atoms or molecules, electrons placed in orbits deeper than the outer orbit can be set free, making the atoms very unstable (radicals).

Radiation events can be categorized not only by their constituents (electrons, protons, neutrons, etc.) but also by their energy. Linear energy transfer (LET) is a physical parameter used to describe either average energy released per unit length of a particle's track, or the amount of energy deposited as the particles traverse a section of tissue. High-LET radiation (neutrons, α-particles) can cause dense ionization along its track, while low-LET radiation (x-rays) produces only a few ionizations sparsely along its track. Radiation interactions with tissues and cells release energy in a biological matter through processes that increase the entropy of the system, often to a limit that the organisms cannot tolerate, resulting in the harmful effects of radiation. This can lead to a great variety of biological consequences, ranging from local recoverable situations to complex irreversible modifications.

BIOLOGICAL EFFECTS OF IR

The biological effects of IR are often associated with absorbed dose values in tissues or cells at risk and other physical factors, such as the radiation type, the relative biological effectiveness of radiation, the spatial and time distribution of the energy deposited, or the total number of cells irradiated. In an attempt to estimate biological effects such as carcinogenesis, the risk is usually considered simply to be proportional to the dose received, although some biological experiments have shown a much more complex relation (9). Radiation damage to cells also depends on the sensitivity of the cells to irradiation. Cells that divide rapidly and/or are not specialized (non-differentiated) seem to be more sensitive at lower doses of radiation than more specialized and less rapidly dividing cells (10).

A biological effect begins with the ionization of atoms. The energy released by IR to a human tissue can abstract electrons from the atoms that constitute the molecules of the tissue. IR can dislocate an electron shared by two atoms forming a molecular bond, and as a result, the bond breaks and the molecule falls apart. In some cases, when IR interacts with cells, it may encounter a crucial part of the cell, such as DNA and chromosomes. However, the cells may stay unharmed by the dose, as the ionization may cause alterations in the structure of the cells similar to those that occur naturally in the cells. Other cells may be damaged, repair the damage, and continue to operate normally. In this case, ionization leads to a cell's structure breakdown, but cells may be capable of repairing the impairment if it is not major. Otherwise, if a damaged cell that does not have sufficient time to repair itself must execute an operation, it will either not be able to fix the damage or perform the repair function incorrectly or incompletely. As a result, the altered cells may no longer be able to reproduce or may reproduce and diverge uncontrollably, forming premalignant tissues. Notably, early experiments on the effect of IR on tissue mast cells had shown an increase in cell populations along with a relative increase in of abnormal types of cells (11). Finally, if radiation has damaged a cell extensively in such a way that it could no longer respond to many of the signals that control cellular growth, it may die through apoptosis.

Cell death can be the ultimate result of cellular and tissue radiation injury (12,13). Knowing the type and mechanisms that lead to radiation-induced cellular death is very important, as killing of target malignant cells is a therapeutic application of irradiation. IR-induced cellular death has been categorized into two main groups/divisions based on the time of decomposition of cells after irradiation, interphase death, and reproductive or mitotic death. *Interphase death* is defined as an irreversible breakdown of cellular metabolism and structure before performing the first mitosis after irradiation, while *reproductive* or *mitotic death* occurs during one or even several divisions after the irradiation. Both types of cellular death can be manifested as apoptosis and/or necrosis (12).

Unlike the effects of low levels of radiation which are more difficult to determine, the biological effects of high levels of radiation exposure are known sufficiently well. Based on how much and how fast a radiation dose is being received, we can group them into acute and chronic doses. Acute radiation doses are relatively large doses of 0.1 Gy or greater, delivered to the entire body during a short period of time. If they are large enough, they may lead to observable effects within a few hours or weeks. Sometimes acute doses are responsible for the appearance of a pattern of clinically recognizable signs and symptoms (syndromes). On the other hand, chronic doses are relatively small doses of radiation (in the range of a few mGy or a maximum of 10 cGy) received over a prolonged period of time. As a result, the organism has time to repair the radiation-induced damage, due to the small percentage of cells affected, and therefore it is easier to tolerate a chronic dose than an acute one without excluding though the possibility of long-term effects like chromosomal instability in these low-dose exposed cells and their neighboring 'bystander' ones.

TYPES OF RADIATION EFFECTS

It is considered that two distinct types of biological effects exist: stochastic and nonstochastic (deterministic) effects. The term *deterministic* (causally determined by preceding events) is not very accurate, as it has been recognized that both early and late tissue reactions are not necessarily predetermined and that they can change after irradiation, but it is still employed, due to its historical use in many recommendations and standards (14). In general, stochastic effects (mainly cancer) are caused by nonlethal mutations in cells, whereas deterministic effects are caused by cellular death. The effects that can be diagnosed in a medical examination within a few weeks after radiation exposure, including skin burns and reduction in the levels of the blood cells, are deterministic effects. Such effects occur when the number of dead cells, due to a high radiation dose, exceeds the capacity for replacement, that is, when the equilibrium state between production and death is perturbed by an excess of the latter. As a result, the target organs and tissues stop working properly, which eventually leads to biological modifications with serious consequences.

To recognize deterministic effects, a given threshold dose has been established above which they appear. This dose is high enough, so it does not include doses from natural radiation or from occupational exposure. On the other hand, stochastic effects develop at low-dose radiation and there is not a threshold dose. Although the existence of a low-dose threshold does not seem implausible for radiation-related cancers of certain tissues, it cannot favor the existence of a universal threshold (15). It appears that the frequency of the effect's appearance in the exposed population is usually dose-dependent and increasing linearly with the dose at least above 0.1 Gy, but there is not such a relation between the severity of the effect and the radiation dose. In its 2007 recommendations (16) the International Commission on Radiological Protection (ICRP) concluded that for the induction of cancer at low doses, the use of a simple proportional relationship between increased dose and increased risk is a scientifically plausible assumption.

The investigation of acute and longer-term effects of low doses of IR on cells and tissues has exceptional importance in radiotherapy as well as in the management of accidental radiation exposure. Considering that the pathological processes begin immediately after irradiation and that the clinical and histological features may not be identifiable for weeks, months or even years after exposure, post-irradiation events of *in vivo* systems upon radiation exposition can also be classified as acute, consequential or late effects, according to the time before appearance of symptoms (17). Acute- and delayed-phase responses are also observed *in vitro* (in cell cultures).

Direct and nearly immediate cellular toxicity caused by free radical–mediated damage to cellular DNA leads to acute injuries, which occur within a few hours or days after irradiation, while delayed radiation complications may develop many years after the radiation exposure. The effects may be long lasting (i.e., they can be observed up to 45 cell doublings after radiation exposure) as a consequence of the radiation-induced genomic instability. In some cases, acute reactions persist into the late period, and the resulting chronic lesions add to the overall damage. These lesions are defined as consequential late effects (18). At the cellular level the radiation response may be manifested in irreversible changes such as mutations, especially inaction of DNA repair, cell-cycle control and apoptotic genes, malignant transformation, development of abnormal cell forms and the death of cells, or as minor reversible structural and functional alterations of biological systems. Both the irreversible and reversible damage of cells are manifested at the subcellular level as structural and functional changes in cell organelles (19).

DNA DAMAGE: CHROMOSOMAL INSTABILITY

The most sensitive target for radiation effects is the DNA molecule. The effects of radiation on DNA are considered as the final process of a number of physical and chemical events, initiated by the ionizing interaction and followed by biological alterations of various types. IR induces a plethora of types of DNA damages, but the identity of some specific lesions which are responsible for cell death, mutation, and carcinogenesis, remains uncertain. Incompletely or incorrectly repaired DNA lesions are the prime source of lethal and mutagenic effects of IR. However, recently, several *nontargeted* effects (i.e., bystander effects), in which cells that are not directly exposed to radiation show the same effects as exposed cells, have been identified and are under intense research.

IR produces many types of DNA damages, including DNA base alterations, DNA–DNA cross-links, single- (SSBs) or double-stranded breaks (DSBs) (6), and other non-DSB clustered DNA lesions consisting of different lesions at close proximity to each other, usually within 10 to 20 base pairs (20). The radiation-induced chromosomal aberrations that are usually being observed are gaps within the chromatid strands and breaks in chromosomes. Chromosome abnormalities, such as gaps, translocations, and deletions were observed after whole-body irradiation, by high-resolution scanning electron microscopy (SEM) in canine chromosomes, which are morphologically similar to human chromosomes (21). Radiation energy is the main cause of such lesions, which can develop as a consequence of radiation-induced loss of genome stability in the surviving cells (22).

Most of the important biological effects of IR are a direct consequence of the rejoining of two DSBs. When a DSB occurs, the next cellular division may develop permanent genetic damage, in the form of chromosomal aberrations. In most of these DSBs, the broken pieces reunite with the original formation, and there may be a point mutation (the result of damage to one point on one gene) at the site of the original break. In some breaks, however, the fragments do not reunite, resulting in an aberration. Conversion of DNA damage into chromosome damage in response to cell-cycle regulation and specifically chromatin condensation after irradiation is a very important feature of chromosome dynamics, which can be the actual determining factor in the induction of chromosomal breaks, especially at low doses (23). The interchange of the fragments among broken chromosomes and the production of abnormal ones is another important consequence of chromosomal breakage, and cells with such aberrant chromosomes usually have impaired reproductive capacity, which causes serious problems in the next mitosis (cellular division). Many of the chromosomal abnormalities are lethal, as the cell fails to complete mitosis either the next time it divides or within the next few divisions, and there are other abnormalities, which allow the cell to divide, but they may contribute to the production of cancerous cells many cell generations later.

CLUSTERED DNA DAMAGES

Clustered DNA damages—two or more closely spaced lesions (single- or double-stranded breaks, abasic sites, oxidized bases) on opposing strands that present a remarkable complexity—are regarded as critical lesions, capable of producing mutagenic or even lethal effects in a cell (Figure 18.1). The concept of clustered DNA damage was first conceived to account for the elevated rates of lethality observed in irradiated cells (24,25). Interestingly, even low doses (0.1 to 1 Gy) of high-LET IR seem to be able to cause clustered damages in human cells (26,27). The origins of these multiple damage areas in the DNA are traced to a single radiation track, which is accompanied by numerous energy depositions rather than several track events (28,29).

The complexity of clustered DNA lesions is regarded as an inhibiting factor during error correction. It has been hypothesized that IR-induced lesions formed within these DNA damage sites would be less readily repaired than isolated ones (30). Moreover, clustered DNA lesions seem to have an unfolding catastrophic presence, since experimental evidence has shown that additional DSBs are formed during processing of γ-radiation-induced DNA clustered damage sites (31). As a consequence, studying the possible endpoints of the processes related to clusters is of vast importance. Indeed, Singleton et al. showed significant chromosomal aberrations in normal human cells post α-particle irradiation and subsequent cluster formation, denoting the possible induction of genomic instability and carcinogenesis (32). Recent studies by various groups seem to agree on the importance of unrepaired clustered DNA lesions, especially when it comes to chromosomal instability (33,34).

SIGNIFICANT CONTRIBUTION OF INFLAMMATION IN THE CHRONIC LATE EFFECTS OF RADIATION

Radiation treatment for cancer has been seen unequivocally as a double-edged sword; on one hand, there is therapeutic gain to some extent, since the rapidly proliferating cancer cells are targeted. On the other hand, normal tissue is inevitably irradiated as well. This is mainly the reason a fractionated regime is often preferred, considering that normal cells will have a certain amount of time to recover, so they may be

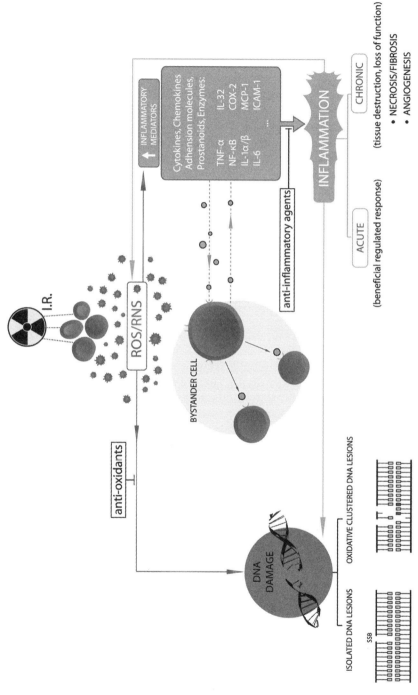

FIGURE 18.1 Clustered damaged DNA areas are regarded as capable of producing mutagenic or even lethal effects in a cell. (*See insert for color representation of the figure.*)

sparcd. Even though this notion is generally correct, radiation side effects do appear in normal tissue and can be characterized as acute or chronic.

This response diversity is best illustrated in the following cases. Research has shown a brain, liver, and intestine cell response to x-rays only 3 hours post-irradiation in murine models, using nuclear factor-κB (NF-κB) as a sensor (35). Apart from its well-documented role in orchestrating the immune response, NF-κB is also implicated in inflammation and oncogenic processes (36). Furthermore, a 67-year-old woman was reported to develop radiation-induced lung fibrosis 10 years after radiation treatment completion (37). Bearing this information in mind, along with other case reports pointing in the same direction (38,39), it is quite safe to assume that radiation-induced injury and inflammation in particular have taken the form of a clinical disease (Table 18.1 and Figure 18.1).

The modulation of a healthy natural response such as inflammation and its development to a chronic unresolved reaction is possible through a sustained, dys-regulated acute inflammation that is no longer self-limiting and remains without treatment. Yet there is another form of chronic inflammation that may evolve with time behind the veil, undetected, with no apparent symptoms. Such a persistent response, regardless of its origin, has been found long ago that is closely linked to cancer induction (40). When it comes to IR, the process can be seen as a vicious circle: radiation is used to treat cancer but can cause chronic inflammation as well; chronic inflammation is undeniably a prelude to cancer precursor sites (damaged, mutated, and unstable cells), and therefore the problematic situation is preserved or even recreated with a secondary tumor induction (41–43). Macrophages and neu-trophils seem to play an important role in chronic inflammation. Macrophages secrete transforming growth factor-β (TGF-β), which in turn attracts neutrophils (44). Both cell types are strategically involved in the inflammatory response and initially not only are they innocuous to the organism, but their action is beyond doubt beneficial when, for instance, wound-healing processes are initiated. Nevertheless, they do have the potential to induce secondary damage to normal tissue if they remain activated for a long period of time (45).

The prolonged presence of inflammatory cells in an organ or tissue can have side effects that vary in histological morphology, severity, and persistence, depending on the area irradiated. For example, radiation-induced inflammation in the lung results in pneumonitis during the acute phase. With time, pneumonitis is replaced by radiation fibrosis and noticeable vascular abnormalities (46). Nevertheless, fibrosis does not necessarily emerge as a result of pneumonitis in a cause-and-effect manner, and its origins are not easily traced. Experiments on fibrosis-susceptible and fibrosis-resistant mice have shown a possible connection between fibrosis and macrophage/lymphocyte activation by specific chemokines (47). As regards the deleterious effects of fibrosis, they can be epitomized by the replacement of healthy tissue with its dysfunctional fibrous counterpart, which has detrimental consequences on tissue flexibility and normal function (48). Although fibrosis was considered irreversible, researchers were able to reduce it when treating patients with pentoxifylline and tocopherol combined (37,49).

Both acute and chronic inflammation can occur in the brain when irradiated for tumors. Acute brain injury usually shows regression or is easily treated, yet this

TABLE 18.1   Overview of Key Radiation-Induced Inflammatory Responses

Radiation type	System	Organ examined	Type of inflammatory response	Ref.
Whole-body ^{12}C, ^{56}Fe-ion irradiation (1–3 Gy)	C57BL/6 mice	Brain	↑ CCR2 (9 months)	(53)
^{137}Cs γ-radiation (10 Gy)	Immortalized murine microglial cells (BV-2)	—	↑ ROS production (1 h) ↑ NF-κB (30 min) ↑ COX-2 (7 h) ↑ MCP-1 (7 h) ↑ IL-1β (24 h) ↑ TNF-α (24 h)	(54)
Abdominal γ-irradiation (10 Gy)	C57BL/6 mice	Intestine	ICAM-1 (24 h)	(58)
γ-Radiation (0–6 Gy)	Human normal lung fibroblasts (HFL-1, MRC-5)	—	↑ COX-2 NF-κB activation ↑ 5-LOX activity (LTB$_4$ production)	(64)
Brain/microglial culture γ-irradiation (5–35 Gy)	C3H/HeN mice	Brain	↑ COX-2 (4 h) TNF-α (4 h) IL-1β (4 h) ↑ ICAM-1 (4 h) MMP-9 (4 h) IL-6 (4 h) MIP-2 (4 h) MCP-1 (4 h) iNOS (24 h)	(83)
Whole-brain γ-irradiation (10 Gy)	F344 × BN rats	Brain	↑ TNF-α (24 h) ↑ IL-1β (24 h) ↑ MCP-1 (24 h)	(84)
Midbrain ^{137}Cs γ-irradiation (25 Gy)	C3Hf/Sed/Kam mice	Brain	↑ TNF-α (2 h) ↑ IL-1α (2 h) ↑ IL-1β (2 h) ↑ ICAM-1	(87)
^{137}Cs γ-radiation (2–10 Gy)	BV-2 cells	—	↑ IL-1β (24 h) ↑ TNF-α (1 h) ↑ COX-2 (7 h) ↑ AP-1 (1 h) ↑NF-κB (1 h) Intracellular ROS generation	(88)
Ionizing radiation (0–6 Gy)	Human umbilical vein endothelial cells (HUVECs)	—	↑ IL-32 (48 h) ↑ COX-2 (1 h) NF-κB activation ↑ cPLA2 ↑ LPC	(97)

(continued)

TABLE 18.1 *(Continued)*

Radiation type	System	Organ examined	Type of inflammatory response	Ref.
Lower lung or whole-lung γ-radiation (10 Gy)	SD rats and C57WT mice	Lung	↑ TNF-α (7 h–20 weeks) production of ROS/RNOS[a]	(98)
Thoracic ^{137}Cs γ-irradiation (5, 12.5 Gy)	C57BL/6 mice	Lung	↑ IL-1α (26 weeks) ↑ TGF-β (8 weeks)	(99)
^{60}Co γ-radiation (2 Gy)	Astrocytes or microglial cells from C3Hf/Sed/Ka mice brains	—	↑ TNF-α (8 h)	(101)

[a]ROS(RNOS) could be produced by the inflammatory cells but possibly also by leakage from damaged mitochondria and/or activation of NAD oxidases or iNOS inside cells (98).

does not apply to chronic radiation-induced brain injury (50). Continued recruitment of microglia, the macrophages of the brain and central nervous system (CNS), seem to hold an important role in the activation of inflammatory mediators (51) and therefore can result to neuroinflammation. Experimental data acquired from chronic neuroinflammation rat model showed susceptibility to cognitive dysfunction (52), suggesting induction of dementia as well. In addition, Rola et al. used heavy ions in whole-body irradiation of rodents and found radiation-induced inflammation unresolved after three months, in the subgranular zone of the hippocampus, with signs of deterioration (53). Moving toward the notion that persistent inflammation is the source of post-irradiation problems and should be inhibited if possible, researchers have found that activation of peroxisome proliferator-activated receptor δ (PPAR-δ), which is a transcription factor expressed primarily in the CNS, can interrupt the radiation-induced pro-inflammatory response in microglia (54).

The intestine is one of the most vulnerable parts of the body when treated with IR. This particular area has been attracting attention for more than 40 years (38). Fibrosis and vascular damage are the principal late effects induced by radiation in the intestine (55). The injury is initiated by the recruitment of leucocytes in the area and the subsequent NF-κB activation (56). NF-κB plays a significant role in both acute and chronic intestinal inflammation, although its action could be characterized as contradictory, depending on the case. During the acute inflammation stage, NF-κB seems to have a beneficial impact, whereas inhibition of the factor, if possible, is the course of action in the case of chronic radiation-induced inflammation (57). Fibrosis and oxidative stress treatment with Cu/Zn-superoxide dismutase (SOD1) (58) and initiation of hyperbaric oxygen therapy to cope with vascular injury (39) have already been used in experiments and show very promising results.

The impact and severity of radiation-induced late effects is associated with many factors. Predisposition in the form of preexisting disease, infection, or inflammatory state has already overburdened the body when irradiation for tumor elimination takes place and the immune system is needed de novo. In the lung, for example, consolidation that resulted from infection is thought to have contributed to the absorbed dose and led to fatal radiation-induced pneumonitis in several cases (59). Another predisposing factor that can modify the risk of radiation injury is genetic susceptibility. Giotopoulos et al. conducted a statistical analysis in a group of 167 breast cancer patients and showed an increased risk for developing radiation-induced fibrosis, as well as other late effects, in patients homozygous for the TGF-β1 variant allele (60). Certainly, genome dependency of chronic inflammatory signaling has been studied *in vitro* in irradiated bone marrow cell cultures (61) and in tumor necrosis factor (TNF)-related apoptosis-inducing ligand (TRAIL)-deficient animals *in vivo* (62).

Radiation-induced chronic inflammation is obviously a side effect of vast importance that requires treatment as soon as possible. Situations difficult to resolve, such as fibrosis and neoangiogenesis, can be avoided this way. The concept of halting inflammation by inhibiting inflammatory mediators (especially after irradiation) seems quite appealing, and it is not uncommon for a lot of researchers to suggest this idea as a possible solution (63,64). Anti-inflammatory agents, combined with conventional methods such as radiation and chemotherapy, seem to present a promising approach. In addition, modification of diet habits with supplementation such as dietary flaxseed has been shown to have a protective and anti-inflammatory action (65).

INFLAMMATION AND RADIATION-INDUCED NONTARGETED EFFECTS

Since IR was introduced in diagnosis and therapy, a theory emerged, supporting the seemingly obvious; the side effects of radiation were caused by direct damage to the DNA. Yet a plethora of data inconsistent with this notion started to appear and gradually formed a conceptual shift in our understanding. As it seems, damage sustained by cytoplasm and not necessarily the nucleus itself during α-particle irradiation *in vitro* can result in mutagenicity. In fact, this type of cell irradiation potentially poses a greater threat than direct nucleus traversal, since a particle hit causes implications but not survival issues (66). In addition, the importance and radiation vulnerability of organelles situated in the cytoplasm is shown by studies in silico as well (67) and the results show good agreement with experiment. However, a comparison between *in vitro* and *in vivo* approaches should be made since the outcome may vary greatly. Data showing less damage accumulated in cells *in vivo* than *in vitro* (68) could be explained by divergence of the defense mechanism(s) in and out of the organism (45).

Nontargeted effects are classified into two main groups. Nontargeted effects can actually appear in cells that are not irradiated directly but are descendants of irradiated ancestors many cell divisions ago. In this case, a nontargeted effect appearance is regarded as radiation-induced genomic instability, since the progeny of the irradiated cells is likely to carry mutations and a considerable accumulation of chromosomal abnormalities. Naturally, nontargeted effects are also present when unirradiated cells show a variety of responses consistent with the irradiation pattern. This is typical for cells in the vicinity of directly irradiated ones, and there is an underlying signal-transferring pathway that makes communication between cells possible (61,68–70). This type of effect is referred to as a *radiation-induced bystander effect*. This type of effect has particular interest when studying the impact of low-dose high-LET radiation, notably in space travel and domestic radon gas emission (10,71).

Apparently, radiation-induced genomic instability is interwoven with inflammation. Lorimore et al. have studied mice susceptible to radiation-induced myeloid leukemia (CBA/Ca) and resistant mice as well (C57BL/6). An *in vivo* γ-irradiated bone marrow–conditioned medium was used to support the growth of normal unirradiated murine cells. Apart from an evident reliance upon predisposing genetic factors, at least when studying specific endpoints, the results showed an elevated expression of TNF-α and the presence of oxidative stress, as a result of the proinflammatory activation of macrophages. In other words, macrophages can induce chromosomal instability as a result of an inflammatory response relying greatly on genotype (70). In a recent study, the same group of researchers exhibited the implication of cyclooxygenase-2 (COX-2) in the process (61), therefore linking pro-inflammatory cytokine signaling with delayed effects of IR and, notably, genomic instability. This concept is supported by another *in vivo* experiment by Mukherjee et al., where the same murine models were used, along with an anti-inflammatory drug introduced in the mice diet. Treated mice showed a quite limited expression of chromosomal instability (72), thus sustaining the strong relation between genomic instability and radiation-induced inflammatory mechanisms. Certainly, recruitment of inflammatory mediators other than the ones studied is possible (61,73), considering that one cytokine can trigger the activation of others, and the result takes the form of an avalanche (Figure 18.1).

As regards radiation-induced bystander effects, Rastogi et al. used the aforementioned CBA/Ca and C57BL/6 models to indicate the involvement of COX-2 pathway in the bystander response. Interestingly, radiation-induced bystander effects appeared in both genotypes (74), in contrast to chromosomal instability, which was expressed only by CBA/Ca cells (70). An *in vitro* study by Zhou et al. also manifested the role of COX-2 in the bystander response, by irradiating normal human lung fibroblasts with α-particles. The causality between COX-2 pathway and bystander response was tested by inserting a special COX-2 inhibitor. Indeed, the COX-2 inactivation caused a dramatic decrease in the bystander effect (75). Shao et al. conducted an experiment using a high-LET particle microbeam to irradiate human fibroblast cells. The researchers observed formation of micronuclei, small nuclei created because of failure to integrate chromosomes or parts of them in the nucleus of the daughter cell. A combination of ROS scavengers and intercellular communication inhibitors resulted in an almost catholic restrain of the bystander effect (71).

Radiation-induced genomic instability and bystander effects seem to have a close connection, so that it is not always possible to distinguish one from another. Furthermore, the mechanisms characterizing these two nontargeted effects apparently coincide and they are linked in a cause-and-effect manner, while it is sometimes unclear which one actually causes the other (45).

RADIATION-INDUCED UP-REGULATION OF INFLAMMATORY MEDIATORS

The development and progression of radiation-induced late effects are caused, in part, by an acute and chronic oxidative stress (76). Chronic inflammation and organ dysfunction come as a result of radiation-induced increases in ROSs and RNSs, while an acute inflammatory response is observed, involving increased production of inflammatory mediators, such as cytokines, interleukins (ILs), and chemokines. The associated chronic overproduction of cytokines and growth factors may result in fibrosis and/or necrosis (77), while there is evidence that inflammation is also correlated with cancer development or progression (2,78). Radiation-induced inflammation and the resulting production of ROSs or RNSs are considered to play a significant role in normal tissue response to radiation damage (79,80). As a result, the dose of radiation that can be delivered safely to cancer patients with solid tumors has to be limited, in order to reduce radiation-induced late normal tissue injury.

ROSs or RNSs can be produced by inflammatory cells, but also by leakage from damaged mitochondria and by activation of NAD oxidases or iNOSs inside the cells (Figure 18.1 and Table 18.1). Increased production of ROSs, which results in oxidation of DNA and activation of pro-inflammatory mediators, has been observed *in vitro* and *in vivo*. Specifically, expression of pro-inflammatory cytokines (TNF-α, TGF-β, IL-6, IL-1α/-1β) and intercellular adhesion molecule-1 (ICAM-1), was maintained persistently from 12 h to 1 week in an IR-irradiated mouse model (64), while endothelial ICAM-1 expression was determined in mice 24 h after they received sham irradiation or abdominal irradiation with 10 Gy (58). ICAM-1 has a central role in the process of intestinal leukocyte recruitment in response to abdominal irradiation. Drugs that can either languish the amplitude of the inflammatory response or reduce chronic oxidative stress of late radiation-induced normal tissue injury, such as PPAR-γ agonists and antioxidants/antioxidant enzymes, have been tested in animal models, and their effectiveness in reducing radiation-induced late effects indicates that radiation-induced chronic injury may be prevented and/or treated (79).

CENTRAL NERVOUS SYSTEM

The acute-phase molecular response of the brain to IR involves expression of inflammatory gene products. Brain responses are often considered to be slow to develop after irradiation, due to the appearance of severe clinical or radiological manifestations of brain necrosis, ranging from months to years (81). However, less severe acute symptoms may occur with large fractions soon after brain radiotherapy. Partial or whole-brain irradiation is often required for both primary and metastatic brain cancer treatment, but the radiation dose that can be delivered to the tumor safely is limited by the risk of toxicity to the adjacent normal brain tissue. Radiation-induced brain injury can lead to severe deficits several months to years after irradiation (82). Although it was considered that the brain was not responsive to inflammation, *in vivo* studies suggest that there is a significant increase in pro-inflammatory mediators within hours of irradiating the rodent brain (53,83–85).

The exact mechanisms underlying radiation-induced late effects may still be unclear, but oxidative stress and inflammation are thought to play a significant role in radiation-induced normal tissue injury. Brain irradiation induces regionally specific alterations in cytokine gene and protein expression, and IL-1α/-1β and TNF-α have been reported to stimulate proliferation of astrocytes and microglial cells (86,87). They induce expression of ICAM-1 on astrocytes, microglial cells, and brain microvascular endothelial cells, which may promote autoimmune responses. Stimulation of astrocytes and microglial cells by a variety of signals leads to synthesis of IL-1α/-1β, IL-6, and TNF-α, which may cause cellular cascades and clinical symptoms. The results of an experiment in mice show that TNF-α and IL-1α/-1β were induced consistently in the brain following irradiation, in addition to an acute inflammatory response. Their production may be part of an adaptive survival response by tissues to oxidative stress that, in the case of IR, could have both beneficial and detrimental effects (85).

Microglia, the immune cells of the brain, are key mediators of neuroinflammation. They can become activated by a variety of stimuli and release a host of pro-inflammatory cytokines, chemokines, and ROSs/RNSs. Irradiating microglia cells *in vitro* leads to an increase in pro-inflammatory mediators such as the cytokines TNF-α, IL-1β, and IL-6 and the chemokines monocyte chemoattractant protein (MCP-1) and ICAM-1 (83,88). The exact role of these pro-inflammatory mediators in radiation-induced brain injury is not clearly defined; however, increased levels of cytokines have been connected with some pathological conditions of the CNS, such as Alzheimer's and Parkinson's disease (89,90), while TNF-α has been identified in brain lesions in multiple sclerosis (91). Irradiating BV-2 cells, a murine microglial cell line, with a single 2- to 10-Gy dose, resulted in increased ROS production 1 h post-irradiation, while an increase in COX-2 and MCP-1 levels was observed 7 h post-irradiation and an increase in IL-1β and TNF-α message levels was determined 24 h post-irradiation (54).

Other reports have also demonstrated that irradiating BV-2 cells leads to an increase in COX-2 protein and IL-1β and TNF-α message levels, while an increase in the DNA-binding activity of redox-regulated pro-inflammatory transcription factors activator protein-1 (AP-1) and NF-κB was also observed (88,92). In addition to that, the results of studies on COX-2 expression in the mouse brain following IR *in vivo* and in murine glial cell cultures *in vitro* indicated that COX-2 is significantly induced in brain as well as in astrocyte and microglial cultures by radiation injury (64,83). When cells are exposed to a number of stresses, such as IR, induction of compensatory activations of multiple intracellular signaling pathways takes place. Pathways activated by IR or other type of stress include downstream of death receptors such as the transcription factor NF-κB. (93).

Irradiation increases significantly activation of AP-1, NF-κB, and cAMP response element-binding protein (CREB). Whole-brain irradiation induces regionally specific pro-inflammatory environments through activation of AP-1, NF-κB, and CREB and overexpression of TNF-α, IL-1β, and MCP-1 in rat brain and may contribute to unique pathways for radiation-induced impairments in tissue function (84). MCP-1 and its receptor CCR2 play significant roles in neuroinflammation (94). CCR2 is a chemokine receptor and is involved in monocyte infiltration in inflammatory diseases, such as rheumatoid arthritis, as well as in the inflammatory response against tumors. There is evidence of an apparent persistent inflammatory response after particle irradiation, when CCR2 was found to be present in C57BL/6 mouse brain 9 months after exposure to high-LET irradiation (53).

After irradiation, increased intracellular ROS generation leads to activation of the stress-activated kinases PKCα, MEK1/2, and ERK1/2 and the pro-inflammatory transcription factors NF-κB and AP-1. Activation of these transcription factors increases the expression of some pro-inflammatory mediators, such as COX-2, MCP-1, IL-1β, and TNF-α, which contribute to the inflammatory phenotype of microglia (54). Experimental data suggest that in the BV-2 cells the radiation-induced increase in TNF-α and COX-2 is regulated primarily by the AP-1 pathway, while IL-1β expression is controlled by both AP-1 and NF-κB pathways. Overall, these results highlight the importance of these two transcription factors in mediating the microglial proinflammatory response following irradiation (88).

CIRCULATORY SYSTEM

IL-32 is a cytokine that is detected in human hematopoietic cells. It induces the expression of various cytokines, such as TNF-α and IL-8, in monocytes and activates the NF-κB in lymphocytes. IL-32 expression is regulated by inflammatory cytokines, so it has been considered to play a role in inflammatory/autoimmune diseases, such as rheumatoid arthritis (95,96). IR treatment has been shown to increase IL-32 expression dramatically in vascular endothelial cells through multiple pathways, and thus neutralization of IL-32 function may indicate ways to control IR-induced inflammation. In addition, irradiation of human umbilical vein endothelial cells led to an induction of IL-32 expression through NF-κB activation, through induction of cPLA2 and LPC, and induction of COX-2 and subsequent conversion of arachidonic acid to prostacyclin (97).

LUNG

Radiation-induced inflammation and production of ROSs affect normal tissue response significantly. Continuing ROS production in the tissue contributes to DNA damage that have been observed at late times in lung fibroblasts, over prolonged periods (98). In normal lung fibroblasts, IR-induced ROSs activated the PI3K/Akt, NF-κB pathway, leading to COX-2-mediated prostaglandin E2 (PGE2) production, and the FLAP/5-LOX pathway, resulting in LTB4 (a leukotriene involved in inflammation) production (64). *In vivo* studies have shown early overproduction of both pro-inflammatory (IL-1α, TNF-α) and pro-fibrogenic (TGF-β1) cytokines during radiotherapy, and have associated this sustained production of cytokines with the development of late and acute expression of radiation-induced lung damage, such as radiation pneumonitis (RP) (99). TNF-α is an inducible multifunctional cytokine known to be involved in many aspects of immune regulation and inflammation. Radiation has been shown to increase TNF-α message and/or protein production by tumor cells (100,101), and the intracellular production of TNF-α in the irradiated cells may cause lethality that is more significant than killing caused by the direct effects of IR alone. Cells that produce TNF-α after irradiation may also enhance radiation damage in adjacent cells.

Experimental results suggest that TNF-α plays an important role in the induction of radiation-induced pneumonitis. Whole-lung irradiation of Sprague–Dawley rats suggested a cyclic pattern of increased TNF-α expression, following irradiation and showed an early peak at about 7 h, a second at 5 to 12 days, and a third at 4 to 20 weeks. The inflammatory response seemed to be organ-wide, as this pattern of TNF-α expression was the same in the upper region of the lung and in the lower irradiated region (98). IL-6 is a multifunctional cytokine that acts as both pro-inflammatory and anti-inflammatory cytokine and is closely correlated with the regulation of immunological and inflammatory response. During a three-dimensional conformal radiation therapy in patients with non-small-cell lung cancer, IL-6 levels remained significantly higher in RP patients compared to those in non-RP patients, with a significant peak at 2 weeks of treatment (102).

CONCLUDING REMARKS

Regarding exposure of humans to IR, there is sufficient evidence for the carcinogenicity of x- and γ-radiation. According to a recent IARC monograph (103), these types of radiation can undoubtedly cause cancer of the salivary gland, esophagus, stomach, colon, lung, bone, basal cell of the skin, female breast, kidney, urinary bladder, brain and CNS, thyroid, and leukemia (excluding chronic lymphocytic leukemia). It must also be mentioned that positive associations have been observed between x- and γ-radiation and cancer of the rectum, liver, pancreas, ovary, and in prostate, and non-Hodgkin's lymphoma and multiple myeloma. In-utero exposure to x- and γ-radiation causes cancer. In this chapter we focused on and discussed more extensively evidence on the involvement and suggested role(s) of inflammation in acute and, especially, chronic radiation-induced cell and tissue injury. The induction of chronic oxidative stress, DNA damage, and inflammation is accepted as a possible major contributor of the malignant transformation and initiation of secondary primary cancers in radiotherapy-treated cancer patients. Future directions should involve *in vivo* or *in vitro* studies that explicitly explore mechanistic pathways of the late radiation effects, especially the development of secondary primary cancers and the mechanisms that may modulate the toxicity of the exposure like for example the inflammation status and the immune system response after treatment. In this case, several modulating factors, such as natural antioxidant or nonsteroidal anti-inflammatory drug supplementation, may prove beneficial to patients augmenting the life-saving effects of radiation.

Acknowledgments

This study has been partially supported by A. Georgakilas's funding from EU grant MC-CIG-303514, COST ACTION CM1202, "Biomimetic Radical Chemistry" and the European Union (European Social Fund–ESF) and Greek national funds through the Operational Program "Education and Lifelong Learning" of the National Strategic Reference Framework (NSRF)—Research Funding Program: THALES grant MIS 379346. We would like to thank Dr. Mike E. Robbins and designer Andreas A. Karoutzos for their invaluable help.

REFERENCES

1. Scott BR, Walker DM, Tesfaigzi Y, Schollnberger H, Walker V. Mechanistic basis for nonlinear dose–response relationships for low-dose radiation-induced stochastic effects. Nonlinearity Biol Toxicol Med 2003;1:93–122.
2. Coussens LM, Werb Z. Inflammation and cancer. Nature 2002;420:860–867.
3. Kryston TB, Georgiev A, Georgakilas AG. Role of oxidative stress and DNA damage in human carcinogenesis. Mutat Res 2011;711:193–201.
4. Nowsheen S, Aziz K, Kryston TB, Ferguson NF, Georgakilas AG. The interplay between inflammation and oxidative stress in carcinogenesis. Curr Mol Med 2012;12:672–680.
5. Prise KM, O'Sullivan JM. Radiation-induced bystander signalling in cancer therapy. Nature Reviews: Cancer 2009;9:351–360.
6. Lett JT. Damage to cellular DNA from particulate radiations, the efficacy of its processing and the radiosensitivity of mammalian cells: emphasis on DNA double strand breaks and chromatin breaks. Radiat Environ Biophys 1992;31:257–277.
7. Cramp WA, Yatvin MB, Harms-Ringdahl M. Recent developments in the radiobiology of cellular membranes. Acta Oncol 1994;33:945–952.
8. Daniniak N, Tann BJ. Utility of biological membranes as indicators for radiation exposure: alterations in membrane structure and function over time. Stem Cells 1995;13:142–152.
9. Mayneord WV, Clarke RH. Carcinogenesis and radiation risk: a biomathematical reconnaissance. Br J Radiol 1975;12:1–112.
10. Hada M, Georgakilas AG. Formation of clustered DNA damage after high-LET irradiation: a review. J Radiat Res 2008;49:203–210.
11. Dutta-Choudhuri R, Roy H. Effects of ionizing radiations on tissue mast cells. Int J Appl Radiat Isot 1959;4:268–269.
12. Harms-Ringdahl M, Nicotera P, Radford IR. Radiation induced apoptosis. Mutat Res 1996;366:171–179.
13. Hendry JH, West CM. Apoptosis and mitotic cell death: their relative contributions to normal-tissue and tumour radiation response. Int J Radiat Biol 1997;71:709–719.
14. Preston RJ. Radiation effects. Ann ICRP 2012;41:4–11.
15. International Commission on Radiation Protection. Low dose extrapolation of radiation-related cancer risk. Ann ICRP 2005;35.
16. International Commission on Radiation Protection. The 2007 recommendations of the International Commission on Radiological Protection. Ann ICRP 2007;37.
17. Rubio CA, Jalnäs M. Dose-time-dependent histological changes following irradiation of the small intestine of rats. Digest Dis Sci 1996;41:392–401.
18. Dörr W, Hendry JH. Consequential late effects in normal tissues. Radiother Oncol 2001;61:223–231.
19. Somosy Z. Radiation response of cell organelles. Micron 2000;31:165–181.
20. Georgakilas AG. Processing of DNA damage clusters in human cells: current status of knowledge. Mol Biosyst 2008;4:30–35.
21. Niiro GK, Seed TM. SEM of canine chromosomes: normal structure and the effects of whole-body irradiation. Scanning Microsc 1988;2:1593–1598.
22. Morgan WF, Day JP, Kaplan MI, McGhee EM, Limoli CL. Genomic instability induced by ionizing radiation. Rad Res 1996;146:247–258.
23. Terzoudi GI, Pantelias GE. Conversion of DNA damage into chromosome damage in response to cell cycle regulation of chromatin condensation after irradiation. Mutagenesis 1997;12:271–276.

24. Ward JF. Some biochemical consequences of the spatial distribution of ionizing radiation produced free radicals. Radiat Res 1981;86:185–195.

25. Ward JF. Biochemistry of DNA lesions. Radiat Res 1985;104:S103–S111.

26. Sutherland B, Bennett PV, Sidorkina O, Laval J. DNA damage clusters induced by ionizing radiation in isolated DNA and in human cells. Proc Natl Acad Sci USA 2000;97:103–108.

27. Goodhead DT, Thacker J, Cox R. Weiss Lecture: Effects of radiations of different qualities on cells: molecular mechanisms of damage and repair. Int J Radiat Biol 1993;63:543–556.

28. Lomax ME, Gulston MK, O'Neill P. Chemical aspects of clustered DNA damage Induction by ionising radiation. Radiat Prot Dosimetry 2002;99:63–68.

29. Sutherland BM, Bennett PV, Sidorkina O, Laval J. Clustered damages and total lesions induced in DNA by ionizing radiation: oxidized bases and strand breaks. Biochemistry 2000;39:8026–8031.

30. Nikjoo H, O'Neill P, Wilson EW, Goodhead D. Computational approach for determing the spectrum of DNA damage induced by ionizing radiation. Radiat Res 2001;156:577–583.

31. Gulston M, de Lara C, Jenner T, Davis E, O'Neill P. Processing of clustered DNA damage generates additional double-strand breaks in mammalian cells post-irradiation. Nucleic Acids Res 2004;32:1602–1609.

32. Singleton BK, Griffin CS, Thacker J. Clustered DNA damage leads to complex genetic changes in irradiated human cells. Cancer Res 2002;62:6263–6269.

33. Hair JM, Terzoudi GI, Hatzi VI, Lehockey KA, Srivastava D, Wang W, et al. BRCA1 role in the mitigation of radiotoxicity and chromosomal instability through repair of clustered DNA lesions. Chem Biol Interact 2010;188: 350–358.

34. Asaithamby A, Hu B, Chen DJ. Unrepaired clustered DNA lesions induce chromosome breakage in human cells. Proc Natl Acad Sci USA 2011;108:8293–8298.

35. Chang CT, Lin H, Ho TY, Li CC, Lo HY, Wu SL, et al. Comprehensive assessment of host responses to ionizing radiation by nuclear factor-kappaB bioluminescence imaging-guided transcriptomic analysis. PLoS One 2011;6:e23682.

36. Dolcet X, Llobet D, Pallares J, Matias-Guiu X. NF-κB in development and progression of human cancer. Virchows Arch 2005;446:475–482.

37. Delanian S. Striking regression of radiation-induced fibrosis by a combination of pentoxifylline and tocopherol. Br J Radiol 1998;71:892–894.

38. DeCosse JJ, Rhodes RS, Wentz WB, Reagan JW, Dworken HJ, Holden WD. The natural history and management of radiation induced injury of the gastrointestinal tract. Ann Surg 1969;170: 369–384.

39. Warren DC, Feehan P, Slade JB, Cianci PE. Chronic radiation proctitis treated with hyperbaric oxygen. Undersea Hyper Med 1997;24:181–184.

40. Balkwill F, Mantovani A. Inflammation and cancer: Back to Virchow? Lancet 2001;357:539–545.

41. Schonfeld S, Bhatti P, Brown E, Linet M, Simon S, Weinstock R, et al. Polymorphisms in oxidative stress and inflammation pathway genes, low-dose ionizing radiation, and the risk of breast cancer among US radiologic technologists. Cancer Causes Control 2010;21:1857–1866.

42. Multhoff G, Radons J. Radiation, inflammation, and immune responses in cancer. Front Oncol 2012;2:58.

43. Gallet P, Phulpin B, Merlin JL, Leroux A, Bravetti P, Mecellem H, et al. Long-term alterations of cytokines and growth factors expression in irradiated tissues and relation with histological severity scoring. PLoS One 2011;6:e29399.

44. Nathan C. Points of control in inflammation. Nature 2002;420:846–852.

45. Wright EG, Coates PJ. Untargeted effects of ionizing radiation: implications for radiation pathology. Mutat Res 2006;597:119–132.

46. Stone HB, Coleman CN, Anscher MS, McBride WH. Effects of radiation on normal tissue: consequences and mechanisms. Lancet Oncol 2003;4:529–536.

47. Johnston CJ, Williams JP, Okunieff P, Finkelstein JN. Radiation-induced pulmonary fibrosis: examination of chemokine and chemokine receptor families. Radiat Res 2002;157:256–265.

48. Ferrero-Miliani L, Nielsen OH, Andersen PS, Girardin SE. Chronic inflammation: importance of NOD2 and NALP3 in interleukin-1beta generation. Clin Exp Immunol 2007;147:227–235.

49. Delanian S, Porcher R, Balla-Mekias S, Lefaix JL. Randomized, placebo-controlled trial of combined pentoxifylline and tocopherol for regression of superficial radiation-induced fibrosis. J Clin Oncol 2003;21:2545–2550.

50. Greene-Schloesser D, Robbins ME, Peiffer AM, Shaw EG, Wheeler KT, Chan MD. Radiation-induced brain injury: a review. Front Oncol 2012;2:73.

51. Kim SU, de Vellis J. Microglia in health and disease. J Neurosci Res 2005;81:302–313.

52. Rosi S, Ramirez-Amaya V, Hauss-Wegrzyniak B, Wenk GL. Chronic brain inflammation leads to a decline in hippocampal NMDA-R1 receptors. J Neuroinflamm 2004;1:12.

53. Rola R, Sarkissian V, Obenaus A, Nelson GA, Otsuka S, Limoli CL, et al. High-LET radiation induces inflammation and persistent changes in markers of hippocampal neurogenesis. Radiat Res 2005;164:556–560.

54. Schnegg CI, Kooshki M, Hsu F-C, Sui G, Robbins ME. PPARδ prevents radiation-induced proinflammatory responses in microglia via transrepression of NF-κB and inhibition of the PKCα/MEK1/2/ERK1/2/AP-1 pathway. Free Radic Biol Med 2012;52:1734–1743.

55. Stone HB, Coleman NC, Anscher MS, McBride WH. Effects of radiation on normal tissue: consequences and mechanisms. Lancet Oncol 2003;4:529–536.

56. Molla M, Panes J. Radiation-induced intestinal inflammation. World J Gastroenterol 2007;13:3043–3046.

57. Eckmann L, Nebelsiek T, Fingerle AA, Dann SM, Mages J, Lang R, et al. Opposing functions of IKKbeta during acute and chronic intestinal inflammation. Proc Natl Acad Sci USA 2008;105:15058–15063.

58. Mollà M, Gironella M, Salas A, Closa D, Biete A, Gimeno M, et al. Protective effect of superoxide dismutase in radiation-induced intestinal inflammation. Int J Radiat Oncol Biol Phys 2005;61:1159–1166.

59. Movsas B, Raffin TA, Epstein AH, Link CJ, Jr. Pulmonary radiation injury. Chest 1997;111:1061–1076.

60. Giotopoulos G, Symonds RP, Foweraker K, Griffin M, Peat I, Osman A, et al. The late radiotherapy normal tissue injury phenotypes of telangiectasia, fibrosis and atrophy in breast cancer patients have distinct genotype-dependent causes. Br J Cancer 2007;96:1001–1007.

61. Lorimore SA, Mukherjee D, Robinson JI, Chrystal JA, Wright EG. Long-lived inflammatory signaling in irradiated bone marrow is genome dependent. Cancer Res 2011;71:6485–6491.

62. Finnberg N, Klein-Szanto AJ, El-Deiry WS. TRAIL-R deficiency in mice promotes susceptibility to chronic inflammation and tumorigenesis. J Clin Invest 2008;118:111–123.

63. Deorukhkar A, Krishnan S. Targeting inflammatory pathways for tumor radiosensitization. Biochem Pharmacol 2010;80:1904–1914.

64. Yang HJ, Youn H, Seong KM, Yun YJ, Kim W, Kim YH, et al. Psoralidin, a dual inhibitor of COX-2 and 5-LOX, regulates ionizing radiation (IR)-induced pulmonary inflammation. Biochem Pharmacol 2011;82:524–534.

65. Lee JC, Krochak R, Blouin A, Kanterakis S, Chatterjee S, Arguiri E, et al. Dietary flaxseed prevents radiation-induced oxidative lung damage, inflammation and fibrosis in a mouse model of thoracic radiation injury. Cancer Biol Ther 2009;8:47–53.

66. Wu LJ, Randers-Pehrson G, Xu A, Waldren CA, Geard CR, Yu Z, et al. Targeted cytoplasmic irradiation with alpha particles induces mutations in mammalian cells. Proc Natl Acad Sci USA 1999;96:4959–4964.

67. Kuncic Z, Byrne HL, McNamara AL, Guatelli S, Domanova W, Incerti S. In silico nanodosimetry: new insights into nontargeted biological responses to radiation. Comput Math Methods Med 2012;2012:147252.

68. Watson GE, Pocock DA, Papworth D, Lorimore SA, Wright EG. In vivo chromosomal instability and transmissible aberrations in the progeny of haemopoietic stem cells induced by high- and low-LET radiations. Int J Radiat Biol 2001;77:409–417.

69. Hei TK, Zhou H, Chai Y, Ponnaiya B, Ivanov VN. Radiation induced non-targeted response: mechanism and potential clinical implications. Curr Mol Pharmacol 2011;4:69–105.

70. Lorimore SA, Chrystal JA, Robinson JI, Coates PJ, Wright EG. Chromosomal instability in unirradiated hemaopoietic cells induced by macrophages exposed in vivo to ionizing radiation. Cancer Res 2008;68:8122–8126.

71. Shao C, Furusawa Y, Kobayashi Y, Funayama T, Wada S. Bystander effect induced by counted high-LET particles in confluent human fibroblasts: a mechanistic study. FASEB J 2003;17:1422–1427.

72. Mukherjee D, Coates PJ, Lorimore SA, Wright EG. The *in vivo* expression of radiation-induced chromosomal instability has an inflammatory mechanism. Radiat Res 2012;177:18–24.

73. Fiers W. Tumor necrosis factor: characterization at the molecular, cellular and *in vivo* level. FEBS Lett 1991;285:199–212.

74. Rastogi S, Coates PJ, Lorimore SA, Wright EG. Bystander-type effects mediated by long-lived inflammatory signaling in irradiated bone marrow. Radiat Res 2012;177:244–250.

75. Zhou H, Ivanov VN, Gillespie J, Geard CR, Amundson SA, Brenner DJ, et al. Mechanism of radiation-induced bystander effect: role of the cyclooxygenase-2 signaling pathway. Proc Natl Acad Sci USA 2005;102:14641–14646.

76. Zhao W, Diz DI, Robbins ME. Oxidative damage pathways in relation to normal tissue injury. Br J Radiol 2007;80:S23–S31.

77. Denham JW, Hauer-Jensen M. The radiotherapeutic injury: a complex "wound." Radiother Oncol 2002;63:129–145.

78. Mantovani A, Allavena P, Sica A, Balkwill F. Cancer-related inflammation. Nature 2008;454:436–444.

79. Zhao W, Robbins MEC. Inflammation and chronic oxidative stress in radiation-induced late normal tissue injury: therapeutic implications. Curr Med Chem 2009;16:130–143.

80. Robbins ME, Zhao W. Chronic oxidative stress and radiation-induced late normal tissue injury: a review. Int J Radiat Biol 2004;80:251–259.

81. Fike JR, Cann CE, Turowski K, Higgins RJ, Chan AS, Phillips TL, et al. Radiation dose response of normal brain. Int J Radiat Oncol Biol Phys 1988;14:63–70.

82. Johannesen TB, Lien HH, Hole KH, Lote KH. Radiological and clinical assessment of long-term brain tumour survivors after radiotherapy. Radiother Oncol 2003;69:169–176.

83. Kyrkanides S, Moore AH, Olschowka JA, Daeschner JC, Williams JP, Hansen JT, et al. Cyclooxygenase-2 modulates brain inflammation-related gene expression in central nervous system radiation injury. Mol Brain Res 2002;104:159–169.

84. Lee WH, Sonntag WE, Mitschelen M, Yan H, Lee YW. Irradiation induces regionally specific alterations in pro-inflammatory environments in rat brain. Int J Radiat Biol 2010;86:132–144.

85. Hong JH, Chiang CS, Campbell IL, Sun JR, Withers HR, McBride WH. Induction of acute phase gene expression by brain irradiation. Int J Radiat Oncol Biol Phys 1995;33:619–626.

86. Merrill JE. Effects of interleukin-1 and tumor necrosis factor-alpha on astrocytes, microglia, oligodendrocytes, and glial precursors *in vitro*. Dev Neurosci 1991;13:130–137.

87. Selmaj KW, Farooq M, Norton WT, Raine CS, Brosnan CF. Proliferation of astrocytes *in vitro* in response to cytokines: a primary role for tumor necrosis factor. J Immunol 1990;144:129–135.

88. Ramanan S, Kooshki M, Zhao W, Hsu FC, Robbins ME. PPARα ligands inhibit radiation-induced microglial inflammatory responses by negatively regulating NF-κβ and AP-1 pathways. Free Radic Biol Med 2008;45:1695–1704.

89. Griffin WS, Sheng JG, Royston MC, Gentleman SM, McKenzie JE, Graham DI, et al. Glial-neuronal interactions in Alzheimer's disease: the potential role of a 'cytokine cycle' in disease progression. Brain Pathol 1998;8:65–72.

90. Mogi M, Harada M, Narabayashi H, Inagaki H, Minami M, Nagatsu T. Interleukin (IL)-1 beta, IL-2, IL-4, IL-6 and transforming growth factor-alpha levels are elevated in ventricular cerebrospinal fluid in juvenile parkinsonism and Parkinson's disease. Neurosci Lett 1996;211:13–16.

91. Hofman FM, Hinton DR, Johnson K, Merrill JE. Tumor necrosis factor identified in multiple sclerosis brain. J Exp Med 1989;170:607–612.

92. Dong X, Dong J, Zhang R, Fan L, Liu L, Wu G. Anti-inflammatory effects of tanshinone IIA on radiation-induced microglia BV-2 cells inflammatory response. Cancer Biother Radiopharm 2009;24:681–687.

93. Dent P, Yacoub A, Contessa J, Caron R, Amorino G, Valerie K, et al. Stress and radiation-induced activation of multiple intracellular signaling pathways. Radiat Res 2003;159:283–300.

94. Gerard C, Rollins BJ. Chemokines and disease. Nat Immunol 2001;2:108–115.

95. Kim SH, Han SY, Azam T. Interleukin-32: a cytokine and inducer of TNFalpha. Immunity 2005;22:131–142.

96. Joosten LA, Netea MG, Kim SH. IL-32, a proinflammatory cytokine in rheumatoid arthritis. Proc Natl Acad Sci USA 2006;103:3298–3303.

97. Kobayashi H, Yazlovitskaya EM, Lin PC. Interleukin-32 positively regulates radiation-induced vascular inflammation. Int J Radiat Oncol Biol Phys 2009;74:1573–1579.

98. Hill RP, Zaidi A, Mahmood J, Jelveh S. Investigations into the role of inflammation in normal tissue response to irradiation. Radiother Oncol 2011;101:73–79.

99. Rubin P, Johnston CJ, Williams JP. A perpetual cascade of cytokines postirradiation leads to pulmonary fibrosis. Int J Radiat Oncol Biol Phys 1995;33:99–109.

100. Hallahan DE, Spriggs DR, Beckett MA, Kufe DW, Weichselbaum RR. Increased tumor necrosis factor alpha mRNA after cellular exposure to ionizing radiation. Proc Natl Acad Sci USA 1989;86:10104–10107.

101. Chiang CS, McBride WH. Radiation enhances tumor necrosis factor alpha production by murine brain cells. Brain Res 1991;566:265–269.

102. Arpin D, Perol D, Blay JY, Falchero L, Claude L, Vuillermoz-Blas S, et al. Early variations of circulating interleukin-6 and interleukin-10 levels during thoracic radiotherapy are predictive for radiation pneumonitis. Clin Oncol 2005;23:8748–8756.

103. International Agency for Research on Cancer. Working Group on the Evaluation of Carcinogenic Risks to Humans X- and γ-Radiation. IARC Monogr Eval Carcinog Risks Hum 2012;100D:103–229.

CHAPTER 19

PHOTOCARCINOGENESIS AND INFLAMMATION

Chikako Nishigori

There is ample evidence demonstrating that solar ultraviolet (UV) light induces human skin cancers. First, epidemiological studies have demonstrated a negative correlation between the latitude of residence and incidence and mortality rates of skin cancer in homogeneous populations (1,2). Second, skin cancer can be produced in mice by UV irradiation; the action spectrum of photocarcinogenesis falls into UVB (280 to 320 nm) (3,4). Third, patients with genetic disorders that lead to deficiencies in repairing UV-induced DNA damage are prone to develop cancers in sun-exposed areas of the skin (5).

Accumulation of DNA lesions in several genes, such as oncogenes and tumor suppressor genes, plays a crucial role in carcinogenesis. UV irradiation causes DNA lesions, which are thought to be responsible for sunlight-induced skin cancers. Indeed, even in actinic keratosis, precancerous lesions of squamous cell caricinoma (SCC), genetic alterations can be observed (6).

UVB-induced skin cancer development (photocarcinogenesis) is a complex multistage process that involves initiation, promotion, and progression. Each of these processes is mediated by various cellular, biochemical, and molecular changes. The pyrimidine photoproducts that result from direct DNA damage induced by UV are an important cause of skin cancer. DNA damage induces mutations that lead to the up- or down-regulation of signal transduction pathways, cell-cycle dysregulation, and depletion of antioxidant defenses (6,7). In addition to these changes, UVB-induced immunosuppression also plays an important role in photocarcinogenesis (8,9).

Chronic inflammation is induced by various environmental factors, including sunlight. Epidemiological and experimental studies have indicated that chronic inflammation is closely involved in human carcinogenesis. Epidemiological studies have shown that acute intense exposure to UV rays is also a major risk factor for developing skin cancer, implying that intermittent severe inflammation caused by acute sunburn is an important factor in the development of skin cancers.

Cancer and Inflammation Mechanisms: Chemical, Biological, and Clinical Aspects, First Edition.
Edited by Yusuke Hiraku, Shosuke Kawanishi, and Hiroshi Ohshima.
© 2014 John Wiley & Sons, Inc. Published 2014 by John Wiley & Sons, Inc.

DNA DAMAGE CAUSED BY SUNLIGHT

UV wavelengths from the Sun that reach the surface of the Earth are UVA (320 to 400 mm) and part of UVB (280 to 320 nm) because UV wavelengths below 300 nm are absorbed by the ozone layer. Irradiation with UVA or UVB results in the production of reactive oxygen species (ROSs). However, because the absorbance spectrum of DNA extends well into the UVB range, DNA absorbs more UVB photon energy directly than UVA. Therefore, UVB acts on DNA by exciting the nucleobases directly, resulting in the formation of dimeric photoproducts at dipyrimidine sites in an oxygen-independent manner. Comparatively, UVA and visible light tend to participate in the formation of ROSs in the presence of photosensitizers (10,11) irrespective of the involvement of oxygen. UVB not only produces pyrimidine dimers and (6-4) photoproducts (6-4 PPs) by direct excitation, but also produces ROSs in the presence of photosensitizers, which cause oxidative DNA damage. Previous studies have suggested that dipyrimidine photoproducts are the major molecule among UV-induced DNA lesions involved in cytotoxicity and mutagenesis (12).

On the other hand, ROSs cause various biological effects via the redox-signaling pathway (13) and oxidative DNA damage and thus play a role in carcinogenesis (14). Several types of modified DNA are produced by ROSs. Among them, 8-hydroxydeoxyguanosine (8-OHdG) has been established as a sensitive marker of oxidative DNA damage. It can be generated in the presence of a photosensitizer either electron transfer or by hydroxy radicals and singlet oxygen ($^1O_2$) with oxygen molecules. 8-OHdG is generated in fibroblasts derived from human skin in response to exposure to artificial light sources such as sunlamps (UVB) or monochromatic radiation ranging from 312 nm up to near visible (434 nm). Kvam and Tyrrell analyzed the spectrum for the yield of oxidative damage in confluent, nongrowing, primary skin fibroblasts, and they determined that UVA and near-visible radiation cause almost all of the guanine oxidation (10). Ito et al. reported that the photodynamic action of riboflavin caused the formation of 8-OHdG in double-stranded DNA and that no enhancing effect by D_2O was observed (15), suggesting that $^1O_2$ molecules were not involved. The estimated ratio of 8-OHdG yield to total guanine loss indicated that the photoexcited riboflavin specifically induced 8-OHdG formation at the guanine residues located 5′ to another guanine by electron transfer. The guanine base in genomic DNA is highly susceptible to oxidative stress, as it possesses the lowest oxidation potential of all the bases. Besaratinia reported that UVA irradiation at a dose of 18 J/cm^2 produced significant levels of 8-oxo-2′-deoxyguanosine (8-oxodG) in DNA, and UVA-induced mutations were characterized by statistically significant increases in G $\rightarrow$ T transversions and small tandem base deletions (16). Drobetsky et al. used a mutation detection system involving adenine phosphoribosyltransferase genes to demonstrate that T$\rightarrow$G transversions, which are a generally rare class of mutation, are induced at high frequencies (up to 50%) in UVA-exposed cells (17). These studies indicate that UVA, which constitutes more than 90% of the UV light that reaches the surface of the Earth, is primarily responsible for ROS generation.

On the other hand, recent work has shown that cyclobutane pyrimidine dimers are produced at higher yield than 8-oxoguanine (8-oxoG) after exposure to UVA in rodent and human skin cells (18–20). We found that the diuretic agent hydrochloroth-iazide (HCT) significantly enhanced the production of thymine dimers (T< >Ts) in a wide range of UVA wavelengths, and this enhancement occurred independent of the presence of oxygen (21). This finding indicates that excited HCT molecules function as UVA-absorbing chromophores that transfer energy to adjacent pyrimidines, resulting in the formation of T< >Ts.

UV-INDUCED DNA LESIONS AND MUTATIONS IN SKIN CANCERS

The maximum action spectrum of UV-induced carcinogenesis in animal experiments is within the UVB range, with the peak at 293 nm (22). Direct absorption of UV energy by DNA leads to the formation of pyrimidine photoproducts. Formation of dipyrimidine photoproducts can lead to UV signature mutations, which have been demonstrated in UV irradiated bacteria and mammalian shuttle vectors (23,24). UV signature mutations are known to be a *transition* type mutation, is defined as a change from one pyrimidine (cytosine or thymine) or purine (guanine or adenine) to the other at dipyrimidine sequence site. Many studies have analyzed the molecular changes observed in skin cancers. Ziegler et al. reported that *p53* mutations were detected in nonmelanoma skin cancers from Caucasian people in much higher frequencies, approximately 50 to 90%, than in internal malignancies (6). These mutations are predominated by the C:G → T:A transition type at dipyrimidine sites: namely, the UV signature mutations. Several other reports have demonstrated that the types of mutations that are not considered to be caused by dipyrimidine photoproducts are frequently observed in human skin cancers developed in sun-exposed body sites (25) and UVB-induced murine skin cancer (26), indicating that oxidative DNA damage also plays a role to some extent in cancer development, although dipyrimidine photoproducts remain to be the major players in photocarcinogenesis.

Patients with xeroderma pigmentosum (XP) are deficient in nucleotide excision repair, and their greatly increased susceptibility to skin cancer in sun-exposed areas highlights the role of UV-induced DNA damage in carcinogenesis. XP-A, the most severe type of XP, induces severe photosensitivity, which is obvious within 12 months of age, and the onset of both nonmelanoma skin cancers (NMSCs) and malignant melanomas (MMs) as early as at 10 years of age (21). Most *p53* mutations in skin cancers from XP patients are UV-signature mutations CC → TT base substitutions (27–30). The spectra were similar to those detected in NMSCs of sun-exposed areas of the body in non-XP patients. Comparatively, the frequency of *ras* gene mutations was far less in the same samples. We could detect only one mutation at codon 61 in the Ki-*ras* gene, and the other four mutations found occurred at codons 6, 7, 15, and 16 in the Ha-*ras* gene. The latter mutations are not common and possess a low transforming ability, which implies that DNA damage caused by sunlight rarely affects the crucial sites of the *ras* gene that would lead to the activation of *ras* (31).

A comparative analysis of the mutation frequency and pattern of the *p53* gene in 49 NMSCs, which consisted of 16 basal cell carcinomas (BCCs) and nine squamous cell carcinomas (SCCs) from sun-exposed areas, and 16 BCCs and eight SCCs from covered areas, indicated that the mutation frequency in the *p53* gene in these two groups were similar at 48% and 46%, respectively. However, the mutational type significantly differed between these two groups; UV signature mutations were detected in 50% (6/12) of tumors that had developed in sun-exposed areas but in less than 20% (2/11) of tumors from covered areas, and this difference was statistically significant (9). This finding indicates that the *p53* mutations detected in the NMSCs that developed in sun-exposed areas are UV signature mutations and that *p53* mutations play a significant role in photocarcinogenesis. These results also suggest that UV light plays a role in inducing mutations in tumor-related genes.

UV-INDUCED INFLAMMATION AND SKIN CANCERS

The most obvious and well-known effect of sun exposure is sunburn (erythema). In humans, slight erythema starts several hours after sun exposure on the site and becomes edematous. Usually, the peak of erythema and swelling is reached 12 to 24 h after exposure and subsides within a few days with pigmentation and, occasionally, fine scales observed. The sunburn reaction depends on several factors, including UV dose, UV wavelength, and skin type. A histological study of moderate doses of UVB irradiation to human skin revealed that dermal neutrophils appeared immediately after irradiation and increased in number from the onset of erythema (0.5 to 4 h) to a maximum at 14 h, dwindling thereafter up to 48 h, thereby indicating that infiltration of neutrophils is a key step in the initiation of acute skin reactions mediated by UVB (32). Following UV absorption by cellular molecules, photochemical reactions occur, and these reactions are responsible for initiating sunburn reactions.

Devary et al. (33) suggested that the UV response is initiated at or near the plasma membrane, and the response may be elicited by oxidative stress caused by UV radiation. Indeed, there is plenty of evidence that various antioxidants attenuate erythema or edema mediated by UVB (34–36). Furthermore, low levels of oxidants can modify cell signaling via redox regulation, and these signal modifications have functional consequences (37). UV irradiation triggers sequential molecular responses, thereby activating cell surface growth factors and pro-inflammatory cytokine receptors. Subsequently, these receptors induce various protein kinase signal transduction pathways that up-regulate expression and functional activation of the nuclear transcription factor AP-1 and thus cause diverse gene expression (38).

Mice treated with the specific p38 mitogen-activated protein kinase (p38 MAPK) inhibitors SB202190 and SB242235 showed marked inhibition of acute sunburn inflammation and apoptosis (39,40). Furthermore, Giri et al. showed that ROS generation followed by the development of cutaneous inflammation after photosensitization acts as a tumor promoter (41). Moore et al. reported that tumor necrosis factor-α (TNF-α)-deficient mice are resistant to skin carcinogenesis (42). They also reported that the frequency of DNA adduct formation and *c-Ha-ras* mutations was the same in wild-type and TNF-$\alpha^{-/-}$ epidermis after 7,12-dimethylbenz[*a*]anthracene (DMBA) treatment. These results imply that this process is independent of initiation.

The pro-inflammatory cytokine TNF-α is a critical mediator of tumor promotion, which acts via the protein kinase C (PKC)-α and AP-1-dependent pathways (43). Persistent oxidative stress in cancer (44) may also cause activation of transcription factors and protooncogenes such as *c-fos* and *c-jun* (45) as well as genetic instability. Such stress may also contribute to the maintenance of malignant characteristics of neoplasms.

We have shown previously that large amounts of 8-OHdG as well as 4-hydroxy-2-nonenal-modified proteins and 3-nitro-L-tyrosine are formed in mice skin after 10 minimal erythema dose (MED) exposures three times per week for 2 weeks (total dose: 100.8 kJ/m^2). In addition, significantly increased 3-nitro-L-tyrosine modifications were detected in the 10 MED-irradiated groups (46); 3-nitro-L-tyrosine suggested the presence of nitric oxide (NO)-mediated oxidative damage in chronic inflammation. Peroxynitrite (ONOO$^-$) is generated by the reaction of NO with superoxide ($O_2^{\bullet-}$), which is released by infiltrating neutrophils and macrophages (47). Immunohistochemical studies of chronically UV-irradiated skin specimens revealed that 8-OHdG is produced not only in the nuclei of epidermal keratinocytes but also in the inflammatory cells. *In vivo* repair kinetics of 8-OHdG demonstrate a slower time to recover to the basal level than for pyrimidine dimers, which suggests an important role of the persistent inflammatory response in oxidative stress (46).

DNA DAMAGE AND INFLAMMATION

We demonstrated previously the high susceptibility to UV-induced skin cancer in *Ogg1* knockout mice, which are deficient in glycosylase/AP lyase, which plays a role in the removal of 8-oxoG from DNA (14). However, unexpectedly, the *p53* mutations found in skin cancers produced in *Ogg1* knockout mice were mainly CC → TT UV signature mutations, not GC → TA transversions, which generally identify the presence of 8-oxoG (48). Furthermore, the gene profiling studies of *Ogg1* knockout mouse skin revealed that the inflammatory response pathway-related genes such as *cxcl1* and *il-6* were the most affected of the up-regulated genes in *Ogg1* knockout mice after UVB exposure (49). In fact, we observed much more neutrophil infiltrate after UV exposure in *Ogg1* knockout mice, indicating that DNA damage (i.e., accumulation of 8-oxoG) caused by the deficiency of the Ogg1 protein leads to mutations. Moreover, the presence of DNA damage itself causes the up-regulation of inflammatory response genes, and it is largely involved in photocarcinogenesis in *Ogg1* knockout mice (14). ROSs caused by direct UV exposure and associated inflammatory reactions caused by UV exposure induce accumulation of oxidative DNA damage, such 8-oxoG in the skin, implying that oxidative DNA damage induced by sunlight plays an important role in the development of skin cancers.

Rodier et al. reported that irreparably large radiation doses induce DNA double-stranded breaks and increase IL-6 secretion (50). We compared *IL-6* levels after 250 mJ/cm^2 of UVB irradiation in wild-type and *Ogg1* knockout mice by real-time PCR (RT-PCR) and found that *Ogg1* knockout mice express significantly higher levels of *IL-6* at 3 and 24 h after UVB exposure than those of wild-type mice. We also performed RT-PCR for inflammatory soluble factors such as *Mmp2* and *Tnfα* but found no significant difference between the two *Ogg1* genotypes. Our data indicate that IL-1β and IL-6 are the most important candidate cytokines to induce inflammatory conditions associated with 8-oxoG accumulation following UVB exposure.

DIFFERENCE BETWEEN CHEMICALLY INDUCED CANCER AND PHOTOCARCINOGENESIS

Phosphoinositide-specific phospholipase C (PLC), which is an effector of *Ras*, hydrolyzes the membrane phospholipid phosphatidylinositol 4,5-bisphosphate to generate diacylglycerol, which, in turn, activates PKC. *PLCε* knockout ($PLCε^{-/-}$) mice exhibit marked resistance to tumor formation in a two-stage skin chemical carcinogenesis protocol using DMBA as an initiator and the phorbol ester 12-*O*-tetradecanoylphorbol-13-acetate (TPA) as a promoter (51). *PLCε* knockout mice also showed reduced inflammatory responses after TPA treatment, suggesting that PLCε positively regulates skin chemical carcinogenesis and the inflammatory response. However, surprisingly, chronic exposure to UVB induced both a greater number and variety of types of skin tumors in *PLCε* $^{-/-}$ mice than in wild-type (*PLCε* $^{+/+}$) and *PLCε* heterozygous (*PLCε* $^{+/-}$) mice (52). Pathophysiological examination revealed that the frequency of UVB-induced cell death in the skin cells of *PLCε* $^{-/-}$ mice was reduced compared with the $PLCε^{+/+}$ mice, thereby indicating a crucial role of PLCε in regulating UVB-induced cell death as well as the inflammatory response. It also indicates the importance of UV-induced cell death in the prevention of photocarcinogenesis.

Low levels of oxidants can modify cell-signaling proteins, and these modifications have functional consequences. Gopalakrishna and Jaken demonstrated that various antioxidant preventive agents inhibit PKC-dependent cellular responses and speculated that PKC is a logical candidate for redox modification by oxidants and antioxidants that may determine their cancer-promoting and anticancer activities, respectively (53).

MELANOMA AND UV-INDUCED INFLAMMATION

Recently, much attention has been paid to the occurrence of melanoma formation and UV-induced inflammation. It is generally accepted that continual inflammation increases the risk of cancer, which is consistent with the finding that excessive intense intermittent sun exposure is one of the most important risk factors for melanoma. Noonan et al. reported that a single dose of burning UV radiation to neonates, but not to adults, of the hepatocyte growth factor/scatter factor transgenic mice is necessary and sufficient to induce melanoma with a high incidence (54). This also provides experimental grounds for the epidemiological evidence that childhood sunburn is a major risk factor for the development of melanomas (55,56). Zaidi et al. reported an unanticipated role for interferon-γ (IFN-γ) in promoting melanocytic cell survival/immunoevasion (57). They reported that the expression profile of neonatal skin melanocytes isolated following a melanomagenic UVB dose bear distinct, persistent interferon response signature, including genes associated with immuno-evasion. UVB-induced melanocyte activation, characterized by aberrant growth and migration, was abolished by antibody-mediated systemic blockade of IFN-γ, but not type I interferons. Admixed recruited skin macrophages, which produce IFN-γ, enhanced transplanted melanoma growth; IFN-γ blockade abolished macrophage-enhanced melanoma growth and survival. These findings explain partly why inter-mittent burning doses of sun exposure is the highest risk factor for melanomagenesis.

Whether UVA or UVB is more dangerous in the development of melanoma is still controversial. Using a fish melanoma model, Setlow et al. reported that UVA is responsible for melanoma development (58), whereas Mitchell et al. reported that UVB, but not UVA, causes melanoma by using a *Xiphophorus* hybrid melanoma model (59). Noonan et al. also reported that UVB, but not UVA, induced melanoma in a mice melanoma model (60). Both NMSCs and melanomas are induced by solar light, but there are some differences. Melanocytes show resistance to UVB-induced apoptosis. Consequently, melanocytes survive after acute sunburn with large amounts of DNA damage, whereas keratinocytes tend to be apoptotic after large doses of UV exposure. These differences explain the discrepancy between chemical carcinogen-esis and UV carcinogenesis observed using the PLCε knockout mice. Indeed, the most frequently developed body site of the superficial-spreading type of melanoma, which is the most common type of MM among the Caucasian population, is the trunk or thigh, which is often exposed intensively to the sun when sunbathing. Often, we see sunburn freckles in patients with superficial spreading type MM (SSM). Eumelanin protects the skin from UV, whereas pheomelanin acts as a photosensi-tizer and causes oxidative DNA damage in melanocytes. Eumelanin predominates in mouse melanocytes, whereas various compositions of eumelanin and pheomelanin are observed in human melanocytes, depending on the skin type. Therefore, careful consideration is necessary when comparing the results in mice with those in humans (61).

THE ROLE OF UVA IN PHOTOCARCINOGENESIS

Until recently, studies on UV-induced carcinogenesis have focused on UVB-induced DNA mutations. However, recently the role of UVA in photocarcinogenesis has received much greater attention. One reason for this is the involvement of UVA-induced ROSs in the development of melanomas. The second reason is that many studies revealed that UVA generates not only ROSs but also cyclobutane pyrimidine dimers *in vivo*. It was demonstrated recently that cyclobutane pyrimidine dimers are produced at higher yields than 8-oxoG after UVA exposure in rodent and human skin cells (18,19); therefore, there was a paradigm shift regarding the theory of photocarcinogenesis. A recent series of studies demonstrating that UVA produces T<>Ts at much higher levels than other types of pyrimidine dimers and does not form 6-4 PPs (20) explains the mutation spectrum of the relevant genes in cancers of sun-exposed areas of the skin in humans (7,27,30). Van Kranen et al. reported that UVA-induced murine skin cancers are less frequent than UVB-induced murine skin cancers and that the incidence of *p53* alterations in UVA-induced tumors is lower than that in UVB-induced tumors (62). These facts imply that UVB is more mutagenic and carcinogenic than UVA. An *in vivo* study analyzing the action spectrum for photocarcinogenesis in a mice model revealed that UVA is partly responsible for photocarcinogenesis (22). UVA seems to exert cancer-promoting properties besides DNA damage to cause DNA mutations. Many of the carcinogenic functions of UVA have been attributed to its production of ROSs and subsequent induction of the inflammatory signaling pathway. However, it should be noted that UVA is also capable of inducing T<>Ts in human skin.

DEVELOPMENT OF SKIN CANCER IN GENETIC DISORDERS

In mammalian cells, DNA lesions are repaired by several repair systems, such as nucleotide excision repair (NER), base excision repair, and mismatch repair. If the DNA lesions remain at the replication fork, homologous recombination or the translesional DNA synthesis system works. However, if the DNA damage is too extensive to repair via these systems, it may cause mutations. Bulky DNA lesions causing DNA conformational change, such as UV-induced pyrimidine photoproducts, are removed by NER. Deficiency in NER leads to three human genetic diseases: XP, cockayne syndrome (CS), and trichothiodystrophy. Of these, only patients with XP (subdivided into XP-A through XP-G based on the responsible genes) are predisposed to skin cancers. The majority of *p53* mutations in skin cancers from XP patients are CC → TT base substitutions, or UV-signature mutations (27–29). Some investigators have suggested that repair insufficiencies lead to oxidative stress in addition to the accumulation of pyrimidine dimers in XP cells (63,64).

Nevoid basal cell carcinoma syndrome (NBCCS) is an autosomal dominant disease characterized by tumorigenesis, such as multiple BCCs, odontogenic keratocysts, and developmental abnormalities (65). Germline mutations in *PTCH* have been identified in patients with NBCCS (66). In addition, the site-specific distribution of the BCCs in NBCCS patients indicates that UV exposure plays a role in the development of BCCs in NBCCS (67). In the general population, approximately 10% of BCCs occur on the trunk versus 35% among all NBCCS cases. The site-specific distribution of BCCs in Japanese NBCCS indicates a similar tendency (68), suggesting that NBCCS patients are sensitive to intermittent intense sun exposure because people receive occasional intermittent intense sun exposure at this anatomical site. Skin fibroblasts from patients with NBCCS were hypersensitive to death by UVB but not UVC radiation, in comparison with skin fibroblasts from normal individuals; these patients also did not have impaired pyrimidine dimer–removal systems (68,69). A previous report shows that the removal of 8-OHdG in fibroblasts after UVB exposure is slightly impaired in NBCCS cells compared to that in normal cells, which implies that oxidative stress plays a role in the development of BCCs and other tumors in NBCCS (68). Aszterbaum et al. reported that UV and ionizing radiation enhance the growth of BCCs and trichoblastomas in patched heterozygous knockout mice (70).

UV-INDUCED IMMUNE SUPPRESSION

The immune system plays an important role in UV carcinogenesis by contributing to host resistance against skin cancer development. However, UV radiation circumvents immune surveillance against skin cancers by modulating the immune response in a way that favors tumor development. Skin cancers induced by UV radiation are highly antigenic and can therefore be recognized by the immune system. This is apparent with UV-induced murine skin cancers, many of which are immunologically rejected upon transplantation into normal syngenic mice (71). The exceptionally high incidence of skin cancer, particularly squamous cell carcinoma, on the sun-exposed skin of immunosuppressed renal transplant patients (72), suggests that UV-induced human skin cancers are also highly antigenic. However, despite the potential for immunological control, skin cancer occurs with high frequency in susceptible, sun-exposed populations. Previous studies, mainly those using murine models, have provided an explanation for this paradox by demonstrating that UV radiation not only transforms cells by inducing mutations but also interferes with host immunity against the developing skin tumors. These studies demonstrate that UV irradiation of the skin produces both local immune suppression that inhibits immune functions within the irradiated skin and systemic immune suppression against antigens introduced at a critical time after exposure to UV radiation. Modulation of immune responses initiated at nonirradiated sites is now known to involve soluble mediators,

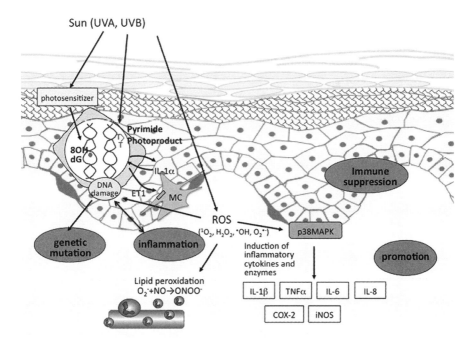

FIGURE 19.1   Schematic summary of photocarcinogenesis detailed in the text. (*See insert for color representation of the figure.*)

at least some of which are produced by UV-irradiated keratinocytes (73). Different mediators appear to be involved in the modulation of delayed-type hypersensitivity (DTH) and the contact hypersensitivity (CHS) response. In particular, IL-10 antibodies prevent UV-induced suppression of DTH responses, whereas TNF-α antibodies inhibit UV-induced suppression of CHS (73). UV-induced pyrimidine dimers have been suggested to trigger cytokine production by epidermal cells, which is involved in UV-induced specific immune suppression (74). Furthermore, lipid peroxidation by UV-induced prostaglandin formation may also play a role in immune suppression (75,76).

In summary, both directly and indirectly, UV induces DNA lesions that cause genetic mutations and trigger inflammation and immune suppression which allow tumor growth. Both UV itself and UV-induced inflammation lead to ROS generation. These ROSs also cause DNA lesions and enhance mutation frequency. Furthermore, lipid peroxidation caused by UV and ROSs is involved in immune suppression (Figure 19.1).

REFERENCES

1. Armstrong BK, Kricker A. The epidemiology of UV induced skin cancer. J Photochem Photobiol B 2001;63:8–18.
2. Hu S, Ma F, Collado-Mesa F, Kirsner RS. UV radiation, latitude, and melanoma in US Hispanics and blacks. Arch Dermatol Res 2004;140:819–824.
3. Das M, Bickers DR, Santella RM, Mukhtar H. Altered patterns of cutaneous xenobiotic metabolism in UVB-induced squamous cell carcinoma in SKH-1 hairless mice. J Invest Dermatol 1985;84:532–536.
4. Nishigori C, Tanaka M, Moriwaki S, Imamura S, Takebe H. Accelerated appearance of skin tumors in hairless mice by repeated UV irradiation with initial intense exposure and characterization of the tumors. Jpn J Cancer Res 1992;83:1172–1178.
5. Nishigori C, Moriwaki S, Takebe H, Tanaka T, Imamura S. Gene alterations and clinical characteristics of xeroderma pigmentosum group A patients in Japan. Arch Dermatol 1994;130:191–197.
6. Ziegler A, Jonason AS, Leffell DJ, Simon JA, Sharma HW, Kimmelman J, et al. Sunburn and p53 in the onset of skin cancer. Nature 1994;372:773–776.
7. Nishigori C, Hattori Y, Toyokuni S. Role of reactive oxygen species in skin carcinogenesis, Antioxid Redox Signal 2004;6:561–570.
8. Loser K, Apelt J, Voskort M, Mohaupt M, Balkow S, Schwarz T, et al. IL-10 controls ultraviolet-induced carcinogenesis in mice. J Immunol 2007;179:365–371.
9. Nishigori C. Cellular aspects of photocarcinogenesis. Photochem Photobiol Sci 2006;5:208–214.
10. Kvam E, Tyrrell RM. Induction of oxidative DNA base damage in human skin cells by UV and near visible radiation. Carcinogenesis 1997;18:2379–2384.
11. Rosen JE, Prahalad AK, Schlüter G, Chen D, Williams GM. Quinolone antibiotic photodynamic production of 8-oxo-7,8-dihydro-2-deoxyguanosine in cultured liver epithelial cells. Photochem Photobiol 1997;65:990–996.
12. Ellison MJ, Childs JD. Pyrimidine dimers induced in *Escherichia coli* DNA by ultraviolet radiation present in sunlight. Photochem Photobiol 1981;34:465–469.
13. Ono R, Masaki T, Dien S, Yu X, Fukunaga A, Yodoi J, Nishigori C. Suppressive effect of recombinant human thioredoxin on ultraviolet light-induced inflammation and apoptosis in murine skin. J Dermatol 2012;39:843–851.
14. Kunisada M, Sakumi K, Tominaga Y, Budiyanto A, Ueda M, Ichihashi M, et al. 8-Oxoguanine formation induced by chronic ultraviolet B exposure makes *ogg1* knockout mice susceptible to skin carcinogenesis. Cancer Res 2005;65:6006–6010.

15. Ito K, Inoue S, Yamamoto K, Kawanishi S. 8-Hydroxydeoxyguanosine formation at the 5' site of 5'-GG-3' sequences in double-stranded DNA by UV radiation with riboflavin. J Biol Chem 1993;268:13221–13227.

16. Besaratinia A, Synold TW, Xi B, Pfeifer GP. G-to-T transversions and small tandem base deletions are the hallmark of mutations induced by ultraviolet A radiation in mammalian cells. Biochemistry 2004;43:8169–8177.

17. Drobetsky EA, Turcotte J, Châteauneuf A. A role for ultraviolet A in solar mutagenesis. Proc Natl Acad Sci USA 1995;92:2350–2354.

18. Douki T, Reynaud-Angelin A, Cadet J, Sage E. Bipyrimidine photoproducts rather than oxidative lesions are the main type of DNA damage involved in the genotoxic effect of solar UVA radiation. Biochemistry 2003;43:9221–9226.

19. Courdavault S, Baudouin C, Charveron M, Favier A, Cadet J, Douki T. Larger yield of cyclobutane dimers than 8-oxo-7,8-dihydroguanine in the DNA of UVA-irradiated human skin cells. Mutat Res 2004;556:135–142.

20. Mouret S, Baudouin C, Charveron M, Favier A, Cadet J, Douki T. Cyclobutane pyrimidine dimers are predominant DNA lesions in whole human skin exposed to UVA radiation. Proc Natl Acad Sci USA 2006;103:13765–13770.

21. Kunisada M, Masaki T, Ono R, Morinaga H, Nakano E, Yogianti F, et al. Hydrochlorothiazide enhances UVA-induced DNA damage. Photochem Photobiol 2013;89:649–654.

22. de Gruijl F. Action spectrum for photocarcinogenesis. Recent Results Cancer Res 1995;139:21–30.

23. Miller J. Mutagenic specificity of ultraviolet light. J Mol Biol 1985;182:45–65.

24. Brash DE, Rudolph JA, Simon JA, Lin A, McKenna GJ, Baden HP, et al. A role for sunlight in skin cancer: UV-induced *p53* mutations in squamous cell carcinoma. Proc Natl Acad Sci USA 1991;88:10124–10128.

25. Pierceall WE, Goldberg LH, Tainsky MA, Mukhopadhyay T, Ananthaswamy HN. *Ras* gene mutation and amplification in human nonmelanoma skin cancers. Mol Carcinog 1991;4:196–202.

26. Nishigori C, Wang S, Miyakoshi J, Sato M, Tsukada T, Yagi T, et al. Mutations in *ras* genes in cells cultured from mouse skin tumors induced by ultraviolet irradiation. Proc Natl Acad Sci USA 1994;91:7189–7193.

27. Dumaz N, Drougard C, Sarasin A, Daya-Grosjean L. Specific UV-induced mutation spectrum in the *p53* gene of skin tumors from DNA-repair-deficient xeroderma pigmentosum patients. Proc Natl Acad Sci USA 1993;90:10529–10533.

28. Sato M, Nishigori C, Zghal M, Yagi T, Takebe H. Ultravioletspecific mutations in *p53* gene in skin tumors in xeroderma pigmentosum patients. Cancer Res 1993;53:2944–2946.

29. Spatz A, Giglia-Mari G, Benhamou S, Sarasin A. Association between DNA repair-deficiency and high level of *p53* mutations in melanoma of xeroderma pigmentosum. Cancer Res 2001;61:2480–2486.

30. Matsumura Y, Sato M, Nishigori C, Zghal M, Yagi T, Imamura S, et al. High prevalence of mutations in the *p53* gene in poorly differentiated squamous cell carcinomas in xeroderma pigmentosum patients. J Invest Dermatol 1995;105:399–401.

31. Sato M, Nishigori C, Lu Y, Zghal M, Yagi T, Takebe H. Far less frequent mutations in *ras* genes than in the *p53* gene in skin tumors of xeroderma pigmentosum patients. Mol Carcinogenesis 1994;11:98–105.

32. Hawk JL, Murphy GM, Holden CA. The presence of neutrophils in human cutaneous ultraviolet-B inflammation. Br J Dermatol 1988;118:27–30.

33. Devary Y, Gottlieb RA, Smeal T, Karin M. The mammalian ultraviolet response is triggered by activation of src tyrosine kinases. Cell 1992;71:1081–1091.

34. Köpcke W, Krutmann J. Protection from sunburn with beta-carotene: a meta-analysis. Photochem Photobiol 2008;84:284–288.

35. Eberlein-König B, Placzek M, Przybilla B. Protective effect against sunburn of combined systemic ascorbic acid (vitamin C) and *d*-alpha-tocopherol (vitamin E). J Am Acad Dermatol 1998;38:45–48.

36. Katiyar SK, Perez A, Mukhtar H. Green tea polyphenol treatment to human skin prevents formation of ultraviolet light B–induced pyrimidine dimers in DNA. Clin Cancer Res 2000;6:3864–3869.

37. Sun Y, Oberley LW. Redox regulation of transcriptional activators. Free Radic Biol Med 1996;21:335–348.

38. Rittié L, Fisher GJ. UV-light-induced signal cascades and skin aging. Ageing Res Rev 2002;1:705–720.

39. Hildesheim J, Rania TA, Fornace AJ. p38 Mitogen-activated protein kinase inhibitor protects the epidermis against the acute damaging effects of ultraviolet irradiation by blocking apoptosis and inflammatory responses. J Invest Dermatol 2004;122:497–502.

40. Kim AL, Labasi JM, Zhu Y, Tang X, McClure K, Gabel CA, et al. Role of p38 MAPK in UVB-induced inflammatory responses in the skin of SKH-1 hairless mice. J Invest Dermatol 2005;124: 1318–1325.

41. Giri U, Sharma SD, Abdulla M, Athar M. Evidence that *in situ* generated reactive oxygen species act as a potent stage I tumor promoter in mouse skin. Biochem Biophys Res Commun 1995;209:698–705.

42. Moore RJ, Owens DM, Stamp G, Arnott C, Burke F, East N, et al. Mice deficient in tumor necrosis factor-alpha are resistant to skin carcinogenesis. Nat Med 1999;5:828–831.

43. Arnott CH, Scott KA, Moore RJ, Hewer A, Phillips DH, Parker P, et al. Tumour necrosis factor-alpha mediates tumour promotion via a PKC alpha- and AP-1-dependent pathway. Oncogene 2002;21:4728–4738.

44. Toyokuni S, Okamoto K, Yodoi J, Hiai H. Persistent oxidative stress in cancer. FEBS Lett 1995;358:1–3.

45. Crawford D, Zbinden I, Amstad P, Cerutti P. Oxidant stress induces the proto-oncogenes c-*fos* and c-*myc* in mouse epidermal cells, Oncogene 1988;3:27–32.

46. Hattori Y, Nishigori C, Tanaka T, Uchida K, Nikaido O, Osawa T, et al. 8-Hydroxy-2′-deoxyguanosine is increased in epidermal cells of hairless mice after chronic ultraviolet B exposure. J Invest Dermatol 1996;107:733–737.

47. Kaur H, Halliwell B. Evidence for nitric oxide–mediated oxidative damage in chronic inflammation: nitrotyrosine in serum and synovial fluid from rheumatoid patients. FEBS Lett 1994;350:9–12.

48. Yogianti F, Kunisada M, Ono R, Sakumi K, Nakabeppu Y, Nishigori C. Skin tumours induced by narrowband UVB have higher frequency of *p53* mutations than tumours induced by broadband UVB independent of *Ogg1* genotype. Mutagenesis 2012;27:637–643.

49. Kunisada M, Yogianti F, Sakumi K, Ono R, Nakabeppu Y, Nishigori C. Increased expression of versican in the inflammatory response to UVB- and reactive oxygen species–induced skin tumorigenesis. Am J Pathol 2011;179:3056–3065.

50. Rodier F, Coppé JP, Patil CK, Hoeijmakers WA, Muñoz DP, Raza SR, et al. Persistent DNA damage signalling triggers senescence-associated inflammatory cytokine secretion. Nat Cell Biol 2009;11:973–979.

51. Bai Y, Edamatsu H, Maeda S, Saito H, Suzuki N, Satoh T, et al. Crucial role of phospholipase Cepsilon in chemical carcinogen-induced skin tumor development. Cancer Res 2004;64:8808–8810.

52. Oka M, Edamatsu H, Kunisada M, Hu L, Takenaka N, Dien S, et al. Enhancement of ultraviolet B–induced skin tumor development in phospholipase Cε-knockout mice is associated with decreased cell death. Carcinogenesis 2010;31:1897–1902.

53. Gopalakrishna R, Jaken S. Protein kinase C signaling and oxidative stress. Free Radic Biol Med 2000;28:1349–1361.

54. Noonan FP, Recio JA, Takayama H, Duray P, Anver MR, Rush WL, et al. Neonatal sunburn and melanoma in mice. Nature 2001;413:271–272.

55. Autier P, Doré JF. Influence of sun exposures during childhood and during adulthood on melanoma risk. EPIMEL and EORTC Melanoma Cooperative Group. European Organisation for Research and Treatment of Cancer. Int J Cancer 1998;77:533–537.

56. Whiteman DC, Whiteman CA, Green AC. Childhood sun exposure as a risk factor for melanoma: a systematic review of epidemiologic studies. Cancer Causes Control 2001;12:69–82.

57. Zaidi MR, Davis S, Noonan FP, Graff-Cherry C, Hawley TS, Walker RL, et al. Interferon-γ links ultraviolet radiation to melanomagenesis in mice. Nature 2011;469:548–553.

58. Setlow RB, Grist E, Thompson K, Woodhead AD. Wavelengths effective in induction of malignant melanoma. Proc Natl Acad Sci USA 1993;90:6666–6670.

59. Mitchell DL, Fernandez AA, Nairn RS, Garcia R, Paniker L, Trono D, et al. Ultraviolet A does not induce melanomas in a *Xiphophorus* hybrid fish model. Proc Natl Acad Sci USA 2010;107:9329–9334.

60. Noonan FP, Zaidi MR, Wolnicka-Glubisz A, Anver MR, Bahn J, Wielgus A, et al. Melanoma induction by ultraviolet A but not ultraviolet B radiation requires melanin pigment. Nat Commun 2012;3:884.

61. Hill HZ, Hill GJ. UVA, pheomelanin and the carcinogenesis of melanoma. Pigment Cell Res 2000;13 Suppl 8:140–144.

62. van Kranen HJ, de Laat A, van de Ven J, Wester PW, de Vries A, Berg RJ, et al. Low incidence of p53 mutations in UVA (365-nm)-induced skin tumors in hairless mice. Cancer Res 1997;57:1238–1240.

63. Reardon JT, Bessho T, Kung HC, Bolton PH, Sancar A. *In vitro* repair of oxidative DNA damage by human nucleotide excision repair system: possible explanation for neurodegeneration in xeroderma pigmentosum patients. Proc Natl Acad Sci USA 1997;94:9463–9468.

64. Satoh MS, Jones CJ, Wood RD, Lindahl T. DNA excision-repair defect of xeroderma pigmentosum prevents removal of a class of oxygen free radical–induced base lesions. Proc Natl Acad Sci USA 1993;90:6335–6339.

65. Gorlin RJ. Nevoid basal-cell carcinoma syndrome. Medicine (Baltim) 1987;66:98–113.

66. Hahn H, Wicking C, Zaphiropoulous PG, Gailani MR, Shanley S, Chidambaram A, et al. Mutations of the human homolog of *Drosophila* patched in the nevoid basal cell carcinoma syndrome. Cell 1996;85:841–851.

67. Goldstein AM, Bale SJ, Peck GL, DiGiovanna JJ. Sun exposure and basal cell carcinomas in the nevoid basal cell carcinoma syndrome. J Am Acad Dermatol 1993;29:34–41.

68. Nishigori C, Arima Y, Matsumura Y, Matsui M, Miyachi Y. Impaired removal of 8-hydroxydeoxyguanosine induced by UVB radiation in naevoid basal cell carcinoma syndrome cells. Br J Dermatol 2005;153 (Suppl 2):52–56.

69. Applegate LA, Goldberg LH, Ley RD, Ananthaswamy HN. Hypersensitivity of skin fibroblasts from basal cell nevus syndrome patients to killing by ultraviolet B but not by ultraviolet C radiation. Cancer Res 1990;50:637–641.

70. Aszterbaum M, Epstein J, Oro A, Douglas V, LeBoit PE, Scott MP, et al. Ultraviolet and ionizing radiation enhance the growth of BCCs and trichoblastomas in patched heterozygous knockout mice. Nat Med 1999;5:1285–1291.

71. Kripke ML. Antigenicity of murine skin tumors induced by ultraviolet light. J Natl Cancer Inst 1974;53:1333–1336.

72. Glover MT, Niranjan N, Kwan JT, Leigh IM. Non-melanoma skin cancer in renal transplant recipients: the extent of the problem and a strategy for management. Br J Plast Surg 1994;47:86–89.

73. Rivas JM, Ullrich SE. The role of IL-4, IL-10, and TNF-α in the immune suppression induced by ultraviolet radiation. J Leukocyte Biol 1994;56:769–775.

74. Nishigori C, Yarosh DB, Ullrich SE, Vink AA, Bucana CD, Roza L, et al. Evidence that DNA damage triggers interleukin 10 cytokine production in UV-irradiated murine keratinocytes. Proc Natl Acad Sci USA 1996;93:10354–10359.

75. Shreedhar V, Giese T, Sung VW, Ullrich SE. A cytokine cascade including prostaglandin E2, IL-4, and IL-10 is responsible for UV-induced systemic immune suppression. J Immunol 1998;160:3783–3789.

76. Halliday GM. Inflammation, gene mutation and photoimmunosuppression in response to UVR-induced oxidative damage contributes to photocarcinogenesis. Mutat Res 2005;571:107–120.

CHAPTER 20

CHEMOPREVENTION OF COLORECTAL CANCER BY ANTI-INFLAMMATORY AGENTS

Michihiro Mutoh, Mami Takahashi, and Keiji Wakabayashi

In recent years, colorectal cancer has increasingly become a major cause of cancer mortality in advanced countries, including Japan. Therefore, elucidation of the mechanisms of colorectal carcinogenesis and the search for chemopreventive agents are important and urgent tasks. Since chronic inflammatory status and associated changes, such as elevation of production of cytokines and growth factors, appear to predispose to cancer development in any site of the body, the metabolic pathways that are switched on under such conditions might be good targets for chemopreventive agents.

In human colorectal cancer tissue, overexpression of enzymes associated with inflammation, such as inducible cyclooxygenase (COX)-2 and inducible nitric oxide synthase (iNOS) have been reported (1,2). Thus, it is suggested that their reaction products, prostaglandin E_2 (PGE_2) and nitric oxide (NO), might contribute to the development of colorectal cancer. To date, several mechanisms involved in colorectal neoplasia have been clarified. K-*ras* mutations contribute to the induction of hyperplastic changes (3). Mutated K-*ras* activates the mitogen-activated protein kinase (MAPK) and phosphoinositide-3 kinase (PI3K)/Akt pathways and results in cyclin D1 and COX-2 overexpression, which in turn may induce iNOS expression in the presence of inflammatory stimuli (4,5). Overexpressed COX-2 produces excess prostaglandins (PGs) and causes cell proliferation and inhibition of apoptosis, to some extent mediated by PGE_2 receptor subtypes EP1, EP2, and EP4 (6).

Adenomatous polyposis coli (APC) or *β-catenin* mutations appear to be involved in the generation of dysplastic lesions (3), stabilizing β-catenin protein in the cytoplasm and activating β-catenin/Tcf signaling to up-regulate target genes, such as cyclin D1 (7). β-Catenin alteration is suggested to be involved in increased expression of iNOS (4). NO produced by iNOS causes DNA damage and neovascularization, which promotes carcinogenesis. Moreover, NO itself could induce COX-2 expression (8).

Cancer and Inflammation Mechanisms: Chemical, Biological, and Clinical Aspects, First Edition.
Edited by Yusuke Hiraku, Shosuke Kawanishi, and Hiroshi Ohshima.
© 2014 John Wiley & Sons, Inc. Published 2014 by John Wiley & Sons, Inc.

With elucidation of the mechanisms of colorectal carcinogenesis, a great deal of interest has been concentrated on anti-inflammatory agents, including nonsteroidal anti-inflammatory drugs (NSAIDs), which act by inhibiting COX enzymes. Screening of anti-inflammatory agents as potential colorectal cancer chemopreventive agents has been carried out using several *in vivo* animal models. The majority feature application of azoxymethane (AOM), a very potent carcinogen that induces colorectal cancers at high incidence in rats and mice. Short-term treatment with AOM results in the development of putative preneoplastic aberrant crypt foci (ACF). Such biomarker lesions are thought to be useful surrogates for tumors in assessing the effects of agents capable of preventing carcinogenesis in the colon (9,10). Furthermore, the *Apc* gene–deficient mouse, an animal model of human familial adenomatous polyposis (FAP) characterized by large numbers of intestinal polyps because of a truncation mutation in the *Apc*, is also a useful model to evaluate cancer chemopreventive agents. Indeed, many FAP model mice, such as Apc^{1309} (C57BL/6J$^{Apc/ApcD1309}$) (mutation at codon 1309; develop ca.35 polyps/mouse), Apc^{Min} (*Min*) (mutation at codon 850; develop ca.100 polyps/mouse), Apc^{D716} (mutation at codon 716; develop ca.300 polyps/mouse), and Apc^{1638} (mutation at codon 1638; develop ca.10 polyps/mouse) strains are now used world wide (11–15).

In this chapter we aim to provide a summary of this field of research with attention to possible mechanisms of action and potential application of anti-inflammatory agents for practical cancer prevention.

ANTI-INFLAMMATORY AGENTS TARGETING CYCLOOXYGENASE

A large number of epidemiological studies have indicated that NSAIDs can reduce the risk of colorectal cancer. For example, people who take aspirin regularly demonstrate at most a 40% reduction in the relative risk of colorectal cancer and associated mortality (16). Furthermore, celecoxib, a COX-2-selective inhibitor, and indomethacin and sulindac, conventional NSAIDs (Figure 20.1) that inhibit both COX-1 and COX-2, can actually cause regression of existing colorectal polyps in patients with FAP (17–19). Although there are several molecular mechanisms assumed to be involved in the reduction of colorectal cancer by NSAIDs, such as inactivation of Akt, activation of AMP-activated protein kinase (AMPK), and inhibition of transcription factor nuclear factor-κB (NF-κB), the most likely possibility is linked directly to their inhibition of COX (20–22).

Prostanoid synthesis starts with release of arachidonic acid (AA) from cell membrane phospholipids, mediated primarily via the action of phospholipase A$_2$ (6). COX then catalyzes the conversion of AA to PGG$_2$ and under the influence of peroxidase activity of the enzyme, this is rapidly converted to PGH$_2$. There are two isoforms of COX: the constitutive enzyme, COX-1, present in many cells and tissues, and the inducible enzyme, COX-2, produced in response to growth factors, mitogens, and pro-inflammatory cytokines. PGH$_2$ is additionally isomerized to PGE$_2$, PGD$_2$, PGF$_2$, PGI$_2$, and thromboxane A$_2$ by their respective PG synthases (6). Nonenzymatic

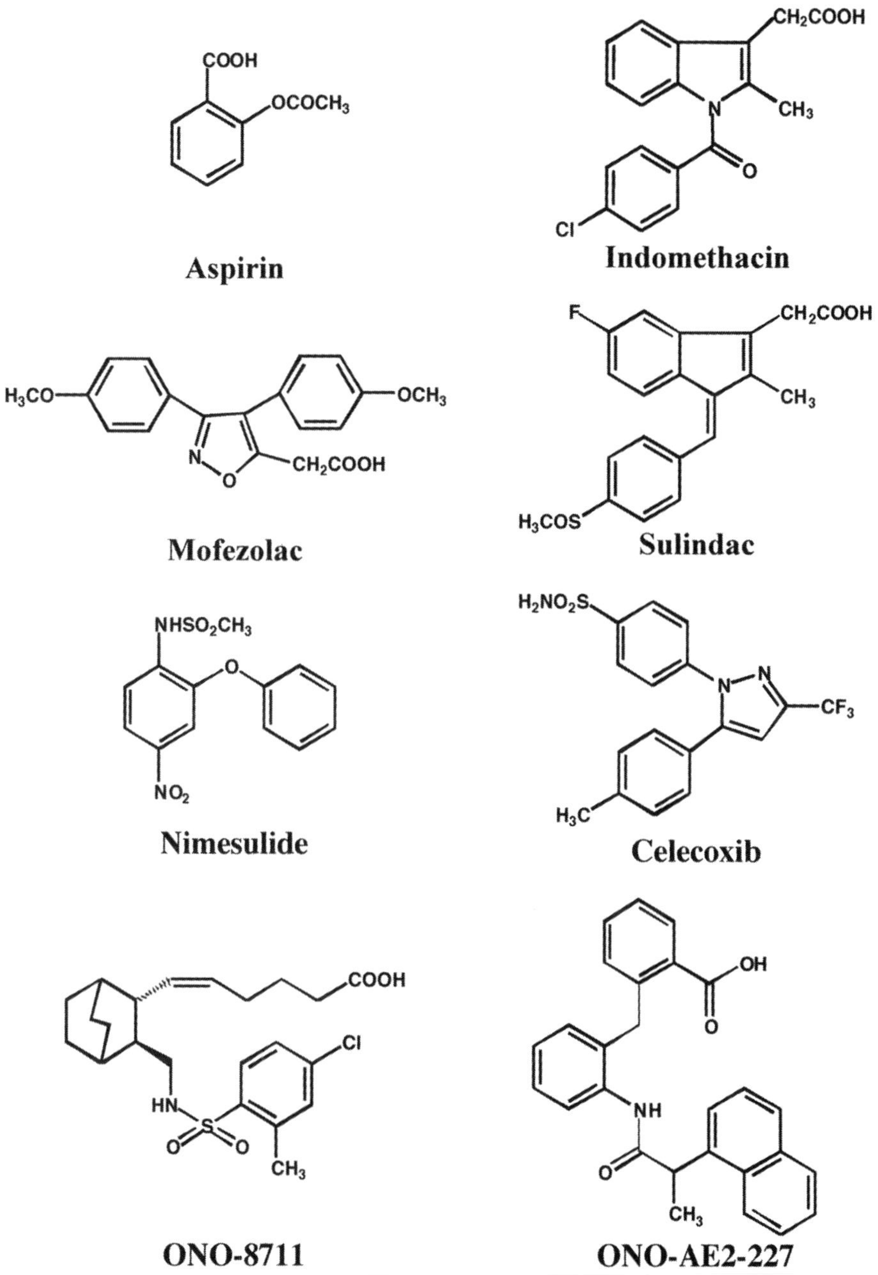
COOH
OCOCH3
Aspirin
CH2COOH
N
CH3
O
Cl
Indomethacin
H3CO
OCH3
N
O
CH2COOH
Mofezolac
F
CH2COOH
CH3
H3COS
Sulindac
NHSO2CH3
O
NO2
Nimesulide
H2NO2S
N
N
CF3
H3C
Celecoxib
COOH
Cl
HN
S
O
O
CH3
ONO-8711
OH
O
NH
O
CH3
ONO-AE2-227

FIGURE 20.1 Structures of COX-1 and COX-2 inhibitors.

dehydration of PGD_2 results in generation of PGJ_2, 12-PGJ_2, and 15-deoxy-$\Delta^{12,14}$-PGJ_2 (15-Δ-PGJ_2).

As PGE_2 synthesis is elevated in colon cancer, it is likely that PGE_2 would enhance carcinogenesis more than other prostanoids. In fact, in an AOM-induced colorectal carcinogenesis model in F344 rats, administration of PGE_2 enhanced colon carcinogenesis through induction of cell proliferation and reduction of apoptosis (23). It is conceivable that both COX isoforms may play important roles in colorectal carcinogenesis. Dietary administration of 1200 ppm mofezolac, [3,4-di(4-methoxyphenyl)-5-isoxazolyl acetic acid], a COX-1-selective inhibitor (Figure 20.1), was found to reduce the number of ACF per rat treated with AOM. Treatment with the same dose of mofezolac reduced the number of intestinal polyps in APC^{1309} mice to 59% of that in the untreated control mice (24). Dietary administration of 400 ppm nimesulide, a COX-2-selective inhibitor (Figure 20.1), was found to reduce the number of intestinal polyps in *Min* mice (25). Furthermore, homologous genetic disruption of either COX-1 or COX-2 markedly reduced polyp formation in *Min* mice (26).

There are four PGE_2 receptor subtypes, EP1 to 4, and assessment of mRNA expression has demonstrated up-regulation of EP1 and EP2 and down-regulation of EP3 in AOM-induced rat and mouse colon cancers. EP4 mRNA is consistently expressed in normal mucosa and tumors (27). Using PGE receptor subtype knockout mice, the roles of these receptors in colon carcinogenesis have been investigated (27–29). EP1 receptor selective antagonists, ONO-8711, {6-[(2*S*,3*S*)-3-(4-chloro-2-methylphenylsulfonylaminomethyl)-bicyclo[2.2.2]octan-2-yl]-5Z-hexenoic acid}, and the EP4 receptor–selective antagonist, ONO-AE2-227, 2-[2-{2-(1-naphthyl)propanoylamino}phenyl]methylbenzoic acid, inhibited development of AOM-induced ACF in male C57BL/6J mice (Figure 20.1). Moreover, when *Min* mice were given 500 ppm ONO-8711 in the diet, the number of intestinal polyps was reduced significantly, to 57% of that in the untreated control mice (27). Deficiencies of EP1 and EP4 also caused a decrease in ACF formation in the colons of mice treated with AOM (28,29). Sonoshita et al. reported that double knockout of *Apc* and *EP2* genes decreased intestinal polyp development (30).

In contrast, deficiency of EP3 was found to enhance colon tumor formation after exposure to AOM (27). The available observations suggest that EP1, EP2, and EP4 are promotive in colon carcinogenesis, while EP3 could play suppressive roles, particularly in late stages. Of note, deficiencies of other specific membrane receptors—DP for PGD_2, FP for PGF_2, IP for PGI_2, and TP for thromboxane A_2—did not decrease ACF formation after AOM treatment (29).

ANTI-INFLAMMATORY AGENTS TARGETING iNOS

NO is an essential mediator of physiological processes in the digestive tract, maintaining the mucosa and regulating blood flow and peristalsis (31). However, over-production of NO contributes to tissue damage, colon cancer cell growth, and DNA deamination (32,33). L-Arginine is converted to L-citrulline and NO by NOS, neuronal (nNOS), inducible (iNOS), and endothelial (eNOS) NOS isoforms. Thus, L-arginine

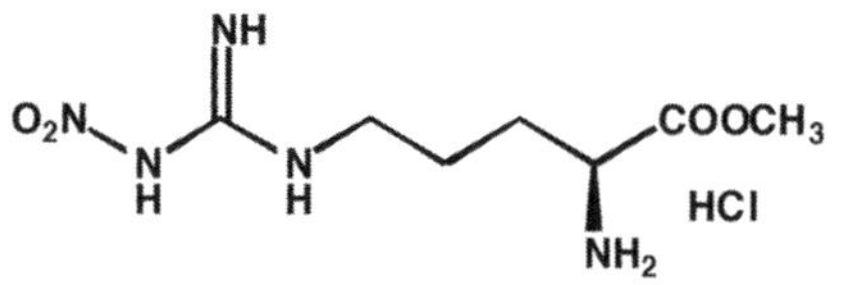

L-NAME

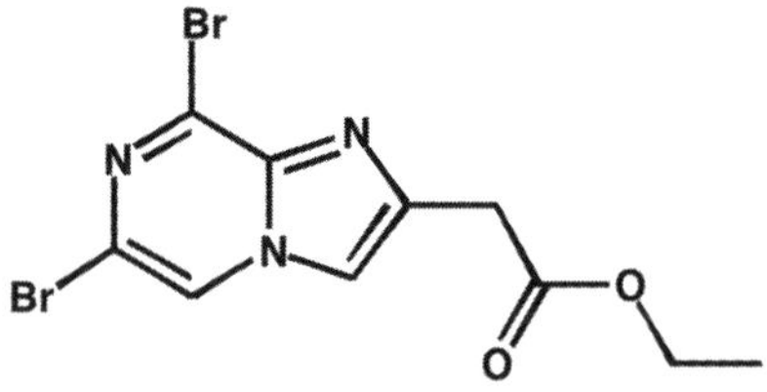

SG-51

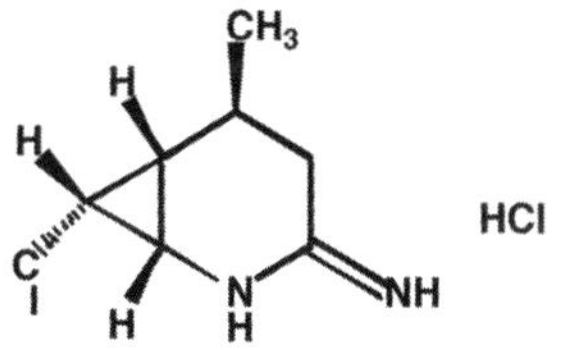

ONO-1714

FIGURE 20.2 Structures of NOS inhibitors.

analogs, $N(\omega)$-nitro-L-arginine methyl ester (L-NAME), have attracted attention as possible inhibitors (Figure 20.2). iNOS expression is barely detectable in normal colon epithelial or stromal cells. However, it is found in lesions in which β-catenin alterations are observed, [i.e., human colon adenomas and adenocarcinomas (32,33)], and in AOM-induced rat colon dysplastic ACF and tumors (4).

Administration of a specific iNOS inhibitor, ONO-1714, {(1S,5S,6R,7R)-7-chloro-3-imino-5-methyl-2-azabicyclo[4.1.0]heptane hydrochloride}(Figure 20.2), at doses of 10, 20, 50, and 100 ppm, reduced the number of AOM-induced ACFs in the F344 rats up to 53% of the untreated control value. Moreover, long-term treatment (32 weeks) revealed that 100 ppm ONO-1714 decreased the number of large colon tumors (>3 mm in diameter) (34). These results suggest that iNOS plays roles in both early and late stages of colon carcinogenesis.

In contrast to the normal mucosa, iNOS is also overexpressed in inflamed colonic mucosa, to almost the same extent as in colonic adenocarcinomas, in mice receiving AOM and/or dextran sodium sulfate (DSS) (35,36). An explanation of this expression by inflammatory stimuli could be obtained using IEC-6 rat intestinal epithelial cells transfected with K-*ras* mutant cDNA. In transfected IEC cells, induction of iNOS expression mediated by interleukin-1β (IL-1β) or lipopolysaccharide was elevated markedly compared to the case with transfection of control vector or wild-type K-*ras* cDNA (5). These results suggest that activating mutations of K-*ras* caused by the carcinogen AOM are associated with up-regulation of iNOS expression in the presence of inflammatory stimuli. It has been reported that ONO-1714 effectively inhibited DSS-induced large bowel carcinogenesis in *Min* mice. Of interest in this context, the suppressive effects of ONO-1714 on the development of large bowel adenocarcinomas were closely correlated with the inhibition of serum triglyceride levels and the inhibition of pro-inflammatory cytokines, tumor necrosis factor-α (TNF-α) and IL-1β, and COX-2 mRNA levels (36).

ANTI-INFLAMMATORY AGENTS TARGETING PPAR-γ

Peroxisome proliferator–activated receptor γ (PPAR-γ) is a member of the ligand-activated nuclear receptor superfamily, which plays key roles in fat metabolism. Moreover, PPAR-γ has been implicated in the pathophysiology of inflammation, obesity, and diabetes (37). Recently, it was shown that activation of PPAR-γ by 15-Δ-PGJ$_2$ or antidiabetic thiazolidinediones exerts antiproliferation, apoptosis, differentiation, and anti-inflammation effects in cancer cells (37). It has been reported that 15-Δ-PGJ$_2$ inhibits NF-κB-dependent gene expression either by functional inactivation of IKK, thereby preventing IκB degradation and nuclear entry of NF-κB, or via direct interference with binding of NF-κB to target DNA sequences.

Administration of pioglitazone, {(±)-5-[4-[2-(5-ethyl-2-pyridyl)ethoxy] benzyl]thiazolidine-2,4-dione monohydrochloride}, a potent PPAR-γ ligand (Figure 20.3), at doses of 100 and 200 ppm in *Apc*^{1309} mice reduced the total numbers of polyps up to 67% of the value in the untreated control group (38). In *Min* mice, treatment with 100 to 1600 ppm for 14 weeks also showed a decrease of intestinal polyps upto 9% of the control number (39). Of note, there exists a PPRE in the promoter region of the *LPL* gene, and pioglitazone treatment induced LPL expression in the liver and intestinal epithelial cells in *Apc*-deficient mice.

NO-1886, 4-[(4-bromo-2-cyanophenyl)carbamoyl]benzylphosphonate (40), is an agent that can induce LPL agonist activity, but unlike bezafibrate and pioglitazone, does not possess PPAR agonistic activity (41) (Figure 20.3). Its administration at doses of 400 and 800 ppm also significantly decreased the total number of intestinal polyps to 48% and 42% of the untreated control value, respectively, in *Min* mice, along with a marked increase in *LPL* mRNA levels in the liver and small intestine (42).

It is interesting that LPL is considered to possess both lipid-modifying and anti-inflammatory functions. It has been reported that LPL suppresses TNF-α- and interferon-γ-evoked inflammation-related gene expression in endothelial cells through inactivation of NF-κB (43). Experiments conducted to clarify the mechanism of NO-1886 influence on colorectal carcinogenesis revealed that the expression levels of TGF-α-induced COX-2 mRNA in human colon cancer cells DLD-1 were reduced. On the other hand, there was no obvious change in the mRNA levels for COX-1 and iNOS. The results were also confirmed by β-gal reporter gene assay in DLD-1 cells (42). Consistent with the *in vitro* data, administration of 400 and

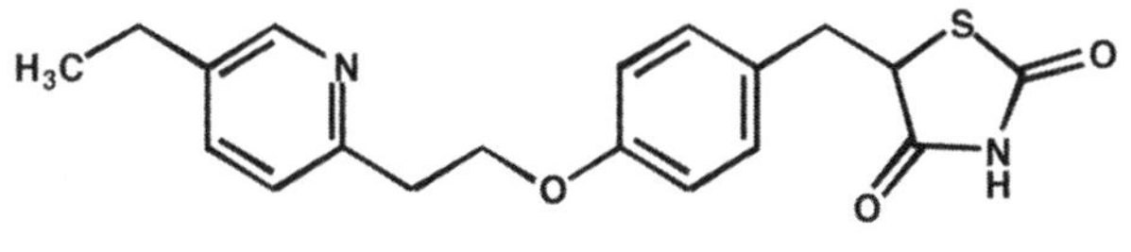

Pioglitazone

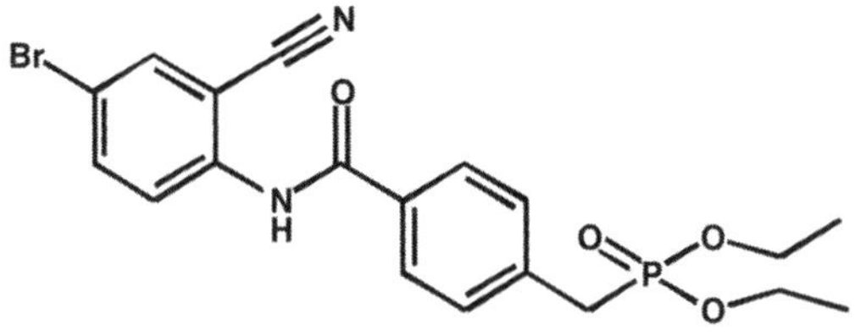

NO-1886

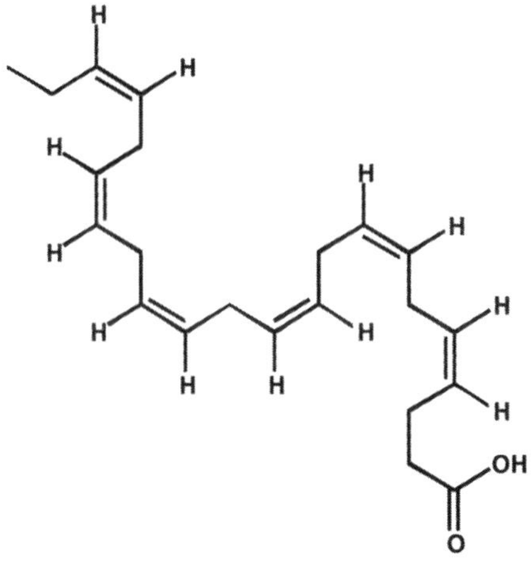

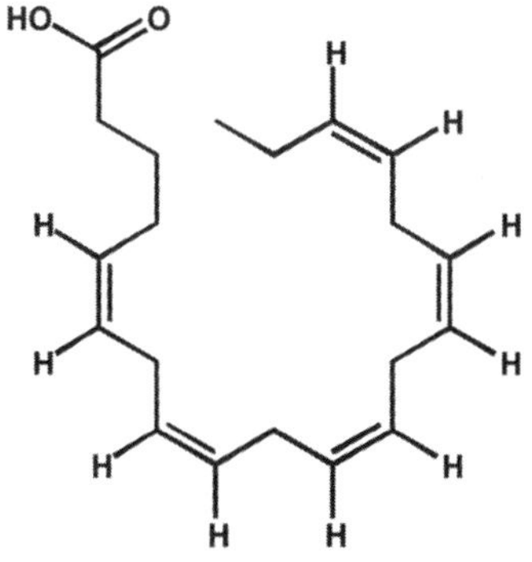

Docosahexaenoic acid   **Eicosapentaenoic acid**

FIGURE 20.3   Structures of lipid-lowering agents.

800 ppm NO-1886 reduced COX-2 mRNA levels in nonpolyp parts of the small intestine of *Min* mice (42).

OTHER CANDIDATES AS CANCER CHEMOPREVENTIVE AGENTS WITH ANTI-INFLAMMATORY POTENTIAL

Other agents from natural compounds that can suppress COX-2 and iNOS are flavonoids and phenolic antioxidants. Putative chemopreventive agents such as catechin, epicatechin, quercetin, kaempferol, genistein, resveratrol, and resorcinol, all having a common resorcin moiety, have been found to suppress COX-2 promoter activity effectively with and without TGF-α stimulation in DLD-1 cells (44,45)

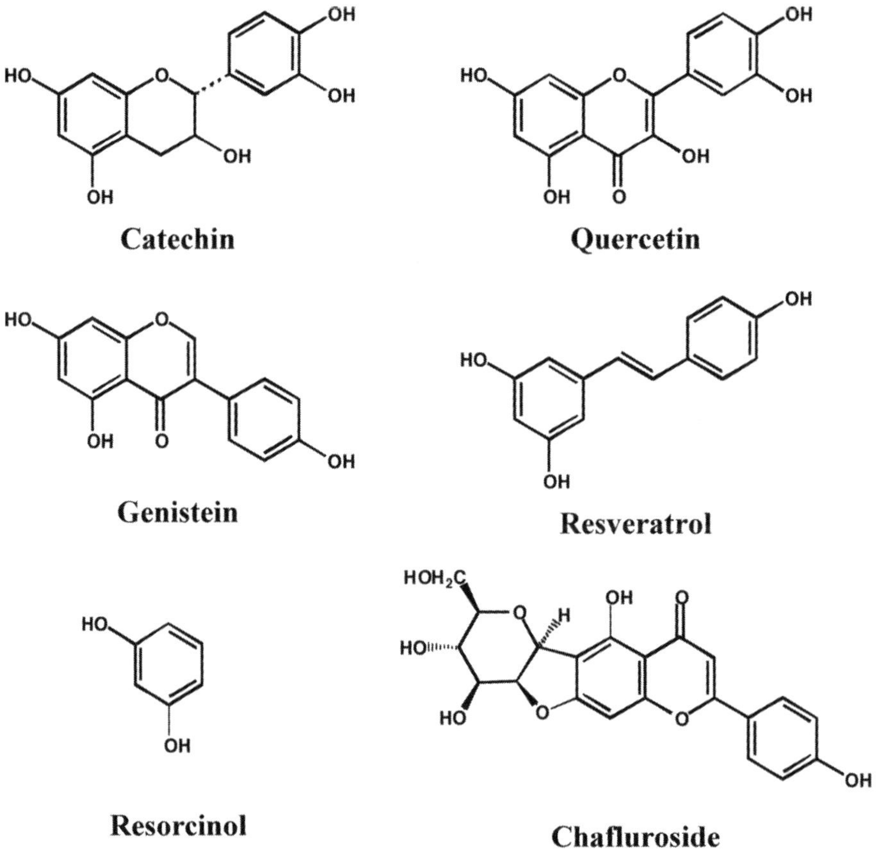

FIGURE 20.4   Structures of natural phenolic antioxidants.

(Figure 20.4). Moreover, ferulic acid derivatives can suppress COX-2 promoter activity, and butyrate reduces iNOS promoter activity (46,47). A new flavone derivative, chafuroside, (2R,3S,4S,4aS,11bS)-3,4,11-trihydroxy-2-(hydroxymethyl)-8-(4-hydroxyphenyl)-3,4,4a,11b-tetrahydro-2H,10H-pyrano[2′,3′:4,5]furo[3,2-g] chromen-10-one, has been isolated as a strong anti-inflammatory compound from oolong tea leaves and found to exert strong suppressive effects on intestinal tumorigenesis (48) (Figure 20.4). Administration of 10 ppm chafuroside reduced AOM-induced ACF formation and total numbers of polyps in the *Min* mice to 56% of the untreated control value.

A major component of fish oil, docosahexaenoic acid (DHA) lowered the numbers of moderately differentiated adenocarcinomas developing in the middle and distal colon after AOM treatment in F344 rats compared to untreated controls, with significant reductions in levels of PGE_2 and AA in the blood plasma (49) (Figure 20.3). DHA also exerts inhibitory effects on intestinal polyp development in $Apc^{\Delta 716}$ mice (50).

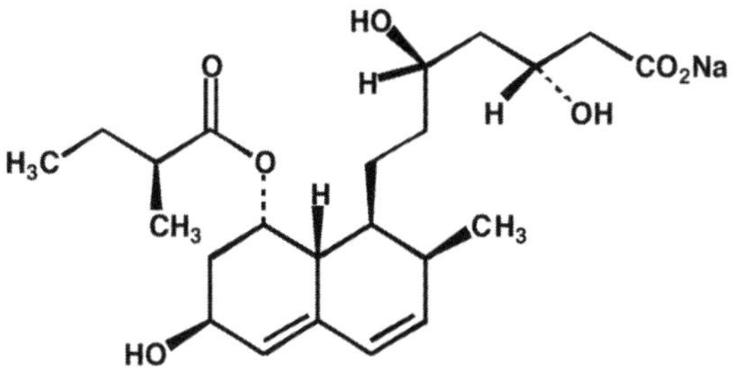

Pravastatin

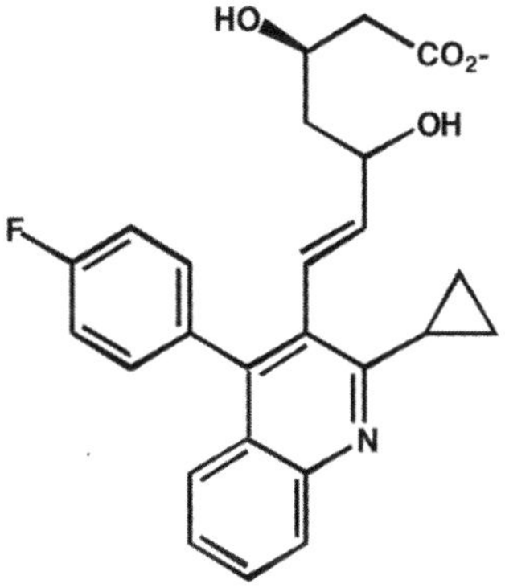

Pitavastatin 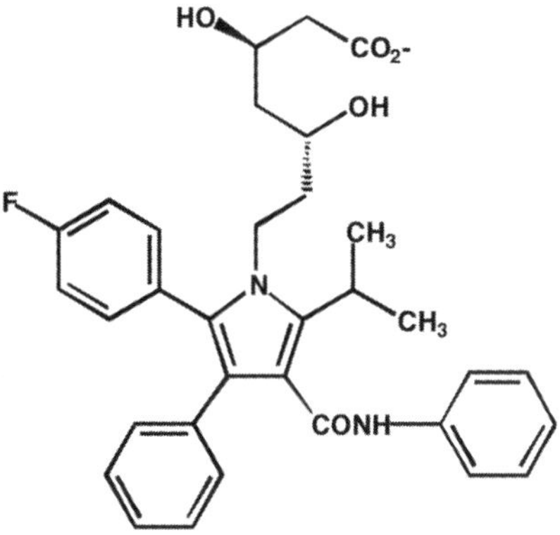   **Atorvastatin**

FIGURE 20.5   Structures of statins.

Synthetic agents that can suppress COX-2 and iNOS include statins, 3-hydroxy-3-methylglutaryl-coenzyme A (HMG-CoA) reductase inhibitors, which are commonly used for the treatment of hypercholesterolemia. Pravastatin and atorvastatin have been reported to suppress 1,2-dimethylhydrazine (DMH) or AOM-induced colon cancer development in mice and rats, respectively (51,52). In addition, atorvastatin at a dose of 100 ppm reduced the incidence of small intestinal polyps in *Min* mice to 70% of the value in untreated control mice (53). Furthermore, 10 ppm pitavastatin, (+)-monocalcium bis{(3R,5S,6E)-7-[2-cyclopropyl-4-(4-fluorophenyl)-3-quinolyl]-3,5-dihydroxy-6-heptenoate}, a novel lipophilic statin, decreased the incidence of colon adenomas or adenocarcinomas induced by AOM/DSS treatment in ICR mice to about 78% of that in the untreated control group (54) (Figure 20.5). In addition to the main function of statins, which is inhibition of the synthesis of mevalonate, they also suppress inflammation. Treatment of *Min* mice with pitavastatin at dose of 40 ppm decreased the total number of polyps to 65.8% of the untreated control value (55). Serum levels of total cholesterol and triglycerides were slightly reduced, and those of IL-6, leptin, and monocyte chemoattractant protein-1 (MCP-1) were decreased. mRNA expression levels of COX-2, IL-6, iNOS, MCP-1, and plasminogen activator inhibitor-1 (Pai-1) were reduced significantly in intestinal nonpolyp parts by pitavastatin treatment (55).

An epidemiological study demonstrated that chronic use of statins for more than five years significantly reduced the risk of colorectal cancer by 47% (56). On the other hand, other epidemiological studies using first-generation statins did not provide fully consistent results (57,58). Discrepancies could be explained by suboptimal administration of the drugs and their characteristics. Atorvastatin and pitavastatin, third-generation statins, possess strong serum lipid-lowering and anti-inflammatory potential. Further epidemiological data using third-generation statins are desired to evaluate chemopreventive effects against colorectal cancer.

FUTURE PROSPECTS

Combination Effects

In the future, use of anti-inflammatory drugs with other chemopreventive agents might find clinical application. Methods targeting several molecules could allow reduction of the dosage of the individual agents, resulting in lowered side effects. Moreover, targeting several molecules could provide synergistic effects. The combinations aspirin + α-difluoromethylornithine, the selective iNOS inhibitor L-N(6)-(1-iminoethyl)lysine tetrazole amide (SC-51) + the selective COX-2 inhibitor celecoxib, atorvastatin + sulindac, and atorvastatin + naproxen (59–61) have already been shown to reduce AOM-induced colon ACF/cancer formation (Figures 20.1 and 20.5). Moreover, the combinations ONO-8711 + ONO-AE2-227 and mofezolac + nimesulide were found to reduce intestinal polyp formation in *Apc*-deficient mice (62,63) (Figure 20.1).

Targeting Obesity

Recently, there is increasing consensus that obesity should be thought of as a pre-inflammatory condition. Because hypertrophy of fatty tissue results in an increase and changed profiles of adipocytokines, low-grade inflammation is evoked. It has become clear that factors such as insulin resistance, dyslipidemia, and subsequent adipocytokine imbalance could be involved in the promotion of colorectal carcinogenesis. Animal experiments have shown that some adipocytokines may play important roles not only in progression to malignancy, but also in very early stages of colorectal cancer development.

Obese mice such as the KK-A^y strain were revealed to be highly susceptible to induction of colon ACF and development of colorectal carcinomas by treatment with AOM (64). In addition to severe hyperinsulinemia and hypertriglyceridemia, the KK-A^y mouse exhibits abdominal obesity and resulting elevation of serum adipocytokines/cytokines, such as IL-6, leptin, and Pai-1, compared with values for lean C57BL/6J mice. In the visceral fat tissue, significant overexpression of pro-inflammatory cytokine mRNAs, such as for IL-6, leptin, MCP-1, Pai-1, and TNF-α was confirmed; in contrast, that for adiponectin was decreased. Thus, consequent adipocytokine imbalance is suggested to contribute to the promotion of colon carcinogenesis.

As correction of this balance must be considered as a mean of cancer prevention, it might be important to develop selective adipocytokine-regulated drugs and search for agents from drugs with few side effects. Administration of a Pai-1 inhibitor, SK-216, at 25- to 100-ppm doses in *Min* mice, characterized by high levels of serum Pai-1, was found to reduce serum Pai-1 and hepatic Pai-1 mRNA levels, and decreased total numbers of intestinal polyps significantly, up to 56% of the untreated group value (65). Thus, adipocytokines such as adiponectin and Pai-1 are considered to be key molecular targets for cancer chemopreventive approaches.

Clinical Trials

Among candidate substances, sulindac has been studied frequently in the clinical setting (66–68), being reported to reduce the number and size of colorectal adenomas in a double-blind randomized trial (66). However, it has been also reported that sulindac may cause serious side-effects at a dose of more than 100 mg/day, at least in Japanese FAP patients (67). COX-2 selective inhibitors, including celecoxib and rofecoxib, were hoped to be ideal cancer chemopreventive agents, because they cause little damage to the gastric mucosa (19, 69), but recent studies have revealed that they may cause cardiotoxicity (70,71). Another trial using natural fish oil, omega-3 polyunsaturated fatty acid eicosapentaenoic acid (EPA) (Figure 20.3), showed promising reduction in rectal polyp number and size in FAP patients (72).

Meanwhile, aspirin, one of the conventional NSAIDs, is again attracting attention as a chemopreventive agent. Widespread and long-term use of aspirin for cardiovascular disease prevention allowed the accumulation of abundant evidence regarding its safety profile, and the dual benefit for patients with significant risk factors for both cardiovascular disease and colorectal cancer is an obvious advantage.

The CAPP2 randomized trial in 1000 Lynch syndrome gene carriers found almost a 60% reduction in new cancer development at about five years after randomization (73). In this experiment aspirin (600 mg/day) was used for a minimum of two years (73). This finding suggests that follow-up for several years after randomized trials is necessary to evaluate the effects of aspirin, and potentially also for other colon cancer chemopreventive agents. It is clearly desirable that more data be accumulated, specifically for Asian populations, to better assess chemoprevention of colon cancer by aspirin in the future.

REFERENCES

1. Eberhart CE, Coffey RJ, Radhika A, Giardiello FM, Ferrenbach S, Dubois RN. Up-regulation of cyclooxygenase 2 gene expression in human colorectal adenomas and adenocarcinomas. Gastroenterology 1994;107:1183–1188.
2. Ambs S, Merriam WG, Bennett WP, Felley-Bosco E, Ogunfusika MO, Oser SM, et al. Frequent nitric oxide synthase-2 expression in human colon adenomas: implication for tumor angiogenesis and colon cancer progression. Cancer Res 1998;58:334–341.

3. Jen J, Powell SM, Papadopoulos N, Smith KJ, Hamilton SR, Vogelstein B, Kinzler KW. Molecular determinants of dysplasia in colorectal lesions. Cancer Res 1994;54:5523–5526.

4. Takahashi M, Wakabayashi K. Gene mutations and altered gene expression in azoxymethane-induced colon carcinogenesis in rodents. Cancer Sci 2004;95:475–480.

5. Takahashi M, Mutoh M, Shoji Y, Kamanaka K, Naka M, Maruyama T, et al. Transfection of K-rasAsp12 cDNA markedly elevates IL-1β and lipopolysaccharide-mediated inducible nitric oxide synthase expression in rat intestinal epithelial cells. Oncogene 2003;22:7667–7676.

6. Mutoh M, Takahashi M, Wakabayashi K. Roles of prostanoids in colon carcinogenesis and their potential targeting for cancer chemoprevention. Curr Pharm Des 2006;12:2375–2382.

7. Vogelstein B, Kinzler KW. Cancer genes and the pathways they control. Nat Med 2004;10:789–799.

8. Salvemini D, Misko TP, Masferrer JL, Seibert K, Currie MG, Needleman P. Nitric oxide activates cyclooxygenase enzymes. Proc Natl Acad Sci USA 1993;90:7240–7244.

9. Boone CW, Steele VE, Kelloff GJ. Screening for chemopreventive (anticarcinogenic) compounds in rodents. Mutat Res 1992;267:251–255.

10. Olivo S, Wargovich MJ. Inhibition of aberrant crypt foci by chemopreventiveagents. *In Vivo* 1998;12:159–166.

11. Quesada CF, Kimata H, Mori M, Nishimura M, Tsuneyoshi T, Baba S. Piroxicam and acarbose as chemopreventive agents for spontaneous intestinal adenomas in APC gene 1309 knockout mice. Jpn J Cancer Res 1998;89:392–396.

12. Moser AR, Pitot HC, Dove WF. A dominant mutation that predisposes to multiple intestinal neoplasia in the mouse. Science 1990;247:322–324.

13. Oshima M, Oshima H, Kitagawa K, Kobayashi M, Itakura C, Taketo M. Loss of Apc heterozygosity and abnormal tissue building in nascent intestinal polyps in mice carrying a truncated Apc gene. Proc Natl Acad Sci USA 1995;92:4482–4486.

14. Fodde R, Edelmann W, Yang K, van Leeuwen C, Carlson C, Renault B, et al. A targeted chain-termination mutation in the mouse Apc gene results in multiple intestinal tumors. Proc Natl Acad Sci USA 1994;91:8969–8973.

15. McCart AE, Vickaryous NK, Silver A. Apc mice: models, modifiers and mutants. Pathol Res Pract 2008;204:479–490.

16. Thun MJ, Namboodiri MM, Heath CW Jr. Aspirin use and reduced risk of fatal colon cancer. N Engl J Med 1991;325:1593–1596.

17. Giardiello FM, Yang VW, Hylind LM, Krush AJ, Petersen GM, Trimbath JD, et al. Primary chemo-prevention of familial adenomatous polyposis with sulindac. N Engl J Med 2002;346:1054–1059.

18. Akasu T, Yokoyama T, Sugihara K, Fujita S, Moriya Y, Kakizoe T. Peroral sustained-release indomethacin treatment for rectal adenomas in familial adenomatous polyposis: a pilot study. Hepatogastroenterology 2002;49:1259–1261.

19. Steinbach G, Lynch PM, Phillips RK, Wallace MH, Hawk E, Gordon GB, et al. The effect of celecoxib, a cyclooxygenase-2 inhibitor, in familial adenomatous polyposis. N Engl J Med 2000;342: 1946–1952.

20. Kulp SK, Yang YT, Hung CC, Chen KF, Lai JP, Tseng PH, et al. 3-Phosphoinositide-dependent protein kinase-1/Akt signaling represents a major cyclooxygenase-2-independent target for celecoxib in prostate cancer cells. Cancer Res 2004;64:1444–1451.

21. Hawley SA, Fullerton MD, Ross FA, Schertzer JD, Chevtzoff C, Walker KJ, et al. The ancient drug salicylate directly activates AMP-activated protein kinase. Science 2012;336:918–922.

22. Loveridge CJ, MacDonald AD, Thoms HC, Dunlop MG, Stark LA. The proapoptotic effects of sulindac, sulindac sulfone and indomethacin are mediated by nucleolar translocation of the RelA(p65) subunit of NF-kappaB. Oncogene 2008;27:2648–2655.

23. Kawamori T, Uchiya N, Sugimura T, Wakabayashi K. Enhancement of coloncarcinogenesis by prostaglandin E$_2$ administration. Carcinogenesis 2003;24:985–990.

24. Kitamura T, Kawamori T, Uchiya N, Itoh M, Noda T, Matsuura M, et al. Inhibitory effects of mofezolac, a cyclooxygenase-1 selective inhibitor, on intestinal carcinogenesis. Carcinogenesis 2002;23:1463–1466.

25. Nakatsugi S, Fukutake M, Takahashi M, Fukuda K, Isoi T, Taniguchi Y, et al. Suppression of intestinal polyp development by nimesulide, a selective cyclooxygenase-2 inhibitor, in Min mice. Jpn J Cancer Res 1997;88:1117–1120.

26. Chulada PC, Thompson MB, Mahler JF, Doyle CM, Gaul BW, Lee C, et al. Genetic disruption of Ptgs-1, as well as Ptgs-2, reduces intestinal tumorigenesis in Min mice. Cancer Res 2000;60:4705–4708.

27. Shoji Y, Takahashi M, Kitamura T, Watanabe K, Kawamori T, Maruyama T, et al. Downregulation of prostaglandin E receptor subtype EP3 during colon cancer development. Gut 2004;53: 1151–1158.

28. Watanabe K, Kawamori T, Nakatsugi S, Ohta T, Ohuchida S, Yamamoto H, et al. Role of the prostaglandin E receptor subtype EP1 in colon carcinogenesis. Cancer Res 1999;59:5093–5096.

29. Mutoh M, Watanabe K, Kitamura T, Shoji Y, Takahashi M, Kawamori T, et al. Involvement of prostaglandin E receptor subtype EP4 in colon carcinogenesis. Cancer Res 2002;62:28–32.

30. Sonoshita M, Takaku K, Sasaki N, Sugimoto Y, Ushikubi F, Narumiya S, et al. Acceleration of intestinal polyposis through prostaglandin receptor EP2 in ApcΔ716 knockout mice. Nat Med 2001;7:1048–1051.

31. Kolios G, Valatas V, Ward SG. Nitric oxide in inflammatory bowel disease: a universal messenger in an unsolved puzzle. Immunology 2004;113:427–437.

32. Hussain SP, Hofseth LJ, Harris CC. Radical causes of cancer. Nat Rev Cancer 2003;3:276–285.

33. Hofseth LJ, Hussain SP, Wogan GN, Harris CC. Nitric oxide in cancer and chemoprevention. Free Radic Biol Med 2003;34:955–968.

34. Takahashi M, Mutoh M, Shoji Y, Sato H, Kamanaka Y, Naka M, et al. Suppressive effect of an inducible nitric oxide inhibitor, ONO-1714, on AOM-induced rat colon carcinogenesis. Nitric Oxide 2006;14:130–136.

35. Kohno H, Suzuki R, Sugie S, Tanaka T. β-Catenin mutations in a mouse model of inflammation-related colon carcinogenesis induced by 1,2-dimethylhydrazine and dextran sodium sulfate. Cancer Sci 2005;96:69–76.

36. Tanaka T, Kohno H, Suzuki R, Hata K, Sugie S, Niho N, et al. Dextran sodium sulfate strongly promotes colorectal carcinogenesis in $Apc^{Min/+}$ mice: inflammatory stimuli by dextran sodium sulfate results in development of multiple colonic neoplasms. Int J Cancer 2006;118:25–34.

37. Na HK, Surh YJ. Peroxisome proliferator–activated receptor gamma (PPARgamma) ligands as bifunctional regulators of cell proliferation. Biochem Pharmacol 2003;66:1381–1391.

38. Niho N, Takahashi M, Kitamura T, Shoji Y, Itoh M, Noda T, et al. Concomitant suppression of hyperlipidemia and intestinal polyp formation in Apc-deficient mice by peroxisome proliferator-activated receptor ligands. Cancer Res 2003;63:6090–6095.

39. Niho N, Takahashi M, Shoji Y, Takeuchi Y, Matsubara S, Sugimura T, et al. Dose-dependent suppression of hyperlipidemia and intestinal polyp formation in Min mice by pioglitazone, a PPAR gamma ligand. Cancer Sci 2003;94:960–964.

40. Tsutsumi K, Inoue Y, Shima A, Iwasaki K, Kawamura M, Murase T. The novel compound NO-1886 increases lipoprotein lipase activity with resulting elevation of high-density lipoprotein cholesterol, and long-term administration inhibits atherogenesis in the coronary arteries of rats with experimental atherosclerosis. J Clin Invest 1993;92:411–417.

41. Doi M, Kondo Y, Tsutsumi K. Lipoprotein lipase activator NO-1886 (ibrolipim) accelerates the mRNA expression of fatty acid oxidation–related enzymes in rat liver. Metabolism 2003;52:1547–1550.

42. Niho N, Mutoh M, Takahashi M, Tsutsumi K, Sugimura T, Wakabayashi K. Concurrent suppression of hyperlipidemia and intestinal polyp formation by NO-1886, increasing lipoprotein lipase activity in Min mice. Proc Natl Acad Sci USA 2005;102:2970–2974.

43. Kota RS, Ramana CV, Tenorio FA, Enelow RI, Rutledge JC. Differential effects of lipoprotein lipase on tumor necrosis factor-alpha and interferon-gamma-mediated gene expression in human endothelial cells. J Biol Chem 2005;280:31076–31084.

44. Mutoh M, Takahashi M, Fukuda K, Matsushima-Hibiya Y, Mutoh H, Sugimura T, et al. Suppression of cyclooxygenase-2 promoter-dependent transcriptional activity in colon cancer cells by chemopreventive agents with a resorcin-type structure. Carcinogenesis 2000;21:959–963.

45. Mutoh M, Takahashi M, Fukuda K, Komatsu H, Enya T, Matsushima-Hibiya Y, et al. Suppression by flavonoids of cyclooxygenase-2 promoter-dependent transcriptional activity in colon cancer cells: structure–activity relationship. Jpn J Cancer Res 2000;9:686–691.

46. Hosoda A, Ozaki Y, Kashiwada A, Mutoh M, Wakabayashi K, Mizuno K, et al. Syntheses of ferulic acid derivatives and their suppressive effects on cyclooxygenase-2 promoter activity. Bioorg Med Chem 2002;10:1189–1196.

47. Sasahara Y, Mutoh M, Takahashi M, Fukuda K, Tanaka N, Sugimura T, et al. Suppression of promoter-dependent transcriptional activity of inducible nitric oxide synthase by sodium butyrate in colon cancer cells. Cancer Lett 2002;177:155–161.

48. Niho N, Mutoh M, Sakano K, Takahashi M, Hirano S, Nukaya H, et al. Inhibition of intestinal carcinogenesis by a new flavone derivative, chafuroside, in oolong tea. Cancer Sci 2006;97:248–251.

49. Takahashi M, Fukutake M, Isoi T, Fukuda K, Sato H, Yazawa K, et al. Suppression of azoxymethane-induced rat colon carcinoma development by a fish oil component, docosahexaenoic acid (DHA). Carcinogenesis 1997;18:1337–1342.

50. Oshima M, Takahashi M, Oshima H, Tsutsumi M, Yazawa K, et al. Effects of docosahexaenoic acid (DHA) on intestinal polyp development in Apc delta 716 knockout mice. Carcinogenesis 1995;16:2605–2607.

51. Narisawa T, Fukaura Y, Tanida N, Hasebe M, Ito M, Aizawa R. Chemopreventive efficacy of low dose of pravastatin, an HMG-CoA reductase inhibitor, on 1,2-dimethylhydrazine-induced colon carcinogenesis in ICR mice. Tohoku J Exp Med 1996;180:131–138.

52. Reddy BS, Wang CX, Kong AN, Khor TO, Zheng X, Steele VE, et al. Prevention of azoxymethane-induced colon cancer by combination of low doses of atorvastatin, aspirin, and celecoxib in F344 rats. Cancer Res 2006;66:4542–4546.

53. Swamy MV, Patlolla JMR, Steele VE, Kopelovich L, Reddy BS, Rao CV. Chemoprevention of familial adenomatous polyposis by low dose of atorvastatin and celecoxib given individually and in combination to APC^{min} mice. Cancer Res 2006;66:7370–7377.

54. Yasui Y, Suzuki R, Miyamoto S, Tsukamoto T, Sugie S, Kohno H, et al. A lipophilic statin, pitavastatin, suppresses inflammation-associated mouse colon carcinogenesis. Int J Cancer 2007;121:2331–2339.

55. Teraoka N, Mutoh M, Takasu S, Ueno T, Yamamoto M, Sugimura T, et al. Inhibition of intestinal polyp formation by pitavastatin, a HMG-CoA reductase inhibitor. Cancer Prev Res (Phila) 2011;4:445–453.

56. Poynter JN, Gruber SB, Higgins RD, Almog R, Bonner JD, Rennert HS, et al. Statins and the risk of colorectal cancer. N Engl J Med 2005;352:2184–2192.

57. ALLHAT Officers and Coordinators for the ALLHAT Collaborative Research Group. The anti-hypertensive and lipid-lowering treatment to prevent heart attack trial: major outcomes in moderately hypercholesterolemic, hypertensive patients randomized to pravastatin vs usual care—the antihypertensive and lipid-lowering treatment to prevent heart attack trial (ALLHAT-LLT). JAMA 2002;288:2998–3007.

58. Downs JR, Clearfield M, Weis S, Whitney E, Shapiro DR, Beere PA, et al. Primary prevention of acute coronary events with lovastatin in men and women with average cholesterol levels: results of AFCAPS/TexCAPS-Air Force/Texas coronary atherosclerosis prevention study. JAMA 1998;279:1615–1622.

59. Li H, Schut HA, Conran P, Kramer PM, Lubet RA, Steele VE, et al. Prevention by aspirin and its combination with alpha-difluoromethylornithine of azoxymethane-induced tumors, aberrant crypt foci and prostaglandin E2 levels in rat colon. Carcinogenesis 1999;20:425–430.

60. Rao CV, Indranie C, Simi B, Manning PT, Connor JR, Reddy BS. Chemopreventive properties of a selective inducible nitric oxide synthase inhibitor in colon carcinogenesis, administered alone or in combination with celecoxib, a selective cyclooxygenase-2 inhibitor. Cancer Res 2002;62:165–170.

61. Suh N, Reddy BS, DeCastro A, Paul S, Lee HJ, Smolarek AK, et al. Combination of atorvastatin with sulindac or naproxen profoundly inhibits colonic adenocarcinomas by suppressing the p65/β-catenin/cyclin D1 signaling pathway in rats. Cancer Prev Res (Phila) 2011;4:1895–1902.

62. Kitamura T, Itoh M, Noda T, Tani K, Kobayashi M, Maruyama T, et al. Combined effects of prostaglandin E receptor subtype EP1 and subtype EP4 antagonists on intestinal tumorigenesis in adenomatous polyposis coli gene knockout mice. Cancer Sci 2003;94:618–621.

63. Kitamura T, Itoh M, Noda T, Matsuura M, Wakabayashi K. Combined effects of cyclooxygenase-1 and cyclooxygenase-2 selective inhibitors on intestinal tumorigenesis in adenomatous polyposis coli gene knockout mice. Int J Cancer 2004;109:576–580.

64. Teraoka N, Mutoh M, Takasu S, Ueno T, Nakano K, Takahashi M, et al. High susceptibility to azoxymethane-induced colorectal carcinogenesis in obese KK-A^y mice. Int J Cancer 2011;129:528–535.

65. Mutoh M, Niho N, Komiya M, Takahashi M, Ohtsubo R, Nakatogawa K, et al. Plasminogen activator inhibitor-1 (Pai-1) blockers suppress intestinal polyp formation in Min mice. Carcinogenesis 2008;29:824–829.

66. Giardiello FM, Hamilton SR, Krush AJ, Piantadosi S, Hylind LM, Celano P, et al. Treatment of colonic and rectal adenomas with sulindac in familial adenomatous polyposis. N Engl J Med 1993;328:1313–1316.

67. Ishikawa H, Akedo I, Suzuki T, Narahara H, Otani T. Adverse effects of sulindac used for prevention of colorectal cancer. J Natl Cancer Inst 1997;89:1381.

68. Kim B, Giardiello FM. Chemoprevention in familial adenomatous polyposis. Best Pract Res Clin Gastroenterol 2011;25:607–622.

69. Higuchi T, Iwama T, Yoshinaga K, Toyooka M, Taketo MT, Sugihara K. A randomized double blind, placebo-controlled trial of the effects of rofecoxib, a selective cyclooxygenase-2 inhibitor, on rectal polyps in familial adenomatous polyposis patients. Clin Cancer Res 2003;9:4756–4760.

70. Solomon SD, McMurray JJ, Pfeffer MA, Wittes J, Fowler R, Finn P, et al. [Adenoma Prevention with Celecoxib (APC) Study Investigators]. Cardiovascular risk associated with celecoxib in a clinical trial for colorectal adenoma prevention. N Engl J Med 2005;352:1071–1080.

71. Bresalier RS, Sandler RS, Quan H, Bolognese JA, Oxenius B, Horgan K, et al. [Adenomatous Polyp Prevention on Vioxx (APPROVe) Trial Investigators]. Cardiovascular events associated with rofecoxib in a colorectal adenoma chemoprevention trial. N Engl J Med 2005;352:1092–1102.

72. West NJ, Clark SK, Phillips RK, Hutchinson JM, Leicester RJ, Belluzzi A, et al. Eicosapentaenoic acid reduces rectal polyp number and size in familial adenomatous polyposis. Gut 2010;59:918–925.

73. Burn J, Gerdes AM, Macrae F, Mecklin JP, Moeslein G, Olschwang S, et al. (CAPP2 Investigators). Long-term effect of aspirin on cancer risk in carriers of hereditary colorectal cancer: an analysis from the CAPP2 randomised controlled trial. Lancet 2011;378:2081–2087.

CHAPTER 21

NUTRACEUTICALS AND COLON CANCER PREVENTION

Deepak Poudyal and Lorne J. Hofseth

The American Cancer Society (ACS) has listed colorectal cancer (CRC) as the third most common cancer in men and the second in women. In the United States, ACS estimated 141,210 new cases of colorectal cancer and 49,380 deaths in 2011. According to the second expert report, around 1 million cases of colon and rectal cancer were recorded in 2002, accounting for around 9% overall cancer, of which nearly 530,000 deaths were recorded in the same year, making it the fourth most common cause of death from cancer (1). Approximately 95% of colorectal cancers are adenocarcinoma; the rest include mucinous carcinomas and adenosquamous carcinomas (2). CRC often produces symptoms at an early stage, making it treatable through the detection and removal of the precancerous lesions, meaning that survival rates are relatively high (1). The risk of CRC increases with age. Modifiable factors associated with increased risk of CRC are obesity, physical inactivity, a diet high in red or processed meat, heavy alcohol consumption, and smoking (3). Colorectal cancer is one of the two major cancers whose risk is commonly agreed to be modified primarily by food and nutrition (1).

In 1914, Peyton Rous observed that the restriction of food consumption delayed the development of tumor metastasis in mice (4). In the 1930s, exploration of the role of diet in human cancers began, and even at that stage evidence emerged of the capacity of a higher intake of plant foods to reduce the risk of cancer (5). Nutrients and food constituents have the ability either to inhibit several events that lead to cancer or contribute to cancer development by altering DNA itself or by altering the translation of the genetic message in DNA (1). One landmark study published in the 1980s concluded: "It is highly likely that the United States will eventually have the option of adopting a diet that reduces its incidence of cancer by approximately one third, and it is absolutely certain that another one third could be prevented by abolishing smoking" (6). Cancer development, or carcinogenesis, requires a series of cellular changes starting from tumor initiation, promotion, and progression to form a tumor mass. The capacity of the cell to achieve effective cancer prevention or repair is dependent on the extracellular microenvironment, including the availability of energy

Cancer and Inflammation Mechanisms: Chemical, Biological, and Clinical Aspects, First Edition.
Edited by Yusuke Hiraku, Shosuke Kawanishi, and Hiroshi Ohshima.
© 2014 John Wiley & Sons, Inc. Published 2014 by John Wiley & Sons, Inc.

and the presence of appropriate macro- and micronutrients (1). The panel for the second expert report has found the evidence convincing that diets high in vegetables, dietary factors [high in starch, non-starch polysaccharides (fiber), and carotenoids, all found in foods of plant origin] may decrease the risk of CRC, whereas alcohol and consumption of diets high in red meat and sugar probably increase the risk of this cancer (1).

PATHOGENESIS OF COLON CANCER

The earliest model of pathogenesis of colorectal cancer, the adenoma–carcinoma hypothesis (7), is now widely accepted. The adenoma–carcinoma sequence is a complex multistep process in which cells accumulate a change in multiple genes that control cell growth and differentiation, resulting in a neoplastic phenotype (8). CRC could arise due to either hereditary or non-hereditary conditions. Between 5 and 10% of colorectal cancer is a consequence of recognized hereditary conditions, the two major ones being familial adenomatous polyposis (FAP) and hereditary nonpolyposis colon cancer (HNPCC) (9). The remainder include nonhereditary (sporadic) CRC (SCC) and long standing Inflammatory bowel disease [IBD: ulcerative colitis (UC) or Crohn's disease (CD)] [reviewed in (10)]. Two pathways that lead to CRC are the gatekeeper and caretaker pathways. The gatekeeper pathway is involved in 85% of SCC and also with FAP, which involves the disruption of genes in the adenomatous polyposis coli (APC) pathway that regulate growth (11). Fifteen percent of SCC and HNPCC involve caretaker pathways, where genes that maintain genetic stability are disrupted (11). Three major types of colon cancer are discussed below.

Hereditary Colon Cancer

Patients with a familial risk, those who have two or more first- or second-degree relatives (or both) with CRC, make up approximately 20% of all patients with CRC, whereas approximately 5 to 10% of the total annual burden of CRC is mendelian in nature; that is, it is inherited in an autosomal dominant manner (9).

HNPCC   Hereditary nonpolyposis colorectal cancer, also called Lynch syndrome, accounts for approximately 5 to 8% of all CRC patients (12). HNPCC tumors exhibit microsatellite instability (MSI), occuring mostly in the right colon (proximal end), have diploid DNA, and carry characteristic mutations (TGF-β type II receptor, BAX), behave indolently (13), and harbor mutations in the mismatch-repair (MMR) genes (12). The average age at the onset of HNPCC is approximately 45 years (12). Accelerated carcinogenesis occurs in HNPCC. That is, a tiny colonic adenoma may emerge as a carcinoma within two to three years, compared with 8 to 10 years in the general population (9).

FAP   Familial adenomatous polyposis tumors exhibit chromosomal instability, occur mostly in the left side of the colon (distal origin), have aneuploid DNA, carry characteristic mutations in the *APC*, *p53*, and K-*ras* genes and behave aggressively

(13). Multiple colonic adenomas occur at an early age in patients with familial adenomatous polyposis, occasionally during the preteen years, and proliferate throughout the colon, with malignant degeneration in most patients by the age of 40 to 50 years (9).

Nonhereditary Colon Cancer (Sporadic)

The two main types of genomic instability that contribute to colon carcinogenesis are chromosomal instability (CIN) and microsatellite instability (MSI), accounting for 85 and 15% of SCC, respectively [reviewed in (10)]. Loss of *APC* function is typically an early event in SCC pathogenesis, classifying the APC gene as a "gatekeeper" of the colon (10). Tumors that arise via the CIN/tumor suppressor gene pathway are typically microsatellite stable (MSS), and the remaining 15% of sporadic CRCs arise through the MSI pathway (10). Compared with MSS sporadic colon cancers, MSI sporadic colon cancers are more likely to be diploid (normal DNA content), located in the proximal colon, mucinous, poorly differentiated, show lymphocytic infiltration, and are associated with a more favorable prognosis (14). The major carcinogenic pathways that lead to SCC (i.e., CIN, MSI, and hypermethylation) also occur in colitis-associated colorectal cancers (CAC) [reviewed in (10)]. The earlier average age at the onset of SCC is approximately 63 years (12).

Colitis-Associated Colorectal Cancer

Cancers in the setting of IBD are believed to occur by a progression from no dysplasia to indefinite dysplasia to low-grade dysplasia (LGD) to high-grade dysplasia (HGD) to carcinoma (10). The frequency of CIN (85%) and MSI (15%) in CAC is roughly the same as in SCC (15). For the most part, in IBD, neoplastic lesions arise within areas of the mucosa that have been involved with colonic inflammation (10). This might explain the fact that during inflammation, the healing of UC by reepithelization of colonic mucosa leads to abnormal cell growth, resulting in neoplastic lesions (10).

CHEMOPREVENTIVE MECHANISMS BY NUTRACEUTICALS

Prevention doesn't mean elimination—it means reduction in occurrence, such that at any age fewer people have cancer than otherwise would be the case (1). During colon tumor progression, specific molecular processes have been targeted for chemopreventive intervention, including chronic inflammation, proliferation and differentiation signaling, apoptosis, cell surface growth factor receptors, angiogenesis, and metastasis [reviewed in (16)]. Because therapies including surgery, chemotherapy, radiotherapy, and molecular-targeted therapy are still limited to advanced forms of tumors, the role of dietary substances for both prevention and control of colon cancer through chemopreventive strategies is being investigated [modified from (16)]. At the end of the multistage carcinogenesis pathway, mutated cells will bear some or all of the six hallmarks of the cancer: 1. growth signal autonomy, 2. evasion of growth inhibitory signals, 3. evasion of apoptosis, 4. unlimited replication 5. sustained angiogenesis, and 6. invasion and metastasis (17).

Molecular Targets of Nutraceutical Chemoprevention

Same nutraceutical agents can induce or suppress various pathways during cancer chemoprevention. Nutraceuticals interfere with various molecular pathways that are involved in carcinogenesis. Several key players are involved in triggering cancer progression, thus making them a perfect target during chemoprevention. The overall goal of nutraceuticals is to counteract the hallmarks of cancer: induction of cell-cycle arrest and apoptosis, inhibition of angiogenesis and metastasis, and modulation of signal transduction pathways. The target pathways and molecules that are affected by nutraceuticals are illustrated in Tables 21.1 to 21.4 and described briefly in the sections that follow.

p53 Family Members The tumor suppressor p53 plays a pivotal role in controlling cell cycle, apoptosis, genomic integrity, and DNA repair in response to various genotoxic stresses (18). After activation, p53 can bind to regulatory DNA sequences and activate the expression of target genes, which can be grouped into four categories: cell cycle inhibition (p21, reprimo, cyclin G1, GADD45, 14-3-3), apoptosis (PERP, NOXA, PUMA, p53AIP1, ASPP1/2, Fas, BAX, PIDD), genetic stability (p21, DDB2, MSH2, XPC), and inhibition of angiogenesis (TSP1, Maspin, BAI1, GD-AIF) [reviewed in (18)]. Hence, p53 is one of the most sought after targets of chemopreventive nutraceuticals.

Nuclear Factor-κB (NF-κβ) NF-κβ is a transcription factor consisting of closely related proteins that generally exist as dimers and bind to a common DNA sequence within the promoter/enhancer of target genes, called the κβ site, to promote transcription of target genes [such as COX-2, iNOS, Bcl-2, Bcl-x(L), cyclin D1, MMP, and VEGF] through the recruitment of coactivators and corepressors (18). One of the chemopreventive aspects is inhibition of NF-κβ pathways by nutraceuticals through blocking one of the five members of NF-κβ [p50, p52, p65(Rel A), c-Rel, and Rel B], upstream activators such as TNF-α receptors, and finally, the target genes.

Signal Transducers and Activators of Transcription Pathway (STAT) Constitutive activation of STAT3 and STAT5 led by activation of tyrosine kinases has been linked to multiple myelomas, lymphomas, leukemia, and several solid tumors (19). Inhibition of STAT3 phosphorylation could be an important mechanism during cancer chemoprevention by nutraceuticals.

Cyclooxygenase-2 (COX-2) COX-2 is an inducible prostaglandin G/H synthase that is involved in prostaglandin (PG) synthesis. Overproduction of COX-2 and PG production from free arachidonic acid have been implicated in colon carcinogenesis (20). COX-2-mediated increased PGE2 levels have been believed to enhance tumor

TABLE 21.1 MicroRNA and Nutraceuticals in Colon Cancer Prevention[a]

MiRs	Regulation	Modulating dietary agents	Main regulated functions	Experimental models/cells	Ref.[b]
miR-663	Up	Resveratrol	Targets canonical TGF-β signaling, PTEN, DICER1	SW480	(1)
miR-146	Up	Polypherolic extract	Inflammation	Human colon fibroblast, CCD-18Co cells	(2)
Let-7d, miR-15, miR-107, miR-191, miR-324	Up	PUFA	Cell proliferation, apoptosis	Sprague–Dawley rats (AOM$^+$)	(3)
miR-21	Down	Curcumin	Cell proliferation, Invasion, and metastasis	HCT-116	(4)
miR-155	Down	Quercetin, isorhamnetin	Inflammation	RAW264.7 cells (LPS$^+$) and C57BL/6 mice	(5)
miR-106b	Down	PUFA	Cell proliferation	HCT-116	(6)
miR-29b	Up	Hexane extract of American ginseng	NF-κβ pathway, inflammation	HCT-116	UD (D.P., L.H.)

[a] AOM, azoxymethane; LPS, lipopolysaccharide; UD, unpublished data.

[b]
1. Tili E, Michaille JJ, Alder H, Volinia S, Delmas D, Latruffe N, et al. Resveratrol modulates the levels of microRNAs targeting genes encoding tumor-suppressors and effectors of TGFbeta signaling pathway in SW480 cells. Biochem Pharmacol 2010;80:2057–2065.
2. Noratto GD, Kim Y, Talcott ST, Mertens-Talcott SU. Flavonol-rich fractions of yaupon holly leaves (*Ilex vomitoria*, Aquifoliaceae) induce microRNA-146a and have anti-inflammatory and chemopreventive effects in intestinal myofibroblast CCD-18Co cells. Fitoterapia 2011;82:557–569.
3. Davidson LA, Wang N, Shah MS, Lupton JR, Ivanov I, Chapkin RS. n-3 Polyunsaturated fatty acids modulate carcinogen-directed non-coding microRNA signatures in rat colon. Carcinogenesis. 2009;30:2077–2084.
4. Mudduluru G, George-William JN, Muppala S, Asangani IA, Kumarswamy R, Nelson LD, et al. Curcumin regulates miR-21 expression and inhibits invasion and metastasis in colorectal cancer. Biosci Rep. 2011;31:185–197.
5. Boesch-Saadatmandi C, Loboda A, Wagner AE, Stachurska A, Jozkowicz A, Dulak J, et al. Effect of quercetin and its metabolites isorhamnetin and quercetin-3-glucuronide on inflammatory gene expression: role of miR-155. J Nutr Biochem 2011;22:293–299.
6. Hu S, Dong TS, Dalal SR, Wu F, Bissonnette M, Kwon JH, et al. The microbe-derived short chain fatty acid butyrate targets miRNA-dependent p21 gene expression in human colon cancer. PLoS One 2011;6:e16221.

305

TABLE 21.2 Colon Cancer Prevention in *In Vitro* Cell model by Nutraceuticals

Nutraceuticals or dietary agents	Main regulated function	Targets	Models/cells	Refs.[a]
FLAVONOIDS				
1. Quercetin	Inhibits cell growth	β-Catenin	SW480	(1)
		EFGR and MAPK pathway	HT29	(2)
	Induces apoptosis	p53, p21, and AMPK	HT29	(3)
	Induces autophagy	Oncogenic RAS		(4)
2. Apigenin	Induces apoptosis	APC	HT29	(5)
	Cell-cycle arrest	APC, decreased CDC-42 and cyclin B1	HT29, SW480, Caco2	(5,6)
3. Hydroxylated polymethoxy flavones (PMFs) (citrus origin)	Inhibits cellular proliferation and induces apoptosis	p21Cip1/Waf1, CDK-2, CDK-4, phospho-Rb, Mcl-1, caspase3 and -8, PARP	HT29, HCT116	(7)
4. Epigallocatechin-3-gallate (EGCG)	Antiproliferation and senescence	Down-regulation of EGFR via MAPK, inhibition of telomerase and topoisomerase I	SW480, HCT116, HT29	(8–10)

[a]
1. Park CH, Chang JY, Hahm ER, Park S, Kim HK, Yang CH. Quercetin, a potent inhibitor against beta-catenin/Tcf signaling in SW480 colon cancer cells. Biochem Biophys Res Commun 2005;328:227–234.
2. Fridrich D, Teller N, Esselen M, Pahlke G, Marko D. Comparison of delphinidin, quercetin and (−)-epigallocatechin-3-gallate as inhibitors of the EGFR and the ErbB2 receptor phosphorylation. Mol Nutr Food Res 2008;52:815–822.
3. Kim HJ, Kim SK, Kim BS, Lee SH, Park YS, Park BK, et al. Apoptotic effect of quercetin on HT-29 colon cancer cells via the AMPK signaling pathway. J Agric Food Chem 2010;58:8643–8650.
4. Psahoulia FH, Moumtzi S, Roberts ML, Sasazuki T, Shirasawa S, Pintzas A. Quercetin mediates preferential degradation of oncogenic Ras and causes autophagy in Ha-RAS-transformed human colon cells. Carcinogenesis 2007;28:1021–1031.
5. Chung CS, Jiang Y, Cheng D, Birt DF. Impact of adenomatous polyposis coli (APC) tumor supressor gene in human cokon cancer cell lines on cell cycle arrest by apigenin. Mol Carcinog 2007;46:773–782.
6. Wang W, Heideman L, Chung CS, Pelling JC, Koehler KJ, Birt DF. Cell-cycle arrest at G2/M and growth inhibition by apigenin in human colon carcinoma cell lines. Mol Carcinog 2000;28:102–110.
7. Qiu P, Dong P, Guan H, Li S, Ho CT, Pan MH, et al. Inhibitory effects of 5-hydroxy polymethoxyflavones on colon cancer cells. Mol Nutr Food Res 2010;54 (Suppl 2):S244–S252.
8. Adachi S, Shimizu M, Shirakami Y, Yamauchi J, Natsume H, Matsushima-Nishiwaki R, et al. (−)-Epigallocatechin gallate downregulates EGF receptor via phosphorylation at Ser1046/1047 by p38 MAPK in colon cancer cells. Carcinogenesis 2009;30:1544–1552.
9. Naasani I, Oh-Hashi F, Oh-Hara T, Feng WY, Johnston J, Chan K, et al. Blocking telomerase by dietary polyphenols is a major mechanism for limiting the growth of human cancer cells *in vitro* and *in vivo*. Cancer Res 2003;63:824–830.
10. Berger SJ, Gupta S, Belfi CA, Gosky DM, Mukhtar H. Green tea constituent (−)-epigallocatechin-3-gallate inhibits topoisomerase I activity in human colon carcinoma cells. Biochem Biophys Res Commun 2001;288:101–105.

TABLE 21.2 (Continued)

Nutraceuticals or dietary agents	Main regulated function	Targets	Models/cells	Refs.[a]
5. Delphinidin	Inhibits invasion and metastasis	Suppression of ERK1/2, Akt and NF-κβ signaling pathways	HCT116, SW620	(11,12)
	Induces apoptosis	Activation of caspase cascade, increasing Bax protein, decreasing Bcl-2 protein, and suppression of NF-κβ pathways	HCT116	(13)
6. Cyanidin	Cytotoxicity and antimetastasis	Inhibition of glutathione reductase, depleting glutathione, and accumulating cellular ROS	LoVo/ADR	(14)
	Antiproliferative and anti-inflammatory	Reduction of iNOS and COX-2 expression	HT29	(15)
7. Pro-anthocyanidin	Cytotoxicity and antimetastasis	Inhibition of glutathione reductase, depleting glutathione, and accumulating cellular ROS	LoVo/ADR	(14)
	Induces apoptosis	Down-regulation of PI3K and Akt signaling pathways	Caco2	(16)
8. Fisetin	Induces apoptosis	Inhibition of COX-2 and Wnt/EGFR/NF-κβ signaling pathways	HT29	(17)

11. Larsen CA, Dashwood RH. (−)-Epigallocatechin-3-gallate inhibits Met signaling, proliferation, and invasiveness in human colon cancer cells. Arch Biochem Biophys 2010;501:52–57.

12. Zhou F, Zhou H, Wang T, Mu Y, Wu B, Guo DL, et al. Epigallocatechin-3-gallate inhibits proliferation and migration of human colon cancer SW620 cells in vitro. Acta Pharmacol Sin 2012;33:120–126.

13. Yun JM, Afaq F, Khan N, Mukhtar H. Delphinidin, an anthocyanidin in pigmented fruits and vegetables, induces apoptosis and cell cycle arrest in human colon cancer HCT116 cells. Mol Carcinog 2009;48:260–270.

14. Cvorovic J, Tramer F, Granzotto M, Candussio L, Decorti G, Passamonti S. Oxidative stress-based cytotoxicity of delphinidin and cyanidin in colon cancer cells. Arch Biochem Biophys 2010;501:151–157.

15. Kim JM, Kim JS, Yoo H, Choung MG, Sung MK. Effects of black soybean [Glycine max (L.) Merr.] seed coats and its anthocyanidins on colonic inflammation and cell proliferation in vitro and in vivo. J Agric Food Chem 2008;56:8427–8433.

16. Engelbrecht AM, Mattheyse M, Ellis B, Loos B, Thomas M, Smith R, et al. Proanthocyanidin from grape seeds inactivates the PI3-kinase/PKB pathway and induces apoptosis in a colon cancer cell line. Cancer Lett 2007;258:144–153.

17. Suh Y, Afaq F, Johnson JJ, Mukhtar H. A plant flavonoid fisetin induces apoptosis in colon cancer cells by inhibition of COX2 and Wnt/EGFR/NF-kappaB-signaling pathways. Carcinogenesis 2009;30:300–307.

307

TABLE 21.2 (Continued)

Nutraceuticals or dietary agents	Main regulated function	Targets	Models/cells	Refs.[a]
9. Genistein	Antiproliferation	Dephosphorylation and nuclear retention of FOXO3 (active state), inhibition β-catenin mediated WNT signaling through increasing sFRP2 gene expression	DLD1	(18,19)
	Induces apoptosis	Regulation of p21Waf1 and Bax/Bcl-2 expression	HT29	(20)
10. Silibinin	Antiproliferative	Inhibition of cyclin-CDK promoter activity, increases p21 and p27 protein expression	HCT116	(21)
	Induces apoptosis	TRAIL mediated	SW480, SW620	(22,23)
POLYPHENOLICS				
1. Curcumin (diferuloylmethane)	Induces apoptosis	Oxidative stress, cleavage of PARP, caspase3 and reduction in Bcl-xL level, production of ROS and Ca^{2+}, down-regulation of PPAR δ, 14-3-3ε, and VEGF, inhibition of proteasome, TRAIL-induced ROS-mediated DR-5 up-regulation	HCT116, HT29, HCT116, $p53^{-/-}$, Colo205, HCT116, SW480	(24–29)

18. Qi W, Weber CR, Wasland K, Savkovic SD. Genistein inhibits proliferation of colon cancer cells by attenuating a negative effect of epidermal growth factor on tumor suppressor FOXO3 activity. BMC Cancer 2011;11:219.

19. Zhang Y, Chen H. Genistein attenuates WNT signaling by up-regulating sFRP2 in a human colon cancer cell line. Exp Biol Med (Maywood) 2011;236:714–722.

20. Yu Z, Li W, Liu F. Inhibition of proliferation and induction of apoptosis by genistein in colon cancer HT-29 cells. Cancer Lett 2004;215:159–166.

21. Hogan FS, Krishnegowda NK, Mikhailova M, Kahlenberg MS. Flavonoid, silibinin, inhibits proliferation and promotes cell-cycle arrest of human colon cancer. J Surg Res 2007;143:58–65.

22. Kauntz H, Bousserouel S, Gosse F, Raul F. Silibinin triggers apoptotic signaling pathways and autophagic survival response in human colon adenocarcinoma cells and their derived metastatic cells. Apoptosis 2011;16:1042–1053.

23. Kauntz H, Bousserouel S, Gosse F, Raul F. The flavonolignan silibinin potentiates TRAIL-induced apoptosis in human colon adenocarcinoma and in derived TRAIL-resistant metastatic cells. Apoptosis 2012;17:797–809.

24. Watson JL, Hill R, Yaffe PB, Greenshields A, Walsh M, Lee PW, et al. Curcumin causes superoxide anion production and p53-independent apoptosis in human colon cancer cells. Cancer Lett 2010;297:1–8.

25. Moragoda L, Jaszewski R, Majumdar AP. Curcumin induced modulation of cell cycle and apoptosis in gastric and colon cancer cells. Anticancer Res 2001;21:873–878.

TABLE 21.2 (Continued)

Nutraceuticals or dietary agents	Main regulated function	Targets	Models/cells	Refs.[a]
	Induces autophagy and senescence	Through SA-β-galactosidase activation	HCT116	(30)
	Antiproliferation	Inhibition Wnt/β-catenin pathway by decreasing transcriptional coactivator p300, suppression of EGFR gene expression, and modulation of Akt/mTOR signaling	SW480, HCT116, DLD1, HCT15	(31–33)
	Inhibits invasion	Down-regulation of COX-2, MMP-2 expression via inhibition of p65	Colo205	(34)

26. Su CC, Lin JG, Li TM, Chung JG, Yang JS, Ip SW, et al. Curcumin-induced apoptosis of human colon cancer colo 205 cells through the production of ROS, Ca2+ and the activation of caspase-3. Anticancer Res 2006;26:4379–4389.

27. Wang JB, Qi LL, Zheng SD, Wang HZ, Wu TX. Curcumin suppresses PPARdelta expression and related genes in HT-29 cells. World J Gastroenterol 2009;15:1346–1352.

28. Milacic V, Banerjee S, Landis-Piwowar KR, Sarkar FH, Majumdar AP, Dou QP. Curcumin inhibits the proteasome activity in human colon cancer cells *in vitro* and *in vivo*. Cancer Res 2008;68:7283–7292.

29. Jung EM, Lim JH, Lee TJ, Park JW, Choi KS, Kwon TK. Curcumin sensitizes tumor necrosis factor–related apoptosis-inducing ligand (TRAIL)–induced apoptosis through reactive oxygen species-mediated upregulation of death receptor 5 (DR5). Carcinogenesis 2005;26:1905–1913.

30. Mosieniak G, Adamowicz M, Alster O, Jaskowiak H, Szczepankiewicz AA, Wilczynski GM, et al. Curcumin induces permanent growth arrest of human colon cancer cells: link between senescence and autophagy. Mech Ageing Dev 2012;133:444–455.

31. Ryu MJ, Cho M, Song JY, Yun YS, Choi IW, Kim DE, et al. Natural derivatives of curcumin attenuate the Wnt/beta-catenin pathway through down-regulation of the transcriptional coactivator p300. Biochem Biophys Res Commun 2008;377:1304–1308.

32. Chen A, Xu J, Johnson AC. Curcumin inhibits human colon cancer cell growth by suppressing gene expression of epidermal growth factor receptor through reducing the activity of the transcription factor Egr-1. Oncogene 2006;25:278–287.

33. Johnson SM, Gulhati P, Arrieta I, Wang X, Uchida T, Gao T, et al. Curcumin inhibits proliferation of colorectal carcinoma by modulating Akt/mTOR signaling. Anticancer Res 2009;29:3185–3190.

34. Su CC, Chen GW, Lin JG, Wu LT, Chung JG. Curcumin inhibits cell migration of human colon cancer colo 205 cells through the inhibition of nuclear factor kappa B /p65 and down-regulates cyclooxygenase-2 and matrix metalloproteinase-2 expressions. Anticancer Res 2006;26:1281–1288.

TABLE 21.2 (Continued)

Nutraceuticals or dietary agents	Main regulated function	Targets	Models/cells	Refs.[a]
2. Resveratrol (3,5,3'-trihydroxystilbene)	Antiproliferative	Inhibition of IGF-1R, Akt/Wnt pathway, inhibition of polyamine synthesis and increased polyamine catabolism through cFOS activity, G2 arrest through inhibition of CDK7 and p34CDC2 kinase, reduced telomerase activity, targets p38MAPK and PPAR γ and SSAT activity	Caco2, HT29, WiDr, Caco2, HCT116	(35–39)
	Induces apoptosis	Activation p53 proteins, ROS-triggered autophagy, AMPK signaling	HT29, SW480	(35,40,41)
	Inhibits metastasis	Inhibition VEGF and MMP-9 expression	Lovo	(42)
3. Pterostilbene (dimethyl ether analog of resveratrol)	Antiproliferative	Decrease in c-Myc and cyclin D1 expression, suppression of iNOS and COX-2 expression through down-regulation of p38 cascade	HT29	(43)

35. Vanamala J, Reddivari L, Radhakrishnan S, Tarver C. Resveratrol suppresses IGF-1 induced human colon cancer cell proliferation and elevates apoptosis via suppression of IGF-1R/Wnt and activation of p53 signaling pathways. BMC Cancer 2010;10:238.

36. Wolter F, Turchanowa L, Stein J. Resveratrol-induced modification of polyamine metabolism is accompanied by induction of c-Fos. Carcinogenesis 2003;24:469–474.

37. Liang YC, Tsai SH, Chen L, Lin-Shiau SY, Lin JK. Resveratrol-induced G2 arrest through the inhibition of CDK7 and p34CDC2 kinases in colon carcinoma HT29 cells. Biochem Pharmacol 2003;65:1053–1060.

38. Fuggetta MP, Lanzilli G, Tricarico M, Cottarelli A, Falchetti R, Ravagnan G, et al. Effect of resveratrol on proliferation and telomerase activity of human colon cancer cells in vitro. J Exp Clin Cancer Res 2006;25:189–193.

39. Ulrich S, Loitsch SM, Rau O, von Knethen A, Brune B, Schubert-Zsilavecz M, et al. Peroxisome proliferator-activated receptor gamma as a molecular target of resveratrol-induced modulation of polyamine metabolism. Cancer Res 2006;66:7348–7354.

40. Miki H, Uehara N, Kimura A, Sasaki T, Yuri T, Yoshizawa K, et al. Resveratrol induces apoptosis via ROS-triggered autophagy in human colon cancer cells. Int J Oncol 2012;40:1020–1028.

41. Hwang JT, Kwak DW, Lin SK, Kim HM, Kim YM, Park OJ. Resveratrol induces apoptosis in chemoresistant cancer cells via modulation of AMPK signaling pathway. Ann NY Acad Sci 2007;1095:441–448.

42. Wu H, Liang X, Fang Y, Qin X, Zhang Y, Liu J. Resveratrol inhibits hypoxia-induced metastasis potential enhancement by restricting hypoxia-induced factor-1 alpha expression in colon carcinoma cells. Biomed Pharmacother 2008;62:613–621.

43. Paul S, Rimando AM, Lee HJ, Ji Y, Reddy BS, Suh N. Anti-inflammatory action of pterostilbene is mediated through the p38 mitogen-activated protein kinase pathway in colon cancer cells. Cancer Prev Res (Phila) 2009;2:650–657.

TABLE 21.2 (Continued)

Nutraceuticals or dietary agents	Main regulated function	Targets	Models/cells	Refs.[a]
CAROTENOIDS				
1. β-carotene	Antiproliferative	Induction of apoptosis	HT29	(44)
2. Lycopene	Antiproliferative	Suppression of Akt activation, decreased β-catenin and cyclin B1 expression and increased nuclear level of p27, inactivation of Ras signaling	HT29, HCT116, HT29	(45,46)
	Inhibits invasion	Suppression of MMP-7 and Akt pathway	HT29	(47)
ISOTHIOCYANATES				
1. Phenethyl isothiocyanate (PEITC)	Antiproliferative	Decrease in cyclin (A, D, and E) expression by activating p38MAPK	HT29	(48)
	Induces apoptosis	Mitochondrial caspase cascade and activation of JNK pathway	HT29	(49)
2. Benzyl isothiocyanate (BITC)	Antimetastasis	Down-regulation of MMP gene expression via suppression of PKC, MAPK, and NK-κβ levels	HT29	(50)

44. Briviba K, Schnabele K, Schwertle E, Blockhaus M, Rechkemmer G. Beta-carotene inhibits growth of human colon carcinoma cells *in vitro* by induction of apoptosis. Biol Chem 2001;382:1663–1668.

45. Tang FY, Shih CJ, Cheng LH, Ho HJ, Chen HJ. Lycopene inhibits growth of human colon cancer cells via suppression of the Akt signaling pathway. Mol Nutr Food Res 2008;52:646–654.

46. Palozza P, Colangelo M, Simone R, Catalano A, Boninsegna A, Lanza P, et al. Lycopene induces cell growth inhibition by altering mevalonate pathway and Ras signaling in cancer cell lines. Carcinogenesis 2010;31:1813–1821.

47. Lin MC, Wang FY, Kuo YH, Tang FY. Cancer chemopreventive effects of lycopene: suppression of MMP-7 expression and cell invasion in human colon cancer cells. J Agric Food Chem 2011;59:11304–11318.

48. Cheung KL, Khor TO, Yu S, Kong AN. PEITC induces G1 cell cycle arrest on HT-29 cells through the activation of p38 MAPK signaling pathway. AAPS J 2008;10:277–281.

49. Hu R, Kim BR, Chen C, Hebbar V, Kong AN. The roles of JNK and apoptotic signaling pathways in PEITC-mediated responses in human HT-29 colon adenocarcinoma cells. Carcinogenesis 2003;24:1361–1367.

50. Lai KC, Huang AC, Hsu SC, Kuo CL, Yang JS, Wu SH, et al. Benzyl isothiocyanate (BITC) inhibits migration and invasion of human colon cancer HT29 cells by inhibiting matrix metalloproteinase-2/-9 and urokinase plasminogen (uPA) through PKC and MAPK signaling pathway. J Agric Food Chem 2010;58:2935–2942.

TABLE 21.2 (*Continued*)

Nutraceuticals or dietary agents	Main regulated function	Targets	Models/cells	Refs.[a]
INDOLES				
1. 3,3′-Diindolylmethane (DIM)	Cell-cycle arrest	Inhibition of CDK2 activity and increase in protein levels of p21 and p27	HT29	(51)
	Induces apoptosis	Proteasomal degradation of class I HDACs and induction of DNA damage	HT29	(52)
RETINOLS				
1. Retinol	Antimetastasis	Decreases in MMP-2 and -9 activity, proteasomal degradation of β-catenin, and suppression of PI3K activity	HCT116, SW620	(53–55)
POLYUNSATURATED FATTY ACID (PUFA)				
1. ω-3-PUFA docosahexaenoic acid (DHA)	Antiproliferative	Suppression of PGE2 production and COX-2 expression	LS174T	(56)
	Induces apoptosis	Down-regulation of Akt signaling, proteasomal degradation of β-catenin	Caco2, SW480, HCT116	(57,58)

51. Choi HJ, Lim do Y, Park JH. Induction of G1 and G2/M cell cycle arrests by the dietary compound 3,3′-diindolylmethane in HT-29 human colon cancer cells. BMC Gastroenterol 2009;9:39.

52. Li Y, Li X, Guo B. Chemopreventive agent 3,3′-diindolylmethane selectively induces proteasomal degradation of class I histone deacetylases. Cancer Res 2010;70:646–654.

53. Park EY, Wilder ET, Lane MA. Retinol inhibits the invasion of retinoic acid–resistant colon cancer cells *in vitro* and decreases matrix metalloproteinase mRNA, protein, and activity levels. Nutr Cancer 2007;57:66–77.

54. Dillard AC, Lane MA. Retinol increases beta-catenin-RXRalpha binding, leading to the increased proteasomal degradation of beta-catenin and RXRalpha. Nutr Cancer 2008;60:97–108.

55. Park EY, Wilder ET, Chipuk JE, Lane MA. Retinol decreases phosphatidylinositol 3-kinase activity in colon cancer cells. Mol Carcinog 2008;47:264–274.

56. Habbel P, Weylandt KH, Lichopoj K, Nowak J, Purschke M, Wang JD, et al. Docosahexaenoic acid suppresses arachidonic acid–induced proliferation of LS-174T human colon carcinoma cells. World J Gastroenterol 2009;15:1079–1084.

57. Toit-Kohn JL, Louw L, Engelbrecht AM. Docosahexaenoic acid induces apoptosis in colorectal carcinoma cells by modulating the PI3 kinase and p38 MAPK pathways. J Nutr Biochem 2009;20:106–114.

58. Calviello G, Resci F, Serini S, Piccioni E, Toesca A, Boninsegna A, et al. Docosahexaenoic acid induces proteasome-dependent degradation of survivin and apoptosis in human colorectal cancer cells not expressing COX-2. Carcinogenesis 2007;28:1202–1209.

TABLE 21.2 *(Continued)*

Nutraceuticals or dietary agents	Main regulated function	Targets	Models/cells	Refs.[a]
2. Conjugated Linoleic acid (CLA)	Antiproliferative	Reduction in phosphorylation of ERK1/2 and downstream c-myc	Caco2	(59)
	Anti-inflammatory	Decrease in expression of TNF-α, IL-6, and IL-12	Caco2	(60)
ORIENTAL HERBAL MEDICINE EXTRACT				
1. Ginseng Extract	Antiproliferative	Increased in p53 and p21 expression	HCT116	(61)
2. Ginsenosides (active component)	Induces apoptosis	Decrease in NF-κβ activity, AMPK signaling	HT29	(62,63)
3. *Ginkgo biloba* extract	Induces cell-cycle arrest	Increase in caspase3 activity, increase in p53, and reduction bcl-2 mRNA	HT29	(64)

59. Bocca C, Bozzo F, Gabriel L, Miglietta A. Conjugated linoleic acid inhibits Caco-2 cell growth via ERK-MAPK signaling pathway. J Nutr Biochem 2007;18:332–340.

60. Reynolds CM, Loscher CE, Moloney AP, Roche HM. *Cis-9, trans-11*-conjugated linoleic acid but not its precursor *trans*-vaccenic acid attenuate inflammatory markers in the human colonic epithelial cell line Caco-2. Br J Nutr 2008;100:13–17.

61. King ML, Murphy LL. Role of cyclin inhibitor protein p21 in the inhibition of HCT116 human colon cancer cell proliferation by American ginseng (*Panax quinquefolius*) and its constituents. Phytomedicine 2010;17:261–268.

62. Kim SM, Lee SY, Yuk DY, Moon DC, Choi SS, Kim Y, et al. Inhibition of NF-kappaB by ginsenoside Rg3 enhances the susceptibility of colon cancer cells to docetaxel. Arch Pharm Res 2009;32:755–765.

63. Yuan HD, Quan HY, Zhang Y, Kim SH, Chung SH. 20(*S*)-Ginsenoside Rg3-induced apoptosis in HT-29 colon cancer cells is associated with AMPK signaling pathway. Mol Med Rep 2010;3:825–831.

64. Chen XH, Miao YX, Wang XJ, Yu Z, Geng MY, Han YT, et al. Effects of *Ginkgo biloba* extract EGb761 on human colon adenocarcinoma cells. Cell Physiol Biochem 2011;27:227–232.

TABLE 21.3   Colon Cancer Chemoprevention by Nutraceuticals, Animal Studies

Nutraceuticals	Chemopreventive mechanisms	Animal experimental model	Refs.[a]
FLAVONOIDS			
1. Quercetin	Reduces polyp number and size distribution by reducing macrophage infiltration	Apc$^{Min/+}$ mice	(1)
	Suppresses development of preneoplastic lesion and their proliferation (2), decreases AOM-induced colonic neoplasia (3)	C57BL/KsJ-db/db (AOM+) (2); CF1 mice (AOM$^+$) (3)	(2,3)
	Reduces the number of ACFs	Rats (AOM$^+$) (4,5)	(4, 5)
2. Apigenin	Reduces number of ACF by reducing the ornithine decarboxylase activity	CF1 mice (AOM$^+$)	(6)
3. EGCG	Reduces tumor multiplicity	Apc$^{Min/+}$ mice (7)	(7)
	Inhibits angiogenesis by blocking the induction of VEGF	Xenograft model in athymic BALB/c nude (HT29$^+$) mice	(8)
	Reduces the number of ACF through reduction of p-GSK-3β, β-catenin, COX-2 and cyclin D1 level	C57BL/KsJ-db/db (AOM$^+$) mice	(9)

[a] 1. Murphy EA, Davis JM, McClellan JL, Carmichael MD. Quercetin's effects on intestinal polyp multiplicity and macrophage number in the Apc(Min/+) mouse. Nutr Cancer 2011;63:421–426.

2. Miyamoto S, Yasui Y, Ohigashi H, Tanaka T, Murakami A. Dietary flavonoids suppress azoxymethane-induced colonic preneoplastic lesions in male C57BL/KsJ-db/db mice. Chem Biol Interact 2010;183:276–283.

3. Deschner EE, Ruperto J, Wong G, Newmark HL. Quercetin and rutin as inhibitors of azoxymethanol-induced colonic neoplasia. Carcinogenesis 1991;12:1193–1196.

4. Choi SY, Park JH, Kim JS, Kim MK, Aruoma OI, Sung MK. Effects of quercetin and beta-carotene supplementation on azoxymethane-induced colon carcinogenesis and inflammatory responses in rats fed with high-fat diet rich in omega-6 fatty acids. Biofactors 2006;27:137–146.

5. Volate SR, Davenport DM, Muga SJ, Wargovich MJ. Modulation of aberrant crypt foci and apoptosis by dietary herbal supplements (quercetin, curcumin, silymarin, ginseng and rutin). Carcinogenesis 2005;26:1450–1456.

6. Au A, Li B, Wang W, Roy H, Koehler K, Birt D. Effect of dietary apigenin on colonic ornithine decarboxylase activity, aberrant crypt foci formation, and tumorigenesis in different experimental models. Nutr Cancer 2006;54:243–251.

7. Bose M, Hao X, Ju J, Husain A, Park S, Lambert JD, et al. Inhibition of tumorigenesis in ApcMin/+ mice by a combination of (−)-epigallocatechin-3-gallate and fish oil. J Agric Food Chem 2007;55:7695–7700.

8. Jung YD, Kim MS, Shin BA, Chay KO, Ahn BW, Liu W, et al. EGCG, a major component of green tea, inhibits tumour growth by inhibiting VEGF induction in human colon carcinoma cells. Br J Cancer 2001;84:844–850.

9. Shimizu M, Shirakami Y, Sakai H, Adachi S, Hata K, Hirose Y, et al. (−)-Epigallocatechin gallate suppresses azoxymethane-induced colonic premalignant lesions in male C57BL/KsJ-db/db mice. Cancer Prev Res (Phila) 2008;1:298–304.

TABLE 21.3 *(Continued)*

Nutraceuticals	Chemopreventive mechanisms	Animal experimental model	Refs.[a]
4. Naringenin	Reduces the number of ACFs and increases the apoptosis of luminal surface colonocytes (10)	Male Sprague–Dawley rats (AOM[+])	(10)
5. Limonin (limonoids, citrus fruit)	Reduces the ACF number by suppressing proliferation and elevating apoptosis through anti-inflammatory activities (11)	Male Sprague–Dawley rats (AOM[+]) (11); male Fisher F344 rats (AOM[+]) (12)	(11,12)
6. Genistein	Reduces colonic inflammation and COX-2 gene expression	Wistar rats (TNBS[+])	(13)
7. Soy isoflavonones	Prevents development of early colon neoplasia by suppressing Wnt/β-catenin signaling	Male Sprague–Dawley rats (AOM[+])	(14)
8. Anthocyanins (tart cherry)	Suppresses ACF and COX-2 expression	Male Sprague–Dawley rats (DMH[+])	(15)
	Inhibits intestinal tumor development by reducing tumor number and volume	Apc[Min/+]	(16)
9. Pro-anthocyanidin	Reduces ACF formation and colonic cell proliferation by lowering PCNA expression and inducing apoptosis	Fisher F344 rats (AOM[+])	(17)

10. Leonardi T, Vanamala J, Taddeo SS, Davidson LA, Murphy ME, Patil BS, et al. Apigenin and naringenin suppress colon carcinogenesis through the aberrant crypt stage in azoxymethane-treated rats. Exp Biol Med (Maywood) 2010;235:710–717.

11. Vanamala J, Leonardi T, Patil BS, Taddeo SS, Murphy ME, Pike LM, et al. Suppression of colon carcinogenesis by bioactive compounds in grapefruit. Carcinogenesis 2006;27:1257–1265.

12. Tanaka T, Maeda M, Kohno H, Murakami M, Kagami S, Miyake M, et al. Inhibition of azoxymethane-induced colon carcinogenesis in male F344 rats by the citrus limonoids obacunone and limonin Carcinogenesis 2001;22:193–198.

13. Seibel J, Molzberger AF, Hertrampf T, Laudenbach-Leschowski U, Diel P. Oral treatment with genistein reduces the expression of molecular and biochemical markers of inflammation in a rat model of chronic TNBS-induced colitis. Eur J Nutr 2009;48:213–220.

14. Zhang Y, Li Q, Zhou D, Chen H. Genistein, a soya isoflavone, prevents azoxymethane-induced up-regulation of WNT/beta-catenin signalling and reduces colon pre-neoplasia in rats. Br J Nutr 2012:1–10.

15. Min WK, Sung HY, Choi YS. Suppression of colonic aberrant crypt foci by soy isoflavones is dose-independent in dimethylhydrazine-treated rats. J Med Food 2010;13:495–502.

16. Kang SY, Seeram NP, Nair MG, Bourquin LD. Tart cherry anthocyanins inhibit tumor development in Apc(Min) mice and reduce proliferation of human colon cancer cells. Cancer Lett 2003;194:13–19.

17. Nomoto H, Iigo M, Hamada H, Kojima S, Tsuda H. Chemoprevention of colorectal cancer by grape seed proanthocyanidin is accompanied by a decrease in proliferation and increase in apoptosis. Nutr Cancer 2004;49:81–88.

TABLE 21.3 (Continued)

Nutraceuticals	Chemopreventive mechanisms	Animal experimental model	Refs.[a]
10. Silibinin	Inhibits ACF formation: through decreasing iNOS, COX-2, and cyclin D1 expression (18); down-regulating IL-1β, TNF-α, MMP7 expression (19)	Male Fisher F344 rats (AOM[+]) (18); Wistar rats (AOM[+]) (19)	(18,19)
	Decreases intestinal polyps via suppression of β-catenin/GSK3β signaling, suppression of cyclin, c-Myc, and cytokine production	Apc[Min/+] mice	(20,21)
	Polyphenolics		
1. Curcumin	Reduces total number of colonic premalignant lesions by activating AMP-activated kinase (AMPK), decreasing COX-2 and NF-κβ activity (attenuating chronic inflammation and improving adipocytokine imbalance)	C57BL/KsJ-db/db (db/db) obese mice (AOM[+])	(22)
	Reduces the incidence and multiplicity of noninvasive and invasive adenocarcinoma (23–25); also reduced COX-2 metabolism (26)	Male F344 rats (AOM[+]) (23, 26); CF1 mice (AOM[+]) (24); C57BL/6J [Min/+] mice (25)	(23–26)
	Decreases the percent of MDSC and IL-6 levels in tumor tissues	Mouse colon cancer allograft model	(27)

18. Velmurugan B, Singh RP, Tyagi A, Agarwal R. Inhibition of azoxymethane-induced colonic aberrant crypt foci formation by silibinin in male Fisher 344 rats. Cancer Prev Res (Phila) 2008;1:376–384.

19. Kauntz H, Bousserouel S, Gosse F, Marescaux J, Raul F. Silibinin, a natural flavonoid, modulates the early expression of chemoprevention biomarkers in a preclinical model of colon carcinogenesis. Int J Oncol 2012;41:849–854.

20. Rajamanickam S, Kaur M, Velmurugan B, Singh RP, Agarwal R. Silibinin suppresses spontaneous tumorigenesis in APC min/+ mouse model by modulating beta-catenin pathway. Pharm Res 2009;26:2558–2567.

21. Rajamanickam S, Velmurugan B, Kaur M, Singh RP, Agarwal R. Chemoprevention of intestinal tumorigenesis in APCmin/+ mice by silibinin. Cancer Res 2010;70:2368–2378.

22. Kubota M, Shimizu M, Sakai H, Yasuda Y, Terakura D, Baba A, et al. Preventive effects of curcumin on the development of azoxymethane-induced colonic preneoplastic lesions in male C57BL/KsJ-db/db obese mice. Nutr Cancer 2012;64:72–79.

23. Kawamori T, Lubet R, Steele VE, Kelloff GJ, Kaskey RB, Rao CV, et al. Chemopreventive effect of curcumin, a naturally occurring anti-inflammatory agent, during the promotion/progression stages of colon cancer. Cancer Res 1999;59:597–601.

TABLE 21.3 (*Continued*)

Nutraceuticals	Chemopreventive mechanisms	Animal experimental model	Refs.[a]
2. Resveratrol	Suppresses DSS induced colitis via down-regulation of p38, PGES-1, iNOS, and COX-2	C57BL/6 Mice	(28)
	Suppresses colitis-associated colon cancer by reducing tumor incidence and multiplicity (29), decreases ACF formation (30,31)	C57BL/6 (AOM + DSS) (29); Rats (DMH[+]) (30,31)	(29–31)
	Modulates early and late events of colon carcinogenesis and maintain colonic mucosal integrity by inhibiting COX-2, ODC, caspase 3 and hsp70 protein	Rats (DMH[+])	(32)
	Down-regulates genes required for cell-cycle progression and proliferation (cyclin D1, D2, DP-1 transcription factor, and Y-box binding protein) and up-regulates gene for recruitment and activation of immune cells (CTLAg-4, LIFR, MCP3) and inhibits TGF-β	Apc[Min/+] mice	(33)

24. Huang MT, Lou YR, Ma W, Newmark HL, Reuhl KR, Conney AH. Inhibitory effects of dietary curcumin on forestomach, duodenal, and colon carcinogenesis in mice. Cancer Res 1994;54:5841–5847.

25. Perkins S, Verschoyle RD, Hill K, Parveen I, Threadgill MD, Sharma RA, et al. Chemopreventive efficacy and pharmacokinetics of curcumin in the min/+ mouse, a model of familial adenomatous polyposis. Cancer Epidemiol Biomark Prev 20C2;11:535–540.

26. Rao CV, Rivenson A, Simi B, Reddy BS. Chemoprevention of colon carcinogenesis by dietary curcumin, a naturally occurring plant phenolic compound. Cancer Res 1995;55:259–266.

27. Tu SP, Jin H, Shi JD, Zhu LM, Suo Y, Lu G, et al. Curcumin induces the differentiation of myeloid-derived suppressor cells and inhibits their interaction with cancer cells and related tumor growth. Cancer Prev Res (Phila) 2012;5:205–215.

28. Sanchez-Fidalgo S, Cardeno A, Villegas I, Talero E, de la Lastra CA. Dietary supplementation of resveratrol attenuates chronic colonic inflammation in mice. Eur J Pharmacol 2010;633:78–84.

29. Cui X, Jin Y, Hofseth AB, Pena E, Habiger J, Chumanevich A, et al. Resveratrol suppresses colitis and colon cancer associated with colitis. Cancer Prev Res (Phila) 2010;3:549–559.

30. Sengottuvelan M, Nalini N. Dietary supplementation of resveratrol suppresses colonic tumour incidence in 1,2-dimethylhydrazine-treated rats by modulating biotransforming enzymes and aberrant crypt foci development. Br J Nutr 2006;96:145–153.

31. Sengottuvelan M, Deeptha K, Nalini N. Resveratrol ameliorates DNA damage, prooxidant and antioxidant imbalance in 1,2-dimethylhydrazine induced rat colon carcinogenesis. Chem Biol Interact 2009;181:193–201.

32. Sengottuvelan M, Deeptha K, Nalini N. Influence of dietary resveratrol on early and late molecular markers of 1,2-dimethylhydrazine-induced colon carcinogenesis. Nutrition 2009;25:1169–1176.

33. Schneider Y, Duranton B, Gosse F, Schleiffer R, Seiler N, Raul F. Resveratrol inhibits intestinal tumorigenesis and modulates host-defense-related gene expression in an animal model of human familial adenomatous polyposis. Nutr Cancer 2001;39:102–107.

TABLE 21.3 (Continued)

Nutraceuticals	Chemopreventive mechanisms	Animal experimental model	Refs.[a]
3. Pterostilbene	Reduces ACF formation by decreasing iNOS and COX-2 gene expression and increasing apoptosis in mouse colon (34), down-regulation of Wnt/β-catenin, EGFR, PI3K/Akt and NF-κβ signaling (35)	ICR mice (AOM[+]) (34); male Fisher F344 rats (AOM[+]) (35)	(34,35)
	Carotenoids		
1. β-Carotene	Reduces ACF numbers	Rats (AOM[+])	(4)
2. Lycopene	Attenuates inflammatory status of colitis by lowering the level of myeloperoxidase activity	Male Wistar rats (TNBS[+])	(36)
	Suppressed growth, progression and migration of colorectal tumors by up-regulation of p21 CIP1/WAF1 proteins and suppression of PCNA, COX-2, PGE2 and phosphorylation of ERK1/2 protein and reduced plasma level of MMP-9	Female BALB/cAnN-Foxn1 nude xenograft model (HT-29)	(37)
3. Astaxanthin	Reduces tumor invasion by decreasing expression of MMP-2/9, Akt, ERK-2, NF-κβ, and COX-2	Wistar rats (DMH[+])	(38)

34. Chiou YS, Tsai ML, Wang YJ, Cheng AC, Lai WM, Badmaev V, et al. Pterostilbene inhibits colorectal aberrant crypt foci (ACF) and colon carcinogenesis via suppression of multiple signal transduction pathways in azoxymethane-treated mice. J Agric Food Chem 2010;58:8833–8841.

35. Paul S, DeCastro AJ, Lee HJ, Smolarek AK, So JY, Simi B, et al. Dietary intake of pterostilbene, a constituent of blueberries, inhibits the beta-catenin/p65 downstream signaling pathway and colon carcinogenesis in rats. Carcinogenesis 2010;31:1272–1278.

36. Reifen R, Nur T, Matas Z, Halpern Z. Lycopene supplementation attenuates the inflammatory status of colitis in a rat model. Int J Vitam Nutr Res 2001;71:347–351.

37. Tang FY, Pai MH, Wang XD. Consumption of lycopene inhibits the growth and progression of colon cancer in a mouse xenograft model. J Agric Food Chem 2011;59:9011–9021.

38. Nagendraprabhu P, Sudhandiran G. Astaxanthin inhibits tumor invasion by decreasing extracellular matrix production and induces apoptosis in experimental rat colon carcinogenesis by modulating the expressions of ERK-2, NFκB and COX-2. Invest New Drugs 2011;29:207–224.

TABLE 21.3 (Continued)

Nutraceuticals	Chemopreventive mechanisms	Animal experimental model	Refs.[a]
	Suppresses colitis and associated carcinogenesis by inhibiting occurrence of ulcer, dysplastic crypts, and colonic aderomas by suppressing inflammatory cytokines (TNF-α, IL-1β) and NF-κβ	Male ICR mice (AOM/DSS$^+$)	(39)
4. Lutein	Chemopreventive against the colonic tumors by reducing expression of K-ras, β-catenin, and phosphorylation of PKB in tumors	Rats (DMH$^+$)	(40)
	Isothiocyanates		
1. Phenethyl isothiocyanate	Inhibits tumor incidence and multiplicity by inducing apoptosis (increased cleaved caspases) and cell-cycle arrest (increased p21 expression) (41)	C57BL/6 mice (AOM/DSS$^+$)	(41)
2. Sulforaphane (major isothiocyanate fround in cruciferous vegetables)	Reduces intestinal adenomas through inhibition of Akt, ERK signaling, COX-2 and cyclin D1 protein expression	Apc$^{Min/+}$ (42)	(42)
3. Benzyl isothiocyanate	Reduces incidence and multiplicity of tumors in small intestine and colon	Female ACI/N rats (MAM acetate$^+$)	(43)

39. Yasui Y, Hosokawa M, Mikami N, Miyashita K, Tanaka T. Dietary astaxanthin inhibits colitis and colitis-associated colon carcinogenesis in mice via modulation of the inflammatory cytokines. Chem Biol Interact 2011;193:79–87.

40. Reynoso-Camacho R, Gonzalez-Jasso E, Ferriz-Martinez R, Villalon-Corona B, Loarca-Pina GF, Salgado LM, et al. Dietary supplementation of lutein reduces colon carcinogenesis in DMH-treated rats by modulating K-ras, PKB, and beta-catenin proteins. Nutr Cancer 2011;63:39–45.

41. Cheung KL, Khor TO, Huang MT, Kong AN. Differential in vivo mechanism of chemoprevention of tumor formation in azoxymethane/dextran sodium sulfate mice by PEITC and DBM. Carcinogenesis 2010;31:880–885.

42. Shen G, Khor TO, Hu R, Yu S, Nair S, Ho CT, et al. Chemoprevention of familial adenomatous polyposis by natural dietary compounds sulforaphane and dibenzoylmethane alone and in combination in ApcMin/+ mouse. Cancer Res 2007;67:9937–9944.

43. Sugie S, Okamoto K, Okumura A, Tanaka T, Mori H. Inhibitory effects of benzyl thiocyanate and benzyl isothiocyanate on methylazoxymethanol acetate-induced intestinal carcinogenesis in rats. Carcinogenesis 1994;15:1555–1560.

TABLE 21.3 (*Continued*)

Nutraceuticals	Chemopreventive mechanisms	Animal experimental model	Refs.[a]
	Indoles		
1. 3,3′-Diindolylmethane	Decreases the number of colon tumors and attenuates colonic inflammation	BALB/c mice (AOM/DSS[+])	(44)
	Terpenoids and Saponins		
1. Borneol	Suppresses colitis and decreases proinflammatory cytokines (IL–6 and IL–1β) gene expression	ICR mice (TNBS)	(45)
2. Zerumbone (tropical ginger sesquiterpene)	Decreases frequency of ACFs and adenocarcinoma by reducing NF-κβ and Cox-2 expression	ICR mice (AOM/DSS[+]) (46), F344 rats (AOM[+])	(46,47)
3. Oleanolic acid and ursolic acid (pentacyclic triterpenoid compound)	Reduces frequency of ACFs	Male F344 rats (AOM[+]) (48): male Wistar rats (DMH[+]) (49); Sprague–Dawley rats (AOM[+])	(48–50)
4. Furanolactone columbin (a bitter diterpenoid)	Reduces incidence and multiplicity of colonic adenocarcinoma	Male F344 rats (AOM[+])	(51)
5. Cosfunolide (sesquiterpene)	Reduces incidence and number of ACFs	Male F344 rats (AOM[+])	(52)

44. Kim YH, Kwon HS, Kim DH, Shin EK, Kang YH, Park JH, et al. 3,3′-Diindolylmethane attenuates colonic inflammation and tumorigenesis in mice. Inflamm Bowel Dis 2009;15:1164–1173.

45. Juhas S, Cikos S, Czikkova S, Vesela J, Il'kova G, Hajek T, et al. Effects of borneol and thymoquinone on TNBS-induced colitis in mice. Folia Biol (Praha) 2008;54:1–7.

46. Kim M, Miyamoto S, Yasui Y, Oyama T, Murakami A, Tanaka T. Zerumbone, a tropical ginger sesquiterpene, inhibits colon and lung carcinogenesis in mice. Int J Cancer 2009;124:264–271.

47. Tanaka T, Shimizu M, Kohno H, Yoshitani S, Tsukio Y, Murakami A, et al. Chemoprevention of azoxymethane-induced rat aberrant crypt foci by dietary zerumbone isolated from *Zingiber zerumbet*. Life Sci 2001;69:1935–1945.

48. Janakiram NB, Indranie C, Malisetty SV, Jagan P, Steele VE, Rao CV. Chemoprevention of colon carcinogenesis by oleanolic acid and its analog in male F344 rats and modulation of COX-2 and apoptosis in human colon HT-29 cancer cells. Pharm Res 2008;25:2151–2157.

49. Furtado RA, Rodrigues EP, Araujo FR, Oliveira WL, Furtado MA, Castro MB, et al. Ursolic acid and oleanolic acid suppress preneoplastic lesions induced by 1,2-dimethylhydrazine in rat colon. Toxicol Pathol 2008;36:576–580.

50. Andersson D, Cheng Y, Duan RD. Ursolic acid inhibits the formation of aberrant crypt foci and affects colonic sphingomyelin hydrolyzing enzymes in azoxymethane-treated rats. J Cancer Res Clin Oncol 2008;134:101–107.

51. Kohno H, Maeda M, Tanino M, Tsukio Y, Ueda N, Wada K, et al. A bitter diterpenoid furanolactone columbin from *Calumbae radix* inhibits azoxymethane-induced rat colon carcinogenesis. Cancer Lett 2002;183:131–139.

52. Mori H, Kawamori T, Tanaka T, Ohnishi M, Yamahara J. Chemopreventive effect of costunolide, a constituent of oriental medicine, on azoxymethane-induced intestinal carcinogenesis in rats. Cancer Lett 1994;83:171–175.

TABLE 21.3 (*Continued*)

Nutraceuticals	Chemopreventive mechanisms	Animal experimental model	Refs.[a]
	Polyunsaturated Fatty acid (PUFA)		
1. Eicosapentaenoic acid (EPA)	Reduces tumor incidence and multiplicity	Donyu rats (AOM+) (53), CF1 mouse (AOM+) (54); Apc$^{Min/+}$ (55,56)	(53–56)
	Reduces proliferation index and increases apoptotic index in colonic mucosa	CF1 mouse (AOM+)	(54)
2. Docosahexaenoic acid (DHA)	Reduces ACFs	F344 (DMH+) (57), F344 rats (AOM+) (58)	(57,58)
	Reduces colonic polyp number and size	Apc$^{\Delta716}$	(59)
3. Menhaden oil (both EPA and DHA)	Reduces tumor incidence and multiplicity (60)	Sprague–Dawley rats (DMH+) (60); F344 rats (AOM+) (61); Apc$^{Min/+}$ (7)	(7,60,61)
	Antimetastatic	Sprague–Dawley rats (DMH+) (60)	
4. Fish oil (both EPA and DHA)	Reduces tumor incidence and multiplicity	Sprague–Dawley rats (AOM+) (62); F344 rats (AOM+) (63)	(62,63)
	Reduces ACFs	F344 (AOM+) (64); Wistar (DMH+) (65); Apc$^{Min/+}$ (66)	(64–66)

53. Minoura T, Takata T, Sakaguchi M, Takada H, Yamamura M, Hioki K, et al. Effect of dietary eicosapentaenoic acid on azoxymethane-induced colon carcinogenesis in rats. Cancer Res 1988;48:4790–4794.

54. Deschner EE, Lytle JS, Wong G, Ruperto JF, Newmark HL. The effect of dietary omega-3 fatty acids (fish oil) on azoxymethanol-induced focal areas of dysplasia and colon tumor incidence. Cancer 1990;66:2350–2356.

55. Petrik MB, McEntee MF, Chiu CH, Whelan J. Antagonism of arachidonic acid is linked to the antitumorigenic effect of dietary eicosapentaenoic acid in Apc(Min/+) mice. J Nutr 2000;130:1153–1158.

56. Fini L, Piazzi G, Ceccarelli C, Daoud Y, Belluzzi A, Munarini A, et al. Highly purified eicosapentaenoic acid as free fatty acids strongly suppresses polyps in Apc(Min/+) mice. Clin Cancer Res 2010;16:5703–5711.

57. Takahashi M, Minamoto T, Yamashita N, Yazawa K, Sugimura T, Esumi H. Reduction in formation and growth of 1,2-dimethylhydrazine-induced aberrant crypt foci in rat colon by docosahexaenoic acid. Cancer Res 1993;53:2786–2789.

58. Takahashi M, Fukutake M, Isoi T, Fukuda K, Sato H, Yazawa K, et al. Suppression of azoxymethane-induced rat colon carcinoma development by a fish oil component, docosahexaenoic acid (DHA). Carcinogenesis 1997;18:1337–1342.

59. Oshima M, Takahashi M, Oshima H, Tsutsumi M, Yazawa K, Sugimura T, et al. Effects of docosahexaenoic acid (DHA) on intestinal polyp development in Apc delta 716 knockout mice. Carcinogenesis 1995;16:2605–2607.

60. Nelson RL, Tanure JC, Andrianopoulos G, Souza G, Lands WE. A comparison of dietary fish oil and corn oil in experimental colorectal carcinogenesis. Nutr Cancer. 1988;11:215–220.

61. Reddy BS, Burill C, Rigotty J. Effect of diets high in omega-3 and omega-6 fatty acids on initiation and postinitiation stages of colon carcinogenesis. Cancer Res 1991;51:487–491.

TABLE 21.3 (*Continued*)

Nutraceuticals	Chemopreventive mechanisms	Animal experimental model	Refs.[a]
	Increases colonic apoptotic index through reduction of β-catenin and PPARδ expression	Sprague–Dawley rats (AOM[+])	(67)
	Oriental Herbal Medicines		
1. Ginseng extract	Suppresses colonic inflammation (68)	C57BL/6 (DSS[+]) (68)	(68)
	Reduces tumor incidence and multiplicity (69–71)	C57BL/6 (AOM/DSS[+]) (69,70), F344 rats (DMH[+]) (71)	(69–71)
2. *Ginkgo biloba*	Suppresses colonic inflammation (72,73)	C57BL/6 (DSS[+]) (72), Wistar rats (TNBS[+]) (73)	(72,73)
	Reduces ACF (74)	F344 rats (AOM[+]) (74)	(74)

62. Chang WL, Chapkin RS, Lupton JR. Fish oil blocks azoxymethane-induced rat colon tumorigenesis by increasing cell differentiation and apoptosis rather than decreasing cell proliferation. J Nutr 1998;128:491–497.

63. Singh J, Hamid R, Reddy BS. Dietary fish oil inhibits the expression of farnesyl protein transferase and colon tumor development in rodents. Carcinogenesis 1998;19:985–989.

64. Rao CV, Hirose Y, Indranie C, Reddy BS. Modulation of experimental colon tumorigenesis by types and amounts of dietary fatty acids. Cancer Res 2001;61:1927–1933.

65. Moreira AP, Sabarense CM, Dias CM, Lunz W, Natali AJ, Gloria MB, et al. Fish oil ingestion reduces the number of aberrant crypt foci and adenoma in 1,2-dimethylhydrazine-induced colon cancer in rats. Braz J Med Biol Res 2009;42:1167–1172.

66. Paulsen JE, Elvsaas IK, Steffensen IL, Alexander J. A fish oil derived concentrate enriched in eicosapentaenoic and docosahexaenoic acid as ethyl ester suppresses the formation and growth of intestinal polyps in the *Min* mouse. Carcinogenesis 1997;18:1905–1910.

67. Vanamala J, Glagolenko A, Yang P, Carroll RJ, Murphy ME, Newman RA, et al. Dietary fish oil and pectin enhance colonocyte apoptosis in part through suppression of PPARdelta/PGE2 ard elevation of PGE3. Carcinogenesis 2008;29:790–796.

68. Jin Y, Kotakadi VS, Ying L, Hofseth AB, Cui X, Wood PA, et al. American ginseng suppresses inflammation and DNA damage associated with mouse colitis. Carcinogenesis 2008;29:2351–2359.

69. Poudyal D, Le PM, Davis T, Hofseth AB, Chumanevich A, Chumanevich AA, et al. A hexane fraction of American ginseng suppresses mouse colitis and associated colon cancer: anti-inflammatory and proapoptotic mechanisms. Cancer Prev Res (Phila) 2012;5:685–696.

70. Cui X, Jin Y, Poudyal D, Chumanevich AA, Davis T, Windust A, et al. Mechanistic insight into the ability of American ginseng to suppress colon cancer associated with colitis. Carcinogenesis 2010;31:1734–1741.

71. Li W, Wanibuchi H, Salim EI, Wei M, Yamamoto S, Nishino H, et al. Inhibition by ginseng of 1,2-dimethylhydrazine induction of aberrant crypt foci in the rat colon. Nutr Cancer 2000;36:66–73.

72. Kotakadi VS, Jin Y, Hofseth AB, Ying L, Cui X, Volate S, et al. *Ginkgo biloba* extract EGb 761 has anti-inflammatory properties and ameliorates colitis in mice by driving effector T cell apoptosis. Carcinogenesis 2008;29:1799–1806.

73. Zhou YH, Yu JP, Liu YF, Teng XJ, Ming M, Lv P, et al. Effects of *Ginkgo biloba* extract on inflammatory mediators (SOD, MDA, TNF-alpha, NF-kappaBp65, IL-6) in TNBS-induced colitis in rats. Mediators Inflamm 2006;2006:92642.

74. Suzuki R, Kohno H, Sugie S, Sasaki K, Yoshimura T, Wada K, et al. Preventive effects of extract of leaves of ginkgo (*Ginkgo biloba*) and its component bilobalide on azoxymethane-induced colonic aberrant crypt foci in rats. Cancer Lett 2004;210:159–169.

TABLE 21.4 Human Clinical Studies of Colon Cancer Prevention by Nutraceuticals

Study/trial title	NCT (or ref.)[a]	Study design	Location	Status and phase
QUERCETIN				
1. Combination treatment with curcumin and quercetin in adenomas of FAP	(1)	Non-randomized	Cleveland Clinic, FL	Completed
EPIGALLOCATECHIN-3-GALLATE				
1. Comparison of flavonoid treatment in patients with resected CRC to prevent recurrence	(2)	Prospective cohort	Marien-Hospital Darmstadt, Germany	Completed
2. Green tea extract for nutriprevention of metachronous colon adenomas in the elderly population	(3), NCT01360320	Randomized	University ULM, Germany	Active, phase II
CURCUMIN				
1. Effect of curcumanoids on aberrant crypt foci in the human colon	NCT00176618	Randomized	University of Medicine and Dentistry, NJ	Terminated
2. Curcumin chemoprevention in patients with a high risk of premalignant lesions	(4)	Prospective case study	National Taiwan University College of Medicine	Completed/phase I
3. Use of curcumin in lower gastrointestinal tract in FAP patients	NCT00248053	Non-randomized	Johns Hopkins University, MD	Terminated phase II
4. Curcumin for the chemoprevention of colorectal cancer	NCT00118989	Randomized	University of Pennsylvania	Active, phase II

(continued)

TABLE 21.4 (*Continued*)

Study/trial title	NCT (or Ref.)[a]	Study design	Location	Status and phase
5. Curcumin in preventing colon cancer in smokers with ACF	NCT00365209	Non-randomized	University of California–Irvine	Active, phase II
6. Curcumin for treatment of intestinal adenomas in FAP	NCT00641147	Randomized	Johns Hopkins University, MD	Active, phase II
7. Curcumin, gemcitabine, and celebrex in metastatic colon cancer	NCT00295035	Randomized	Tel-Aviv Sourasky Medical Center, Israel	Unknown, phase III
RESVERATROL				
1. Resveratrol for patients with colon cancer	NCT00256334	Non-randomized	University of California–Irvine	Completed, phase I
2. Resveratrol in treating patients with CRC that can be removed by surgery	NCT00433576	Non-randomized	University of Michigan Cancer center	Completed, phase I
3. Biomarker study for low-dose resveratrol for colon cancer prevention	NCT00578396	Randomized	University of California–Irvine	Unknown, phase I
4. Clinical pharmacology of resveratrol in CRC patients	(5)	Non-randomized	University of Leicester, UK	Completed
β-CAROTENE				
1. β-Carotene for patients with nonactive Crohn's disease	NCT00275418	Randomized	Bnai Zion Medical Center, Israel	Unknown, phase III
2. Vitamins including β-carotene in chemoprevention of cancer including CRC	NCT00270647	Randomized	Brigham and Women's Hospital, MA	Completed, phase II
LYCOPENE				
1. Lycopene decreases insulin-like growth factor-I levels in colon cancer patients	(6)	Randomized	Ben-Gurion University, Israel	Completed, phase I

TABLE 21.4 (*Continued*)

Study/trial title	NCT (or Ref.)[a]	Study design	Location	Status and phase
ω- **PUFA**				
1. Effect of n-3 FA by the consumption of oil-rich fish on reduction of colon cancer and IBD risk	NCT00145015	Randomized	Institute of Food Research, UK	Completed
2. Chemoprevention trial in FAP coli using EPA.	NCT00510692	Randomized	S.L.A. Pharma AG, UK	Completed, phase II/III
3. Effect of two doses of EPA on apoptosis and cell proliferation in colonic mucosa in patients with a history of colonic polyps	NCT00432913	Randomized	S.L.A. Pharma AG, UK	Completed, phase II/III
4. EPA for treatment of CRC liver metastasis	NCT01070355	Randomized	University of Leeds, UK	Completed, phase II
5. Effect of EPA on patients receiving chemotherapy for GI tumor and CRC	NCT01048463	Randomized	Sun Yat-Sen University, China	Active, phase III
6. EPA rich diet in chemoprevention tolerance in advanced colon cancer patients	NCT00398333	Randomized	Hospital Clinic of Barcelona, Spain	Terminated, phase IV
7. ω-3 FA and postoperative complications after colorectal surgery	NCT00488904	Randomized	Aalborg Hospital, Denmark	Completed, phase IV
8. ω-3 FA on functional state and quality of life in malnourished patients with gastroenterological tumors	NCT00168987	Randomized	Charite University, Germany	Completed, phase IV

(*continued*)

TABLE 21.4 (Continued)

Study/trial title	NCT (or Ref.)[a]	Study design	Location	Status and phase
CALCIUM AND VITAMIN				
1. Effect of vitamin D and calcium on genes in colon	NCT00298545	Randomized	Rockefeller University, NY	Completed, phase I
2. Calcium and vitamin D vs. markers of adenomatous polyps	NCT00208793	Randomized	Emory University, GA	Active, phase II
3. Calcium/vitamin D_3, biomarkers, and colon polyp prevention	NCT00399607	Randomized	Emory University, GA	Active, phase III
4. Vitamin D_3/calcium in polyp-prevention study	NCT00153816	Randomized	Dartmouth–Hitchcock Medical Center, NH	Active, phase II/III
VITAMINS				
1. Vitamin D in preventing colon cancer in African-Americans with colon polyps	NCT00870961	Randomized	Northwestern University, IL	Active, phase 0/I
2. Vitamin B (folate) in prevention of CRC in patients at high risk for CRC	NCT00096330	Single group assignment	Roswell Park Cancer Institute, NY	Complete, phase I
3. Vitamin E in treating patients undergoing surgery for CRC	NCT00905918	Randomized	University of Medicine and Dentistry, NJ	Active, phase I
4. DNA changes that affect vitamin D metabolism in patients with CRC receiving vitamin D supplements	NCT00550563	Single group assignment	Roswell Park Cancer Institute, NY	Completed
5. Dietary supplementary vitamin D levels in stage IV CRC patients	NCT01074216	Single group assignment	Memorial Sloan–Kettering Cancer Center, NY	Active, phase II

TABLE 21.4 (*Continued*)

Study/trial title	NCT (or Ref.)[a]	Study design	Location	Status and phase
6. Effect of folate on DNA in colon tissue and blood samples from patients at increased risk of developing colorectal neoplasia	NCT00611000	Random-single blind	Rockefeller University, NY	Completed, phase I
7. Folic acid in preventing colorectal polyps in patients with previous colorectal polyps	NCT00512850	Randomized	Dana-Farber/Brigham and Women's Cancer Center, MA	Active, phase III
CARBOHYDRATES				
1. Role of nondigestible carbohydrates in CRC chemoprevention	NCT01214681	Randomized	Newcastle University, UK	Active, phase I
2. Inositol chemoprevention in colitis-associated dysplasia	NCT01111292	Randomized	Robert H. Lurie Cancer Center, IL	Active, phase I
AMINO ACIDS				
1. Study of effects of L arginine in colitis and colon cancer	NCT01091558	Case–control prospective	Vanderbilt University, TN	Active, phase II

[a]
1. Cruz-Correa M, Shoskes DA, Sanchez P, Zhao R, Hylind LM, Wexner SD, et al. Combination treatment with curcumin and quercetin of adenomas in familial adenomatous polyposis. Clin Gastroenterol Hepatol 2006;4:1035–1038.
2. Hoensch H, Groh B, Edler L, Kirch W. Prospective cohort comparison of flavonoid treatment in patients with resected colorectal cancer to prevent recurrence. World J Gastroenterol 2008;14:2187–2193.
3. Stingl JC, Ettrich T, Muche R, Wiedom M, Brockmoller J, Seeringer A, et al. Protocol for minimizing the risk of metachronous adenomas of the colorectum with green tea extract (MIRACLE): a randomised controlled trial of green tea extract versus placebo for nutriprevention of metachronous colon adenomas in the elderly population. BMC Cancer 2011;11:360.
4. Cheng AL, Hsu CH, Lin JK, Hsu MM, Ho YF, Shen TS, et al. Phase I clinical trial of curcumin, a chemopreventive agent, in patients with high-risk or pre-malignant lesions. Anticancer Res 2001;21:2895–2900.
5. Patel KR, Brown VA, Jones DJ, Britton RG, Hemingway D, Miller AS, et al. Clinical pharmacology of resveratrol and its metabolites in colorectal cancer patients. Cancer Res 2010;70:7392–7399.
6. Walfisch S, Walfisch Y, Kirilov E, Linde N, Mnitentag H, Agbaria R, et al. Tomato lycopene extract supplementation decreases insulin-like growth factor-I levels in colon cancer patients. Eur J Cancer Prev 2007;16:298–303.

promotion by promoting cell proliferation, angiogenesis and apoptotic evasion, stimulating tumor metastasis, and decreasing immune surveillance (21).

Protein Kinase C (PKC)   Modified expression and activity of specific PKC isozymes have been observed in human and rodent colon cancer cells (22). PKC activity was reduced in the mucosa of human colon tumors compared to the normal colon tissue (23–27). In adenomas of APC-mutant min mice, the expression of several PKC isozymes are lost or down-regulated (28). The activation of PKC (through phospholipase C, PLC) might be an integral part of the molecular mechanism involved in the differentiation of intestinal cells promoted by extracellular calcium (22).

Growth Factors and Their Receptors   Deregulation of the PI3K/Akt signaling pathway and its downstream transcription factors are other molecular events that have been implicated in cancer development (29). The active form (phophorylated) of p38 mitogen-activated protein kinase (MAPK) is present mainly in the nuclei of differentiated intestinal cells (30). p38 MAPK interacts and positively regulates the CDX2 transcription factor, which is involved in the control of intestinal cell differentiation (31). Loss of the mouse CDX-2 gene results in the development of polyps in the ileum and colon (22).

Wnt/β-Catenin Pathways   Abnormal activation of the Wnt/β-catenin pathway has been implicated in the development of human colon cancer (20). In normal cells, wild-type adenomatous polyposis Coli (WT APC) protein controls the steady-state levels of β-catenin by targeting it for routine degradation (32,33). Loss of WT APC function results in the translocation of β-catenin to the nucleus, where it interacts with TCF-family transcription factors and activates the transcription of several genes, including those encoding cyclin D1, c-Myc, and antiapoptotic protein survivin, which is overexpressed by colorectal tumors (34).

Matrix Metalloproteinases (MMPs)   MMPs are secreted by inactive zymogens and are activated extracellularly; once activated, they are able to degrade most extracellular matrix components (e.g., collagen, laminin, fibronectin, vitonectin, enactin, proteoglycans), thus enhancing the metastatic potential of cancer cells (35). Degradation of type IV collagen by MMP-2 and MMP-9 often occurs in cancer (35). Increased levels of MMP-9 have been demostrated early in the transition from colon adenoma to adenocarcinoma (36), thus making it an excellent target in colon cancer prevention.

CDK/Cyclins   Loss of cell-cycle regulation is a hallmark of cancer. The eukaryotic cell cycle is regulated by the sequential activation and inactivation of cyclin-dependent kinases (Cdks), which drive cell-cyle progression by forming a Cdk/cyclin complex (assisted by Cdk-activating kinase, CAK); alternatively, Cdk inhibitory subunits (CKIs, such as p21, p27, p16^{INK4}) inactivate the active Cdk/cylin complex [reviewed in (37)]. Thus, the induction of cell-cycle arrest and/or apoptosis is considered to be a promising chemopreventive strategy.

MicroRNA

MicroRNAs (MiRs) are a group of small noncoding endogenous single-stranded RNAs 18 to 25 nucleotides (ca. 22 nt) in length that negatively regulate gene expression by translational inhibition or exonucleolytic messenger RNA (mRNA) decay (38). Several miRs have been found to be either up- or down-regulated in tumors (39–42). Besides the global microRNA expression and signature identification in cancer, several individual microRNAs have evolved as key players in certain malignancies. The latest miRBase, Version 19.0 (August 2012), has 1600 precursors and 2042 mature miRNA sequences annotated in the human genome, and miRNAs are believed to target about one-third of human mRNAs; a single miR can target approximately 200 transcripts simultaneously [modified from (43)]. There are different mechanisms to inhibit mutagenesis and carcinogenesis, and the modulation of miR as an epigenetic response to dietary agents is one such mechanism that has recently evolved as an efficient tool in the cancer chemoprevention study. Table 21.1 summarizes the findings of the studies evaluating modulation of miRNAs by putative colon cancer chemopreventive dietary agents.

Cancer Stem Cells

A subset of tumor cells have been identified that have the ability to self-renew and generate the diverse cells that comprise a tumor, termed *cancer stem cells* (CSCs) (44). CSCs are thought to be resistant to conventional chemotherapies and that this inherent resistance is what leads to relapse in many cancer patients (45). Recent advancements in the identification and isolation of CSCs with the specific cell surface markers ($CD133^+$, $CD44^+/CD24^-$, $Lgr5^+$, $ALDH1^+$, $EpCAM^{hi}/CD44^+$, $CD90^+$) have made it possible to study tissue-specific CSCs (46). Very few studies have been conducted in terms of the chemopreventive effects of nutraceutical or nutrient-derived compounds on colon cancer stem cells. For example, the analog of curcumin, GO-Y030, and curcumin inhibited the STAT3 activation (phosho-STAT3) and cell viability of tumorsphere formation in ($ALDH^+/CD133^+$) colon CSCs (45). GO-Y030 also suppressed the tumor growth of CSCs isolated from SW480 and the HCT116 colon cancer cell line in a mouse model (47). Difluorinated curcumin (CDF), another analog of curcumin, together with conventional chemotherapeutics (5-FU + Ox) could be effective in eliminating chemoresistant colon CSCs by inhibiting growth and inducing apoptosis (48). ω-3 PUFA mediates genes (*Ephrin B1* and *BMP-4*) that regulate the colon stem cell niche and tumor evolution in male Sprague–Dawley rats (AOM^+) (49). There is a huge potential for colon cancer chemoprevention by dietary agents by targeting CSCs. This field of study is yet to be explored, as very little research has been conducted regarding dietary agents and CSCs.

EXPERIMENTAL MODELS OF COLON CARCINOGENESIS IN CHEMOPREVENTION

Colorectal malignancies can develop spontaneously or as a late complication of a chronic inflammatory state (50). For the chemopreventive and therapeutic strategies of CRC, a variety of *in vitro* cellular and *in vivo* animal models have been developed as important tools for study of the development and pathogenesis of disease. CRC development is a long process from normal epithelial cells via aberrant crypts and progressive adenoma stages to carcinomas *in situ* and then metastasis (50). Several cell lines from the different stages of the human colon carcinogenesis have been isolated and developed for the *in vitro* study of colon cancer. Similarly, several animal models with genetic alterations and/or carcinogen induction have been used to study chemoprevention and various stages of colon carcinogenesis.

In Vitro Models

In vitro cellular models permit isolation of specific aspects of tumor biology such that functional analysis of relevant genes or the assessment of effects of endogenous mediators and pharmacological compounds are faster and simpler than in the intact organism (50). Various cell lines are used for the *in vitro* models, such as HCT116, LoVo, HT29, SW420, SW620, DLD1, HCT15, and Caco2. The limitation of *in vitro* models of colon cancer is that it reflects only a specific stage of tumor development, depending on the stage from which the cells originated, and it also lacks the tissue context study of tumor growth and metastasis *in vivo* (50). To address this issue, *in vivo* animal models similar to human CRC have emerged to provide the opportunity to study tumor stages and metastatic processes due to the reliability to induce and establish colon tumors and study chemopreventive features.

Genetically Modified Animals

Genetically modified animals are used primarily for the study of tumor development and rarely for metastasis study. Mutant mice with heterozygous mutations in the APC gene (Min/ΔAPC mouse) develop multiple intestinal neoplasia (Min) and mimic rapid tumor development, as seen in FAP patients (51). Due to the loss of APC allele in the tumors, constitutive activation of the WNT pathway and deregulation in growth of tumor cells occur. However, K-*ras* and *p53* mutation is not seen in Min mice (51). The difference between Min mice and FAP patients is the location of tumor; in Min mice, tumors occur in the small intestine and rarely in the colon, whereas in FAP patients, tumors occur in the small intestine and a large number of adenomatous polyps in the colon and rectum (50). The Apc Min mouse model is used widely for chemopreventive studies of nutrition by adding it to the animal diet.

Construction of a HNPCC animal model has seen less success than that of the FAP model because heterozygous deletion of the MMR genes (*Msh2, Mlh1, Msh6, Msh3, Pms2*, and *Pms1*) are not sufficient to predispose to cancer (50). Homozygous deletion of these MMR genes is used to obtain the disease model for HNPCC-like cancer, but the tumor spetrum seen consists of various lymphomas that are never seen in HNPCC-affected patients (52).

Xenoplant Models

The ability of tumor cells to develop tumors after subcutaneous or intravenous (IV) injection in immunodeficient mouse strains (nude, bg/nu/xid, or SCID mice) has allowed the analysis of human tumors *in vivo* (53). The xenoplant model is often used for the assessment of cytostatic therapeutic compound (50). The model is relevant for tumor metastasis study, but subcutaneous or IV injection of tumor cells neglects the complexity of tumor growth and metastasis and the interaction between tumor and microenvironment, which largely controls the biological characteristics of tumors (50). To overcome this issue and to create a clinically accurate model of human CRC, orthotopic transplantation into the cecum and rectum has been developed where metastasis to liver and lymph node occurs (54). A limitation of this model is the complexity of surgical procedure involved, along with the lack of immune function (50).

Chemically Induced Models

Carcinogen-induced CRC models are a highly reliable way to recapitulate the phases of tumor growth and progression that occur in humans and is frequently used to study the chemopreventive compounds efficacy or their associated risk factors (50). An important benefit of this model is that it is highly reproducible and can readily be tested on animals with different genetic backgrounds.

Heterocyclic Aromatic Amines   After the identification of heterocyclic aromatic amines in grilled or broiled meat, the main representative of this group of carcinogens, 2-amino-3-mehylimidazo [4,5-*f*]quinoline (IQ) and 2-amino-1-methyl-6-phenylimidazo[4,5-*b*] pyridine (PhIP), have been introduced into experimental CRC (55).

Aromatic Amines   Benign and malignant epithelial neoplasms has been observed in rats and rodents with 3,2′-dimethly-4-aminobiphenyl (DMAB), but the tumorigenic ability is less potent (56). The limitation of this model is that multiple injections of DMAB are required and neoplasms can be induced on other tissues (57).

Alkylnitrosoamide Compounds   Methylnitrosourea (MNU) and *N*-methyl-*N*-nitro-*N*-nitrosoguanidine (MNNG) are alkylating carcinogens that induce lesions in the distal colon and rectum in rodent models after intrarectal injection (58). The limitation or weakness of this model is the mode of administration.

Dimethylhydrazine (DMH) and Azoxymethane (AOM)   Repetitive administration of DMH in rodents (59,60) and a single to double dose of AOM injections induces colon tumors with pathological features similar to that seen in sporadic human diseases (61–63). Currently, studies with AOM are preferred because of its practical advantage, such as reproducibility, high potency, simple mode of application, excellent stability in solution, and low price (62). DMH- or AOM-induced tumors share many of the histopathological characteristics of human CRC and they carry frequent mutations in K-ras and β-catenin and some show MSI, indicative of defective MMR system (50). APC and p53 mutations are rare, and the tendency to metastasize is also low (64). The intraperitoneal (IP) injection of AOM (tumor-initiating agent) with DSS (dextran sodium sulfate; tumor-promoting agent) in drinking water (periodic cycle–alternate weeks between water and DSS) (AOM/DSS model) is applicable to the study of tumor progression driven by colitis (62), which causes rapid growth of multiple colon tumors within 10 weeks compared to 30 weeks (for AOM IP injection alone); hence, it also shortens the latency time (62).

DMH is a precursor of methylazoxymethanol (MAM), a carcinogen found in cycad flour (65). Both DMH and AOM require several metabolic activation steps (including N-oxidation and hydroxylation) to induce DNA-reactive adducts. Hydroxylation of AOM results in the formation of the reactive metabolite MAM, which alkylates macromolecules in liver and colon and operates the addition of a methyl group at the O-6 and N-7 positions of guanine (i.e., O-6 methyl deoxyguanosine and N-7-methyl-deoxyguanosine) in the DNA molecule. Methylation at the O-6 position of guanine has been shown to be the primary promutagenic lesions produced by AOM. MAM is activated by (mouse) colon alcoholdehydrogenase (66) and alcohol inducible CYP2E1 (67).

NUTRACEUTICAL AND COLON CANCER

A nutraceutical is a product isolated or purified from foods that are generally sold in a medicinal form not usually associated with foods (68). Nutraceuticals are dietary supplements that contain a concentrated form of bioactive substance derived originally from food (68). The term *nutraceutical* is a combination of "nutrition" and "pharmaceutical," meaning nutrition with a pharmaceutical or heath beneficial value. For simplicity and better understanding, nutraceutical agents have been grouped into the following classes of compounds: flavonoids, polyphenolics, carotenoids, isothiocyanates, saponins, PUFAs, CLAs, oriental herbal medicines. Their effect on *in vitro*, animal and human colon cancer studies are illustrated Tables 21.2, 21.3, and 21.4.

Flavonoids are plant secondary metabolites—present in all terrestrial vascular plants, such as fruits (apples), vegetables (onions), grams, nuts, tea, and wine—that possess anticancer properties (69). Flavonoids have a common phenylchromanone structure (C6–C3–C6) with one or more hydroxyl substituents, including derivatives (refer to Figure 21.1). There are about 4000 isoflavonoid and flavonoid compounds (70). Mammals can only get flavonoids and isoflavonoids through dietary intake (69). The daily intake of flavonoids in the United States and the UK ranges from 20 mg to 1 g (69). Anthocyans (anthocyanins and their aglycons, anthocyanidins) belong to the flavonoid groups of polyphenols, distributed in fruits and vegetables (70).

Hydroxybenzoic acid and hydroxycinnamic acid (e.g., caffeic and ferulic acids) are abundant in food and accounts for nearly one-third of the phenolic compounds in our diet (71). One of the most studied polyphenols is curcumin, a yellow pigment found in turmeric and mustard, with a structure of two ferulic acids linked by methylene, with a β-diketone structure (refer to Figure 21.1) in a highly conjugated system.

Carotenoids are isoprenoid polyene pigments with a long conjugated double-bond chain and nearly bilateral symmetry around the central double bond (72), which give fruits and vegetables such as carrots, cantaloupe, sweet potatoes, and kale their vibrant orange, yellow, and green colors (73). Carotenoids are antioxidants with strong cancer-fighting properties that serve as precursor of vitamin A, except for lutein (73). About 600 carotenoids have been identified, and 40 of them are found in

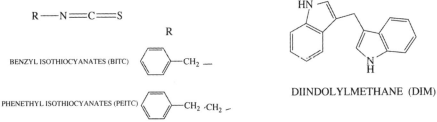

A. FLAVONOIDS

QUERCETIN

	5	6	7	3'	4'	5'
	OH	H	OH	OH	OH	H

B. POLYPHENOLICS

CURCUMIN

C. CAROTENOIDS

BETA-CAROTENE

D. ISOTHIOCYANATES

R—N≡C≡S

	R
BENZYL ISOTHIOCYANATES (BITC)	⟨ ⟩—CH₂ —
PHENETHYL ISOTHIOCYANATES (PEITC)	⟨ ⟩—CH₂ ⋅CH₂ —

E. INDOLES

DIINDOLYLMETHANE (DIM)

F. POLYUNSATURATED FATTY ACIDS (PUFA)

EICOSAPENTAENOIC ACID (EPA)

FIGURE 21.1   Chemical structures of some common nutraceuticals: (A) flavonoids; (B) polyphenolics; (C) carotenoids; (D) isothiocyanates; (E) indoles; (F) polyunsaturated fatty acids.

our daily food (73). Some of the common carotenoids (i.e., β-carotene, lycopene, zeaxanthine, lutein), and their anticancer properties are covered in Tables 21.1 to 21.4.

Isothiocyanates are derived from the hydrolytic breakdown of glucosinolate-sulfur-containing compounds found in cruciferous vegetables, such as broccoli, cabbage, and kale (74). Diindolylmethane (DIM), a metablite of indole-3-carbinol (I3C), is a crucifer-derived phytochemical that is found in some vegetables, including broccoli, cabbage, cauliflower, and brussel sprouts (75). Retinol is vitamin A in the form of retinly palmitate, found in beef, calf, chicken liver, eggs, fish liver oils, and dairy products (whole milk: butter, yogurt, and cheese).

Omega-3 polyunsaturated fatty acids (ω-3 PUFAs) are naturally occurring long-chain fatty acids. The two main ω-3 PUFAs, C20:5 ω-3 EPA and C22:6 ω-3 DHA, are found predominantly in fish oil, such as salmon, mackerel, and sardines (76). Their anticancer properties are illustrated in Tables 21.1 to 21.4.

Human and Clinical Trial Studies Two principal ways of investigating the chemopreventive role of dietary agents against colon cancer are observational (cohort and case–control) and experimental studies [randomized control trials (RCTs)]. Only meta-analysis of several blinded RCTs is considered more reliable when compiling the results. Although a lot of *in vitro* and preclinical studies are done in terms of colon cancer chemopreventive measures by nutraceuticals, only a very few nutraceticals make it to clinical or human studies based on balancing efficacy issues and man-power/monetary issues. It is clear, however, given our recent mechanistic insight in preclincal experiments that human clinical trials are needed to study the use of these relatively safe nutraceuticals to suppress carcinogenesis. Table 21.4 summarizes the nutraceuticals studied in clinical trials as provided by the U.S. National Institutes of Health at Clinicaltrials.gov.

CONCLUDING REMARKS

The effects of nutraceuticals on the hallmarks of cancer, such as apoptosis, cell-cycle regulation, cancer metastasis, signaling pathway, and/or different targets (including oncogenes and tumor suppressors) in the prevention of colorectal cancer have been illustrated in this chapter. The cancer prevention effects of the nutraceutical have been summarized in different study models from *in vitro* cell models to preclinical animal studies to human and clinical trial studies. These studies conclude that nutraceuticals could be a helpful starting point in the design and development of novel colon cancer preventive agents. In the past 30 years, from January 1981 to December 2010, 128 anticancer drugs were introduced in the market; 44 (34%) of which were natural product–derived (77). There is an increase from 25.6% (in 1981) to 50% (in 2010) in natural product and natural product-derived drug entities of all the drugs introduced in the market (77). Two such U.S. Food and Drug Administration (FDA)-approved drugs derived from plants, which have anticancer effects against colon cancer, are Topotecan HCl and Irrinotecan HCl. Both are obtained from *Camptotheca acuminata*, a member of the Nyssaceae plant family. Topotecan HCl was FDA approved in 1996 and is marketed as Hycamtin by GlaxoSmithKline. Irrinotecan HCl was FDA approved in 1997, marketed as Camptosa by Pfizer. Such success stories have attracted a lot of research in the development of new drugs from nutraceutical agents. There is an increasing trend in research and development of new nutraceutical agent since more and more cancer patients are migrating toward the use of such natural compounds, due to fewer side effects than those in commercial synthetic drugs.

REFERENCES

1. Food, Nutrition, Physical Activity, and the Prevention of Cancer: a Global Perspective. Washington, DC: WCRF/AICR; 2007.
2. Kufe DW, Holland JF, Frei E, American Cancer Society. Cancer Medicine, 6th ed. Hamilton, Ontario, Canada: BC Decker; 2003.
3. American Cancer Society. Global Cancer Facts & Figures, 2nd ed. Atlanta: American Cancer Society; 2011.
4. Rous P. The influence of diet on transplanted and spontaneous mouse tumors. J Exp Med 1914;20:433–451.
5. Stocks P, Karn MN. A co-operative study of the habits, home life, dietary and family histories of 450 cancer patients and of an equal number of control patients. Ann Eugen Human Genet 1933;5:30–33.
6. Doll R, Peto R. The causes of cancer: quantitative estimates of avoidable risks of cancer in the United States today. J Natl Cancer Inst 1981;66:1191–1308.
7. Hill MJ, Morson BC, Bussey HJ. Aetiology of adenoma–carcinoma sequence in large bowel. Lancet 1978;1:245–247.
8. Food, Nutrition, Physical Activity, and the Prevention of Cancer: A Global Perspective. Washington, DC: WCRF/AICR; 1997, pp. 216–251.
9. Lynch HT, de la Chapelle A. Hereditary colorectal cancer. N Engl J Med 2003;348:919–932.
10. Itzkowitz SH, Yio X. Inflammation and cancer: IV. Colorectal cancer in inflammatory bowel disease— the role of inflammation. Am J Physiol Gastrointest Liver Physiol 2004;287:G7–17.
11. Weitz J, Koch M, Debus J, Hohler T, Galle PR, Buchler MW. Colorectal cancer. Lancet 2005;365:153–165.
12. Lynch HT, de la Chapelle A. Genetic susceptibility to non-polyposis colorectal cancer. J Med Genet 1999;36:801–818.
13. Smyrk TC, Lynch HT. Microsatellite instability: impact on cancer progression in proximal and distal colorectal cancers. Eur J Cancer. 1999;35:171–172.
14. Jass JR, Whitehall VL, Young J, Leggett BA. Emerging concepts in colorectal neoplasia. Gastroenterology 2002;123:862–876.
15. Willenbucher RF, Aust DE, Chang CG, Zelman SJ, Ferrell LD, Moore DH, 2nd, et al. Genomic instability is an early event during the progression pathway of ulcerative-colitis-related neoplasia. Am J Pathol. 1999;154:1825–1830.
16. Pan MH, Lai CS, Wu JC, Ho CT. Molecular mechanisms for chemoprevention of colorectal cancer by natural dietary compounds. Mol Nutr Food Res. 2011;55:32–45.
17. Hanahan D, Weinberg RA. The hallmarks of cancer. Cell 2000;100:57–70.
18. Amin AR, Kucuk O, Khuri FR, Shin DM. Perspectives for cancer prevention with natural compounds. J Clin Oncol 2009;27:2712–2725.
19. Grandis JR, Drenning SD, Chakraborty A, Zhou MY, Zeng Q, Pitt AS, et al. Requirement of Stat3 but not Stat1 activation for epidermal growth factor receptor–mediated cell growth *in vitro*. J Clin Invest 1998;102:1385–1392.
20. Spychalski M, Dziki L, Dziki A. Chemoprevention of colorectal cancer: A new target needed? Colorectal Dis 2007;9:397–401.
21. Eisinger AL, Prescott SM, Jones DA, Stafforini DM. The role of cyclooxygenase-2 and prostaglandins in colon cancer. Prostaglandins Other Lipid Mediat 2007;82:147–154.

22. Lamprecht SA, Lipkin M. Chemoprevention of colon cancer by calcium, vitamin D and folate: molecular mechanisms. Nat Rev Cancer 2003;3:601–614.

23. Guillem JG, O'Brian CA, Fitzer CJ, Forde KA, LoGerfo P, Treat M, et al. Altered levels of protein kinase C and Ca2+-dependent protein kinases in human colon carcinomas. Cancer Res 1987;47:2036–2039.

24. Kopp R, Noelke B, Sauter G, Schildberg FW, Paumgartner G, Pfeiffer A. Altered protein kinase C activity in biopsies of human colonic adenomas and carcinomas. Cancer Res 1991;51:205–210.

25. Cowell HE, Garrod DR. Activation of protein kinase C modulates cell–cell and cell–substratum adhesion of a human colorectal carcinoma cell line and restores 'normal' epithelial morphology. Int J Cancer 1999;80:455–464.

26. Abraham C, Scaglione-Sewell B, Skarosi SF, Qin W, Bissonnette M, Brasitus TA. Protein kinase C alpha modulates growth and differentiation in Caco-2 cells. Gastroenterology 1998;114:503–509.

27. Scaglione-Sewell B, Abraham C, Bissonnette M, Skarosi SF, Hart J, Davidson NO, et al. Decreased PKC-alpha expression increases cellular proliferation, decreases differentiation, and enhances the transformed phenotype of CaCo-2 cells. Cancer Res 1998;58:1074–1081.

28. Klein IK, Ritland SR, Burgart LJ, Ziesmer SC, Roche PC, Gendler SJ, et al. Adenoma-specific alterations of protein kinase C isozyme expression in Apc(MIN) mice. Cancer Res 2000;60:2077–2080.

29. Bode AM, Dong Z. Targeting signal transduction pathways by chemopreventive agents. Mutat Res 2004;555:33–51.

30. Houde M, Laprise P, Jean D, Blais M, Asselin C, Rivard N. Intestinal epithelial cell differentiation involves activation of p38 mitogen-activated protein kinase that regulates the homeobox transcription factor CDX2. J Biol Chem 2001;276:21885–21894.

31. Lorentz O, Duluc I, Arcangelis AD, Simon-Assmann P, Kedinger M, Freund JN. Key role of the Cdx2 homeobox gene in extracellular matrix-mediated intestinal cell differentiation. J Cell Biol 1997;139:1553–1565.

32. Potter JD. Colorectal cancer: molecules and populations. J Natl Cancer Inst 1999;91:916–932.

33. Vogelstein B, Kinzler KW. The Genetic Basis of Human Cancer. New York: McGraw-Hill; 2001.

34. Gianani R, Jarboe E, Orlicky D, Frost M, Bobak J, Lehner R, et al. Expression of survivin in normal, hyperplastic, and neoplastic colonic mucosa. Hum Pathol 2001;32:119–125.

35. Zucker S, Vacirca J. Role of matrix metalloproteinases (MMPs) in colorectal cancer. Cancer Metastasis Rev 2004;23:101–117.

36. Parsons SL, Watson SA, Collins HM, Griffin NR, Clarke PA, Steele RJ. Gelatinase (MMP-2 and -9) expression in gastrointestinal malignancy. Br J Cancer 1998;78:1495–1502.

37. Johnson PF. Molecular stop signs: regulation of cell-cycle arrest by C/EBP transcription factors. J Cell Sci 2005;118:2545–2555.

38. Kim VN, Nam JW. Genomics of microRNA. Trends Genet 2006;22:165–173.

39. Jin Z, Selaru FM, Cheng Y, Kan T, Agarwal R, Mori Y, et al. MicroRNA-192 and -215 are upregulated in human gastric cancer *in vivo* and suppress ALCAM expression *in vitro*. Oncogene 2011;30:1577–1585.

40. Dahiya N, Sherman-Baust CA, Wang TL, Davidson B, Shih Ie M, Zhang Y, et al. MicroRNA expression and identification of putative miRNA targets in ovarian cancer. PLoS One 2008;3:e2436.

41. Han Y, Chen J, Zhao X, Liang C, Wang Y, Sun L, et al. MicroRNA expression signatures of bladder cancer revealed by deep sequencing. PLoS One 2011;6:e18286.

42. Shimono Y, Zabala M, Cho RW, Lobo N, Dalerba P, Qian D, et al. Downregulation of miRNA-200c links breast cancer stem cells with normal stem cells. Cell 2009;138:592–603.

43. Izzotti A, Cartiglia C, Steele VE, De Flora S. MicroRNAs as targets for dietary and pharmacological inhibitors of mutagenesis and carcinogenesis. Mutat Res 2012.

44. Visvader JE, Lindeman GJ. Cancer stem cells in solid tumours: accumulating evidence and unresolved questions. Nat Rev Cancer 2008;8:755–768.

45. Dean M, Fojo T, Bates S. Tumour stem cells and drug resistance. Nat Rev Cancer 2005;5:275–284.

46. Alison MR, Lim SM, Nicholson LJ. Cancer stem cells: Problems for therapy? J Pathol 2011;223:147–161.

47. Lin L, Liu Y, Li H, Li PK, Fuchs J, Shibata H, et al. Targeting colon cancer stem cells using a new curcumin analogue, GO-Y030. Br J Cancer 2011;105:212–220.

48. Kanwar SS, Yu Y, Nautiyal J, Patel BB, Padhye S, Sarkar FH, et al. Difluorinated-curcumin (CDF): a novel curcumin analog is a potent inhibitor of colon cancer stem-like cells. Pharm Res 2011;28:827–838.

49. Kachroo P, Ivanov I, Davidson LA, Chowdhary BP, Lupton JR, Chapkin RS. Classification of diet-modulated gene signatures at the colon cancer initiation and progression stages. Dig Dis Sci 2011;56:2595–2604.

50. De Robertis M, Massi E, Poeta ML, Carotti S, Morini S, Cecchetelli L, et al. The AOM/DSS murine model for the study of colon carcinogenesis: from pathways to diagnosis and therapy studies. J Carcinog 2011;10:9.

51. Fodde R, Smits R. Disease model: familial adenomatous polyposis. Trends Mol Med 2001;7:369–373.

52. Heyer J, Yang K, Lipkin M, Edelmann W, Kucherlapati R. Mouse models for colorectal cancer Oncogene 1999;18:5325–5333.

53. Garofalo A, Chirivi RG, Scanziani E, Mayo JG, Vecchi A, Giavazzi R. Comparative study on the metastatic behavior of human tumors in nude, beige/nude/xid and severe combined immunodeficient mice. Invasion Metastasis 1993;13:82–91.

54. Hoffman RM. Orthotopic metastatic mouse models for anticancer drug discovery and evaluation: a bridge to the clinic. Invest New Drugs 1999;17:343–359.

55. Tavan E, Cayuela C, Antoine JM, Trugnan G, Chaugier C, Cassand P. Effects of dairy products on heterocyclic aromatic amine-induced rat colon carcinogenesis. Carcinogenesis 2002;23:477–483.

56. Walpole AL, Williams MH, Roberts DC. The carcinogenic action of 4-aminodiphenyl and 3:2′-dimethyl-4-amino-diphenyl. Br J Ind Med 1952;9:255–263.

57. Reddy BS, Ohmori T. Effect of intestinal microflora and dietary fat on 3,2′-dimethyl-4-aminobiphenyl-induced colon carcinogenesis in F344 rats. Cancer Res 1981;41:1363–1367.

58. Narisawa T, Magadia NE, Weisburger JH, Wynder EL. Promoting effect of bile acids on colon carcinogenesis after intrarectal instillation of N-methyl-N'-nitro-N-nitrosoguanidine in rats. J Natl Cancer Inst 1974;53:1093–1097.

59. Deschner EE, Long FC. Colonic neoplasms in mice produced with six injections of 1,2-dimethylhydrazine. Oncology 1977;34:255–257.

60. Thurnherr N, Deschner EE, Stonehill EH, Lipkin M. Induction of adenocarcinomas of the colon in mice by weekly injections of 1,2-dimethylhydrazine. Cancer Res 1973;33:940–945.

61. Tanaka T, Kohno H, Suzuki R, Yamada Y, Sugie S, Mori H. A novel inflammation-related mouse colon carcinogenesis model induced by azoxymethane and dextran sodium sulfate. Cancer Sci 2003;94:965–973.

62. Neufert C, Becker C, Neurath MF. An inducible mouse model of colon carcinogenesis for the analysis of sporadic and inflammation-driven tumor progression. Nat Protoc 2007;2:1998–2004.

63. Suzuki R, Kohno H, Sugie S, Nakagama H, Tanaka T. Strain differences in the susceptibility to azoxymethane and dextran sodium sulfate–induced colon carcinogenesis in mice. Carcinogenesis 2006;27:162–169.

64. Kobaek-Larsen M, Thorup I, Diederichsen A, Fenger C, Hoitinga MR. Review of colorectal cancer and its metastases in rodent models: comparative aspects with those in humans. Comp Med 2000;50:16–26.

65. Laqueur GL. Carcinogenic effects of cycad meal and cycasin, methylazoxymethanol glycoside, in rats and effects of cycasin in germfree rats. Fed Proc 1964;23:1386–1388.

66. Delker DA, McKnight SJ, 3rd, Rosenberg DW. The role of alcohol dehydrogenase in the metabolism of the colon carcinogen methylazoxymethanol. Toxicol Sci 1998;45:66–71.

67. Sohn OS, Fiala ES, Requeijo SP, Weisburger JH, Gonzalez FJ. Differential effects of CYP2E1 status on the metabolic activation of the colon carcinogens azoxymethane and methylazoxymethanol. Cancer Res 2001;61:8435–8440.

68. Nobili S, Lippi D, Witort E, Donnini M, Bausi L, Mini E, et al. Natural compounds for cancer treatment and prevention. Pharmacol Res 2009;59:365–378.

69. Birt DF, Hendrich S, Wang W. Dietary agents in cancer prevention: flavonoids and isoflavonoids. Pharmacol Ther 2001;90:157–177.

70. General procedures and measurement of total phenolics. In: Dey PM, Harborne JB, editors. *Methods in Plant Biochemistry: Plant Phenolics*. London: Academic Press; 1989, pp. 1–28.

71. Yang CS, Landau JM, Huang MT, Newmark HL. Inhibition of carcinogenesis by dietary polyphenolic compounds. Annu Rev Nutr 2001;21:381–406.

72. Britton G. Structure and properties of carotenoids in relation to function. FASEB J 1995;9:1551–1558.
73. Nishino H, Murakosh M, Ii T, Takemura M, Kuchide M, Kanazawa M, et al. Carotenoids in cancer chemoprevention. Cancer Metastasis Rev. 2002;21:257–264.
74. Fahey JW, Zalcmann AT, Talalay P. The chemical diversity and distribution of glucosinolates and isothiocyanates among plants. Phytochemistry 2001;56:5–51.
75. Minich DM, Bland JS. A review of the clinical efficacy and safety of cruciferous vegetable phyto-chemicals. Nutr Rev 2007;65:259–267.
76. Hull MA. Omega-3 polyunsaturated fatty acids. Best Pract Res Clin Gastroenterol 2011;25:547–554.
77. Newman DJ, Cragg GM. Natural products as sources of new drugs over the 30 years from 1981 to 2010. J Nat Prod 2012;75:311–335.

CHAPTER 22

CANCER CHEMOPREVENTION TARGETING COX-2 USING DIETARY PHYTOCHEMICALS

Kyung-Soo Chun and Young-Joon Surh

Carcinogenesis is a complex process that consists of numerous biochemical and molecular events. Chemoprevention is an attempt to use nontoxic natural and synthetic substances or their mixtures to intervene in the relatively early stages of carcinogenesis, before invasive characteristics become manifest. The various stages of carcinogenesis are generally classified as initiation, promotion, and progression. A chemopreventive agent should be able to interfere with one or more phases of multistep carcinogenesis (1). It must be relatively safe as well as effective after long-term administration (2).

There have been multiple lines of emerging evidence, suggesting a cause-and-effect relationship between chronic inflammation and cancer (3,4). Approximately 25% of all cancers are somehow associated with chronic infection and inflammation (5). Although physiological inflammation acts as an adaptive host defense against infection or injury, inadequate resolution of inflammatory responses often leads to various chronic ailments, including cancer (6). Sustained cellular injuries can provoke chronic inflammation, which may cause carcinogenesis. Inflammatory immune cells release numerous pro-inflammatory mediators, such as prostaglandins (PGs), nitric oxide (NO), cytokines, chemokines, and leukotrienes (7). These pro-inflammatory mediators play critical roles in excessive formation of reactive oxygen and nitrogen species (ROSs and RNSs), which can damage such cellular macromolecules as DNA and proteins (8).

PGs are major mediators of inflammation. Some pro-inflammatory PGs, including PGE_2 and $PGF_{2\alpha}$, play roles in carcinogenesis as well as inflammation. Elevated levels of PGs have been observed in various types of human cancers (9,10). In line with this notion, several population-based studies have revealed a significant reduction in the risk of developing colorectal, gastric, esophageal, and breast cancers among people who regularly take nonsteroidal anti-inflammatory drugs (NSAIDs),

Cancer and Inflammation Mechanisms: Chemical, Biological, and Clinical Aspects, First Edition.
Edited by Yusuke Hiraku, Shosuke Kawanishi, and Hiroshi Ohshima.
© 2014 John Wiley & Sons, Inc. Published 2014 by John Wiley & Sons, Inc.

including aspirin (11,12). The cancer chemopreventive properties of NSAIDs have been attributed to their inhibition of cyclooxygenase (COX), or prostaglandin H_2 synthase, which catalyzes the rate-limiting step in the conversion of arachidonic acid to PGs.

COX is the key enzyme in the biosynthesis of the PGs mediating inflammation and other physiological processes. In the early 1990s, COX was demonstrated to exist as two distinct isoforms (13). COX-1, a housekeeping enzyme, is constitutively expressed in nearly all tissues, and mediates physiological responses such as cytoprotection of the stomach, platelet aggregation, and regulation of renal blood flow (14). COX-2 is expressed in cells that mediate inflammation.

Inappropriate up-regulation of COX-2 has been implicated in the pathogenesis of various types of malignancies, including cancers of colon (15), breast (16), stomach (17), pancreas (18), skin (19), urinary bladder (20), and head and neck (21). Possible mechanisms by which COX-2 contributes to carcinogenesis include promotion of cellular proliferation, suppression of cancer cell apoptosis, enhancement of angiogenesis, and invasiveness (22). Therefore, blocking the inappropriate activation or expression of COX-2 is a rational approach for achieving cancer prevention (23).

COX-2 AND CANCER

It is evident that COX-2 plays a crucial role in carcinogenesis. In response to various external stimuli, such as pro-inflammatory cytokines, lipopolysaccharide (LPS), ultraviolet (UV) radiation, ROS, and 12-O-tetradecanoylphorbol-13-acetate (TPA), COX-2 expression is transiently elevated in certain tissues (24). Abnormally overexpressed COX-2 noted in a wide range of cancers exerts anti-apoptotic, proangiogneic, proliferative, and pro-inflammatory actions (25).

In support of the oncogenic functions of COX-2, genetic deletion or pharmacological inhibition of this enzyme abrogates tumorigenesis. The COX-2-null genotype has been associated with reduced tumorigenesis in the colon (26) and the skin (27). In addition, a large number of rodent studies have confirmed the preventive efficacy of selective COX-2 inhibitors in experimentally induced carcinogenesis models (28–32). Epidemiological studies have also shown that long-term use of NSAIDs and selective COX-2 inhibitors (COXIBs) prevents cancer development (33,34). However, prolonged consumption of conventional NSAIDs such as aspirin, which inhibit both COX-1 and COX-2, can cause side effects such as gastric bleeding. This is due largely to their nonselective suppression of COX-1, which has a cytoprotective function. Selective COX-2 inhibitors exhibit lower toxicity while retaining the antineoplastic properties of conventional NSAIDs. Based on these findings, it is conceivable that targeted inhibition of abnormally elevated expression and/or activity of COX-2 may provide one of the most broadly effective and promising therapeutic and preventive approaches in the management of inflammation-associated cancers.

TRANSCRIPTIONAL REGULATION OF COX-2 EXPRESSION

Overexpression of COX-2 can occur as a consequence of both enhanced mRNA stability and/or transcription (35,36). COX-2 mRNA contains an AU-rich element (ARE) in its 3'-untranslated region (37). Several reports have shown that activation of p38 mitogen-activated protein kinase (MAPK) is linked to the stabilization of COX-2 mRNA (38). A variety of transcription factors are involved in the transcriptional regulation of COX-2 (35,36). The promoter region of the COX-2 gene contains a canonical TATA box and various putative transcriptional regulatory elements, such as nuclear factor-κB (NF-κB); CCAAT/enhancer binding protein (C/EBP), which is alternatively known as nuclear factor for interleukin-6 (NF-IL6); polyoma enhancer activator 3 (PEA3); nuclear factor for activating T cells (NFAT); cyclic AMP-response element binding protein (CREB); activator protein-1 (AP-1); activator protein-2 (AP-2); and SP-1 (39). Inappropriate activation of these transcription factors leads to overexpression of COX-2.

The transcription factor NF-κB is one of the most vital components of the link between inflammation and cancer (40). NF-κB has been found to be constitutively overactivated in various tumors, lending support to its oncogenic roles. It is involved in numerous cellular processes during the promotion and progression of carcinogenesis. The role of NF-κB in COX-2 induction has been investigated extensively and is well documented. NF-κB has been shown to act as a positive regulator of LPS-induced COX-2 expression in murine macrophages (41) and human colon adenocarcinoma cells (42). The LPS-induced PGE_2 production in J774 macrophages has been reduced by the antioxidant pyrrolidine dithiocarbamate (PDTC) and the serine protease inhibitor N-$α$-p-tosyl-$τ$-lysine chloromethylketone, which are inhibitors of NF-κB activation (41). Moreover, the expression of COX-2 in IL-1β-stimulated amnion mesenchymal cells was blocked by the NF-κB inhibitor SN50 (43), and an NF-κB decoy abrogated COX-2 expression after NO or LPS treatment in RAW 264.7 macrophages (44).

Another eukaryotic transcription factor, AP-1, also regulates COX-2 expression. AP-1 consists of basic leucine zipper proteins that belong to the Jun and Fos families. COX-2 expression requires the AP-1 proteins, c-Jun and c-Fos, that activate COX-2 transcription by binding to the CRE site located at the 3'-end of the COX-2 promoter in v-*src*-transformed cells (45). In human chondrocytes, okadaic acid increased COX-2 gene transcription through up-regulation of AP-1 and CRE nuclear binding protein (46). In RAW 264.7 macrophages, transient translocation of a dominant-negative c-Jun mutant attenuated NO-induced COX-2 expression (47).

C/EBP transcription factors are also involved in regulating the COX-2 promoter activity. In murine lung tumor–derived cell lines, site-directed mutagenesis of the 3' C/EBP sites inhibited COX-2 promoter activity (48). The overexpression of COX-2 was found to correlate with an increased C/EBPδ level and decreased C/EBPα expression in mouse skin carcinogenesis (49). The COX-2 mRNA induction and promoter activities are impaired in C/EBPβ$^{-/-}$ macrophages and can be restored by the reexpression of C/EBPβ (50). In contrast, COX-2 induction remains completely normal in C/EBPβ-deficient fibroblasts, indicative of the diversity of cell-specific role of C/EBPβ in regulating COX-2 gene expression (50).

It has been shown that the COX-2 induction is mediated through activation of CREB in response to a variety of stimuli, including lysophosphatidylcholine (51), UV (52), and *Helicobacter pylori* infection (53) in various cultured cells. CREB has been reported to play a role in regulating basal COX-2 expression in colon carcinoma cells (36). Furthermore, the induction COX-2 expression is mediated via activation of CREB in activated mast cells (54), UVB-stimulated human keratinocytes (55), and shear stress–stimulated osteoblastic MC3T3-E1 cells (56). Yang and Bleich reported that mutation of the CREB site inhibited COX-2 promoter activity in mouse pancreatic β-cells (57).

CHEMOPREVENTIVE PHYTOCHEMICALS TARGETING COX-2

Numerous anti-inflammatory substances derived from dietary and medicinal plants suppress COX-2 expression/activity and some of them have been reported to inhibit experimentally induced carcinogenesis (vide infra).

Figure 22.1 illustrates chemical structures of some representative chemopreventive phytochemicals targeting COX-2.

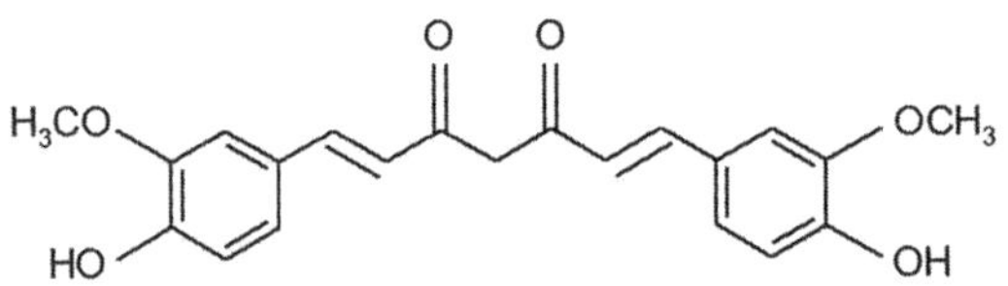

Curcumin

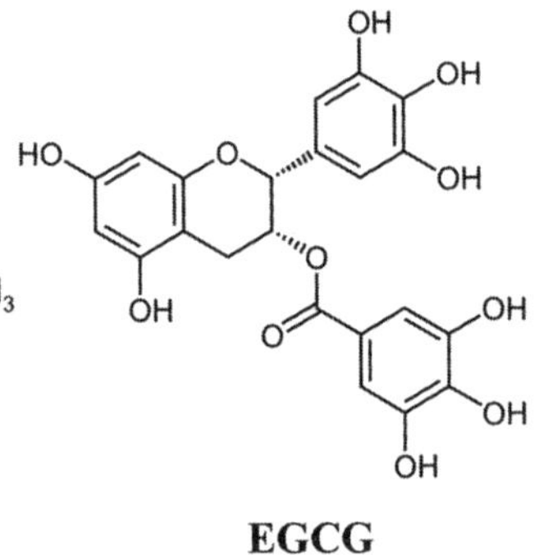

EGCG

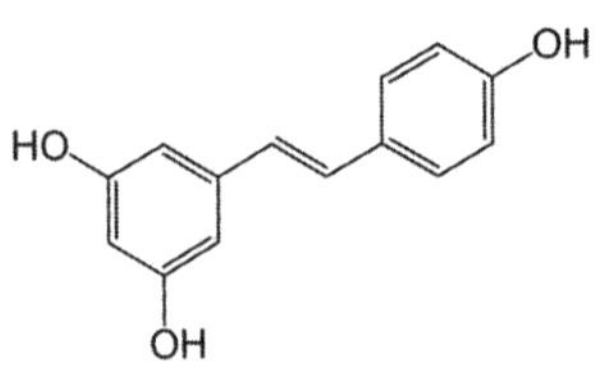

Resveratrol

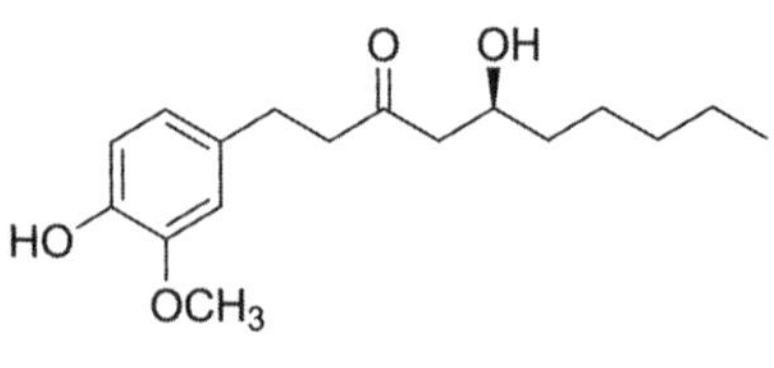

[6]-Gingerol

FIGURE 22.1   Chemical structures of chemopreventive phytochemicals targeting COX-2.

Curcumin

Curcumin is a yellow pigment present in the rhizomes of the Indian spice turmeric (*Curcuma longa* L., Zingiberaceae). The ability of curcumin to inhibit both activity and induced expression of COX-2 has been demonstrated in various cell lines and animal models. Curcumin inhibited the expression of COX-2 and its mRNA transcript in the colon of trinitrobenzenesulfonic acid (TNBS)-treated Sprague–Dawley rats (58), UVB-stimulated HaCaT cells (59), human colon cancer cells (60), TPA-stimulated mouse skin (61), and human pancreatic cancer cells (62). Treatment with curcumin also suppressed the expression of COX-2 protein and mRNA as well as PGE_2 production induced by TPA or bile acid in human gastrointestinal cell lines (63).

The inhibitory effects of curcumin on COX-2 expression are related to its inactivation of AP-1 and NF-κB. Curcumin inhibited TNF-α-induced COX-2 transcription in human colonic epithelial cells by blocking the NF-κB-inducing kinase- and IKK-mediated degradation of IκBα (64). Curcumin also suppressed the expression of COX-2 and generation of PGE_2 in TPA-stimulated mouse skin through suppression of NF-κB activation (61). Curcumin significantly inhibited LPS-mediated induction of COX-2 expression at both mRNA and protein levels in BV2 microglial cells by blocking the activation of NF-κB and AP-1 (65). The inhibitory effect of curcumin on the activation of NF-κB in human myeloid cells stimulated with diverse stimuli, including TNF-α, TPA, and H_2O_2, was reported (66). In another study, curcumin abrogated TNF-induced NF-κB activation in human myeloid leukemia U937 cells by blocking the activation of IKK and AKT (67). Moreover, curcumin diminished TPA-induced expression of the AP-1 component proteins c-Jun and c-Fos in NIH3T3 cells (68) and CD-1 mouse skin *in vivo* (69).

Epigallocatechin Gallate

Epigallocatechin gallate (EGCG), a major green tea polyphenol, has been studied extensively with regard to its cancer preventiue effects. The molecular mechanisms underlying the prevention of inflammation-associated cancer by EGCG have focused on the modulation of COX-2 signaling. Treatment of the human colorectal cancer cells HT-29 and HCA-7 (70) and human colon cancer SW837 cells (71) with EGCG resulted in a significant decrease in the expression of COX-2 protein and mRNA as well as COX-2 promoter activity.

EGCG also suppressed COX-2 expression in response to diverse stimuli, including IL-1β and *N*-nitrosomethylbenzylamine (NMBA) in human epidermal keratinocytes (72), human chondrocytes (73), and esophageal tumors in F344 male rats (74). Moreover, EGCG, given by gavage, inhibited COX-2 expression induced by topically applied TPA in mouse skin (75).

EGCG was found to block the activation of signaling mediated by upstream kinases and transcription factors involved in aberrant COX-2 expression. Several studies demonstrated that EGCG attenuated constitutive activation of NF-κB (70,71) and AP-1 (71) in human colon cancer cells. The activation of AP-1 (76) and NF-κB

(77) in TPA-stimulated mouse epidermal JB6 cells was also inhibited by EGCG. In TPA-treated mouse skin, the DNA binding of NF-κB and CREB was abolished by pretreatment with EGCG (78). The inactivation of NF-κB by EGCG was found to be associated with the inhibition of IKK activity and the suppression of phosphorylation-dependent degradation of IκBα (72). Besides interference with IKK-IκB signaling, the inhibition of upstream MAP kinases by EGCG was attributed to its suppression of NF-κB activation (79,80).

Resveratrol

Resveratrol (*trans*-3,5,4'-trihydroxystilbene) has been isolated from more than 70 plant species, including grapes (81). The compound possesses potent antioxidant and anti-inflammatory activities (81). In a pioneering study, resveratrol blocked the initiation, promotion, and progression stages in a mouse skin carcinogenesis model (82). The blockade of inflammatory signaling pathways involving aberrant COX-2 expression is one of the key molecular mechanisms underlying anti-inflammatory and chemopreventive activities of resveratrol (83,84). Resveratrol has been shown to inhibit COX-2 expression in TPA-treated mouse skin (85), TNBS-stimulated rat colon (86,87), and LPS-, TPA- or H_2O_2-stimulated mouse peritoneal macrophages (88). Moreover, resveratrol diminished the expression of COX-2 mRNA transcript in NMBA-induced esophageal tumors in F344 rats (89). Resveratrol and related stilbene compounds also inhibited the expression of COX-2 and production of PGE_2 in peripheral blood leukocytes stimulated with LPS plus IFN-γ (90).

Resveratrol inhibited NF-κB activation induced by diverse classes of stimuli, such as TPA, LPS, H_2O_2, and okadaic acid in Jurkat-T, HeLa, and glioma cells (91). Resveratrol also suppressed IL-1β-induced activation of NF-κB in acute myeloid leukemia cells (92) and JB6 cells (93). The nuclear translocation and DNA binding of NF-κB were blocked by resveratrol through suppression of phosphorylation and subsequent degradation of IκBα subunits in LPS-stimulated RAW 264.7 cells (94,95). Resveratrol inhibited UVB- and TPA-induced activation of NF-κB in human epidermal keratinocytes (96) and mouse skin (85), respectively, by blocking the activation of IKK as well as phosphorylation and degradation of IκBα. On the other hand, resveratrol abrogated TNF-α-induced activation of NF-κB in U937 cells by suppressing phosphorylation and nuclear translocation of p65, but without affecting IκBα degradation (91). Several studies reported that resveratrol attenuated activation of AP-1 in TPA-treated human mammary epithelial cells (97,98) and mouse skin *in vivo* (84).

Various upstream kinases involved in COX-2 signaling are also the targets of resveratrol in its chemoprevention of inflammation-associated cancers. Pretreatment with resveratrol blocked UVC- and TPA-induced activation of MAPKs and subsequent transcription of the AP-1 reporter gene in HeLa cells (99). Resveratrol inhibited H_2O_2-induced NF-κB activation in HeLa cells partly by blocking activation of PKCμ (100). The differential blockade of the inflammatory signaling by resveratrol appears to be cell type- and stimuli-specific.

[6]-Gingerol

One of the major pungent ingredients of ginger (*Zingiber officinale* Roscoe, Zingiberaceae) is [6]-gingerol, which was found to inhibit epidermal growth factor–induced AP-1 activation and neoplastic transformation in mouse epidermal JB6 cells (101). Topical application of [6]-gingerol inhibited TPA-induced COX-2 expression in mouse skin by blocking key events involved in NF-κB activation, such as degradation of IκBα, nuclear translocation of p65, and interaction of p65 with the coactivator CBP (102). This study also demonstrated that [6]-gingerol diminished NF-κB signaling in mouse skin by blocking TPA-induced phosphorylation of p38 MAPK (102). [6]-Gingerol exerts an inhibitory effect on UVB-induced COX-2 expression through suppression of IκBα phosphorylation in HaCaT cells (103), and a ginger extract containing [6]-gingerol caused a significant decrease in COX-2 expression and PGE_2 production in differentiated U937 cells stimulated with LPS (104).

OTHER POTENTIAL TARGETS FOR INHIBITION OF THE PG SYNTHESIS SIGNALING

Despite reduced gastrointestinal (GI) toxicity, COX–2 selective NSAIDs can cause adverse cardiovascular effects. This was first highlighted by the Vioxx in the Gastrointestinal Outcomes Research (VIGOR) study (105), which resulted in the withdrawal of Rofecoxib from the market (106). The side effects associated with prolonged COX-2 inhibition are attributable to reduced production of thromboxane A_2 (TXA_2) and PGI_2, which leads to imbalance in blood clotting (107). Considering such side effects caused by COX-2-selective inhibitors, a search for other potential targets in the PG synthesis pathway downstream of COX has been needed for better prevention of inflammation-associated carcinogenesis (108). Another attractive strategy for inhibition of PG synthesis is targeting the receptor signaling (109).

The synthesis of PGs is initiated by the liberation of arachidonic acid (AA) from plasma membrane phospholipids by members of the phospholipase A_2 (PLA_2) family. AA is converted immediately into PGH_2 by COX activity at the luminal side of nuclear and ER membranes. PGH_2 functions as an intermediate for the production of PGE_2, PGD_2, $PGF_{2\alpha}$, PGI_2, and TXA_2. These prostanoids have been shown to exert their effects through interaction with specific G protein–coupled receptors: EP (EP1, EP2, EP3, and EP4), DP, FP, IP, and TP, respectively (110).

Recent studies have suggested involvement of EP receptors in cancer. Data from studies utilizing EP receptor–knockout and EP-overexpressing mice as well as receptor-specific agonists and antagonists have indicated an important role for EP2 and EP4 in the formation and growth of several different types of tumors. For example, deficiency in the EP2 receptor reduced the size and the number of intestinal polyps formed in APCD716 mice (111) and inhibited papilloma formation in mouse skin (112). Similarly, homozygous deletion of EP4 receptor, but not EP3 receptor, resulted in a partial decrease in the formation of azoxymethane-induced aberrant crypt foci in mice (113). Furthermore, pharmacological blockade of EP4 by use of specific antagonists diminished carcinogen-induced aberrant crypt foci in wild-type mice and intestinal polyp formation in $Apc^{min+/-}$ mice (113). Deficiency of EP1 also inhibited colon carcinogenesis (114), while transgenic EP1 mice exhibited the accelerated development and progression of skin tumors (115). These studies suggest that the EP2 and EP4 receptors represent potential targets for the chemoprevention of PGE_2-driven tumor development.

POSITIVE FEEDBACK LOOP BETWEEN COX-2 AND EP RECEPTORS

Promoter analysis of the EP2 and EP4 genes identified several consensus sequences responsive to pro-inflammatory stimuli inducing NF-κB, C/EBP, and AP-2, while the promoter region of EP2 also contains Sp1 and MyoD sequences as well as a putative glucocorticoid response element (116,117). Such transcription factors have been known to be activated by PGE_2, and this implies the possibility that COX-2 induction can increase activation as well as induction of EP receptors. Ansari et al. showed that EP2 null mice had a reduced induction of COX-2 expression following TPA treatment compared to wild-type mice (118). Conversely, primary keratinocytes derived from EP2 transgenic mice had increased COX-2 expression after treatment with either TPA or PGE_2 which was mediated through phosphorylation of CREB (118). Steinert et al. also showed that the medullary PGE_2 increased COX-2 expression and activity in a positive feedback loop that involved a cAMP-PKA-CREB-dependent signal transduction (119). EP4 activation could also induce COX-2 expression via the PI3K/AKT pathway, leading to inhibition of GSK-3α activity, which is also inhibited when PKA is activated. This allows translocation of cytosolic β-catenin into the nucleus (120). In the nucleus, β-catenin interacts with the Tcf/Lef family of transcription factors, which results in induction of COX-2 expression. Thus, both EP2 and EP4 induce COX-2, and it is possible that the constitutive up-regulation of COX-2 observed in many epithelial tumors is due to a positive feedback loop involving COX-2 and EP2/EP4 (Figure 22.2).

PHYTOCHEMICALS TARGETING EPs

There are some examples of chemopreventive phytochemicals that target EP receptors. Treatment of human melanoma A375 cells with EGCG resulted in a dose-dependent inhibition of cell migration or invasion of these cells, which was associated with a reduction in the levels of COX-2, PGE_2, EP2, and EP4 (121). EGCG also significantly inhibited the viability and migration of human hepatoma HepG2 cells induced by PGE_2 through upregulation of EP1 expression (122). In melanoma cells, the anticarcinogenic alkaloid berberine was reported to repress expression of EP2, EP4, and COX-2 through inhibition of NF-κB activation (123).

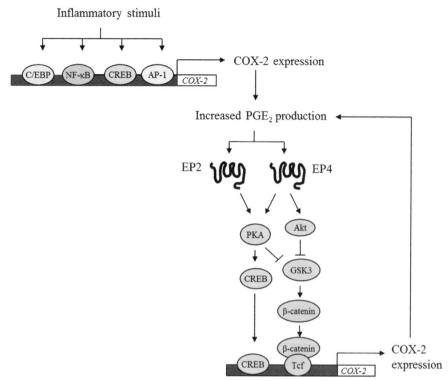

FIGURE 22.2   Intracellular signal transduction involving a positive feedback loop between COX-2 and an EP receptor. (*See insert for color representation of the figure.*)

Moreover, frondoside A, derived from the sea cucumber (*Cucumaria frondosa*), inhibited breast cancer metastasis by antagonizing EP2 and EP4 (124).

CONCLUDING REMARKS

Since abnormal overexpression of COX-2 is implicated in the pathogenesis of various types of human cancers, this inducible enzyme may be a useful target for the evaluation of chemopreventives. Pharmacologic inhibition of COX-2 reduces the formation of various tumors in animals, which supports the cancer chemopreventive potential of selective COX-2 inhibitors. However, given the concerns regarding cardiotoxicity of COX-2 inhibitors, alternative ways of controlling overactivated COX-2 signaling were explored. PGE_2, a major COX-2 product with oncogenic functions, binds to a distinct set of EP receptors (EP1 to 4), and there is an emerging body of literature data suggesting that some, if not all, EP receptors may be therapeutic/preventive targets in the management of inflammation-associated cancer. Future studies will be necessary to identify chemopreventive agents that are effective in repressing either expression or activity of selected EP receptors.

Acknowledgments

This work was supported by Global Core Research Center grant 2011-0030676 from the National Research Foundation, Ministry of Education, Science and Technology of South Korea.

REFERENCES

1. Russo GL. Ins and outs of dietary phytochemicals in cancer chemoprevention. Biochem Pharmacol 2007;74:533–544.
2. Scott EN, Gescher AJ, Steward WP, Brown K. Development of dietary phytochemical chemopreventive agents: biomarkers and choice of dose for early clinical trials. Cancer Prev Res (Phila) 2009;2:525–530.
3. Balkwill F, Mantovani A. Inflammation and cancer: Back to Virchow? Lancet 2001;357:539–545.
4. Kundu JK, Surh YJ. Inflammation: gearing the journey to cancer. Mutat Res 2008;659:15–30.
5. Hussain SP, Harris CC. Inflammation and cancer: an ancient link with novel potentials. Int J Cancer 2007;121:2373–2380.
6. Jackson L, Evers BM. Chronic inflammation and pathogenesis of GI and pancreatic cancers. Cancer Treat Res 2006;130:39–65.
7. Coussens LM, Werb Z. Inflammation and cancer. Nature 2002;420:860–867.
8. Schottenfeld D, Beebe-Dimmer J. Chronic inflammation: a common and important factor in the pathogenesis of neoplasia. CA Cancer J Clin 2006;56:69–83.
9. Bennett A. Prostaglandins: their release, biological effects and relationships to pain and inflammation. Cephalalgia 1986;6 (Suppl 4):17–20.
10. Rigas B, Goldman IS, Levine L. Altered eicosanoid levels in human colon cancer. J Lab Clin Med 1993;122:518–523.
11. Giovannucci E, Egan KM, Hunter DJ, Stampfer MJ, Colditz GA, Willett WC, et al. Aspirin and the risk of colorectal cancer in women. N Engl J Med 1995;333:609–614.
12. Schreinemachers DM, Everson RB. Aspirin use and lung, colon, and breast cancer incidence in a prospective study. Epidemiology 1994;5:138–146.
13. Feng L, Sun W, Xia Y, Tang WW, Chanmugam P, Soyoola E, Wilson CB, Hwang D. Cloning two isoforms of rat cyclooxygenase: differential regulation of their expression. Arch Biochem Biophys 1993;307:361–368.
14. Smith WL, DeWitt DL, Garavito RM. Cyclooxygenases: structural, cellular, and molecular biology. Annu Rev Biochem 2000;69:145–182.
15. Dubois RN. Review article: Cyclooxygenase—a target for colon cancer prevention. Aliment Pharmacol Ther 2000;14 (Suppl 1):64–67.
16. Soslow RA, Dannenberg AJ, Rush D, Woerner BM, Khan KN, Masferrer J, et al. COX-2 is expressed in human pulmonary, colonic, and mammary tumors. Cancer 2000;89:2637–2645.
17. Lim HY, Joo HJ, Choi JH, Yi JW, Yang MS, Cho DY, et al. Increased expression of cyclooxygenase-2 protein in human gastric carcinoma. Clin Cancer Res 2000;6:519–525.
18. Yip-Schneider MT, Barnard DS, Billings SD, Cheng L, Heilman DK, Lin A, et al. Cyclooxygenase-2 expression in human pancreatic adenocarcinomas. Carcinogenesis 2000;21:139–146.
19. Pentland AP, Schoggins JW, Scott GA, Khan KN, Han R. Reduction of UV-induced skin tumors in hairless mice by selective COX-2 inhibition. Carcinogenesis 1999;20:1939–1944.
20. Grubbs CJ, Lubet RA, Koki AT, Leahy KM, Masferrer JL, Steele VE, et al. Celecoxib inhibits N-butyl-N-(4-hydroxybutyl)-nitrosamine–induced urinary bladder cancers in male B6D2F1 mice and female Fischer-344 rats. Cancer Res 2000;60:5599–5602.
21. Chan G, Boyle JO, Yang EK, Zhang F, Sacks PG, Shah JP, et al. Cyclooxygenase-2 expression is up-regulated in squamous cell carcinoma of the head and neck. Cancer Res 1999;59:991–994.
22. Surh YJ, Chun KS, Cha HH, Han SS, Keum YS, Park KK, et al. Molecular mechanisms underlying chemopreventive activities of anti-inflammatory phytochemicals: down-regulation of COX-2 and iNOS through suppression of NFκB activation. Mutat Res 2001;480–481:243–268.

23. Surh YJ. Cancer chemoprevention with dietary phytochemicals. Nat Rev Cancer 2003;3:768–780.
24. Chun KS, Surh YJ. Signal transduction pathways regulating cyclooxygenase-2 expression: potential molecular targets for chemoprevention. Biochem Pharmacol 2004;68:1089–1100.
25. Harris RE. Cyclooxygenase-2 (COX-2) and the inflammogenesis of cancer. Subcell Biochem 2007;42:93–126.
26. Oshima M, Dinchuk JE, Kargman SL, Oshima H, Hancock B, Kwong E, et al. Suppression of intestinal polyposis in $Apc\Delta716$ knockout mice by inhibition of cyclooxygenase 2 (COX-2). Cell 1996;87:803–809.
27. Tiano HF, Loftin CD, Akunda J, Lee CA, Spalding J, Sessoms A, et al. Deficiency of either cyclooxygenase (COX)-1 or COX-2 alters epidermal differentiation and reduces mouse skin tumorigenesis. Cancer Res 2002;62:3395–3401.
28. Wei M, Morimura K, Wanibuchi H, Shen J, Doi K, Mitsuhashi M, et al. Chemopreventive effect of JTE-522, a selective cyclooxygenase-2 inhibitor, on 1, 2-dimethylhydrazine-induced rat colon carcinogenesis. Cancer Lett 2003;202:11–16.
29. Furukawa F, Nishikawa A, Lee IS, Kanki K, Umemura T, Okazaki K, et al. A cyclooxygenase-2 inhibitor, nimesulide, inhibits postinitiation phase of N-nitrosobis(2-oxopropyl)amine–induced pancreatic carcinogenesis in hamsters. Int J Cancer 2003;104:269–273.
30. Narayanan BA, Condon MS, Bosland MC, Narayanan NK, Reddy BS. Suppression of N-methyl-N-nitrosourea/testosterone-induced rat prostate cancer growth by celecoxib: effects on cyclooxygenase-2, cell cycle regulation, and apoptosis mechanism(s). Clin Cancer Res 2003;9:3503–3513.
31. Wilgus TA, Koki AT, Zweifel BS, Rubal PA, Oberyszyn TM. Chemotherapeutic efficacy of topical celecoxib in a murine model of ultraviolet light B–induced skin cancer. Mol Carcinog 2003;38:33–39.
32. Gupta S, Adhami VM, Subbarayan M, MacLennan GT, Lewin JS, Hafeli UO, et al. Suppression of prostate carcinogenesis by dietary supplementation of celecoxib in transgenic adenocarcinoma of the mouse prostate model. Cancer Res 2004;64:3334–3343.
33. Fosslien E. Biochemistry of cyclooxygenase (COX)-2 inhibitors and molecular pathology of COX-2 in neoplasia. Crit Rev Clin Lab Sci 2000;37:431–502.
34. Steinbach G, Lynch PM, Phillips RK, Wallace MH, Hawk E, Gordon GB, et al. The effect of celecoxib, a cyclooxygenase-2 inhibitor, in familial adenomatous polyposis. N Engl J Med 2000;342:1946–1952.
35. Dannenberg AJ, Altorki NK, Boyle JO, Dang C, Howe LR, Weksler BB, et al. Cyclo-oxygenase 2: a pharmacological target for the prevention of cancer. Lancet Oncol 2001;2:544–551.
36. Shao J, Sheng H, Inoue H, Morrow JD, DuBois RN. Regulation of constitutive cyclooxygenase-2 expression in colon carcinoma cells. J Biol Chem 2000;275:33951–33956.
37. Xu K, Robida AM, Murphy TJ. Immediate-early MEK-1-dependent stabilization of rat smooth muscle cell cyclooxygenase-2 mRNA by $G\alpha q$-coupled receptor signaling. J Biol Chem 2000;275:23012–23019.
38. Lasa M, Mahtani KR, Finch A, Brewer G, Saklatvala J, Clark AR. Regulation of cyclooxygenase 2 mRNA stability by the mitogen-activated protein kinase p38 signaling cascade. Mol Cell Biol 2000;20:4265–4274.
39. Kosaka T, Miyata A, Ihara H, Hara S, Sugimoto T, Takeda O, et al. Characterization of the human gene (PTGS2) encoding prostaglandin-endoperoxide synthase 2. Eur J Biochem 1994;221:889–897.
40. Pikarsky E, Porat RM, Stein I, Abramovitch R, Amit S, Kasem S, et al. NFκB functions as a tumour promoter in inflammation-associated cancer. Nature 2004;431:461–466.
41. D'Acquisto F, Iuvone T, Rombola L, Sautebin L, Di Rosa M, Carnuccio R. Involvement of NFκB in the regulation of cyclooxygenase-2 protein expression in LPS-stimulated J774 macrophages. FEBS Lett 1997;418:175–178.
42. Kojima M, Morisaki T, Izuhara K, Uchiyama A, Matsunari Y, Katano M, et al. Lipopolysaccharide increases cyclo-oxygenase-2 expression in a colon carcinoma cell line through NFκB activation. Oncogene 2000;19:1225–1231.
43. Yan X, Wu Xiao C, Sun M, Tsang BK, Gibb W. Nuclear factor κ B activation and regulation of cyclooxygenase type-2 expression in human amnion mesenchymal cells by interleukin-1β. Biol Reprod 2002;66:1667–1671.
44. von Knethen A, Callsen D, Brune B. Superoxide attenuates macrophage apoptosis by NFκB and AP-1 activation that promotes cyclooxygenase-2 expression. J Immunol 1999;163:2858–2866.

45. Xie W, Herschman HR. v-src induces prostaglandin synthase 2 gene expression by activation of the c-Jun N-terminal kinase and the c-Jun transcription factor. J Biol Chem 1995;270:27622–27628.

46. Miller C, Zhang M, He Y, Zhao J, Pelletier JP, Martel-Pelletier J, et al. Transcriptional induction of cyclooxygenase-2 gene by okadaic acid inhibition of phosphatase activity in human chondrocytes: co-stimulation of AP-1 and CRE nuclear binding proteins. J Cell Biochem 1998;69:392–413.

47. von Knethen A, Callsen D, Brune B. NFκB and AP-1 activation by nitric oxide attenuated apoptotic cell death in RAW 264.7 macrophages. Mol Biol Cell 1999;10:361–372.

48. Wardlaw SA, Zhang N, Belinsky SA. Transcriptional regulation of basal cyclooxygenase-2 expression in murine lung tumor-derived cell lines by CCAAT/enhancer-binding protein and activating transcription factor/cAMP response element-binding protein. Mol Pharmacol 2002;62:326–333.

49. Kim Y, Fischer SM. Transcriptional regulation of cyclooxygenase-2 in mouse skin carcinoma cells: regulatory role of CCAAT/enhancer-binding proteins in the differential expression of cyclooxygenase-2 in normal and neoplastic tissues. J Biol Chem 1998;273:27686–27694.

50. Gorgoni B, Maritano D, Marthyn P, Righi M, Poli V. C/EBP β gene inactivation causes both impaired and enhanced gene expression and inverse regulation of IL-12 p40 and p35 mRNAs in macrophages. J Immunol 2002;168:4055–4062.

51. Rikitake Y, Hirata K, Kawashima S, Takeuchi S, Shimokawa Y, Kojima Y, et al. Signaling mechanism underlying COX-2 induction by lysophosphatidylcholine. Biochem Biophys Res Commun 2001;281:1291–1297.

52. Iordanov M, Bender K, Ade T, Schmid W, Sachsenmaier C, Engel K, et al. CREB is activated by UVC through a p38/HOG-1-dependent protein kinase. EMBO J 1997;16:1009–1022.

53. Juttner S, Cramer T, Wessler S, Walduck A, Gao F, Schmitz F, et al. *Helicobacter pylori* stimulates host cyclooxygenase-2 gene transcription: critical importance of MEK/ERK-dependent activation of USF1/-2 and CREB transcription factors. Cell Microbiol 2003;5:821–834.

54. Reddy ST, Wadleigh DJ, Herschman HR. Transcriptional regulation of the cyclooxygenase-2 gene in activated mast cells. J Biol Chem 2000;275:3107–3113.

55. Tang Q, Chen W, Gonzales MS, Finch J, Inoue H, Bowden GT. Role of cyclic AMP responsive element in the UVB induction of cyclooxygenase-2 transcription in human keratinocytes. Oncogene 2001;20:5164–5172.

56. Ogasawara A, Arakawa T, Kaneda T, Takuma T, Sato T, Kaneko H, et al. Fluid shear stress-induced cyclooxygenase-2 expression is mediated by C/EBP β, cAMP-response element-binding protein, and AP-1 in osteoblastic MC3T3-E1 cells. J Biol Chem 2001;276:7048–7054.

57. Yang F, Bleich D. Transcriptional regulation of cyclooxygenase-2 gene in pancreatic beta-cells. J Biol Chem 2004;279:35403–35411.

58. Jiang H, Deng CS, Zhang M, Xia J. Curcumin-attenuated trinitrobenzene sulphonic acid induces chronic colitis by inhibiting expression of cyclooxygenase-2. World J Gastroenterol 2006;12:3848–3853.

59. Cho JW, Park K, Kweon GR, Jang BC, Baek WK, Suh MH, et al. Curcumin inhibits the expression of COX-2 in UVB-irradiated human keratinocytes (HaCaT) by inhibiting activation of AP-1: p38 MAP kinase and JNK as potential upstream targets. Exp Mol Med 2005;37:186–192.

60. Su CC, Chen GW, Lin JG, Wu LT, Chung JG. Curcumin inhibits cell migration of human colon cancer colo 205 cells through the inhibition of nuclear factor κ B/p65 and down-regulates cyclooxygenase-2 and matrix metalloproteinase-2 expressions. Anticancer Res 2006;26:1281–1288.

61. Chun KS, Keum YS, Han SS, Song YS, Kim SH, Surh YJ. Curcumin inhibits phorbol ester-induced expression of cyclooxygenase-2 in mouse skin through suppression of extracellular signal-regulated kinase activity and NFκB activation. Carcinogenesis 2003;24:1515–1524.

62. Li L, Aggarwal BB, Shishodia S, Abbruzzese J, Kurzrock R. Nuclear factor-κB and IκB kinase are constitutively active in human pancreatic cells, and their down-regulation by curcumin (diferuloyl-methane) is associated with the suppression of proliferation and the induction of apoptosis. Cancer 2004;101:2351–2362.

63. Zhang F, Altorki NK, Mestre JR, Subbaramaiah K, Dannenberg AJ. Curcumin inhibits cyclooxygenase-2 transcription in bile acid- and phorbol ester-treated human gastrointestinal epithelial cells. Carcinogenesis 1999;20:445–451.

64. Plummer SM, Holloway KA, Manson MM, Munks RJ, Kaptein A, Farrow S, et al. Inhibition of cyclo-oxygenase 2 expression in colon cells by the chemopreventive agent curcumin involves inhibition of NFκB activation via the NIK/IKK signalling complex. Oncogene 1999;18:6013–6020.

65. Kang G, Kong PJ, Yuh YJ, Lim SY, Yim SV, Chun   W, et al. Curcumin suppresses lipopolysaccharide-induced cyclooxygenase-2 expression by inhibiting activator protein 1 and NFκB bindings in BV2 microglial cells. J Pharmacol Sci 2004;94:325–328.

66. Singh S, Aggarwal BB. Activation of transcription factor NFκB is suppressed by curcumin (diferuloylmethane) [corrected]. J Biol Chem 1995;270:24995–25000.

67. Aggarwal S, Ichikawa H, Takada Y, Sandur SK, Shishodia S, Aggarwal BB. Curcumin (diferuloylmethane) down-regulates expression of cell proliferation and antiapoptotic and metastatic gene products through suppression of IκBα kinase and Akt activation. Mol Pharmacol 2006;69:195–206.

68. Kakar SS, Roy D. Curcumin inhibits TPA induced expression of c-fos, c-jun and c-myc proto-oncogenes messenger RNAs in mouse skin. Cancer Lett 1994;87:85–89.

69. Lu YP, Chang RL, Lou YR, Huang MT, Newmark HL, Reuhl KR, et al. Effect of curcumin on 12-O-tetradecanoylphorbol-13-acetate- and ultraviolet B light–induced expression of c-Jun and c-Fos in JB6 cells and in mouse epidermis. Carcinogenesis 1994;15:2363–2370.

70. Peng G, Dixon DA, Muga SJ, Smith TJ, Wargovich MJ. Green tea polyphenol (−)-epigallocatechin-3-gallate inhibits cyclooxygenase-2 expression in colon carcinogenesis. Mol Carcinog 2006;45:309–319.

71. Shimizu M, Deguchi A, Joe AK, McKoy JF, Moriwaki H, Weinstein IB. EGCG inhibits activation of HER3 and expression of cyclooxygenase-2 in human colon cancer cells. J Exp Ther Oncol 2005;5:69–78.

72. Cui Y, Kim DS, Park SH, Yoon JA, Kim SK, Kwon SB, et al. Involvement of ERK AND p38 MAP kinase in AAPH-induced COX-2 expression in HaCaT cells. Chem Phys Lipids 2004;129:43–52.

73. Ahmed S, Rahman A, Hasnain A, Lalonde M, Goldberg VM, Haqqi TM. Green tea polyphenol epigallocatechin-3-gallate inhibits the IL-1 beta-induced activity and expression of cyclooxygenase-2 and nitric oxide synthase-2 in human chondrocytes. Free Radic Biol Med 2002;33:1097–1105.

74. Li ZG, Shimada Y, Sato F, Maeda M, Itami A, Kaganoi J, et al. Inhibitory effects of epigallocatechin-3-gallate on N-nitrosomethylbenzylamine-induced esophageal tumorigenesis in F344 rats. Int J Oncol 2002;21:1275–1283.

75. Kundu JK, Na HK, Chun KS, Kim YK, et al. Inhibition of phorbol ester-induced COX-2 expression by epigallocatechin gallate in mouse skin and cultured human mammary epithelial cells. J Nutr 2003;133:3805S–3810S.

76. Dong Z, Ma W, Huang C, Yang CS. Inhibition of tumor promoter–induced activator protein 1 activation and cell transformation by tea polyphenols, (−)-epigallocatechin gallate, and theaflavins. Cancer Res 1997;57:4414–4419.

77. Nomura M, Ma W, Chen N, Bode AM, Dong Z. Inhibition of 12-O-tetradecanoylphorbol-13-acetate-induced NFκB activation by tea polyphenols, (−)-epigallocatechin gallate and theaflavins. Carcinogenesis 2000;21:1885–1890.

78. Kundu JK, Surh YJ. Epigallocatechin gallate inhibits phorbol ester-induced activation of NFκB and CREB in mouse skin: role of p38 MAPK. Ann NY Acad Sci 2007;1095:504–512.

79. Ahn SC, Kim GY, Kim JH, Baik SW, Han MK, Lee HJ, et al. Epigallocatechin-3-gallate, constituent of green tea, suppresses the LPS-induced phenotypic and functional maturation of murine dendritic cells through inhibition of mitogen-activated protein kinases and NFκB. Biochem Biophys Res Commun 2004;313:148–155.

80. Vayalil PK, Katiyar SK. Treatment of epigallocatechin-3-gallate inhibits matrix metalloproteinases-2 and -9 via inhibition of activation of mitogen-activated protein kinases, c-jun and NFκB in human prostate carcinoma DU-145 cells. Prostate 2004;59:33–42.

81. Kundu JK, Surh YJ. Molecular basis of chemoprevention by resveratrol: NFκB and AP-1 as potential targets. Mutat Res 2004;555:65–80.

82. Jang M, Cai L, Udeani GO, Slowing KV, Thomas CF, Beecher CW, Fong HH, et al. Cancer chemopreventive activity of resveratrol, a natural product derived from grapes. Science 1997;275:218–220.

83. Aggarwal BB, Bhardwaj A, Aggarwal RS, Seeram NP, Shishodia S, Takada Y. Role of resveratrol in prevention and therapy of cancer: preclinical and clinical studies. Anticancer Res 2004;24:2783–2840.

84. Kundu JK, Chun KS, Kim SO, Surh YJ. Resveratrol inhibits phorbol ester–induced cyclooxygenase-2 expression in mouse skin: MAPKs and AP-1 as potential molecular targets. Biofactors 2004;21:33–39.

85. Kundu JK, Shin YK, Kim SH, Surh YJ. Resveratrol inhibits phorbol ester–induced expression of COX-2 and activation of NFκB in mouse skin by blocking IκB kinase activity. Carcinogenesis 2006;27:1465–1474.

86. Martin AR, Villegas I, Sanchez-Hidalgo M, de la Lastra CA. The effects of resveratrol, a phytoalexin derived from red wines, on chronic inflammation induced in an experimentally induced colitis model. Br J Pharmacol 2006;147:873–885.

87. Martin AR, Villegas I, La Casa C, de la Lastra CA. Resveratrol, a polyphenol found in grapes, suppresses oxidative damage and stimulates apoptosis during early colonic inflammation in rats. Biochem Pharmacol 2004;67:1399–1410.

88. Martinez J, Moreno JJ. Effect of resveratrol, a natural polyphenolic compound, on reactive oxygen species and prostaglandin production. Biochem Pharmacol 2000;59:865–870.

89. Li ZG, Hong T, Shimada Y, Komoto I, Kawabe A, Ding Y, et al. Suppression of N-nitrosomethylbenzylamine (NMBA)-induced esophageal tumorigenesis in F344 rats by resveratrol. Carcinogenesis 2002;23:1531–1536.

90. Richard N, Porath D, Radspieler A, Schwager J. Effects of resveratrol, piceatannol, tri-acetoxystilbene, and genistein on the inflammatory response of human peripheral blood leukocytes. Mol Nutr Food Res 2005;49:431–442.

91. Manna SK, Mukhopadhyay A, Aggarwal BB. Resveratrol suppresses TNF-induced activation of nuclear transcription factors NFκB, activator protein-1, and apoptosis: potential role of reactive oxygen intermediates and lipid peroxidation. J Immunol 2000;164:6509–6519.

92. Estrov Z, Shishodia S, Faderl S, Harris D, Van Q, Kantarjian HM, et al. Resveratrol blocks interleukin-1beta-induced activation of the nuclear transcription factor NFκB, inhibits proliferation, causes S-phase arrest, and induces apoptosis of acute myeloid leukemia cells. Blood 2003;102:987–995.

93. Leonard SS, Xia C, Jiang BH, Stinefelt B, Klandorf H, Harris GK, et al. Resveratrol scavenges reactive oxygen species and effects radical-induced cellular responses. Biochem Biophys Res Commun 2003;309:1017–1026.

94. Cho DI, Koo NY, Chung WJ, Kim TS, Ryu SY, et al. Effects of resveratrol-related hydroxystilbenes on the nitric oxide production in macrophage cells: structural requirements and mechanism of action. Life Sci 2002;71:2071–2082.

95. Tsai SH, Lin-Shiau SY, Lin JK. Suppression of nitric oxide synthase and the down-regulation of the activation of NFκB in macrophages by resveratrol. Br J Pharmacol 1999;126:673–680.

96. Adhami VM, Afaq F, Ahmad N. Suppression of ultraviolet B exposure–mediated activation of NF-kappaB in normal human keratinocytes by resveratrol. Neoplasia 2003;5:74–82.

97. Subbaramaiah K, Chung WJ, Michaluart P, Telang N, Tanabe T, Inoue H, et al. Resveratrol inhibits cyclooxygenase-2 transcription and activity in phorbol ester-treated human mammary epithelial cells. J Biol Chem 1998;273:21875–21882.

98. Subbaramaiah K, Michaluart P, Chung WJ, Tanabe T, Telang N, Dannenberg AJ. Resveratrol inhibits cyclooxygenase-2 transcription in human mammary epithelial cells. Ann NY Acad Sci 1999;889:214–223.

99. Yu R, Hebbar V, Kim DW, Mandlekar S, Pezzuto JM, Kong AN. Resveratrol inhibits phorbol ester and UV-induced activator protein 1 activation by interfering with mitogen-activated protein kinase pathways. Mol Pharmacol 2001;60:217–224.

100. Storz P, Doppler H, Toker A. Activation loop phosphorylation controls protein kinase D–dependent activation of NFκB. Mol Pharmacol 2004;66:870–879.

101. Bode AM, Ma WY, Surh YJ, Dong Z. Inhibition of epidermal growth factor–induced cell transformation and activator protein 1 activation by [6]-gingerol. Cancer Res 2001;61:850–853.

102. Kim SO, Kundu JK, Shin YK, Park JH, Cho MH, Kim TY, et al. [6]-Gingerol inhibits COX-2 expression by blocking the activation of p38 MAP kinase and NFκB in phorbol ester–stimulated mouse skin. Oncogene 2005;24:2558–2567.

103. Kim JK, Kim Y, Na KM, Surh YJ, Kim TY. [6]-Gingerol prevents UVB-induced ROS production and COX-2 expression *in vitro* and *in vivo*. Free Radic Res 2007;41:603–614.

104. Lantz RC, Chen GJ, Sarihan M, Solyom AM, Jolad SD, Timmermann BN. The effect of extracts from ginger rhizome on inflammatory mediator production. Phytomedicine 2007;14:123–128.

105. Malhotra S, Shafiq N, Pandhi P. COX-2 inhibitors: a CLASS act or just VIGORously promoted. MedGenMed 2004;6:6.

106. Graham DJ, Campen D, Hui R, Spence M, Cheetham C, Levy G, et al. Risk of acute myocardial infarction and sudden cardiac death in patients treated with cyclo-oxygenase 2 selective and non-selective non-steroidal anti-inflammatory drugs: nested case–control study. Lancet 2005;365:475–481.

107. Reilly M, Fitzgerald GA. Cellular activation by thromboxane A_2 and other eicosanoids. Eur Heart J 1993;14 (Suppl K):88–93.

108. Cerella C, Sobolewski C, Dicato M, Diederich M. Targeting COX-2 expression by natural compounds: a promising alternative strategy to synthetic COX-2 inhibitors for cancer chemoprevention and therapy. Biochem Pharmacol 2010;80:1801–1815.

109. Hull MA, Ko SC, Hawcroft G. Prostaglandin EP receptors: Targets for treatment and prevention of colorectal cancer? Mol Cancer Ther 2004;3:1031–1039.

110. Hata AN, Breyer RM. Pharmacology and signaling of prostaglandin receptors: multiple roles in inflammation and immune modulation. Pharmacol Ther 2004;103:147–166.

111. Sonoshita M, Takaku K, Sasaki N, Sugimoto Y, Ushikubi F, Narumiya S, et al. Acceleration of intestinal polyposis through prostaglandin receptor EP2 in $Apc\Delta716$ knockout mice. Nat Med 2001;7:1048–1051.

112. Sung YM, He G, Fischer SM. Lack of expression of the EP2 but not EP3 receptor for prostaglandin E2 results in suppression of skin tumor development. Cancer Res 2005;65:9304–9311.

113. Mutoh M, Watanabe K, Kitamura T, Shoji Y, Takahashi M, Kawamori T, et al. Involvement of prostaglandin E receptor subtype EP(4) in colon carcinogenesis. Cancer Res 2002;62:28–32.

114. Watanabe K, Kawamori T, Nakatsugi S, Ohta T, Ohuchida S, Yamamoto H, et al. Role of the prostaglandin E receptor subtype EP1 in colon carcinogenesis. Cancer Res 1999;59:5093–5096.

115. Surh I, Rundhaug JE, Pavone A, Mikulec C, Abel E, Simper M, et al. The EP1 receptor for prostaglandin E2 promotes the development and progression of malignant murine skin tumors. Mol Carcinog 2012;51:553–564.

116. Arakawa T, Laneuville O, Miller CA, Lakkides KM, Wingerd BA, DeWitt DL, et al. Prostanoid receptors of murine NIH 3T3 and RAW 264.7 cells: structure and expression of the murine prostaglandin EP4 receptor gene. J Biol Chem 1996;271:29569–29575.

117. Tsuchiya S, Tanaka S, Sugimoto Y, Katsuyama M, Ikegami R, Ichikawa A. Identification and characterization of a novel progesterone receptor–binding element in the mouse prostaglandin E receptor subtype EP2 gene. Genes Cells 2003;8:747–758.

118. Ansari KM, Sung YM, He G, Fischer SM. Prostaglandin receptor EP_2 is responsible for cyclooxygenase-2 induction by prostaglandin E_2 in mouse skin. Carcinogenesis 2007;28:2063–2068.

119. Steinert D, Kuper C, Bartels H, Beck FX, Neuhofer W. PGE_2 potentiates tonicity-induced COX-2 expression in renal medullary cells in a positive feedback loop involving EP2-cAMP-PKA signaling. Am J Physiol Cell Physiol 2009;296:C75–87.

120. Regan JW. EP2 and EP4 prostanoid receptor signaling. Life Sci 2003;74:143–153.

121. Singh T, Katiyar SK. Green tea catechins reduce invasive potential of human melanoma cells by targeting COX-2, PGE_2 receptors and epithelial-to-mesenchymal transition. PLoS One 2011;6:e25224.

122. Jin J, Chang Y, Wei W, He YF, Hu SS, Wang D, et al. Prostanoid EP1 receptor as the target of (−)-epigallocatechin-3-gallate in suppressing hepatocellular carcinoma cells *in vitro*. Acta Pharmacol Sin 2012;33:701–709.

123. Singh T, Vaid M, Katiyar N, Sharma S, Katiyar SK. Berberine, an isoquinoline alkaloid, inhibits melanoma cancer cell migration by reducing the expressions of cyclooxygenase-2, prostaglandin E(2) and prostaglandin E(2) receptors. Carcinogenesis 2011;32:86–92.

124. Ma X, Kundu N, Collin PD, Goloubeva O, Fulton AM. Frondoside A inhibits breast cancer metastasis and antagonizes prostaglandin E receptors EP4 and EP2. Breast Cancer Res Treat 2012;132:1001–1008.

CHAPTER 23

REGULATION OF INFLAMMATION-ASSOCIATED INTESTINAL DISEASES WITH PHYTOCHEMICALS

Akira Murakami

Inflammation is a complex, highly sequential series of events that is provoked by a variety of stimuli, including pathogens, noxious mechanical and chemical agents, and autoimmune responses. The subsequent cascade of events is characterized by such signs and symptoms as redness, swelling, heat, and pain. The inflammatory response occurs in the vascularized connective tissue, including plasma, circulating cells, blood vessels, and intracellular and extracellular components. This corresponds with increased microvascular caliber, enhanced vascular permeability, leukocyte recruitment, and release of inflammatory mediators, such as interleukin (IL)-1β, IL-6, tumor necrosis factor (TNF)-α, nitric oxide (NO), and prostaglandin (PG) E_2. Inflammation is the primary process through which the body repairs tissue damage and defends itself against stimuli. In a physiological condition, the regulated response protects against further injury and clears damaged tissue, whereas in a pathological situation, inflammation can result in tissue destruction and lead to organ dysfunction.

Of importance, a considerable proportion of chronic inflammatory diseases display an overlap with the onset and development of cancer, such as in cases of ulcerative colitis (UC) and Crohn's disease (colorectal cancer), reflux esophagitis, Barrett's esophagus (esophageal carcinoma), and hepatitis (hepatocellular carcinoma) (1). Further, inflammation has been recognized to play a pivotal role in insulin resistance, obesity, and diabetes (2), as well as in brain and myocardial infarctions that originate from vascular atherosclerosis (3). As described below, mechanisms underlying intestinal bowel diseases (IBDs), such as UC and colorectal cancer (CRC), remain to be fully elucidated. Also, it has been suggested that regulation of inflammatory conditions with food and drugs that are designed from a mechanistic viewpoint should provide great benefit for health promotion and disease prevention. In particular, food phytochemicals such as flavonoids have attracted much attention as promising agents capable of regulating inflammatory states. In this chapter I describe briefly possible mechanisms of action underlying inflammation-associated intestinal diseases as well as their regulation with food phytochemicals.

DEXTRAN SULFATE SODIUM–INDUCED MOUSE COLITIS MODEL

IBD is characterized by a pronounced infiltration of neutrophils into colonic lesions, accompanied by epithelial cell necrosis and ulceration. Although infection, environmental factors, heredity, and immunological abnormalities have been proposed as causes (4), the precise pathogenesis of IBD is poorly understood. In general, an appropriate or optimal animal model of IBD should display certain key characteristics, as the gut should exhibit morphological alterations, inflammation, signs and symptoms, as well as a pathophysiology and course similar or identical to those seen in humans (5). It is also recommended that the animal model have a well-defined genetic background in addition to a well-characterized immune system, so that reagents that are readily available and accessible for experimentation can be used along with well-defined criteria for successful management and/or manipulation.

The dextran sulfate sodium (DSS)–induced model exhibits many symptoms similar to those seen human UC, including diarrhea, bloody feces, body weight loss, mucosal ulcerations, and shortening of the colorectum (6) (Figure 23.1). In addition, DSS-induced colitis animal models have a number of advantages over others, such as simple experimental methods, reproducibility of the time course of development and severity of colitis among individual mice, and relative uniformity of the lesions induced (7). Therefore, this model is thought to be reliable for studying the pathogenesis of UC and testing drugs for treatment, although the precise mechanism of colitis induction by DSS remains unclear. Various mucosal inflammatory mediators have been investigated for their contribution to the pathogenesis of IBD. Recent studies have reported that pro-inflammatory cytokines, including IL-1β, TNF-α, IL-6,

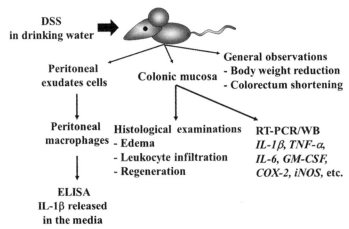

FIGURE 23.1 A DSS-administered mouse is a suitable animal model for investigating the mechanism of colitis and for identifying promising agents with anti-inflammatory properties.

and granulocyte macrophage-colony-stimulating factor (GM-CSF), were increased in the colonic mucosa of active IBD patients. Importantly, they correlate with the severity of inflammation, which suggests that these cytokines may play important roles in the induction and amplification of inflammatory responses, as well as in the healing of intestinal tissue injury (8). In addition, increased production of inducible NO synthase (iNOS) and cyclooxygenase (COX)-2 have been found in IBD (9) as well as in models of intestinal inflammation (10), while a prostanoid was reported to be protective against the development of UC (11). Thus, the role of COX in UC development remains controversial. Therefore, precise elucidation of the complex network of various cytokines and other chemical mediators is required for an understanding of how local inflammatory responses are regulated.

CXCL16 AS THE CELL SURFACE RECEPTOR FOR DSS

As mentioned above, IL-1β is one of the key pro-inflammatory cytokines that play a pivotal role in the onset of experimental colitis and human IBD; however, the molecular mechanisms underlying its production remain unclear. We reported previously that DSS caused a post-translational enhancement of IL-1β secretion from murine peritoneal macrophage (pMϕ), presumably through reactive oxygen species (ROS) generation, the resulting activation of both the p38 mitogen-activated protein kinase (MAPK) and extracellular signal-regulated kinases (ERK)1/2 pathways, and finally, caspase1 activation (12). More important, CXCL16 was identified as the cell surface receptor of DSS, which is involved in IL-1β production (12) (Figure 23.2). A scavenger receptor that binds phosphatidylserine and oxidized lipoprotein

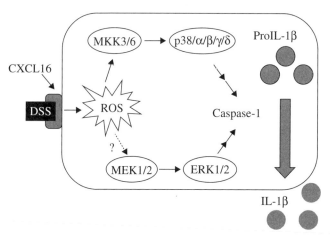

FIGURE 23.2 Proposed molecular events involved in the DSS-induced IL-1β production pathway in murine pMϕ. DSS may be recognized by CXCL16, after which ROSs are intracellularly generated by an unknown mechanism. This leads to activation of the MKK3/6 and p38 MAPK pathways for caspase-1 activation, which is responsible for the conversion of proIL-1β to biologically active IL-1β. Alternatively, DSS treatment may induce the activation of MEK1/2 and ERK1/2 pathways, which might also be associated with caspase1 and IL-1β release.

(SR-PSOX) has been reported to recognize negative-charged molecules, such as oxidized LDL and DSS, in phorbol ester–stimulated THP-1 human monocytes based on their uptake (13). It is a transmembrane type of chemokine, expressed selectively on antigen-presenting cells such as dendritic cells and macrophages, and was identified as the ligand for an orphan G-protein-coupled chemokine receptor/CXC chemokine receptor 6 (14). It should be noted that SR-PSOX was indicated to be biochemically and functionally identical to CXCL16 (15). Intriguingly, orally administered DSS was previously reported to be taken up by macrophages in inflamed mucosa, spleens, and mesenteric lymph nodes in mice with chronic colitis (6). Further, Mietus-Snyder et al. reported that oxidative stress may up-regulate scavenger receptor expression, thereby promoting the uptake of LDL in macrophages (16), whereas we did not see any increase in CXCL16 expression levels following DSS treatment. In any case, CXCL16 may be the key molecule for further investigation of mechanisms of DSS-induced colitis.

REGULATION OF COLITIS WITH NATURALLY OCCURRING FLAVONOIDS

Most of the current therapies for IBD involve treatment with glucocorticosteroids and 5-aminosalicylic acid; however, they display limited beneficial action (17). Immunosuppressive drugs have also been used to control severe illness, regardless of the more serious complications and toxic side effects associated with them (18). On the other hand, flavonoids are plant secondary metabolites ubiquitously distributed throughout the plant kingdom, and numerous reports have shown their antioxidative and anti-inflammatory activities in cellular and rodent models. They are also known to be inhibitors of several enzymes that are activated in certain inflammatory conditions (19), while a variety of cell types associated with the immune system are down-regulated by certain flavonoids *in vitro* (20). Further, most flavonoids show potent antioxidative/radical scavenging effects (21). Rutin (RUT: 3-*O*-rhamnosylglucosylquercetin, Figure 23.3) occurs widely in various foods, including buckwheat, parsley, tomatoes, and apricots, and is one of the most common naturally occurring flavonoids, with a wide range of biological activities (22). It and its aglycone, quercetin (QER: 3,3′,4′,5,7-pentahydroxyflavone, Figure 23.3), have been reported to exert numerous biochemical and pharmacological activities, such as free-radical scavenging (23), suppression of cellular immune and inflammatory responses, and anticarcinogenic activities in rodents (24).

We found previously that a diet containing 0.1% RUT but not QER attenuated DSS-induced body weight loss and shortening of the colorectum and dramatically improved colitis histologic scores (25) (Table 23.1). Further, DSS-induced increases in colonic mucosal IL-1β levels were blunted significantly in RUT- but not QER-fed mice, while dietary RUT attenuated the expressions of IL-1β and IL-6 mRNA in colonic mucosa. As for dose dependency, 0.01%, but not 0.001%, dietary RUT significantly reduced mucosal IL-1β levels. RUT was previously reported to attenuate 2,4,6-trinitrobenzene sulfonic acid–induced colitis in rats, presumably by mitigating intestinal oxidative stress (26). Galvez et al. also found that oral

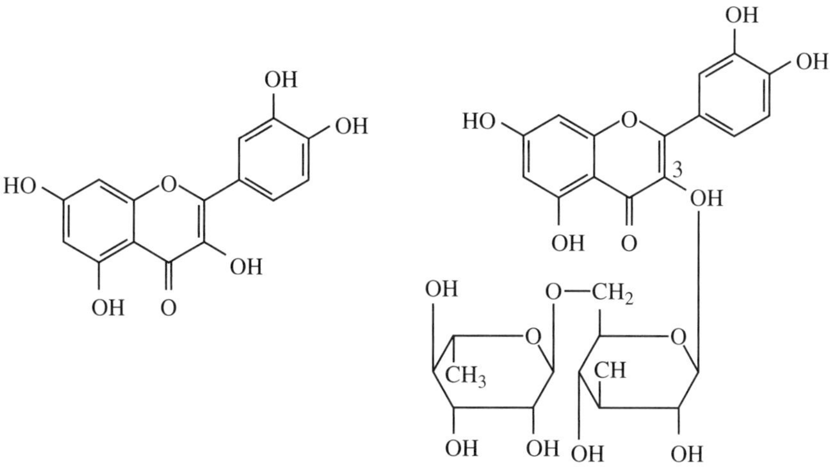

Quercetin   **Rutin**

FIGURE 23.3   Chemical structures of quercetin and rutin.

administration of rutin suppressed colonic damage and inflammation associated with acetic acid–induced colitis in rats (27). However, those studies presented scant experimental data regarding the molecular mechanisms underlying RUT-suppressed colitis. Dietary RUT was profoundly effective for attenuating DSS-induced colitis in mice, which probably occurred through attenuation of pro-inflammatory gene

TABLE 23.1   Suppressive Effects of Dietary Rutin and Quercetin on DSS-Induced Colitis

	0.1% Rutin	0.1% Quercetin
Pro-inflammatory mRNA		
IL-1β	Decreased	NE[a]
IL-6	Decreased	NE
GM-CSF	NE	NE
iNOS	NE	NE
COX-1	NE	NE
COX-2	NE	NE
TNF-α	NE	NE
IL-1β protein	Decreased[b]	NE[c]
Histological scores		
Ulcer	Decreased	NE
Edema	Decreased	Decreased
Leukocyte infiltration	Decreased	NE
Colorectum shortening	Suppressed	NE

[a]Not effective.
[b]Rutin was effective at the dose of 0.01% but not 0.001%, and 0.1% rutin decreased IL-1β even in a therapeutic protocol.
[c]At doses of 0.001, 0.01, and 0.1%.
Source: Kwon et al. (25).

expression, especially that of IL-1β and IL-6 (25). Another important observation was that oral administration of RUT 3 days after beginning DSS treatment reversed colitis significantly, as shown by suppression of both colorectum shortening and IL-1β production. To our knowledge, there are only a few reports of natural compounds that can reverse experimental colitis when used in a therapeutic protocol (28).

Oxidative stress has been implicated in the pathogenesis of DSS-induced colitis. Tardieu et al. reported that DSS increased the level of an oxidative DNA damage biomarker, 8-oxo-7,8-dihydro-2′-deoxyguanosine, in rat colonic mucosa (29). Sustained production of ROSs during colonic inflammation may overwhelm the endogenous antioxidant defense system, and in accordance with that notion, there are several independent reports of decreased antioxidant levels in patients with IBD (30). Thus, a strategy based on the use of antioxidants for compensating the dysregulated antioxidative defense system in inflamed intestines is reasonable and may be effective for controlling IBD. Most flavonoids, which occur widely throughout the plant kingdom (31), are well characterized as distinct antioxidative agents, with one such compound being RUT (32). On the other hand, the polyphenolic QER has been shown to act as a pro-oxidant in some models when used at high concentrations. Laughton et al. reported that certain phenolic compounds, including QER, enhanced generation of the hydroxyl radical (33). Smith et al. also reported that QER, as well as related flavonoids, had the potential to serve as accelerative agents for DNA degradation via free-radical formation (34), while dietary QER induced focal areas of dysplasia in the colon of 22% of normal mice (35). In contrast to those observations, the pro-oxidative nature of RUT has yet to be shown, suggesting that the presence of glycosyl moiety is a notable determinant for exertion of its activity. To strengthen that hypothesis, a free hydroxyl group at the 3-position, which is present in QER but masked in RUT, has been shown to be a prerequisite for pro-oxidative capability (36). Meanwhile, RUT has been shown to be metabolized mainly to phenylacetic acids such as 3-hydroxyphenylacetic acid in the human colon (37), although we could not determine whether RUT itself or its metabolized product(s) are responsible for attenuation of colitis. On the other hand, as it has been reported that RUT is absorbed from colon (38), it is possible that its ability to ameliorate DSS colitis results from an action within the colonic tissue itself or via RUT (or its metabolites) in the plasma. There are significant differences among the mouse, rat, and human gut with respect to the mechanisms by which they utilize or exclude luminal flavonoids.

OBESITY-ASSOCIATED CRC

Numerous epidemiological studies have provided abundant evidence that environmental factors, rather than genetic variations between populations, are of prime importance in the etiology of CRC (39). One of the most influential factors in coronary heart disease, stroke, and cancer, especially in industrialized countries, is obesity, whose prevalence has increased markedly over the past two decades (40). The results of case–control and prospective studies suggest that obesity is a strong risk factor for colorectal cancer, especially in men (41). More recently, a prospective population-based study of about 90,000 subjects conducted by the American Cancer Society confirmed that obesity is associated directly with an increased risk of death from colon cancer (42). Furthermore, findings from a metaanalysis study, which included information from 70,000 cases of CRC, showed that people with a body mass index (BMI) $\geq$ 30 kg/m^2 have about a 20% greater risk of developing CRC then do those considered to be of normal weight (BMI < 25 kg/m^2).

Although the underlying mechanisms of obesity-related CRC are not fully understood, a number of possibilities explaining these observations have been proposed. Among pathophysiological and biological mechanisms, the two most studied candidate systems are the insulin–(insulin-like growth factor) IGF-1 axis and adipocytokines. It is well established that BMI is correlated with serum insulin level, and many obese persons are insulin-resistant. Insulin resistance is at the core of many candidate systems, and interactions between several systems cause CRC. Insulin activation of insulin receptors triggers an intracellular signaling cascade in both the ERK and phosphatidylinositol 3-kinase (PI3K) pathways, which are involved in cell proliferation and survival. Conventionally, insulin appears to be mitogenic only at supra-physiological levels, and its main proliferative effects are probably through IGF-1 receptors (43). Thus, it is considered that the most important role of insulin is to reduce the levels of insulin-like growth factor binding proteins (IGFBPs), specifically IGFBP-1 and -2 (44).

Until the discovery of adipocytokines, adipose tissue was thought to have only passive functions, such as serving as an energy storage depot and mechanical barrier. Adipocytokines are a group of adipose tissue–secreted hormones that were initially reported in the early 1990s when leptin was described (45). Later, it was shown that leptin, resistin, plasminogen activator inhibitor-1, TNF-α, and IL-6 had positive relationships to adiposity (46). Since they have crucial roles in immune regulation, vascular function, and adipocyte metabolism, adipocytokines are considered to be central players in the pathogenesis of metabolic syndrome, a cluster of clinical symptoms that includes obesity and insulin resistance. Consequently, regulation of body weight and obesity-related CRC is rapidly becoming a critical concern for public health experts and medical researchers worldwide (47).

Leptin and adiponectin are the most abundantly produced adipocytokines, and their levels in the circulation largely reflect the amount of fat and distribution of white adipose tissue (WAT) in the body. Therefore, they are the most intensively studied adipocytokines in regard to their hormonal roles in carcinogenesis. Leptin, a 16-kDa protein encoded by the *ob* gene, was discovered in 1994 and shown to be a regulator of body weight and energy balance that acts in the hypothalamus (45). C57BL/KsJ-*db/db* (*db/db*) mice have a defect in the leptin receptor (Ob-R) gene (48), which leads to leptin regulatory impairments of food intake, resulting in hyperinsulinemia, hyperglycemia, and hyperleptinemia in subjects with extreme obesity (49). Although this hormone is produced primarily by WAT, it is interesting that its expression has also been detected in other tissues, including those in the gastrointestinal tract (50). Most findings presented about leptin are in regard to its involvement in energy balance and obesity development, and results indicate that the protein regulates nearly every step in the transition from normal weight to obesity, including food intake, absorption, utilization, and storage (51). In humans, leptin levels are proportional to BMI and are elevated in obese persons (52). In fact, two prospective studies of the relationship of serum leptin with CRC demonstrated significant associations, while a colonoscopy study showed associations with adenoma in men but not in women (53). Meanwhile, leptin is a pleiotropic hormone, and several *in vitro* studies conducted with various cancer cell lines have shown that it can act as a mitogenic, anti-apoptotic, and tumorigenic factor (54). These activities of leptin are mediated through the transmembrane receptor Ob-R. In mammalian tissue, at least six splicing variants with different cytoplasmic domains have been described, with the long (Ob-Rb) and short (Ob-Ra) forms two of the major isoforms present in mammalian cells (55). Ob-Ra is more abundant in peripheral tissues and is expressed primarily in the hypothalamus (56). However, several recent studies have reported that Ob-Rb was found in a variety of cancer cells, including colon cancer. It is known that Ob-Rb can mediate multiple signaling pathways, such as PI3K/protein kinase B (Akt) and nuclear factor (NF)-κB, in addition to Janus-activated kinase 2 (JAK2)/signal transducers and activators of transcription 3 (STAT3), as well as MAPK/ERK (40). In addition, it was recently reported that Ob-Ra mainly activates MAPK and seems to be responsible for mitogenic activity (57).

Most studies on the relationship of obesity and colorectal carcinogenesis have used obese animals (e.g., *db/db*, *ob/ob* or high-fat-diet-consumption mice). However, it is difficult to determine which factor is involved in colon carcinogenesis in obese animals, because in addition to several adipocytokines, there are an untold number of altered physiological factors in obese persons, and they interact with each other in complex ways. In view of these considerations, we investigated the serum leptin level in an AOM/DSS-induced colon carcinogenesis model using nonobese wild-type mice (58). Of note, the weight of the epididymal fat pad and mesenteric fat, as well as serum leptin levels in mice given AOM and/or DSS, were increased markedly compared to untreated mice. In addition, leptin protein production in the epididymal fat pad of AOM- and DSS-treated mice was 4.7-fold greater than that of untreated controls. Furthermore, the insulin-signaling molecules Akt, S6, MEK1/2, and ERK1/2 were concomitantly activated in the epididymal fat of AOM- and DSS-treated mice, while such treatments also increased the levels of insulin and IGF-1. Taken together, higher levels of insulin and IGF-1 promote insulin signaling in epididymal fat, thereby increasing serum leptin levels, which may play a crucial role in both obesity-related and obesity-independent colon carcinogenesis (58) (Figure 23.4). In addition, leptin mRNA expression and protein production in mouse epididymal fat pads is increased significantly in chemically induced colon carcinogenesis model mice, although such findings have been reported with other types of obese animals (59). Moreover, substantial activation of the insulin signaling pathway in the epididymal fat pads was found as well as elevation of serum levels of insulin and IGF-1. Thus, hyperleptinemia observed in mice treated with AOM and DSS may be due, at least in part, to an increased visceral fat mass where leptin production is increased, and to the fact that leptin production in adipocytes is driven by elevated levels of insulin and IGF-1 in serum.

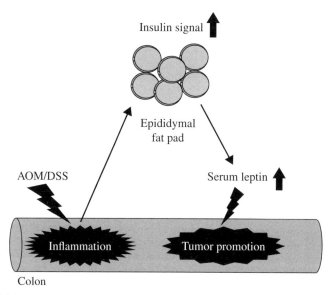

FIGURE 23.4   The higher levels of serum insulin and IGF-1 promote insulin signaling in epididymal fat, which may lead to an increased amount of visceral fat tissue and a resulting elevation of serum leptin level. These events may play critical roles in the development of CRC not only in obese but also in nonobese persons.

SUPPRESSIVE EFFECTS OF NOBILETIN ON INFLAMMATION-ASSOCIATED CRC

We recently assessed the effects of flavonoids on serum leptin levels and tumor formation in an AOM/DSS-induced colon carcinogenesis model (60–62). Nobiletin (Figure 23.5), which markedly suppressed leptin secretion from 3T3-L1 mouse pre-adipocytes, abolished colonic malignancy and dramatically decreased serum leptin levels in mice (60–62). In addition, chrysin, which did not affect leptin secretion from 3T3-L1 cells, did little to suppress tumor formation and serum leptin levels. These results suggest that serum leptin levels are related to colon carcinogenesis and that dietary nobiletin suppresses carcinogenesis partly through a mechanism

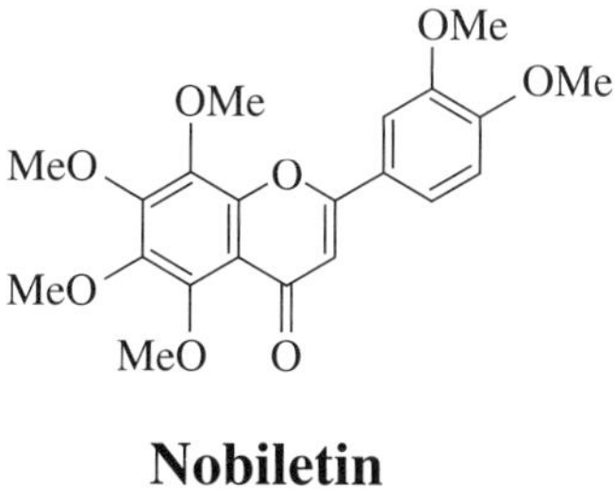

Nobiletin

FIGURE 23.5   Chemical structure of nobiletin.

involving leptin. Although additional studies are necessary to confirm this, synthetic drugs or food ingredients targeting leptin secretion and its actions may be useful in the future prevention of obesity-associated colorectal cancer development. We demonstrated that flavonoids, especially nobiletin, concomitantly suppressed hyper-leptinemia and colon tumor development, whereas other adipocytokines (e.g., TNF-α, IL-6, adiponectin) and triglycerides did not significantly alter development. It is well known that obese animals with elevated leptin levels (e.g., wild-type mice fed a high-fat diet and *db/db* mice) are highly susceptible to chemically induced carcinogenesis (63). Collectively, it can be hypothesized that leptin plays an important role not only in colitis- but also in obesity-related, colon carcinogenesis, and flavonoids are able to inhibit that, at least in part, through regulation of leptin levels.

Recently, obesity has become a focus point in investigations to identify dietary and lifestyle factors related to increased risk of colorectal cancer (64). Metabolic stress resulting from obesity has been shown to be associated with increased levels of oxidative stress (65), inflammatory cytokines, insulin (66), and lipids (67). Of interest, Niho et al. revealed a hyperlipidemic state in *Apc* gene-deficient mice used as a model of familial adenomatous polyposis patients (68) compared to their wild-type counterparts. Our findings regarding the serum level of triglycerides are similar, although the difference did not reach statistical significance (58). This discrepancy may be attributeed to the differences in the genetic backgrounds of the mice used and/or the experimental protocols. Obesity is driven by WAT, from which excess or reduced levels of adipocytokines are secreted (69). Adiponectin is the most abundant cytokine in adipocytes and has been reported to have antidiabetic and anti-inflammatory properties (70), and low levels of adiponectin have been shown to be associated with an increased risk of colorectal cancer in humans (71). Several classical pro-inflammatory cytokines (e.g., TNF-α and to a large extent IL-6) are also secreted from adipocytes (72) and may participate in the regulation of obesity (73). In addition, epidemiological studies have revealed the roles of TNF-α and IL-6 in the onset of several types of cancer (74). However, no results have been published regarding the hormonal role of leptin in chemically induced carcinogenesis in rodents.

As described above, leptin is a pleiotropic hormone, and several *in vitro* studies showed that leptin can act as a mitogenic, anti-apoptotic, and tumorigenic factor with various cancer cell lines (54). In our previous study, the mean body and epididymal fat weights in AOM/DSS-treated mice were greater than those in control mice, while nobiletin feeding decreased those values (58). These results raise the possibility that the elevation of serum leptin levels seen in carcinogenesis model mice is in part due to increases in body and fat weights, although the underlying mechanism is still unclear. On the other hand, several studies have reported findings of great importance: that mesenteric adipose tissue in IBD patients overexpress leptin mRNA (75). Because a DSS-induced colitis animal model is considered to be very reliable and useful for elucidating the mechanism underlying the onset of IBD (6) DSS treatment may be associated with elevated serum leptin levels in AOM/DSS-treated mice. These results, as well as those of several other studies (54), indicate that leptin acts as a mitogenic factor in cultured human colon cancer cells.

Nobiletin, a polymethoxylated flavonoid occurring preferentially in citrus fruit peels, has been reported to inhibit the proliferation of a variety of human cancer cell lines (76) and suppress colon carcinogenesis in rats (77,78). Although several reports have implied the preventive mechanism of nobiletin toward colon carcinogenesis, those results are not definitive. For example, nobiletin inhibited iNOS and COX-2 expression in macrophages (79) and reduced PGE_2 levels in rat colonic mucosa treated with AOM (77). Present results showed that dietary nobiletin decreased body and epididymal fat weights, which had been increased by treatment with AOM/DSS. These effects may contribute to its colon cancer preventive activities, based on the mechanisms of action described above. In accordance with this notion, Saito et al. recently reported that nobiletin enhanced both the differentiation and lipolysis of adipocytes via activation of signaling cascades mediated by cAMP and cAMP-responsive element binding protein (80).

CONCLUDING REMARKS

In conclusion, dietary RUT, even at a low dose, was found to attenuate the production of critical pro-inflammatory mediator genes such as IL-1β, thereby ameliorated DSS-induced colitis in mice. This flavonoid also notably ameliorated colitis under a therapeutic protocol. Because RUT is a common phytochemical found in a variety of fruits and vegetables, the feasibility of a therapeutic approach for IBD and CRC using a RUT-supplemented diet may be reasonable and promising. However, further mechanistic and toxicity studies to explore the effectiveness of this phytochemical for preventing and treating IBD and colorectal carcinogenesis are expected. In addition, CXCL 16 was identified as the cell surface receptor for DSS, which may be a significant target for anti-inflammatory drugs. On the other hand, the higher levels of serum insulin and IGF-1 promote insulin signaling in the epididymal fat, which may lead to an increased amount of visceral fat tissue and a resulting elevation of serum leptin level. These events may play critical roles in the development of CRC not only in obese but also in nonobese persons. Our data suggest that the serum levels of leptin are related to colon carcinogenesis and that dietary nobiletin suppresses carcinogenesis partly through regulation of this hormone. Although additional studies are necessary to confirm this speculation, synthetic drugs or food ingredients targeting leptin secretion and activities may be useful for regulating obesity-associated colorectal cancer development.

Acknowledgments

The author thanks previous graduate students Ki Han Kwon and Shingo Miyamoto, and Takuji Tanaka (Tohkai Cytopathology Institute), for their notable contributions to our work cited in this chapter. This study was partially supported by a Grant-in-Aid for Cancer Research from the Ministry of Health, Labor and Welfare of Japan.

REFERENCES

1. Coussens LM, Werb Z. Inflammation and cancer. Nature 2002;420:860–867.
2. Dandona P, Aljada A, Bandyopadhyay A. Inflammation: the link between insulin resistance, obesity and diabetes. Trends Immunol 2004;25:4–7.

3. Libby P, Ridker PM. Inflammation and atherosclerosis: role of C-reactive protein in risk assessment. Am J Med 2004;116 (Suppl 6A):9S–16S.

4. Sartor RB. Pathogenesis and immune mechanisms of chronic inflammatory bowel diseases. Am J Gastroenterol 1997;92:5S–11S.

5. Jurjus AR, Khoury NN, Reimund JM. Animal models of inflammatory bowel disease. J Pharmacol Toxicol Methods 2004;50:81–92.

6. Okayasu I, Hatakeyama S, Yamada M, Ohkusa T, Inagaki Y, Nakaya R. A novel method in the induction of reliable experimental acute and chronic ulcerative colitis in mice. Gastroenterology 1990;98:694–702.

7. Murthy SN, Cooper HS, Shim H, Shah RS, Ibrahim SA, Sedergran DJ. Treatment of dextran sulfate sodium-induced murine colitis by intracolonic cyclosporin. Dig Dis Sci 1993;38:1722–1734.

8. Lauritsen K, Laursen LS, Bukhave K, Rask-Madsen J. *In vivo* effects of orally administered prednisolone on prostaglandin and leucotriene production in ulcerative colitis. Gut 1987;28:1095–1099.

9. Rachmilewitz D, Stamler JS, Bachwich D, Karmeli F, Ackerman Z, Podolsky DK. Enhanced colonic nitric oxide generation and nitric oxide synthase activity in ulcerative colitis and Crohn's disease. Gut 1995;36:718–723.

10. Morteau O, Morham SG, Scllon R, Dieleman LA, Langenbach R, Smithies O, et al. Impaired mucosal defense to acute colonic injury in mice lacking cyclooxygenase-1 or cyclooxygenase-2. J Clin Invest 2000;105:469–478.

11. Hirata I, Murano M, Nitta M, Sasaki S, Toshina K, Maemura K, et al. Estimation of mucosal inflammatory mediators in rat DSS-induced colitis: possible role of PGE(2) in protection against mucosal damage. Digestion 2001;63 (Suppl 1):73–80.

12. Kwon KH, Ohigashi H, Murakami A. Dextran sulfate sodium enhances interleukin-1 beta release via activation of p38 MAPK and ERK1/2 pathways in murine peritoneal macrophages. Life Sci 2007;81:362–371.

13. Shimaoka T, Nakayama T, Kume N, Takahashi S, Yamaguchi J, Minami M, et al. Cutting edge: SR–PSOX/CXC chemokine ligand 16 mediates bacterial phagocytosis by APCs through its chemokine domain. J Immunol 2003;171:1647–1651.

14. Shimaoka T, Nakayama T, Fukumoto N, Kume N, Takahashi S, Yamaguchi J, Minami M, et al. Cell surface-anchored SR-PSOX/CXC chemokine ligand 16 mediates firm adhesion of CXC chemokine receptor 6-expressing cells. J Leukoc Biol 2004;75:267–274.

15. Shimaoka T, Nakayama T, Hieshima K, Kume N, Fukumoto N, Minami M, et al. Chemokines generally exhibit scavenger receptor activity through their receptor-binding domain. J Biol Chem 2004;279:26807–26810.

16. Mietus-Snyder M, Gowri MS, Pitas RE. Class A scavenger receptor up-regulation in smooth muscle cells by oxidized low density lipoprotein: enhancement by calcium flux and concurrent cyclooxygenase-2 up-regulation. J Biol Chem 2000;275:17661–17670.

17. Podolsky DK. Inflammatory bowel disease (2). N Engl J Med 1991;325:1008–1016.

18. Shanahan F. Inflammatory bowel disease: immunodiagnostics, immunotherapeutics, and ecotherapeutics. Gastroenterology 2001;120:622–635.

19. Havsteen B. Flavonoids, a class of natural products of high pharmacological potency. Biochem Pharmacol 1983;32:1141–1148.

20. Middleton E, Kandasami C. Effects of flavonoids on immune and inflammatory cell functions. Biochem Pharmacol 1992;43:1167–1179.

21. Mora A, Payá M, Ríos JL, Alcaraz MJ. Structure–activity relationships of polymethoxyflavones and other flavonoids as inhibitors of non-enzymic lipid peroxidation. Biochem Pharmacol 1990;40:793–797.

22. Erlund I, Kosonen T, Alfthan G, Mäenpää J, Perttunen K, Kenraali J, et al. Pharmacokinetics of quercetin from quercetin aglycone and rutin in healthy volunteers. Eur J Clin Pharmacol 2000;56:545–553.

23. Kandaswami C, Middleton E. Free radical scavenging and antioxidant activity of plant flavonoids. Adv Exp Med Biol 1994;366:351–376.

24. Deschner EE, Ruperto J, Wong G, Newmark HL. Quercetin and rutin as inhibitors of azoxymethanol-induced colonic neoplasia. Carcinogenesis 1991;12:1193–1196.

25. Kwon KH, Murakami A, Tanaka T, Ohigashi H. Dietary rutin, but not its aglycone quercetin, ameliorates dextran sulfate sodium-induced experimental colitis in mice: attenuation of pro-inflammatory gene expression. Biochem Pharmacol 2005;69:395–406.

26. Cruz T, Gálvez J, Ocete MA, Crespo ME, Sánchez de Medina L-H F, et al. Oral administration of rutoside can ameliorate inflammatory bowel disease in rats. Life Sci 1998;62:687–695.

27. Gálvez J, Cruz T, Crespo E, Ocete MA, Lorente MD, Sánchez de Medina F, et al. Rutoside as mucosal protective in acetic acid-induced rat colitis. Planta Med 1997;63:409–414.

28. Tsune I, Ikejima K, Hirose M, Yoshikawa M, Enomoto N, Takei Y, et al. Dietary glycine prevents chemical-induced experimental colitis in the rat. Gastroenterology 2003;125:775–785.

29. Tardieu D, Jaeg JP, Deloly A, Corpet DE, Cadet J, Petit CR. The COX-2 inhibitor nimesulide suppresses superoxide and 8-hydroxy-deoxyguanosine formation, and stimulates apoptosis in mucosa during early colonic inflammation in rats. Carcinogenesis 2000;21:973–976.

30. Buffinton GD, Doe WF. Depleted mucosal antioxidant defences in inflammatory bowel disease. Free Radic Biol Med 1995;19:911–918.

31. Kühnau J. The flavonoids: a class of semi-essential food components: their role in human nutrition. World Rev Nutr Diet 1976;24:117–191.

32. Grinberg LN, Rachmilewitz EA, Newmark H. Protective effects of rutin against hemoglobin oxidation. Biochem Pharmacol 1994;48:643–649.

33. Laughton MJ, Halliwell B, Evans PJ, Hoult JR. Antioxidant and pro-oxidant actions of the plant phenolics quercetin, gossypol and myricetin: effects on lipid peroxidation, hydroxyl radical generation and bleomycin-dependent damage to DNA. Biochem Pharmacol 1989;38:2859–28565.

34. Smith C, Halliwell B, Aruoma OI. Protection by albumin against the pro-oxidant actions of phenolic dietary components. Food Chem Toxicol 1992;30:483–489.

35. Yang K, Lamprecht SA, Liu Y, Shinozaki H, Fan K, Leung D, et al. Chemoprevention studies of the flavonoids quercetin and rutin in normal and azoxymethane-treated mouse colon. Carcinogenesis 2000;21:1655–1660.

36. Kessler M, Ubeaud G, Jung L. Anti- and pro-oxidant activity of rutin and quercetin derivatives. J Pharm Pharmacol 2003;55:131–142.

37. Olthof MR, Hollman PC, Buijsman MN, van Amelsvoort JM, Katan MB. Chlorogenic acid, quercetin-3-rutinoside and black tea phenols are extensively metabolized in humans. J Nutr 2003;133: 1806–1814.

38. Manach C, Morand C, Texier O, Favier ML, Agullo G, Demigné C, et al. Quercetin metabolites in plasma of rats fed diets containing rutin or quercetin. J Nutr 1995;125:1911–1922.

39. Flood DM, Weiss NS, Cook LS, Emerson JC, Schwartz SM, Potter JD. Colorectal cancer incidence in Asian migrants to the United States and their descendants. Cancer Causes Control 2000;11: 403–411.

40. Garofalo C, Surmacz E. Leptin and cancer. J Cell Physiol 2006;207:12–22.

41. Kuriyama S, Tsubono Y, Hozawa A, Shimazu T, Suzuki Y, Koizumi Y, et al. Obesity and risk of cancer in Japan. Int J Cancer 2005;113:148–157.

42. McMillan DC, Sattar N, McArdle CS. ABC of obesity: obesity and cancer. BMJ 2006;333:1109–1111.

43. Renehan AG, Frystyk J, Flyvbjerg A. Obesity and cancer risk: the role of the insulin–IGF axis. Trends Endocrinol Metab 2006;17:328–336.

44. Argente J, Caballo N, Barrios V, Muñoz MT, Pozo J, Chowen JA, et al. Disturbances in the growth hormone-insulin-like growth factor axis in children and adolescents with different eating disorders. Horm Res 1997;48 (Suppl 4):16–8.

45. Zhang Y, Proenca R, Maffei M, Barone M, Leopold L, Friedman JM. Positional cloning of the mouse obese gene and its human homologue. Nature 1994;372:425–432.

46. Berggren JR, Hulver MW, Houmard JA. Fat as an endocrine organ: influence of exercise. J Appl Physiol 2005;99:757–764.

47. Kopelman PG. Obesity as a medical problem. Nature 2000;404:635–643.

48. Lee GH, Proenca R, Montez JM, Carroll KM, Darvishzadeh JG, Lee JI, et al. Abnormal splicing of the leptin receptor in diabetic mice. Nature 1996;379:632–635.

49. Frühbeck G, Gómez-Ambrosi J. Rationale for the existence of additional adipostatic hormones. FASEB J 2001;15:1996–2006.

50. Bado A, Levasseur S, Attoub S, Kermorgant S, Laigneau JP, Bortoluzzi MN, et al. The stomach is a source of leptin. Nature 1998;394:790–793.

51. Friedman JM. Leptin, leptin receptors, and the control of body weight. Nutr Rev 1998;56:s38–s46; discussion s54–s75.
52. Hamilton BS, Paglia D, Kwan AY, Deitel M. Increased obese mRNA expression in omental fat cells from massively obese humans. Nat Med 1995;1:953–956.
53. Chia VM, Newcomb PA, Lampe JW, White E, Mandelson MT, McTiernan A, et al. Leptin concentrations, leptin receptor polymorphisms, and colorectal adenoma risk. Cancer Epidemiol Biomark Prev 2007;16:2697–2703.
54. Hardwick JC, Van Den Brink GR, Offerhaus GJ, Van Deventer SJ, Peppelenbosch MP. Leptin is a growth factor for colonic epithelial cells. Gastroenterology 2001;121:79–90.
55. Fei H, Okano HJ, Li C, Lee GH, Zhao C, Darnell R, et al. Anatomic localization of alternatively spliced leptin receptors (Ob-R) in mouse brain and other tissues. Proc Natl Acad Sci USA 1997;94:7001–7005.
56. Bjørbaek C, Uotani S, da Silva B, Flier JS. Divergent signaling capacities of the long and short isoforms of the leptin receptor. J Biol Chem 1997;272:32686–32695.
57. Yamashita T, Murakami T, Otani S, Kuwajima M, Shima K. Leptin receptor signal transduction: OBRa and OBRb of fa type. Biochem Biophys Res Commun 1998;246:752–759.
58. Miyamoto S, Tanaka T, Murakami A. Increased visceral fat mass and insulin signaling in colitis-related colon carcinogenesis model mice. Chem Biol Interact 2010;183:271–275.
59. Lin S, Thomas TC, Storlien LH, Huang XF. Development of high fat diet–induced obesity and leptin resistance in C57Bl/6J mice. Int J Obes Relat Metab Disord 2000;24:639–646.
60. Miyamoto S, Yasui Y, Tanaka T, Ohigashi H, Murakami A. Suppressive effects of nobiletin on hyperleptinemia and colitis-related colon carcinogenesis in male ICR mice. Carcinogenesis 2008;29:1057–1063.
61. Miyamoto S, Yasui Y, Ohigashi H, Tanaka T, Murakami A. Dietary flavonoids suppress azoxymethane-induced colonic preneoplastic lesions in male C57BL/KsJ-db/db mice. Chem Biol Interact 2010;183:276–283.
62. Miyamoto S, Kohno H, Suzuki R, Sugie S, Murakami A, Ohigashi H, et al. Preventive effects of chrysin on the development of azoxymethane-induced colonic aberrant crypt foci in rats. Oncol Rep 2006;15:1169–1173.
63. Jennette JC, Falk RJ. ANCA: Diagnostic markers or pathogenic agents? Bull Rheum Dis 1992;41: 3–6.
64. Calle EE, Kaaks R. Overweight, obesity and cancer: epidemiological evidence and proposed mechanisms. Nat Rev Cancer 2004;4:579–591.
65. Furukawa S, Fujita T, Shimabukuro M, Iwaki M, Yamada Y, Nakajima Y, et al. Increased oxidative stress in obesity and its impact on metabolic syndrome. J Clin Invest 2004;114:1752–1761.
66. Kaaks R, Lukanova A. Energy balance and cancer: the role of insulin and insulin-like growth factor-I. Proc Nutr Soc 2001;60:91–106.
67. Järvinen R, Knekt P, Hakulinen T, Rissanen H, Heliövaara M. Dietary fat, cholesterol and colorectal cancer in a prospective study. Br J Cancer 2001;85:357–361.
68. Niho N, Mutoh M, Komiya M, Ohta T, Sugimura T, Wakabayashi K. Improvement of hyperlipidemia by indomethacin in Min mice. Int J Cancer 2007;121:1665–1669.
69. Wellen KE, Hotamisligil GS. Obesity-induced inflammatory changes in adipose tissue. J Clin Invest 2003;112:1785–1788.
70. Bråkenhielm E, Veitonmäki N, Cao R, Kihara S, Matsuzawa Y, Zhivotovsky B, et al. Adiponectin-induced antiangiogenesis and antitumor activity involve caspase-mediated endothelial cell apoptosis. Proc Natl Acad Sci USA 2004;101:2476–2481.
71. Wei EK, Giovannucci E, Fuchs CS, Willett WC, Mantzoros CS. Low plasma adiponectin levels and risk of colorectal cancer in men: a prospective study. J Natl Cancer Inst 2005;97:1688–1694.
72. Guzik TJ, Mangalat D, Korbut R. Adipocytokines: Novel link between inflammation and vascular function? J Physiol Pharmacol 2006;57:505–528.
73. Borst SE. The role of TNF-alpha in insulin resistance. Endocrine 2004;23:177–182.
74. Yoshimura A. Signal transduction of inflammatory cytokines and tumor development. Cancer Sci 2006;97:439–447.
75. Barbier M, Vidal H, Desreumaux P, Dubuquoy L, Bourreille A, Colombel JF, et al. Overexpression of leptin mRNA in mesenteric adipose tissue in inflammatory bowel diseases. Gastroenterol Clin Biol 2003;27:987–991.

76. Kandaswami C, Perkins E, Drzewiecki G, Soloniuk DS, Middleton E. Differential inhibition of proliferation of human squamous cell carcinoma, gliosarcoma and embryonic fibroblast-like lung cells in culture by plant flavonoids. Anticancer Drugs 1992;3:525–530.

77. Kohno H, Yoshitani S, Tsukio Y, Murakami A, Koshimizu K, Yano M, et al. Dietary administration of citrus nobiletin inhibits azoxymethane-induced colonic aberrant crypt foci in rats. Life Sci 2001;69:901–913.

78. Suzuki R, Kohno H, Murakami A, Koshimizu K, Ohigashi H, Yano M, et al. Citrus nobiletin inhibits azoxymethane-induced large bowel carcinogenesis in rats. Biofactors 2004;22:111–114.

79. Murakami A, Nakamura Y, Torikai K, Tanaka T, Koshiba T, Koshimizu K, et al. Inhibitory effect of citrus nobiletin on phorbol ester-induced skin inflammation, oxidative stress, and tumor promotion in mice. Cancer Res 2000;60:5059–5066.

80. Saito T, Abe D, Sekiya K. Nobiletin enhances differentiation and lipolysis of 3T3-L1 adipocytes. Biochem Biophys Res Commun 2007;357:371–376.

INDEX

Page numbers in **BOLD** refers to major discussion.

1400W, 53

Aberrant crypt foci, 286, 288–289, 292, 294, 314–318, 320–324, 345
Acrolein, 62–63, 76
Actinolite, 223
Activation-induced cytidine deaminase (AID), 3, **119–127**, 217
Acute myeloid leukemia (AML), 11, 13–14, 16–17, 344
Acylated lipopeptide, 108
Adenocarcinoma
  cervical, 166–168
  colon, 66–67, 289, 293, 328, 341
  colorectal, 10, 301
  esophageal, 1, 213–214, 216, 218–219
  lung, 43, 90
  mammary, 29
Adenomatous polyposis coli (APC), 125, 214, 218–219, 285–286, 288, 290, 292, 294, 302–303, 306, 314–317, 319, 321, 328, 330–331, 345, 364
Adiponectin, 294–295, 361, 364
AID, see Activation-induced cytidine deaminase
AIM2, 104–105, 108–109, 238
Akt, 3, 17, 198, 228, 263, 285–286, 307, 309–312, 318–319, 328, 343, 346, 362
Alcohol, 66–67, 69–71, 187, 223, 301–302
Aldehyde dehydrogenase (ALDH), 13, 329
Alkaline phosphatase (ALP), 67–68
Alkylnitrosoamide, 331
ALP, see Alkaline phosphatase
AML, see Acute myeloid leukemia
Amosite, 49, 223

AMP-activated protein kinase (AMPK), 286, 306, 310, 303, 316
Amphibole, 223–224
AMPK, see AMP-activated protein kinase
Anakinra, 110, 112, 230
Angiogenesis, 3, 6, 23, 27, 31, 45, 119, 131, 134, 141, 149–150, 259, 303–304, 314, 328, 340
Anthocyan(s), 332
Anthophyllite, 223
Antigen-presenting cell (APC), 28–29, 33, 95
AOM, see Azoxymethane
AP-1, 24, 84, 227, 237–238, 257, 262–263, 274–275, 341, 343–344
AP-2, 341, 345
APC, see Adenomatous polyposis coli or Antigen-presenting cell
APOBEC, 120
Apurinic site, 43
Arachidonic acid, 65, 67, 75, 80, 153, 155, 157–158, 160, 174, 263, 286, 293, 340, 345
Aromatic amine(s), 331
Asbestos, 1–2, 41, 43–44, 49, 110, **223–230**, 236–239, 241
ASC, 104–109, 112, 238
Aspirin, 228, 286–287, 294–295, 340
ATM, 16, 239–240
Atorvastatin, 293–294
Azoxymethane (AOM), 112, 286, 288–290, 292–294, 305, 314–322, 329, 331–332, 345, 362–365

Bacterial vaginosis, 169
BamHI, 194–197

Cancer and Inflammation Mechanisms: Chemical, Biological, and Clinical Aspects, First Edition.
Edited by Yusuke Hiraku, Shosuke Kawanishi, and Hiroshi Ohshima.
© 2014 John Wiley & Sons, Inc. Published 2014 by John Wiley & Sons, Inc.